Side Reactions in Peptide Synthesis

Side Reactions in Peptide Synthesis

Yi Yang
Chemical Development
Pharmaceutical Drug Development
Ferring Pharmaceuticals A/S
Copenhagen S, Denmark

ELSEVIER

AMSTERDAM • BOSTON • HEIDELBERG • LONDON
NEW YORK • OXFORD • PARIS • SAN DIEGO
SAN FRANCISCO • SINGAPORE • SYDNEY • TOKYO
Academic Press is an Imprint of Elsevier

Academic Press is an imprint of Elsevier
125, London Wall, EC2Y 5AS, UK
525 B Street, Suite 1800, San Diego, CA 92101-4495, USA
225 Wyman Street, Waltham, MA 02451, USA
The Boulevard, Langford Lane, Kidlington, Oxford OX5 1GB, UK

British Library Cataloguing-in-Publication Data
A catalogue record for this book is available from the British Library

Library of Congress Cataloging-in-Publication Data
A catalog record for this book is available from the Library of Congress

ISBN: 978-0-12-801009-9

For information on all Academic Press publications
visit our website at http://store.elsevier.com/

Dedication

Dedicated to my wife, Dan Liu and my son, Qinqin Yang.

Contents

6 Intramolecular Cyclization Side Reactions

7 Side Reactions on Amino Groups in Peptide Synthesis

8 Side Reactions on Hydroxyl and Carboxyl Groups in Peptide Synthesis

Preface

Thanks to their superior properties in terms of high selectivity, enhanced efficacy, and appreciable safety, peptide/peptidomimetic APIs are playing increasingly important roles in the domains of pharmaceuticals and biotech industries by means of hormones, neurotransmitters, growth factors, ion-channel ligands, anti-infectives, and so forth. Simulated by the ever-improving performances of diverse peptide therapeutics, more and more intentions are therefore legitimate to be emphasized on the peptide design and synthesis, endeavored jointly by academic and industrial efforts.

Some inherently specific properties of peptide synthesis relative to those of conventional small molecules could cause rather complicated impurity profiles. Moreover, challenges originated from the impurity formation could be intensified by the fact that certain, if not all, peptide productions lack the intermediary purifying effects in the upstream process prior to chromatographic purification treatment. Needless to mention, the impacts of scaling effects on peptide impurity formation are sometimes tricky to be elucidated. All these intrinsic challenges legitimate the necessity to pay considerable attention to the side reactions that occur in peptide synthesis.

As a first step, the impurity profile of the subject peptide API is supposed to be thoroughly scrutinized, particularly in peptide cGMP production to identify the criticalities of each single impurity against the predefined specifications. The impurities, which are potentially critical to quality or business should be emphasized and paid with peculiar attentions. It is subsequently pertinent to correlate the formation of these critical impurities with the corresponding side reactions and make efforts to elucidate the mechanism of the identified side reactions. Solutions to tackle the relevant side reactions could be designed based on the solid in-depth understanding of the origins and attributes of these side reactions. Different reaction strategies and/or process parameters are supposed to be investigated in order to fit the designed models to the purpose of the impurity suppression. By this means, the critical impurities encountered in peptide production could be diminished or eliminated at the upstream process and alleviate the stress on the purification steps. Hence, the quality of the final peptide API is assured and the process performance could be enhanced accordingly.

It is implied from the aforementioned process optimization procedure that an insight into the peptide side reactions is of crucial importance to the success of peptide API manufacturing. I have written this book to address the most frequent side reactions in peptide synthesis on the basis of a plethora of side reactions encountered in my 10 years of commitment to peptide synthesis research and peptide API production (which is simultaneously unfortunate and fortunate). The side reactions are classified in accordance with their extrinsic properties. Understandably, these categorizations are not absolutely orthogonal and indeed some independence exists. In each chapter of the book, the phenomenon of the side reactions is elucidated. The mechanism of each side reaction is either described or tentatively proposed for further discussion. Finally, diverse possible solutions are suggested in order to tackle the referred side reactions. Abundant literature references are listed for extensive reading.

The systematically organized knowledge behind a plethora of peptide-related side reactions could be sort of help for the colleagues who work in this area, no matter whether they are academic or production oriented. It is especially meaningful for the cGMP production in which a single out-of-specification impurity could ruin the whole production. Detection and analysis of the impurities in peptide synthesis, as well as their corresponding solution is therefore highly accentuated. How to find out the clues from a complicated impurity profile could understandably decide the outcomes of the peptide production. Hopefully this book could be of some help to treat the problems raised by peptide impurities, particularly in the realm of peptide API production and could ultimately assist us to fight relevant diseases.

I wish to express my thanks to Bielefeld University, Lonza AG, Ferring Pharmaceuticals, and GenScript Inc., the former three in particular, that provided me the outstanding platforms to explore the wonderland of peptide. I am grateful to Prof. Dr. Norbert Sewald and Prof. Fernando Albericio. The former introduced me to the peptide realm as a mentor and the latter gave me a lot of instructions in my career. I also appreciate the efforts made by Malik Leila and Jörgen Sjögren for their reviews of the manuscript. Fabrizio Badalassi (my boss) offered me tremendous support during the preparation of the book. Last but not the least, I am obliged to my wife Dan Liu, since her support bestows me all the strength to pursue my dream.

Chapter 1

Peptide Fragmentation/Deletion Side Reactions

Due to the inherent attributes of certain peptide individuals they could undergo a variety of fragmentation processes during synthesis, purification or even storage. Fragmentation could selectively address peptides with characteristic sequences like *N*-terminal *N*-Ac-*N*-alkyl moiety, *N*-acyl-*N*-alkyl-Aib-Xaa- bond, -Asp-Pro-, *N*-terminal His-Pro-Xaa- moiety, *C*-terminal *N*-Me-Xaa, *N*-terminal FITC, thioamide bond and guanidinyl group on Arg side chain, etc. Moreover, utilization of isodipeptide Boc-Ser/Thr(Fmoc-Xaa)-OH as the building block for peptide synthesis could result in the formation of des-Ser/Thr impurity. On top of these specific cases DKP formation could also affect general peptide assembly that leads to the deletion of affected dipeptide moiety from the parental peptide sequence. The occurrence of these fragmentation/deletion side reactions on peptide materials could decrease the manufacturing yield, cause challenges for the down-stream peptide purification, and affect peptide stability upon processing and/or storage. Phenomenon and mechanism of common fragmentation/deletion in peptide synthesis are described in this chapter. Corresponding solutions to minimize these side reactions are proposed.

1.1 ACIDOLYSIS OF PEPTIDES CONTAINING *N*-Ac-*N*-alkyl-Xaa MOTIF

Peptides with a motif of *N*-Ac-*N*-alkyl-Xaa sequence at the *N*-terminus have the distinctively high propensity to suffer from an acidolysis side reaction at the step of acid-mediated peptide cleavage from resin and side chain global deprotection. The *N*-terminal *N*-Ac-*N*-alkyl-Xaa unit might be split from the parental peptide as a 5-member ring derivative, leading to the formation of des-*N*-Ac-*N*-alkyl-Xaa truncated side product.

This kind of side reaction has been detected in the process of the preparation of a series of Arodyn peptides ((acetylated Dyn A) Arodyn 1, 2, 3, 4).[1] It was reasoned that the synthesis of Arodyn 2 resulted in the acidolytic cleavage of *N*-terminal motif *N*-Ac-*N*-Me-Phe during the TFA-mediated global deprotection step (Table 1.1).

The proposed mechanism of the subjected acidolysis side reaction is indicated in Fig. 1.1. It is reasoned in the corresponding investigation that the

Side Reactions in Peptide Synthesis. http://dx.doi.org/10.1016/B978-0-12-801009-9.00001-X

TABLE 1.1 Sequences of Arodyn 1, Arodyn 2, Arodyn 3, Arodyn 4 Peptides

Dyn A(1-11)	H-Tyr-Gly-Gly-Phe-Leu-Arg-Arg-Ile-Arg-Pro-Lys-NH_2
Arodyn 1	Ac-Phe-Phe-Phe-Arg-Leu-Arg-Arg-D-Ala-Arg-Pro-Lys-NH_2
Arodyn 2	Ac-*N*-Me-Phe-Phe-Trp-Arg-Leu-Arg-Arg-D-Ala-Arg-Pro-Lys-NH_2
Arodyn 3	CH_3OCO-*N*-Me-Phe-Phe-Trp-Arg-Leu-Arg-Arg-D-Ala-Arg-Pro-Lys-NH_2
Arodyn 4	*N*-Me-Phe-Phe-Trp-Arg-Leu-Arg-Arg-D-Ala-Arg-Pro-Lys-NH_2

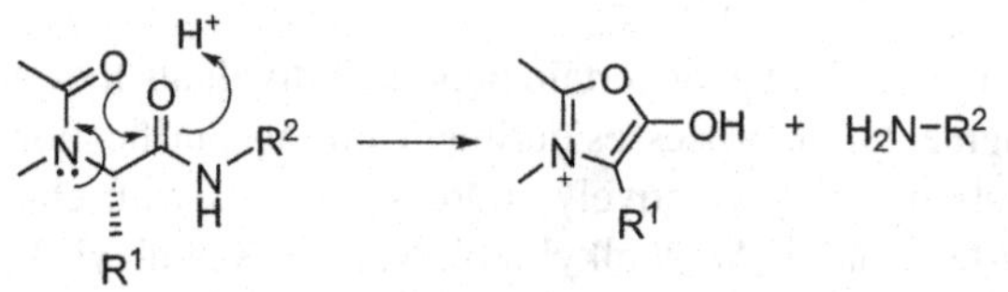

FIGURE 1.1 Proposed mechanism of the acidolytic cleavage of *N*-Ac-*N*-alkyl-Xaa from parental peptide.

occurrence of this side reaction is subject to the actual conditions under which the peptide global deprotection is conducted. It is verified that if the referred reaction is processed at 4°C in the absence of any scavengers the acidolysis of *N*-terminal *N*-Ac-*N*-alkyl-Xaa could be significantly suppressed. No similar impurities with deletion sequences have been detected in the process of Dyn A(1-11) or Arodyn 1 synthesis. The preparation of Arodyn 4 that is devoid of acetyl moiety on its *N*-terminus does not suffer from the concerned acidolysis side reaction upon TFA treatment, accounting for the involvement of the acetyl functional group in the process of *N*-Ac-*N*-alkyl-Xaa acidolysis. Significant *N*-terminus acidolysis side reaction has been invoked in the synthesis of Arodyn 2 in which Ac-*N*-Me-Phe is located on the *N*-terminus compared with the Ac-Phe motif from Arodyn1. This phenomenon is attributed to the presence of *N*-alkyl amino acid residue that favors the advantageous peptide secondary structure facilitating the acidolytic fragmentation of the *N*-terminal residue. In case the *N*-terminal acetyl is replaced by more electron-withdrawing group methyl carbamate, as is the case for Arodyn 3, the subjected acidolysis side reaction on the peptide *N*-terminus would be basically circumvented due to the decrease of the nucleophilicity of the carbonyl oxygen from the methyl carbamate that initiates the ring closure in the acidolytic fragmentation process. It could therefore be deduced from the aforementioned phenomenon that the acidolysis of peptide *N*-terminal *N*-Ac-*N*-alkyl-Xaa motif is induced by the acetyl oxygen nucleophilic attack on the amide bond between the subjected *N*-Ac-*N*-alkyl-Xaa and the neighboring amino acid at its *C*-terminus, facilitated by the advantageous local structure in that the ratio of *cis*-amide bond is significantly increased by the presence of an *N*-alkyl-amino acid residue. Under such circumstances the *N*-acetyl group serves as a nucleophile that initiates the ring closure, and subsequent acidolytic fragmentation of the *N*-Ac-*N*-alkyl-Xaa unit.

1.2 Des-Ser/Thr IMPURITIES INDUCED BY *O*-acyl ISODIPEPTIDE Boc-Ser/Thr(Fmoc-Xaa)-OH AS BUILDING BLOCK FOR PEPTIDE SYNTHESIS

O-acyl isodipeptide derivatives have already found widespread application as effective building blocks in peptide synthesis, particularly for difficult peptide assemblies that are hardly quantitatively realized by the conventional stepwise coupling methods. This methodology takes advantage of the inherent feature of the base-induced reversible intramolecular acyl *O*→*N* shift that involves the ester bond from the Ser/Thr side chain and the α-amino group on the peptide backbone (Fig. 1.2).

The incorporation of the isodipeptide unit into the peptide sequence is intended to disrupt the adverse secondary structure of the subjected peptide that impedes the smooth coupling of the forthcoming amino acid to the elongating peptide chains, particularly for the "difficult couplings." Peptide secondary structures are basically induced and reinforced by diverse molecular interactions such as hydrogen bond, Van der Waals force, hydrophobic interaction, ionic bond, and so forth. The establishment of peptide secondary structure might considerably reduce the flexibility of the affected peptide chains that consequently adversely interferes with the subsequent amino acid couplings during peptide synthesis. This phenomenon is basically regarded as one of the major causes for the nonquantitative amino acid couplings occurred in peptide synthesis that accounts for the generation of peptide impurities with deletion sequences.

O-acyl isodipeptide building blocks[2–4] are utilized in an effort to address this inherent problem in peptide synthesis. The existence of -Xaa-Ser- or -Xaa-Thr- unit in the target peptide sequence is the prerequisite for the employment of

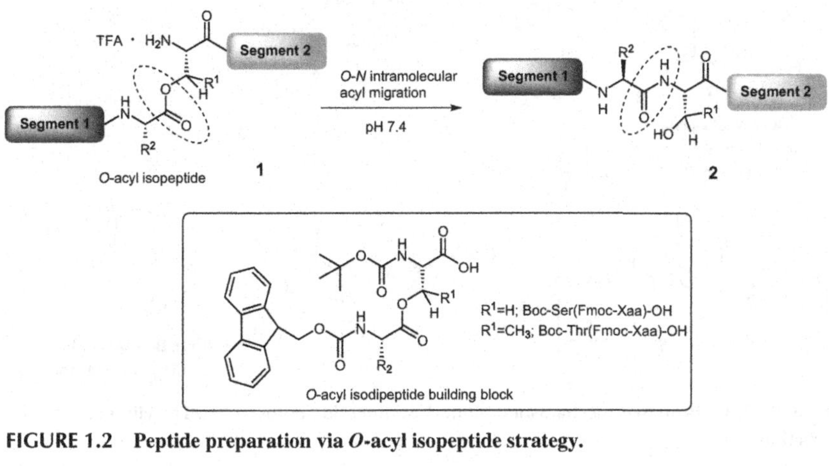

FIGURE 1.2 Peptide preparation via *O*-acyl isopeptide strategy.

O-acyl isodipeptide strategy. The subjected isodipeptide unit is incorporated in the manner of Boc-Ser/Thr(Fmoc-Xaa)-OH building block into the target peptide chains, functioning as the synthon for the natural -Xaa-Ser/Thr- counterpart. The intermediary product containing *O*-acyl isodipeptide structure is depicted as compound **1** in Fig. 1.2. The backbone carboxyl group of the -Xaa- unit is chemically linked with the hydroxyl side chain from Ser/Thr by means of an ester bond (highlighted in a dotted circle). The introduction of *O*-acyl isodipeptide moiety could manifestly disrupt the local peptide secondary structure. The solubility and liquid chromatographic properties of the peptide precursor **1** containing *O*-acyl motif are normally superior to those of its interchangeable *N*-acyl counterpart **2**. These outstanding features of *O*-acyl isopeptide could tremendously facilitate the otherwise challenging chromatographic purification. The purified *O*-acyl isopeptide **1** will be subsequently addressed to the base-catalyzed acyl *O*→*N* shift process that regenerates the natural form of the peptide amide bond via a five-member ring intermediate. The disadvantageous peptide secondary structure that impedes the smooth amino acid coupling is circumvented by this means, significantly facilitating the effective chemical preparation of the target peptide product.

In spite of the successful utility of *O*-acyl isodipeptide strategy manifested in the challenging peptide preparation such as β-amyloid 1-42,[5] it has been detected that this methodology could potentially induce side reactions such as β-elimination which leads to the formation of des-Ser/Thr impurities. The possible mechanism of this side reaction is originated from the formation of active ester Boc-Ser/Thr(Fmoc-Xaa)-OBt **4** derived from the carboxylate activation of its precursor *O*-acyl isodipeptide Boc-Ser/Thr(Fmoc-Xaa)-OH **3** (as depicted in Fig. 1.3). The lifespan of the activated derivative **4** in the reaction system is directly correlated to the kinetics of the subjected acylation reaction. If the referred reaction is proceeding

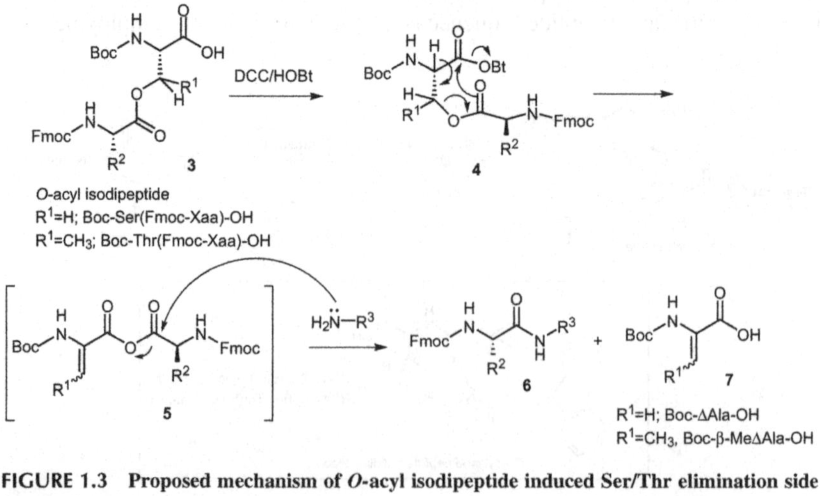

FIGURE 1.3 Proposed mechanism of *O*-acyl isodipeptide induced Ser/Thr elimination side reaction.

sluggishly, Boc-Ser/Thr(Fmoc-Xaa)-OBt **4** will be afforded with sufficient time to deviate from the target intermolecular condensation reaction and undergo intramolecular rearrangement by means of β-elimination, giving rise to the formation of the mixed anhydride 5 from Fmoc-Xaa-OH and Boc-(β-Me)ΔAla-OH, as indicated in Fig. 1.3. As a consequence, the unacylated peptide chain could possibly function with **5** at its two reactive sites, but the anhydride carbonyl at Fmoc-Xaa side is preferred due to the fact that the unsaturated (β-Me)ΔAla side chain unavoidably attenuates the electrophilicity of anhydride carbonyl on the Boc-(β-Me)ΔAla side. The unit of (β-Me)ΔAla is, therefore, excluded from the product structure as Boc-(β-Me)ΔAla-OH **7** upon the nucleophilic attack of the peptide N^{α} on the mixed anhydride **5**, giving rise to the formation of des-Ser/Thr impurity **6**.

In order to verify the proposed mechanism of *O*-acyl isodipeptide-induced deletion side reaction, Boc-Ser(Fmoc-Gly)-OH isodipeptide was incubated in NMP in the presence of DCC (2 equiv.)/HOBt (2 equiv.) for 2 h before 2.2 equiv. benzylamine was charged into the reaction system.[5] The obtained product was analyzed by MS and analytical RP-HPLC, and no Boc-Ser(Fmoc-Gly)-NHBzl was detected while large amount of Fmoc-Gly-NHBzl as well as Boc-ΔAla-NHBzl were located instead. As a matter of fact, the abundance of Fmoc-Gly-NHBzl side-product in the crude material is as high as 80%. In another experiment *O*-acyl isodipeptide Boc-Ser(Fmoc-Gly)-OH was subject to the activation process by 2 equiv. DIC/2 equiv. HOBt in DMF-d_7 for 2 h, ^{1}H-NMR analysis of the obtained product detected 2 types of olefin hydrogen signal which were assigned to E/Z isomers. This result combined with the corresponding MS and RP-HPLC analysis explicitly indicates that Boc-Ser(Fmoc-Gly)-OH has almost been quantitatively converted to the mixed anhydride composed of Fmoc-Gly-OH and Boc-ΔAla-OH within 2 h upon activation by DIC/HOBt.

Moreover, Boc-Ser(Fmoc-Ile)-OH, Boc-Thr(Fmoc-Gly)-OH and Boc-Thr(Fmoc-Ile)-OH were subject to DIC (2 equiv.)/HOBt (2 equiv.) activation in DMF for 2 h, respectively, before 2 equiv. benzylamine was charged into the reaction system to entrap the activated species. Abundant Fmoc-Gly-NHBzl, Fmoc-Ile-NHBzl, Boc-ΔAla/β-MeΔAla-NHBzl were detected as a consequence in the corresponding crude products.[5] All these results have unequivocally verified the susceptibility of *O*-acyl isodipeptide Boc-Ser/Thr(Fmoc-Xaa)-OH to suffer from the undesired rearrangement/deletion side reaction upon activation, while the inclination of this process is seemingly independent on the steric effect of the concerned amino acid.

Another indicative finding towards this side reaction is that when Boc-Ser(Fmoc-Gly)-OH was incubated in $CDCl_3$ in the presence of DIC/HOBt, ^{1}H-NMR of the obtained crude product did not indicate the existence of olefin signals from Boc-ΔAla-OH.[5] This result implies that the inclination of this side reaction of *O*-acyl isodipeptide is considerably influenced by the properties of the solvent. Polar solvents such as DMF or NMP would facilitate this process while unpolar solvents like DCM or $CHCl_3$ could minimize its occurrence. In light of this finding, it is advisable to utilize the unpolar solvents for the activation and coupling of *O*-acyl isodipeptide in order to suppress the deletion side reaction in this process. Moreover, it has been figured out that the types of the coupling reagent additives, such as HOBt, HOAt and HOOBt, would not exert significant impacts on the propensity of this side reaction.

1.3 ACIDOLYSIS OF -*N*-acyl-*N*-alkyl-Aib-Xaa- BOND

Peptide *N*-terminal *N*-Ac-*N*-alkyl-Xaa moiety can not only be addressed to the aforementioned acidolysis process, but is also subjected to the *endo*-peptide bond scission side reaction taken place at the site of -*N*-acyl-*N*-alkyl-Aib-Xaa- sequence upon acid treatment.

The undesired acidolytic fragmentation process on -*N*-acyl-*N*-alkyl-Aib-Xaa- sequence was detected in the preparation of head-to-tail cyclic peptide *cyclo*-[Phe-D-Trp-Lys-Thr-Phe-*N*-Me-Aib].[6] The side chain protected precursor peptide *cyclo*-[Phe-D-Trp-Lys(Boc)-Thr(tBu)-Phe-*N*-Me-Aib] **8** was subjected to TFA/EDT-mediated global deprotection treatment. It is highlighted in Fig. 1.4 that acidolysis at the site of -*N*-Me-Aib-Phe- gives rise to ring disclosure and formation of linear peptide H-Phe-D-Trp-Lys-Thr-Phe-*N*-Me-Aib-OH **9** as well as its thioester counterpart H-Phe-D-Trp-Lys-Thr-Phe-*N*-Me-Aib-SCH_2CH_2SH **10**.

A plethora of peptides containing *N*-Me-Aib residue has been produced, and their crystal structures have been intensively investigated in order to study the mechanism of the acidolysis side reactions occurred at the site of -*N*-Me-Aib-Xaa. The X-ray crystallography analysis of these peptides combining with the kinetics of the acidolysis implies the origins of -*N*-acyl-*N*-alkyl-Aib-Xaa- acidolysis from the aspects of steric effects. It has been discovered in a dedicated investigation[6] that *N*-Me-Aib-containing cyclic peptide *cyclo*-[Phe-Ser(Bzl)-Ser(Bzl)-Phe-*N*-Me-Aib] **11** suffered from acidolytic fragmentation at the site of -*N*-Me-Aib-Phe- upon pure TFA treatment, generating the corresponding

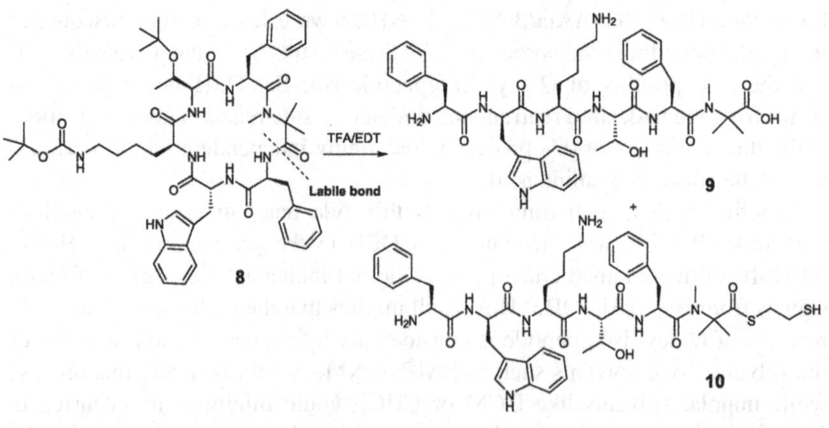

FIGURE 1.4 Acidolysis and ring disclosure of *cyclo*-[Phe-D-Trp-Lys(Boc)-Thr(tBu)-Phe-*N*-Me-Aib].

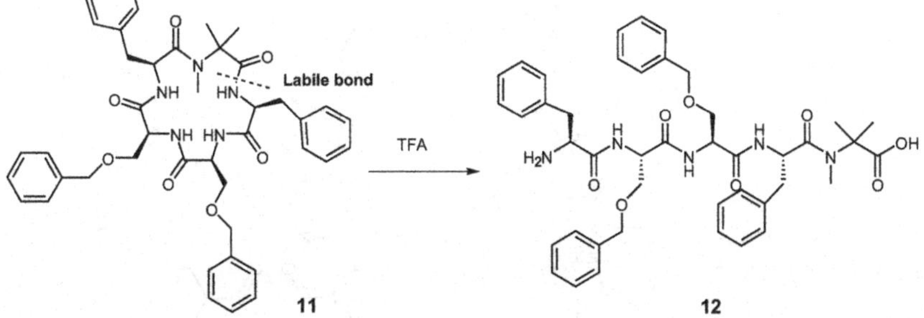

FIGURE 1.5 Acidolysis of peptide *cyclo*-[Phe-Ser(Bn)-Ser(Bn)-Phe-*N*-Me-Aib].

degraded linear peptide H-Phe-Ser(Bzl)-Ser(Bzl)-Phe-*N*-Me-Aib-OH **12** (Fig. 1.5). This ring disclosure process is identified as a pseudo first-order reaction according to its kinetics. The rate of acidolysis is reduced upon the addition of water into the reaction system, while the solvent polarity is significantly decisive for this process in that cyclic peptide **11** underwent a considerably faster acidolysis in TFA/CH_3CN (1:1) ($t_{1/2}$ = 1.1 h) than in TFA/DCM (1:1) ($t_{1/2}$ = 4.1 h). This feature is attributed to the formation of oxazolinium intermediate during acidolysis of the referred -*N*-acyl-*N*-alkyl-Aib-Xaa- peptide bond. Increase of the CH_3CN content will accelerate the kinetics of the ring disclosure.

It was illustrated from an X-ray crystallography study of cyclic peptide **11** that all amide bonds possess ordinary lengths and angles. On the other hand, it was detected that C^{α} of Aib and the carbonyl oxygen from the amino acid preceding Aib residue is spatially in proximity. This would imply that the subjected oxygen atom might be involved in a nucleophilic attack at the -*N*-Me-Aib-Xaa- bond that finally resulted in fragmentation at this site.

The proposed mechanism of the acidolysis of -*N*-acyl-*N*-alkyl-Aib-Xaa- is illustrated in Fig. 1.6 based on the above investigations. Peptide **13** containing

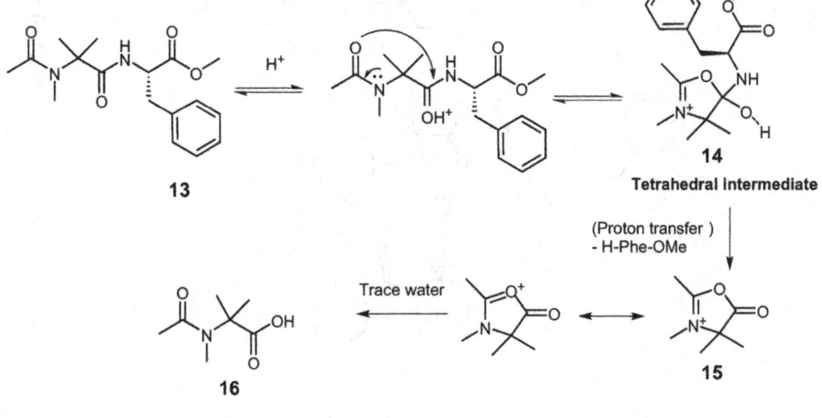

FIGURE 1.6 Proposed mechanism of *N*-acyl-*N*-alkyl-Aib-Xaa- acidolysis.

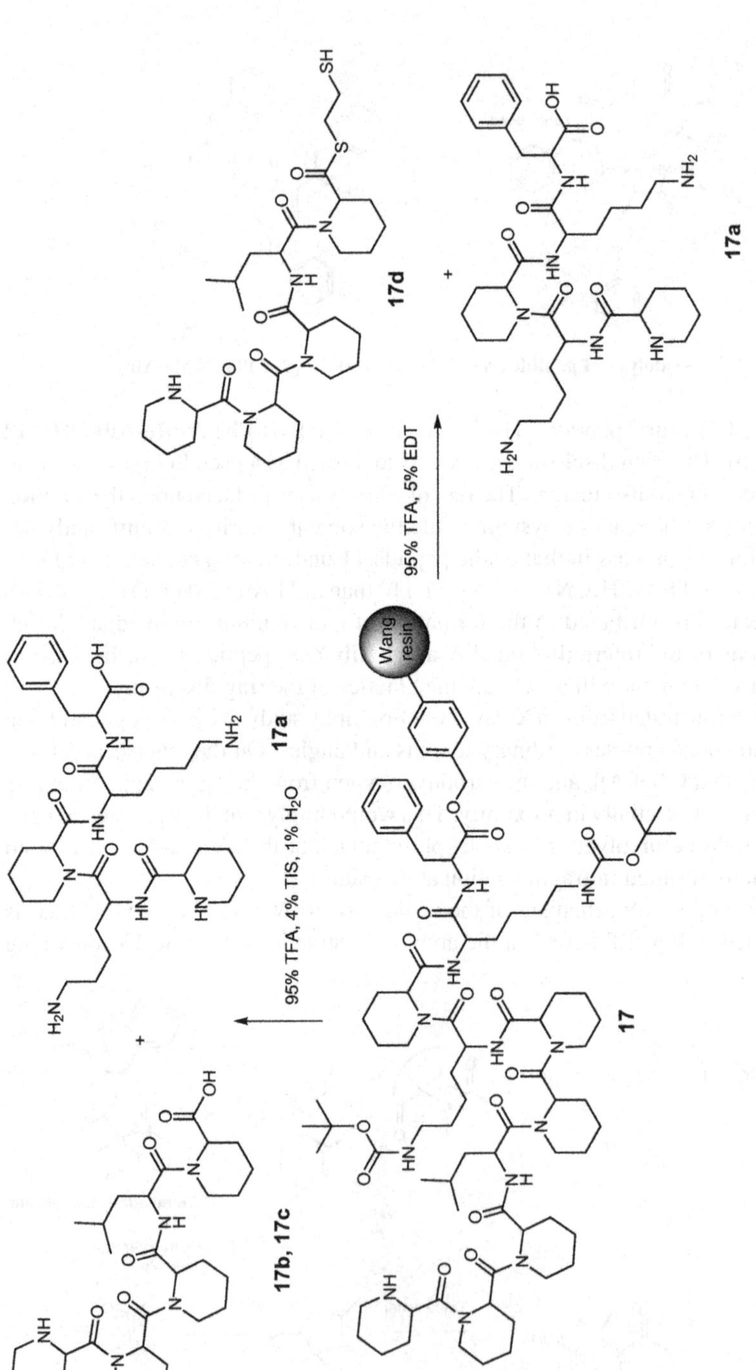

FIGURE 1.7 Acidolysis of Pip-abundant peptide.

N-Ac-*N*-Me-Aib-Xaa- moiety serves as the substrate in this connection. It is readily transformed into a tetrahedral intermediate **14** in acidic milieu. The nitrogen atom from Phe residue in compound **14** does not participate in the conjugation system with *N*-Me-Aib unit, rendering it into a proton acceptor in the acidic condition, dispelling H-Phe-OMe moiety off the intermediate **14** complex, and giving rise to the formation of oxo-oxazolinium derivative **15**. The latter is rapidly hydrolyzed into *N*-Ac-*N*-Me-Aib-OH **16**, finalizing the acidolysis process.

Similar side reactions have also been identified in the preparation of various Pip-abundant peptide derivatives.[7] Treatment of peptidyl resin **17** by 95% TFA/4% TIS/1% H_2O led to the formation of peptide fragments: **17a**, **17b**, and **17c** (Fig. 1.7). It was verified by MS and RP-HPLC that the amide bond between Pip^5 and Pip^6 was subjected to the fragmentation in this process, resulting in the formation of degraded peptide fragments: **17a**, **17b**, and **17c** – the two latter derivatives are diastereomers since the concerned acidolysis at the amide bond between Pip^5 and Pip^6 simultaneously induces configuration conversion on Pip^5-C^{α}. Meanwhile, treatment of **17** by 95% TFA/5% EDT released fragments **17a** and **17d**, the latter is the corresponding thioester of **17b/c** derivatives.

1.4 ACIDOLYSIS OF -Asp-Pro- BOND

It is known that the -Asp-Pro- peptide bond is labile under acidic conditions, such as in TFA,[8] HF,[9] formic acid,[10] and acetic acid.[11] Acidolysis of -Asp-Pro- peptide bond may not only take place in HF-mediated peptide side chain global deprotection reaction but also in weak acidic milieu (pH = 4).[12] The mechanism of this process (see also Fig. 1.8) is basically akin to that of aspartimide formation in that the amide nitrogen atom from Pro backbone attacks the carboxyl side chain of the preceding Asp residue, forming an instable cationic imide intermediate **18**[8] readily hydrolyzed into peptidyl fragments **19** and **20** whose *C*- and *N*-terminus are occupied by the subjected Asp and Pro, respectively.

In another separate investigation[13] protein E298D eNOS has been identified to suffer from acidolytic fission at -Asp^{298}-Pro^{299}- sequence, giving rise to 100 and 35 kDa fragments, while its native protein counterpart eNOS (Glu^{298})

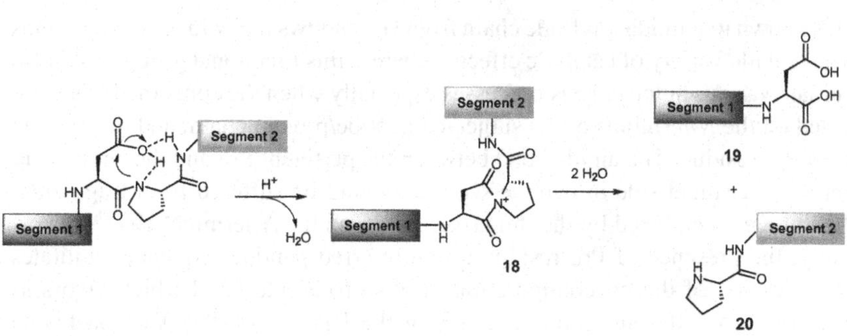

FIGURE 1.8 Proposed mechanism of -Asp-Pro- acidolysis.

is exempted from acidolysis under the same conditions. This distinctive contrast accounts for the notorious susceptibility of -Asp-Pro- to undergo acidolytic fragmentation, highly probable via the imide intermediate formation step.

The local peptide/protein secondary structure around -Asp-Pro- sequence plays a critically important role in dictating the readiness of the acidolysis side reaction. It has been discovered[12] that cellulosomal scaffoldin protein unit cohesin2-CBD undergoes fragmentation in a buffer solution at pH 4 and the acidolysis site is exactly -Asp-Pro- sequence. While this protein contains three -Asp-Pro- moieties located at -Asp^{40}-Pro^{41}-, -Asp^{50}-Pro^{51}-, -Asp^{57}-Pro^{58}-, respectively, only the -Asp^{57}-Pro^{58}- unit suffers from the acidolysis, and the other 2 remain intact upon the treatment. It is subsequently disclosed that the labile -Asp^{57}-Pro^{58}- sequence is located at a relatively rigid turn structure motif synergically stabilized by multiple hydrogen bonds.[14,15] The crystal structure of the parental protein[12] indicates that the carboxyl side chain of Asp^{50} does not lie in close proximity to Pro,^{51} whereas Asp^{40} and Asp^{57} side chains are spatially closer to Pro^{41} and Pro,^{58} respectively. Moreover, the oxygen atoms from Asp^{40} and Asp^{57} carboxyl side chains are noncovalently paired with reciprocal nitrogen atoms from Asn^{42} and Asn^{59} amide side chains, respectively, by means of hydrogen bond. This spatial alignment brings the carboxyl side chain from Asp, and the backbone amide on the neighboring Pro into proximity, and locks the local moiety into an advantageous conformation that promotes both the imide intermediate generation and the subsequent hydrolysis.

Some peptides such as Herpes simplex virion-originated peptide might suffer from -Asp-Pro- cleavage during FAB-MS analysis.[16] Meanwhile, when the labile -Asp-Pro- unit in the referred peptide was replaced by -Asn-Pro- the stability of the modified peptide could be considerably enhanced under FAB-MS analysis conditions, and no -Asn-Pro- fragmentation was detected. Laser irradiation might induce -Asp-Pro- fission as well.[17]

1.5 AUTODEGRADATION OF PEPTIDE *N*-TERMINAL H-His-Pro-Xaa- MOIETY

It is known that imidazolyl side chain from His endows many functional proteins with a wide variety of catalytic effects, whereas this functional group could also initiate various autocatalysis processes especially when the concerned His is located on the *N*-terminus of the subjected peptide/protein chain, and neighbored by a Pro residue. The amide bond between the pertinent Pro and the amino acid on its *C*-terminal side in peptide sequence could be suffered from fragmentation process catalyzed by the imidazole group on the *N*-terminal His.[18] Apparently, the presence of Pro residue in the referred peptide sequence facilitates the adoption of the *cis*-configuration of His-Pro amide bond which favors as a consequence the nucleophilic attack of the His-N^{α} on -Pro-Xaa- backbone amide, whereas the nucleophilicity of His-N^{α} is strengthened through the effect

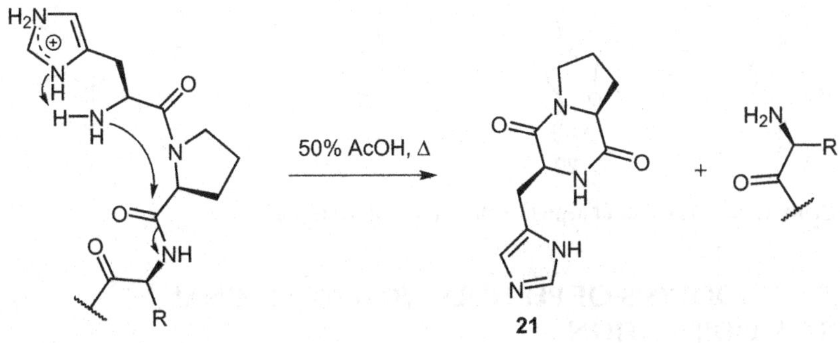

FIGURE 1.9 Proposed mechanism of His-mediated peptide N-terminal His-Pro-Xaa fragmentation.

of deprotonation exerted by His imidazolyl side chain. This His-dictated autocatalysis process gives rise to the cleavage of His-Pro moiety, and ends up with the formation of cyclic derivative His-Pro-diketopiperazine **21**. The proposed mechanism of this process is depicted in Fig. 1.9.

1.6 ACIDOLYSIS OF THE PEPTIDE *C*-TERMINAL *N*-Me-Xaa

Peptides containing *C*-terminal *N*-Me-Xaa residues are basically inherited with decreased stabilities in acidic milieu compared with their counterparts with ordinary non-*N*-alkylated amino acids located at this position. *C*-terminal *N*-Me-Xaa peptides could be subject to acidolytic degradation during acid-mediated peptidyl resin cleavage and/or peptide global deprotection process,[19] leading to the formation of side product with deletion sequence. The mechanism of this side reaction is elucidated in Fig. 1.10. The *C*-terminal *N*-Me-Xaa residue on a peptide would favor the adoption of *cis*-configuration of the amide bond between this residue and its preceding amino acid, facilitating by this means the nucleophilic attack of the *C*-terminal carboxyl group on the subjected amide bond, and giving rise to the formation of an intermediary 5-member ring compound **22**. This intermediate is subsequently rearranged to an anhydride derivative that is in turn hydrolyzed to the peptide side product devoid of the original *C*-terminal amino acid by releasing the subjected *N*-Me-Xaa.

This side reaction normally proceeds slowly, and only a small portion of the material suffers from *C*-terminal *N*-Me-Xaa acidolysis. Moreover, this process affects exclusively peptide acids with C-terminal *N*-Me-Xaa, and if the subjected residue is replaced by the corresponding *N*-Me-Xaa amide this side reaction will be nearly thoroughly suppressed.[20] This phenomenon could be readily rationalized by the reaction mechanism illustrated in Fig. 1.10.

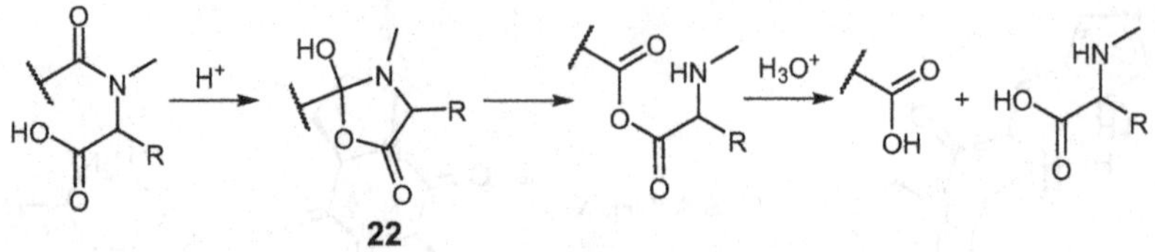

FIGURE 1.10 **Acidolysis of peptide with *C*-terminal *N*-Me-Xaa.**

1.7 ACIDOLYSIS OF PEPTIDES WITH *N*-TERMINAL FITC MODIFICATION

Fluorescent dyes are nowadays routinely utilized as labeling compounds for biomacromolecules. They have been intensively utilized in domains such as fluorescence microscope, flow cytometry, immunofluorescence techniques, and so forth. FITC is one of the most frequently employed fluorescent dyes that could function selectively with amino[21] and/or sulfhydryl[22] functional groups in peptides or proteins, visualizing by this means the affected peptides/proteins under fluorescence. Modifications of target peptides by FITC could be realized in the process of SPPS[23,24] following selective liberation of the amino groups on Lys or Orn side chains,[25] or alternatively, on peptide backbone N^{α} groups.

Side reaction resembling Edman degradation might take place during FITC-mediated modification on peptide N^{α} functional group.[26] Edman degradation, as an intentional method for peptide sequencing, is achieved by the function of the N^{α} group from the target peptide/protein with phenylisothiocynate, and the subsequent acidolysis of the generated phenylthiocarbamoyl derivative into a degraded peptide with a deletion sequence and a split phenylthiohydantoin compound.[27] Peptide modified at its N^{α} functional group by FITC could undergo an equivalent process upon TFA treatment as well. The thiocarbamoyl derivative **23** derived from the reaction between the target peptide and FITC is firstly transformed into a 5-member ring intermediate **24** under acidic condition, that is subsequently split from the parental peptide in the form of fluorescein thiazolinone **25**, and finally rearranged to a stable fluorescein thiohydantoin compound **26**. The mechanism of this process is illustrated in Fig. 1.11. It could be inferred from the proposed mechanism that formation of the 5-member ring intermediate **24** is the key step of the whole process. The spatial proximity between the nitrogen atom from FITC moiety and the amide carbon of the first amino acid on peptide *N*-terminus plays a crucially important role in dictating the propensity of the subjected peptide to undergo the concerned acidolysis side reaction. Compound **23** will be highly susceptible to the transformation into the corresponding 5-member ring intermediate **24**, provided that no spacer is incorporated between FITC moiety and the *N*-terminus of the referred peptide chain. The subsequent fragmentation process will be thus facilitated.

According to the result obtained from a systematic study,[26] the aforementioned degradation process does not take place during FITC-mediated peptide

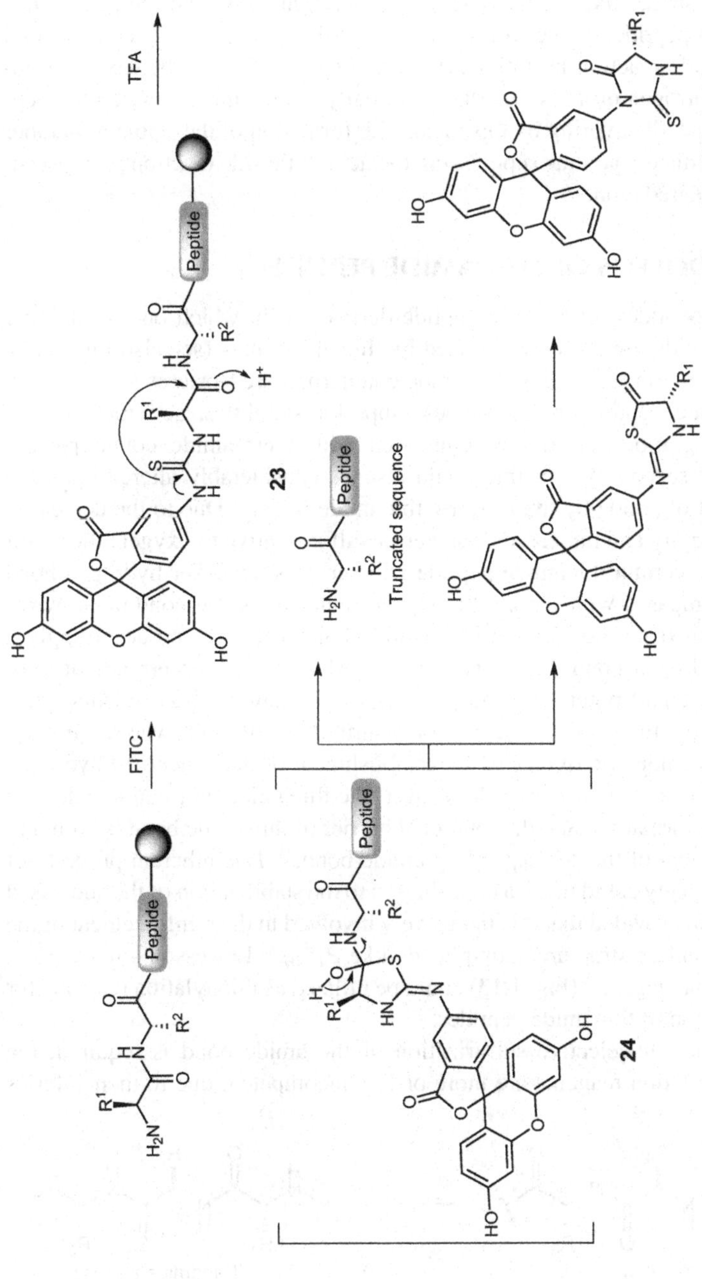

FIGURE 1.11 Mechanism of acidolysis of peptide with N-terminal FITC modification.

N-terminal modification, but at the step of peptide-FITC adduct treatment by TFA. Identical with Edman degradation this side reaction is catalyzed by acid. Normally a spacer like ε-Ahx is squeezed between FITC and the *N*-terminus of the target peptide in an effort to circumvent the occurrence of the undesired acidolysis side reaction by rationally increasing the distance between the nucleophilic nitrogen on FITC and the potentially labile amide bond on the peptide backbone, disfavoring by this means the formation of the stable 5-member ring intermediate, and thus suppressing the acidolytic side reaction reminiscent of Edman degradation.

1.8 ACIDOLYSIS OF THIOAMIDE PEPTIDE

Thioamide peptides refer to the peptide derivatives in which one or multiple backbone amide moieties are replaced by thioamide units (see also Fig. 1.12). Peptide and its corresponding thioamide counterpart are isosteres.

Thioamide peptides might possess superior stabilities, activities or molecular recognition selectivities compared with their amide counterparts.[28] The peptide secondary structure could also be considerably altered upon the replacement of amide by the isosteric thioamide moiety. Due to the decreased electronegativity and increased diameter of sulfur relative to oxygen, the sulfur atom from a certain thioamide peptide will serve as a weaker hydrogen bond acceptor compared with the amide oxygen under the same conditions, while the hydrogen from the subjected thioamide bond will become accordingly a stronger hydrogen bond donor. The overall effects of the occurrence of thioamide bond could potentially either disrupt[29] or stabilize[30] the peptide local secondary structures such as β-turn or γ-turn, depending on whether the addressed thioamide contributes to the establishment of the concerned hydrogen bond as a donor or an acceptor. Moreover, the thioamide unit mainly adopts a Z-planar conformation, and the rotational barrier of thioamide bond is normally higher than that of the corresponding amide bond.[31] This inherent property of thioamide moiety could make a contribution to the stabilization of the addressed peptide chain provided that it is intensively involved in the reinforcement of the peptide secondary structure. Compounds like P_4S_{10},[32] Lawesson's reagent[33,34] or Yokoyama reagent[35] (Fig. 1.13) could be utilized as thionylation reagents for the preparation of thioamide peptides.

When the site-selective sulfurization of the amide bond is required, the above thionylation reagents are more or less incompetent due to their inferior

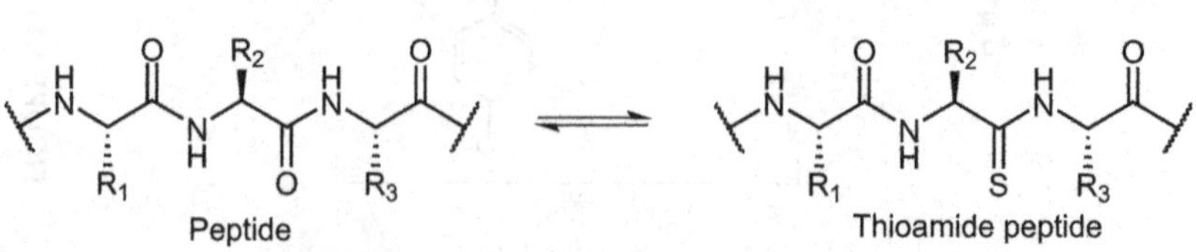

FIGURE 1.12 Normal peptide and thioamide peptide counterpart.

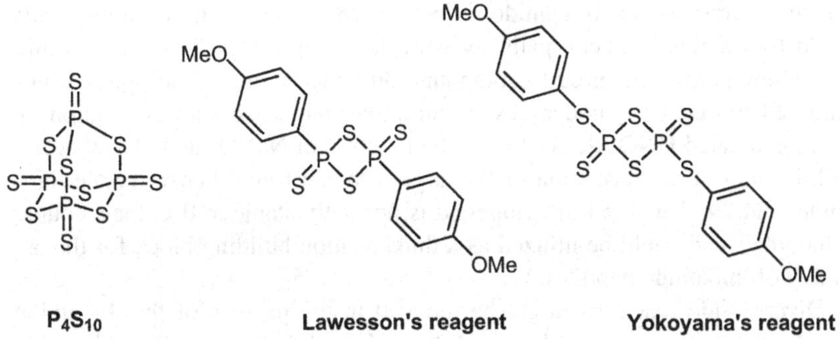

FIGURE 1.13 Thionylation agents P_4S_{10}, Lawesson reagent, and Yokoyama reagent.

selectivities towards multiple amide moieties. Synthetic strategies capable of selectively sulfurizing target amino acids have been developed in an effort to produce the desired thioamide peptides in a site-selective manner. Utility of 6-nitrobenzotriazole thioamino acid stands out as a potent strategy for the preparation of thioamide peptides. The synthetic strategy of the 6-nitrobenzotriazole thioamino acid **29** is illustrated in Fig. 1.14.[36] Boc-amino acid is converted to compound **27** by 4-nitro-1,2-phenylenediamine upon tandem activation and amination of its carboxyl group. The oxygen atom from an amide moiety is more prone to be replaced by sulfur compared to the oxygen atom from an ester unit, and this feature accounts for the strategy of the sulfurization of the amino acid upon its conversion into a potentially active amide instead of an active ester. Amide compound **27** is subsequently addressed to the sulfurization process through the treatment with thionylation reagent P_4S_{10} and is transformed

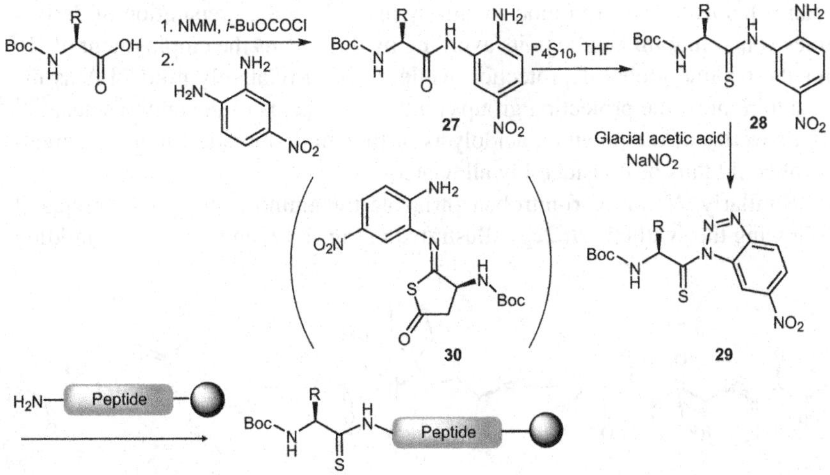

FIGURE 1.14 Scheme of the preparation of 6-nitrobenzotriazole thioamino acid and the thioacylating reaction.

into the corresponding thioamide derivative **28** (all the natural amino acids would follow this synthetic pathway with the exception of Asp which would be predominantly converted to compound **30** under similar conditions). Thioamide **28** subsequently undergoes an intramolecular diazonium cyclization by *in situ* generated HNO_2 derived from the function of $NaNO_2$ and glacial acetic acid, leading to the formation of the target product 6-nitrobenzotriazole thioamino acid **29**. This kind of compound is normally stable at 0°C for a couple of months[36] and could be utilized as a thioacylation building block for the assembly of thioamide peptides.

Diverse side reactions might be induced in the process of the thioamide peptide preparation among which an Edman degradation resembling thioamide acidolysis is frequently detected[37,38] (Fig. 1.15). This acidolytic degradation process normally ends up with the formation of a thiazolone derivative **31** and a peptide *C*-terminal fragment **32** as degradative byproducts. The concerned acidolysis side reaction occurs predominantly in Boc-chemistry at the step of HF-mediated peptide side-chain global deprotection.[39] No unambiguous correlation between the propensity of the acidolysis side reaction and the corresponding peptide sequence has yet been discovered; nevertheless this process is generally more prone to affect the peptide chain with less steric hindrance such as the case for the thioamide derivative of Leucine-enkephalin.[38]

Not only the specific peptide sequence but also the type of acids employed for the peptide side chain global deprotection reaction will affect the lability of thioamide bonds. This assertion is in line with the phenomenon that Clausen et al. observed during the synthesis of thioamide derivative of leucine-enkephalin H-Tyr-Gly-Gly-ψ[CS-NH]-Phe-Leu-OH that no target products were obtained from the HF-mediated global deprotection reaction.[37] In view of the considerable lability of thioamide bond towards strong acids such as HF and TFMSA many chemists resort to Fmoc chemistry instead for the preparation of derivatives containing thioamide units avoiding by this means the employment of HF for the peptide global deprotection while utilizing relatively mild TFA as the acid to remove the protecting groups on the side chains of the amino acids.[40,41] The severity of the potential acidolysis of thioamide moieties upon acid treatment could thus be considerably alleviated.

Similarly, N^{α}-Fmoc-6-nitrobenzotriazole thioamino acids can be prepared following the synthetic strategy illustrated in Fig. 1.14, and utilized as building

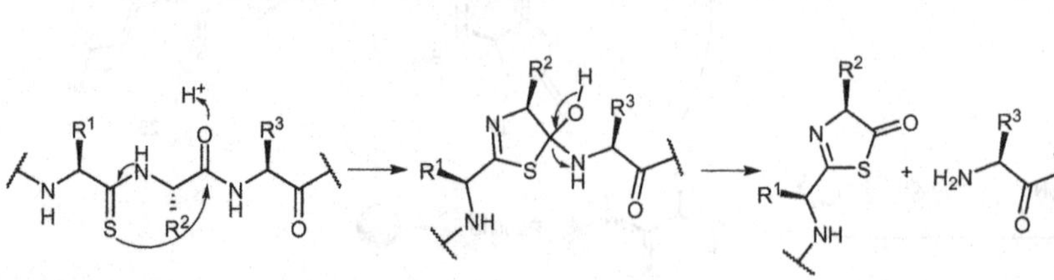

FIGURE 1.15 Mechanism of acidolysis of thioamide peptide.

blocks to construct thioamide peptides via Fmoc-mode SPPS.[40] The degree of thioamide bond acidolysis at the step of TFA-mediated peptide cleavage as well as global deprotection is evidently correlated to the concentration of TFA and the duration of the treatment.[39] It is indicated explicitly that by lowering the TFA concentration, and/or shortening the duration of the treatment of peptide by the subject acid solution the degree of thioamide acidolysis would be accordingly reduced.

Some mild Lewis acids such as $AlCl_3$ or $SnCl_4$ can also be utilized as the reagent to remove N^{α}-Boc protecting groups, which could evidently suppress the acidolysis of thioamide compared with the treatment of the referred peptide by TFA-mediated N^{α}-Boc removal.[42] The reactivity of Lewis acids toward N^{α}-Boc cleavage is apparently dependent on the choice of solvents, e.g., the kinetics of Boc removal by $SnCl_4$ is enhanced in ACN, more than in other ordinary organic solvents.[43] Actually the methodology of N^{α}-Boc cleavage by $SnCl_4$ in ACN was developed in an effort to dedicatedly address the inherent lability of the thioamide bond manifested in ordinary TFA-dictated N^{α}-Boc removal. In addition, it has been discovered that $SnCl_4$ could complex with the liberated N^{α} group[43] – this partially accounts for the enhanced stability of the thioamide bond in the Lewis acid-mediated Boc cleavage environment. $SnCl_4$–peptide complex could be disassociated via the precipitation of peptide methanol solution in ether, or alternatively in the process of liquid chromatographic purification utilizing ACN/H_2O or $MeOH/H_2O$ as an eluent.[43]

N^{α}-protecting groups with enhanced acid lability relative to Boc such as Bpoc or Ddz (Fig. 1.16) could find the utility in thioamide peptide synthesis. The milder N^{α}-protecting group cleavage conditions will correspondingly reduce the extent of potential thioamide acidolysis in this process. It has been verified that Lewis acid solution such as $Mg(ClO_4)_2$/ACN or $ZnCl_2$/THF could be utilized to selectively cleave Bpoc/Ddz protecting groups while the thioamide bonds in the subjected peptides remain largely unaffected under this condition. This methodology could be applied to the SPPS process that utilizes highly acid sensitive solid supports such as Sieber or SASRIN® resin.[44] Bpoc can be thoroughly removed at 50°C by $Mg(ClO_4)_2$/ACN while Boc groups[45] and thioamide bonds are unaffected by this handling. $ZnCl_2/Et_2O$ treatment of peptidyl resin could realize simultaneous peptide cleavage and Boc/OtBu removal.[44] $ZnCl_2$ is not

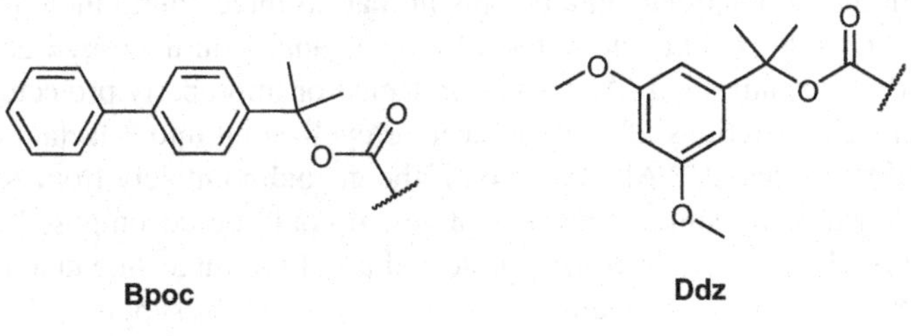

FIGURE 1.16 **Bpoc, Ddz-protecting group.**

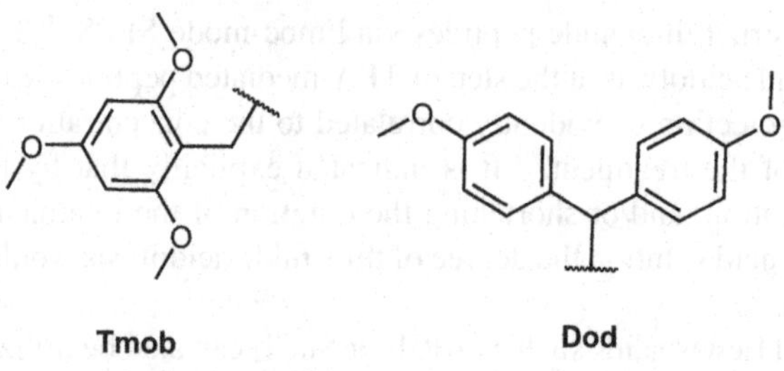

FIGURE 1.17 Tmob, Dod protecting group.

capable of quantitatively cleaving Pmc/Pbf and Trt protecting groups. In view of this limitation orthogonal building blocks such as Arg(Boc)$_2$, Asn(Tmob/Dob) and Gln(Tmob/Dob) (see Fig. 1.17) that are compatible with this strategy are recommended so that the subjected amino acids could be quantitatively regenerated by the treatment of $ZnCl_2$.[44]

SPPS of thioamide peptide employing N^{α}-Bpoc/Ddz protecting strategy is described in Fig. 1.18. Side chain-protected peptides are assembled through Fmoc-chemistry on highly acid-sensitive solid supports such as Sieber or SASRIN® resin prior to the incorporation of thioamino acid that is realized by the employment of preactivated Bpoc-Xaa-6-benzotriazolide as a building block. The following SPPS procedure is conducted under Bpoc-chemistry guidance in that N^{α}-Bpoc-6-benzotriazolide building blocks are incorporated sequentially into the target peptide chain, and the N^{α}-Bpoc-protecting group is cleaved by $Mg(ClO_4)_2$/ACN solution. When the chain assembly is completed the peptidyl resin will then be treated by $ZnCl_2/Et_2O$ with the aim of simultaneously releasing the peptide chains from the solid support and removing Boc/OtBu/Tmob/Dod side chain protecting groups. Pure target thioamide peptide product is finally obtained through RP-HPLC purification. Thioamide peptide with high purity can be prepared with this synthetic strategy, and acidolysis of the labile thioamide bonds will be largely circumvented by this means.

1.9 DEGUANIDINATION SIDE REACTION ON Arg

Arg is to some extent a relatively special amino acid compared to the ordinary side chain mono-functional amino acids in that all three amino moieties from its guanidino side chain could potentially be entangled in undesired acylation during peptide synthesis. If Arg is not protected or improperly protected at its side chain, side reactions like intramolecular cyclization into δ-lactam derivative could be triggered.[46] Alternatively, if the guanidino moiety from Arg side chain is acylated by amino acid derivatives, it could be decomposed into an Orn side product.[47] Arg derivative protected at all the three side chain amino moiety $N^{\delta,\omega,\omega'}$ (Fig. 1.19) could theoretically thoroughly suppress the above side reactions induced by the insufficiently shielded side chain guanidino on

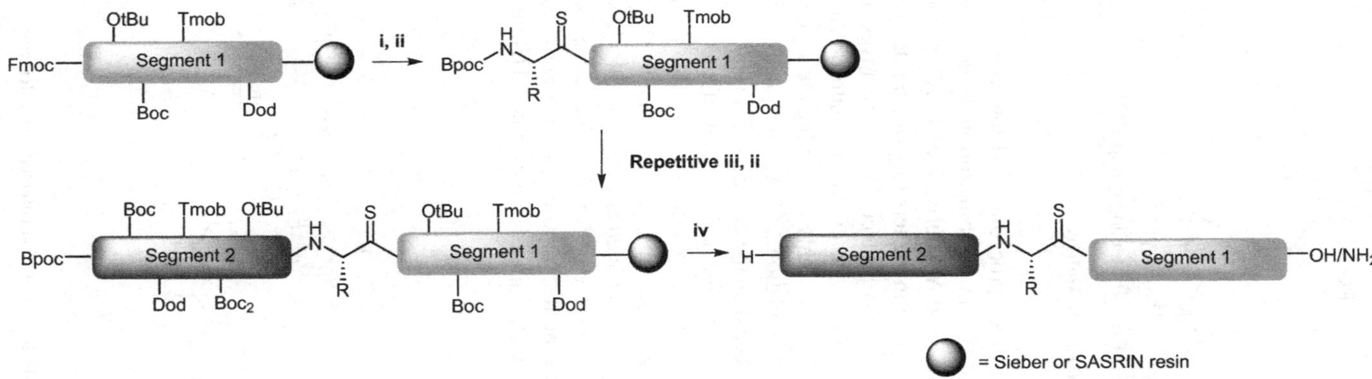

FIGURE 1.18 SPPS of thioamide peptide with Bpoc chemistry: (i) piperidine/DMF; (ii) thioamino acid of Bpoc-Xaa-6-nitrobenzotriazolide/DMF; (iii) $Mg(ClO_4)_2$/ACN; (iv) $ZnCl_2/Et_2O$, DCM.

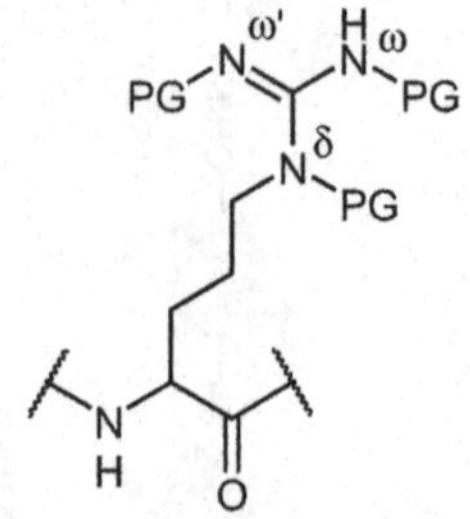

FIGURE 1.19 ***N***$^{\delta,\omega,\omega'}$**-protected Arg derivative.**

Arg, however, due to the tedious procedure and low yield with respect to the preparation of these derivatives, in combination with their significant inferior coupling efficiencies, tri-protected Arg derivative like $N^{\delta,\omega,\omega'}$-tris(Trt)-Arg[48] has not found widespread application in the territory of peptide synthesis.

It has been verified that $N^{\delta,\omega}$-bis-protected Arg derivatives could at least suppress the occurrence of intramolecular cyclization into δ-lactam side product, nevertheless, it is incapable of thoroughly preventing the acylation side reaction on $N^{\omega'}$ and the subsequent degradation.[47,49] As illustrated in Fig. 1.20, if Arg-containing peptide **33** is protected on the Arg guanidino side chain at $N^{\delta,\omega}$ site by carbamate protecting groups (e.g., Boc or Adoc) and left the $N^{\omega'}$ functional group unshielded, the exposed free $N^{\omega'}$ group will be susceptible in the process of the amino acid coupling reaction to the undesired acylation by the activated incoming amino acids, generating over-acylated side product **34** whose Arg guanidino side chain is acylated at the unprotected $N^{\omega'}$ site. This side product is labile under basic condition, and will be decomposed into the corresponding protected Orn derivative **35** and 2-protected amino-5-substituted-1*H*-imidazol-4(5*H*)-one

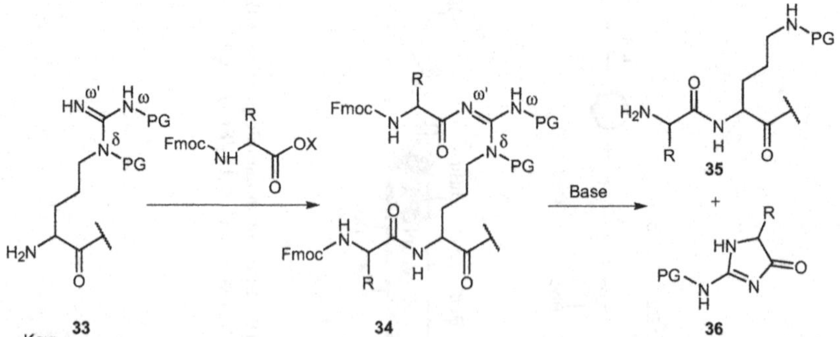

FIGURE 1.20 Scheme of acylation of ***N***$^{\delta,\omega'}$**-protected Arg derivative and base-induced degradation.**

compound **36** upon base treatment like piperidine mediated Fmoc deprotection process.[50] The Orn impurity in the crude could be detected by MS analysis as a derivative with – 42 amu molecular weight compared with the target Arg-containing product. Unlike $N^{\delta,\omega}$-bis-Boc protected derivative $N^{\delta,\omega}$-Arg(Boc)$_2$, $N^{\omega,\omega'}$-bis-protected Arg building block like $N^{\omega,\omega'}$-Arg(Boc)$_2$[51] is exempted from guanidino acylation and subsequent base-induced degradation, but suffers from intramolecular cyclization into δ-lactam instead.[52,53] Moreover, coupling kinetics of Boc-bis-protected Arg is considerably slower than the ordinary Arg derivatives. N^{ω}-mono-protected Fmoc-Arg(Boc)-OH is susceptible to significant extent of guanidino acylation side reaction and subsequent degradation into Orn.[47]

The propensity of an Arg derivative to undergo guanidino acylation and a subsequent degradation process is directly correlated to the protection status of the concerned Arg guanidino side chain. The guanidino side chain from Arg is normally protonated during the process of peptide synthesis due to its strong basicity (pK_a = 12.5) which can largely circumvent the acylation-resembling side reactions.[54,55] Some peptide preparation can be carried out in the presence of side chain-unprotected Arg, and this strategy could be utilized either in solution phase[56] or solid phase peptide synthesis.[57] Since Arg deprotonation takes place at piperidine-mediated Fmoc deprotection steps, this could probably induce the undesired guanidino acylation in the subsequent amino acid coupling reactions, an additional peptidyl resin rinse with 0.25 M HOBt solution is advised to be conducted between the piperidine treatment and the subsequent amino acid coupling,[58] in order to reprotonate the neutralized guanidino side chain on Arg, and minimize the undesired guanidino acylation at the following coupling steps. However, partial deprotonation on guanidino group might take place during the coupling of weakly basic derivatives (e.g., H-Pro-OtBu),[59] which might consequently induce the undesired acylation on the subjected Arg side chain and its degradation to Orn counterpart.[60]

N^{ω}-arylsulfonyl derivatives such as Pmc and Pbf are utilized nowadays as the standard Arg side chain protecting groups. In spite of the incompetence to thoroughly suppress the δ-lactam formation during Arg activation, their overall outstanding performances have established their status as the most intensively employed Arg protecting groups.

Meanwhile, some new types of Arg side chain protecting groups have been invented to address the inherent side reactions. Some candidates such as Suben, Sub and MeSub are displayed in Fig. 1.21.[61] Their striking steric hindrances are beneficial to restrain the side reactions such as δ-lactam formation as well as over-acylation on guanidino group. Their considerably enhanced acidic sensitivities relative to arylsulfonyl-type of protecting groups can be exploited in the way that they could be readily removed in the diluted TFA solution. In spite of the above advantages these new types of Arg side chain protecting groups have not found widespread utility in industrial territories.

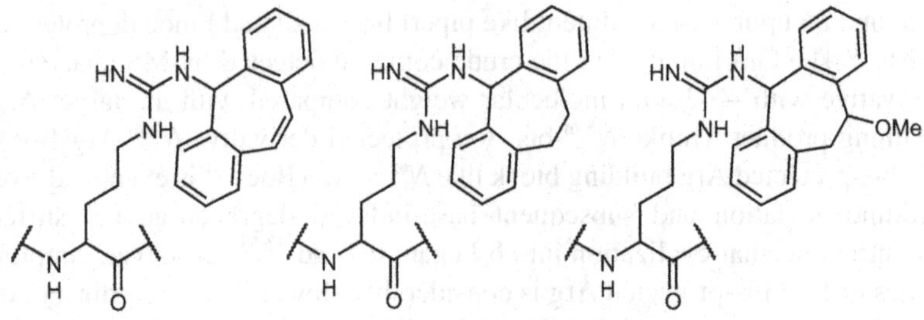

FIGURE 1.21 Arg side chain protecting groups Suben, Sub, and MeSub.

N^{ω}-NO_2 protected Arg derivative Arg(NO_2) could be utilized as an ideal building block in case Arg side chain acylation and the subsequent decomposition side reactions prevail.[62] This merit is attributed to the strong electron-withdrawing effect of NO_2 which can drastically depress the disadvantageous nucleophilicity of N^{ω} and N^{δ} group on the guanidino side chain from the concerned Arg, reducing by this means the possibility of the participation of the guanidino in the above side reaction processes, even though some exceptions do exist.[63,64]

1.10 DKP (2,5-DIKETOPIPERAZINE) FORMATION

DKP (2,5-diketopiperazine) formation is one of the most malignant side reactions in peptide synthesis,[65] which could not only haunt the process of peptide preparation but might also take place even during the storage of peptide materials.[66] Some DKP derivatives play important roles in the domains of drug designs.[67,68] Framework of DKP structure could also be found in a plethora of natural compounds.[69] In spite of the versatility of DKP-related compounds its formation poses significantly serious challenges in peptide synthesis which as consequences induce impurities with deletion sequences. This undesired process might not only drastically reduce the yields of affected peptide production but also introduce unnegligible challenges for the down-stream process such as peptide purification. This side reaction is, therefore, legitimately regarded as one of the most striking problems in peptide synthesis.

The mechanism of DKP formation in the process of peptide synthesis is illustrated in Fig. 1.22. The nucleophilic attack of the N^{α} group from the peptide *N*-terminal amino acid on the carbonyl functionality, either in the form of amide or ester moiety from the second amino acid, gives rise to the fission of the affected amide or ester bond. The *N*-terminal dipeptide is split off the peptide backbone in the form of a 6-member ring derivative diketopiperazine. DKP formation is normally accelerated when the addressed peptide is a depsipeptide (X = O in Fig. 1.22)[70] since hydroxyl derivative is a superior leaving group compared with the amino counterpart. That is to say, depsipeptide is more susceptible to DKP formation facilitated by its inherent ester bond feature.

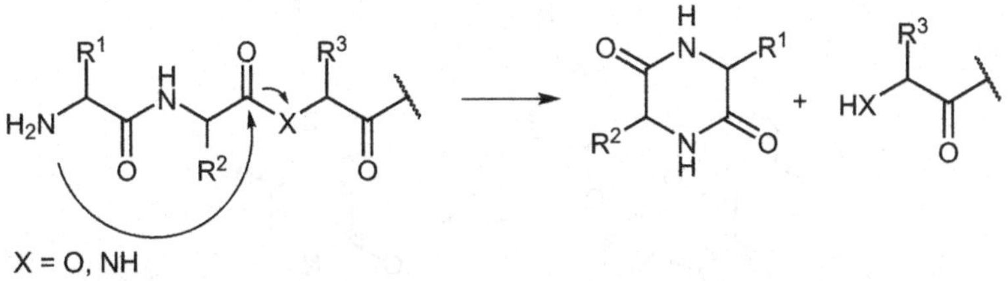

FIGURE 1.22 **Mechanism of DKP formation.**

One of the stimulative effects facilitating the occurrence of DKP formation is the preference of *cis*-configuration adopted by the amide bond between the two constituting amino acids for DKP derivative.[71] This spatial alignment promotes the nucleophilic attack of the N^{α} group on the target amide/ester bond by locking the substrate into an advantageous conformation with respect to DKP formation.[72,73] It is rational to extrapolate from the mechanism of DKP formation that the inclination of this process will be peptide sequence-dependent. That is to say, DKP formation is particularly prone to occur to specific peptide units in which the *cis*-configuration of the amide bond between the two amino acid residues is reinforced by the existence of certain facilitating constituents such as *N*-Me-Xaa,[74] especially on consecutive *N*-Me-Xaa residues.[75] The DKP side reaction could be considerably stimulated under such conditions due to the fact that the occurrence of *N*-Me-Xaa residue in the peptide sequence would promote the adoption of *cis*-configuration by the amide bond on -Xaa_1-*N*-Me-Xaa_2- unit.[76] It is noted that when an *N*-alkyl-Xaa happens to occupy the *C*-terminal location on the parental peptide sequence, the chance of this peptide to undergo DKP formation during solid phase peptide assembly will be drastically enhanced.[77] Besides *N*-alkyl-Xaa-containing derivatives, peptide sequences with alternating L- and D-amino acids are also markedly susceptible to DKP formation,[20] since the derived diketopiperazine compound split off this kind of parental peptide will be stabilized by the featured spatial alignment[65] in that the side-chain functional groups R_1 and R_2 (Fig. 1.22) from the corresponding L- and D-amino acids are resided on the opposite side of the planar DKP ring. Equilibration is favored by this means to the direction of DKP formation. Resembling *N*-Me-Xaa secondary amino acids such as Pro, Hyp[78] and Thz[79] also incline to facilitate DKP formation during peptide synthesis.

It is noted that DKP formation could affect not only the peptides with free N^{α} functional group, but might also take place during the activation of N^{α}-protected dipeptide, or even tripeptide bearing an *N*-alkyl-Xaa, or an amino acid with a secondary amino group like Pro.[75] As depicted in Fig. 1.23 N^{α}-protected dipeptide active ester Z-Gly-Pro-ONp **37** undergoes DKP formation at pH 8, giving rise to N^{α}-Z-protected DKP derivative **38.**[80]

Factors other than peptide sequence affecting DKP formation include the nucleophilicity of the subjected N^{α} functional group. The propensity of DKP formation is understandably proportional to the nucleophilicity of the attacking

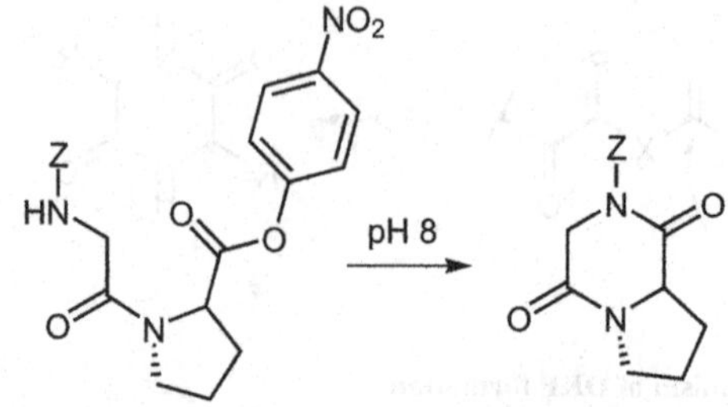

37 38

FIGURE 1.23 **DKP formation from N^{α}-Z-protected active dipeptide.**

N^{α} group. Moreover, the steric effects of the side chains from the DKP-involved amino acids would considerably interfere with the occurrence of DKP formation. Solvent effect is also regarded as one of the prominent factors that exert influences on a DKP side reaction. The impact of solvents on the equilibrium of amide bond *trans→cis* conversion is somehow seemingly irrelevant to DKP formation since the impetus of solvent on amide *trans–cis* equilibrium is trivial.[81] The actual impact of solvents on DKP formation is reflected by their stabilization/destabilization effects on the zwitterionic cyclic intermediate in that solvents with higher abilities to stabilize charged or dipolar solutes by virtue of charge-dipole or dipolar interactions, as well as solvents capable of stabilizing hydrogen donor or acceptor solutes will decrease the kinetics of DKP formation.[82] These kinds of solvents will enlarge the energy difference between the ionic substrate and the transition state that in turn slow-down the DKP formation. Both acid[83] and base[84] could catalyze DKP formation.[82,85] Better leaving groups will facilitate DKP ring closure. For example, in the synthesis of dipeptide ester DKP formation could be accelerated compared with the amide counterpart by virtue of enhanced leaving propensity of alcohol derivative relative to that of amine. If the comparison is narrowed down to dipeptide ester, the steric effect of alcoholic constituent from the subjected ester moiety would evidently affect the extent of DKP formation reflected by the increased susceptibility of dipeptide methyl ester, ethyl ester or benzyl ester to DKP formation compared with their *tert*-butyl ester counterpart, since the enhanced steric hindrance of the latter could effectively impede the process of the intramolecular cyclization and hence DKP formation.[86]

Peptide preparation strategies have been developed accordingly to address the inherent root causes of DKP formation. Since the DKP formation via the cyclization process on the peptide *N*-terminus takes place predominantly during the removal of base-labile N^{α}-protecting group as well as the subsequent acylation reaction, shortening the duration of base treatment on DKP-sensitive peptide chains will effectively restrain the occurrence of the DKP side reaction.[87] Employment of the fluoride-containing substances like TBAF as the reagent for Fmoc cleavage instead of the stimulating piperidine in combination with the utilization of methanol as scavenger to quench the split off nucleophilic DBF compound, might reduce the risk of the sensitive peptide sequence to undergo DKP formation.[88]

Taking advantage of the Trt-Xaa-OH that utilized in place of the ordinary Fmoc-Xaa-OH as building blocks for the preparation of peptides with striking DKP sensitive sequences such derivative would suppress the occurrence of this side reaction at the step of N^{α}-protecting group cleavage.[89] Thanks to its beneficial sensitive acid lability N^{α}-Trt protecting group could be selectively removed under a weakly acidic condition and the subjected N^{α} functional group liberated by this means is protonated, hence its nucleophilicity (that is imperative to invoke the DKP side reaction) would be hence deprived. No separate neutralization step is necessitated under such circumstances since the following amino acid coupling could be conducted in the presence of DIEA which *in situ* neutralizes the protonated N^{α} group, maximizing the utility of DKP restraint by this means.

pNZ (Fig. 1.24) is another valuable N^{α}-protecting group that could be taken advantage of to suppress the DKP side reaction.[90] pNZ is regarded as an alternative to Z which has been widely utilized as the protecting group in peptide synthesis. pNZ could be selectively cleaved off the corresponding peptide precursor by catalytic hydrogenation, or reducing agents such as $SnCl_2$ or $Na_2S_2O_4$ that could reduce nitro compound.[91] The mechanism of pNZ-reductive degradation is depicted in Fig. 1.24. Firstly, pNZ is reduced to *p*-aminobenzyloxycarbonyl that subsequently undergoes spontaneous collapse to afford quinone imine methide and the corresponding carbamate derivative. The latter is converted to amine via decarboxylation process. Similarly, $SnCl_2$ could reduce N^{α}-pNZ derivative to the corresponding amine in the presence of an acidic catalyst. The advantages of pNZ over Fmoc are mainly reflected by its removal condition that is devoid of base treatment, facilitating in this manner the suppression of DKP formation that is catalyzed by bases like piperidine. Resembling N^{α}-Trt strategy the N^{α}-functional group liberated upon pNZ removal could be directly subject to the following acylation reaction without a separate neutralization step, while undergoing *in situ* neutralization-acylation tandem process in the presence of

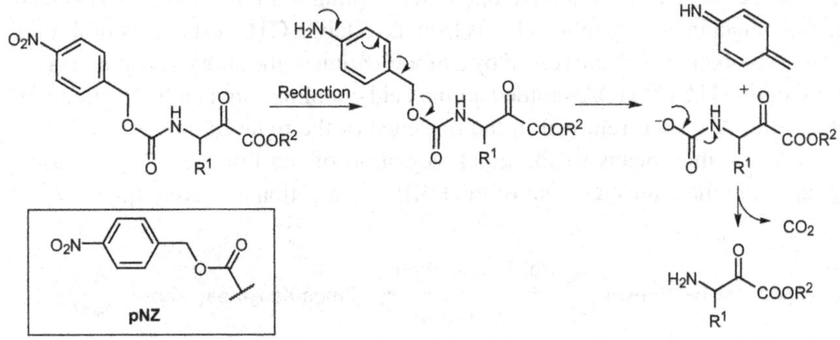

FIGURE 1.24 Mechanism of pNZ cleavage.

an upcoming amino acid coupling reagent and DIEA. The DKP side reaction would be maximally restrained by this means. It is noted, nevertheless, that solid peptide synthesis employing the pNZ strategy should not be carried out in the presence of incompatible solid supports such as CTC resin, since the condition for the pNZ removal might cause peptide premature cleavage from CTC resin and lead to a yield drop as a consequence.[75]

Alloc is another orthogonal N^{α}-protecting group utilized in peptide synthesis. Since Alloc could be removed by Pd(0) catalyst like $Pd(PPh_3)_4$ in the presence of allyl-scavenger,[92] this neutral cleavage condition would consequently lower the extent of DKP formation relative to that occurred in base-mediated Fmoc removal reaction.[75,93] A "tandem Alloc deprotection-amino acid coupling" synthetic strategy has been dedicatedly developed to take advantage of the N^{α}-Alloc cleavage with respect to the DKP formation suppression,[93] addressing the challenging peptide synthesis that is considerably susceptible to DKP side reactions. As is described in Fig. 1.25, if -Xaa_2-Xaa_1- happens to be the sequence that has a high inclination to cyclize and form a DKP derivative during the process of peptide assembly, Xaa_2 could be incorporated into the target peptide chain in the manner of Alloc-Xaa_2-OH to react with Xaa_1-O-Resin and construct the desired Alloc-Xaa_2-Xaa_1- sequence. The N^{α}-Alloc protecting group is subsequently cleaved by $Pd(PPh_3)_4/PhSiH_3$ in the presence of the upcoming amino acid Xaa_3 in its preactivated form as Fmoc-Xaa_3-F. Under such neutral condition the liberated N^{α}-functional group on Xaa_2 could be immediately entrapped by Fmoc-Xaa_3-F, giving rise to the target intermediate Fmoc-Xaa_3-Xaa_2-Xaa_1-O-Resin. This process could maximally shorten the existence of unacylated N^{α}-Xaa_2, lowering by this means the extent of the potential DKP side reaction. This synthetic strategy is established based on the feature that the N^{α}-Alloc protecting group could be removed under neutral condition, as well as a unique property of Fmoc-Xaa-F in that this activated derivative could lead amine acylation reaction even in the absence of DIEA.[94] The combination of the above merits maximally avoids the base employment in the process of DKP-sensitive peptide assembling. The severity of DKP side reaction could by this means be alleviated. This strategy is nonetheless not compatible with CTC resin since the *in situ* formed HF upon the coupling of Fmoc-Xaa-F might cause partial detachment of peptide chains from acid labile CTC resin. This limitation could be somehow circumvented by a modification of the above procedure. Two consecutive HATU/HOAt-guided amino acid couplings are conducted immediately after N^{α}-Alloc removal in the presence of the reduced amount of DIEA, assuring by this means of the quick acylation of the liberated N^{α} functional groups, and the reduced extent of the DKP side reaction as a consequence.[75]

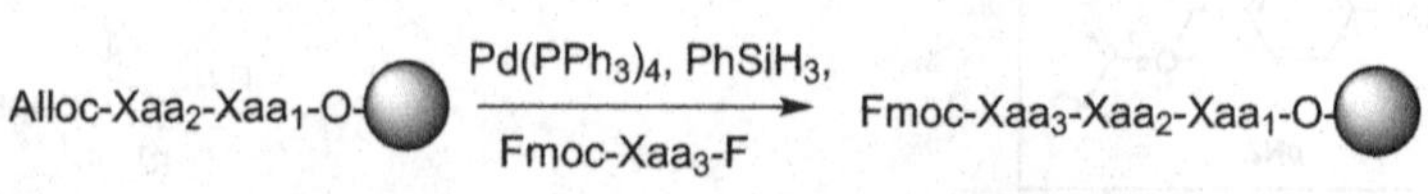

FIGURE 1.25 Tandem Alloc deprotection-amino acid coupling synthetic strategy.

The choice of coupling reagents will affect the extent of the DKP formation. In spite of their relative weaker efficiencies compared with uronium or phosphonium coupling reagents, carbodiimide derivatives like DIC or DCC are preferred for the synthesis of sensitive peptides that are prone to undergo DKP formation at the subjected coupling steps.[75] The preference is attributed to the intrinsic property of carbodiimide derivatives that they could mediate the amino acid couplings in the absence of tertiary bases. DKP side reactions are restrained in this manner by virtue of utilization of carbodiimide coupling reagents for the amino acid couplings at sensitive sequences.

Since DKP formation might prefer to take place on *C*-terminal dipeptide acid sequence, the properties of spacers on the solid support could decisively determine the propensity of the referred peptide to suffer from the DKP formation during SPPS. The steric hindrance of the spacer is of crucial importance in this connection. Compared with a less sterically hindered spacer like HMPA or 4-hyrdoxymethylphenoxy on Wang resin, the 2-chlorotrityl spacer on the CTC resin could effectively suppress the DKP side reaction by virtue of its predominant steric hindrance.[95]

Features of the leaving group such as the steric effect also exerts obvious impacts on DKP formation.[96] More hindered *tert*-butyl ester could replace benzyl ester as the temporary protecting group on *C*-terminal carboxy in solution phase peptide synthesis so that the extent of potential DKP formation could be lowered.

As for the cases of peptide assembly in which DKP formation are extremely liable to occur, employment of the dipeptide building blocks could effectively prevent or alleviate this side reaction. The rationale of this strategy is illustrated in Fig. 1.26. If a peptide with sequence H-Xaa_2-Xaa_3-Xaa_4- **39** is prone to undergo the DKP side reaction to form Xaa_2-Xaa_3 diketopiperazine derivative **42** and a peptide fragment **43** with truncated sequence, the desired acylation of **39** by amino acid building block Fmoc-Xaa_1-OH **40** would not result in the

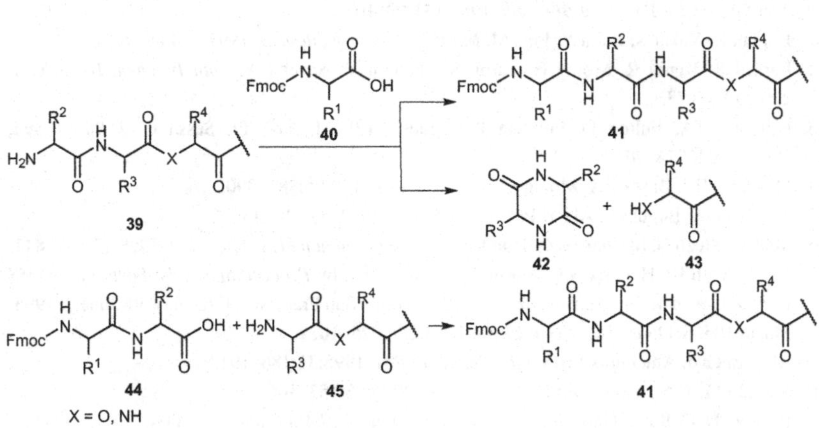

FIGURE 1.26 Avoidance of DKP formation via employment of dipeptide building block.

expected target product Fmoc-Xaa_1-Xaa_2-Xaa_3-Xaa_4- **41**. In order to prevent this uncontrollable side reaction induced by the temporary existence of H-Xaa_2-Xaa_3-Xaa_4- sensitive sequence, new synthetic route is designed in an effort to bypass the formation of the instable intermediate H-Xaa_2-Xaa_3-Xaa_4- that is susceptible to DKP formation. Inspired by this idea the original stepwise coupling strategy could be replaced by a convergent segment condensation method with which dipeptide Fmoc-Xaa_1-Xaa_2-OH **44** is employed as a building block to be assembled to H-Xaa_3-Xaa_4- intermediate **45** to construct the target product Fmoc-Xaa_1-Xaa_2-Xaa_3-Xaa_4- **41**. This synthetic mode could effectively avoid the formation of DKP-sensitive intermediate H-Xaa_2-Xaa_3-Xaa_4- and thus circumvent this severe side reaction. The effects of the strategy of employing dipeptide building blocks have already been verified by many cases of peptide synthesis in which DKP formation are especially liable.[97] Even though this methodology is limited by its inherent shortcomings such as the high tendency of the racemization of dipeptide upon activation. It affords nonetheless an appropriate alternative for the preparation of highly DKP-prone peptides.

REFERENCES

1. Fang W, Bennett MA, Murray TF, Aldrich JV. In: Blondelle SE, ed. *Understanding Biology Using Peptides*. American Peptide Society; 2005:533.
2. Sohma Y, Hayashi Y, Skwarczynski M, et al. *Biopolymers*. 2004;76:344–356.
3. Mutter M, Arunan C, Boyat C, et al. *Angew Chem Int Ed*. 2004;43:4172–4178.
4. Carpino LA, Krause E, Sferdean CD, et al. *Tetrahedron Lett*. 2004;45:7519–7523.
5. Taniguchi A, Yoshiya T, Abe N, et al. *J Pept Sci*. 2007;13:868–874.
6. Creighton CJ, Romoff TT, Bu JH, Goodman M. *J Am Chem Soc*. 1999;121:6786–6791.
7. Rubini C, Osler A, Calderan A, Guiotto A, Ruzza P. *J Pept Sci*. 2008;14:989–997.
8. Piszkiewicz D, Landon M, Smith EL. *Biochem Biophys Res Commun*. 1970;40:1173–1178.
9. Crimmins DL, Mische SM, Denslow ND. *Curr Protoc Prot Sci*. 11.4.1–11.4.11.
10. Tarr GE, Crabb JW. *Anal Biochem*. 1983;131:99–107.
11. Kaplan B, Yakar S, Balta Y, Pras M, Martin B. *J Chromatogr B*. 1997;704:69–76.
12. Lamed R, Kenig R, Morag E, Yaron S, Shoham Y, Bayer EA. *Appl Biochem Biotechnol*. 2001;90:67–73.
13. Fairchild TA, Fulton D, Fontana JT, Gratton J-P, McCabe TJ, Sessa C. *J Biol Chem*. 2001;276:26674–26679.
14. Shimon LJW, Bayer EA, Morag E, et al. *Structure*. 1997;5:381–390.
15. Tavares GA, Béguin P, Alzari PM. *J Mol Biol*. 1997;273:701–713.
16. Mák M, Mezö G, Skribanek Zs, Hundecz F. *Rapid Commun Mass Spectrom*. 1998;12:837–842.
17. Yu W, Vath JE, Huberty MC, Martin SA, Scoble HA. In: *Proceedings of the Forty-first ASMS Conference on Mass Spectrometry and Allied Topic*. San Francisco, CA, May 30–June 4, 1993.
18. Mazur RH, Schlatter JM. *J Org Chem*. 1963;28:1025–1029.
19. Auwera Cvd, Antenunis MJO. *Int J Pept Prot Res*. 1998;31:186–191.
20. Teixidó M, Albericio F, Giralt E. *J Pept Res*. 2005;65:153–166.
21. Boturyn D, Coll J-L, Garanger E, Favrot M-C, Dumy P. *J Am Chem Soc*. 2004;126:5730–5739.
22. Miki M, Dosremedios CG. *J Biochem Tokyo*. 1988;104:232–235.

23. Carrigan C, Imperiali B. *Anal Biochem*. 2005;341:290–298.
24. Song A, Wang X, Zhang J, Marik J, Lebrilla CB, Lam KS. *Bioorg Med Chem Lett*. 2004;14: 161–165.
25. Park SI, Renil M, Vikstrom B, et al. *Lett Pept Sci*. 2002;8:171–178.
26. Julian M, Hernandez A, Maurras A, et al. *Tetrahedron Lett*. 2009;50:260–263.
27. Edman PV. *Nature*. 1956;177:667–668.
28. Hoeg-Jensen T. *Phosphorus Sulfur Silicon Relat Elem*. 1996;108:257–278.
29. Sherman DB, Spatola AF. *J Am Chem Soc*. 1990;112:433–441.
30. Kessler H, Geyer A, Matter H, Göck M. *Int J Pept Protein Res*. 1992;40:25–40.
31. Walter W, Schaumann E. *Chem Ber*. 1971;104:3361–3377.
32. Curphey TJ. *Tetrahedron Lett*. 2002;43:371–373.
33. Cava MP, Levinson MI. *Tetrahedron*. 1985;41:5061–5087.
34. Clausen K, Thorsen M, Lawesson S-O. *Tetrahedron*. 1981;37:3635–3639.
35. Yokoyama M, Hasagawa Y, Hatanaka H, Yakazoe Y, Imamoto T. *Synthesis*. 1984;:827–829.
36. Shalaby MA, Grote CW, Rapoport H. *J Org Chem*. 1996;61:9045–9048.
37. Clausen K, Thorsen M, Lawesson S-O, Spatola AF. *J Chem Soc Perkin Trans*. 1984;1:785–797.
38. Brown DW, Campbell MM, Chambers MS, Walker CV. *Tetrahedron Lett*. 1987;28:2171–2174.
39. Miwa JH, Margarida LA, Meyer AE. *Tetrahedron Lett*. 2001;42:7189–7191.
40. Batjargal S, Wang YJ, Goldberg JM, Wissner RF, Petersson EJ. *J Am Chem Soc*. 2012;134: 9172–9182.
41. Miwa JH, Pallivathucal L, Gowda S, Lee KE. *Org Lett*. 2002;4:4655–4657.
42. Wuts PGM, Greene TW. *Greens's Protective Groups in Organic Synthesis*. Hoboken, NJ: John Wiley, Sons, Inc.; 2007:730.
43. Frank R, Schutkowski M. *Chem Commun*. 1996;:2509–2510.
44. Wildemann D, Drewello M, Fischer G, Schutkowski M. *Chem Commun*. 1999:1809–1810.
45. Stafford JA, Brackeen MF, Karanewsky DS, Valvano NL. *Tetrahedron Lett*. 1993;34: 7873–7876.
46. Cezari MH, Juliano L. *Pept Res*. 1996;9:88–91.
47. Rink H, Sieber P, Raschdorf F. *Tetrahedron Lett*. 1984;25:621–624.
48. Gazis E, Bezas B, Stelakatos GC, Zervas L. In: Young GT, ed. *Peptides 1962*. Oxford: Pergamon; 1963:17.
49. Jaeger G, Geiger R. *Chem Ber*. 1970;103:1727–1747.
50. Isidro-Llobet A, Álvarez M, Albericio F. *Chem Rev*. 2009;109:2455–2504.
51. Lundt BF, Johansen NL, Volund A, Markussen J. *Int J Pept Prot Res*. 1978;12:258–268.
52. Verdini AS, Lucietto P, Fossati G, Giordani C. In: Smith JA, Rivier JE, eds. *Peptides, Chemistry and Biology, Proceedings of the 12th American Peptide Symposium*. Leiden, The Netherlands: ESCOM; 1992:562–563.
53. Verdini AS, Lucietto P, Fossati G, Giordani C. *Tetrahedron Lett*. 1992;33:6541–6542.
54. Du Vigneaud V, Gish DT, Katsoyannis PG, Hess GP. *J Am Chem Soc*. 1958;80:3355–3358.
55. Jones DA, Miculec RA, Mazur RH. *J Org Chem*. 1973;38:2865–2869.
56. Wuensch E, Wendlberger G. *Chem Ber*. 1967;100:160–172.
57. Arzeno HB, Bingenheimer W, Blanchette R, Morgans Jr DJ, Robinson J. *Int J Pept Prot Res*. 1993;41:342–346.
58. Ponsati B, Canas M, Jodes G, Clemente J, Barcadit J. PCT Int. Appl. 2000, WO 2000071570 Al, 2000; *Chem Abstr*. 134:17728.
59. Sureshbabu VV, Narendra N. In: Hughes AB, ed. *Amino Acids, Peptides and Proteins in Organic Chemistry, vol. 4. Protection Reactions, Medicinal Chemistry, Combinatorial Synthesis*. Weinheim: Wiley-VCH Verlag; 2011:43.

60. Atherton E, Cammish LE, Goddard P, Richards JD, Sheppard RC. In: Ragnarsson U, ed. *Peptides 1984*. Stockholm: Almqvist & Wiksell Int; 1985:153.
61. Noda M, Kiffe M. *J Pept Res*. 1997;50:329–335.
62. Bergmann M, Zervas L, Rinke H. *Hoppe-Seyler's Z Physiol Chem*. 1934;224:40–44.
63. Bodanszky M, Sheehan JT. *Chem Ind*. 1960:1268–1269.
64. Paul R, Anderson GW, Callahan FM. *J Org Chem*. 1961;26:3347–3350.
65. Bodanszky M. *Principles of Peptide Synthesis*. Berlin: Sringer-Verlag; 1984:158–201.
66. (a) Battersby JE, Hancock WS, Canova-Davis E, Oeswein J, O'Connor B. *Int J Pept Prot Res*. 1994;44:215–222. (b) Kertescher U, Bienert M, Krause E, Sepetov N, Mehlis B. *Int J Pept Prot Res*. 1993;41:207–211. (c) Steinberg S, Bada JL. *Science*. 1981;213:544–545.
67. Gomes P, Vale N, Moreira R. *Molecules*. 2007;12:2484–2506.
68. Dinsmore CJ, Beshore D. *Tetrahedron*. 2002;58:3297–3312.
69. Witiak DT, Wei Y. *Prog Drug Res*. 1990;35:249–363.
70. Field GB, Noble RL. *Int J Pept Prot Res*. 1990;35:161–214.
71. Grathwohl C, Wüthrich K. *Biopolymers*. 1976;15:2043–2057.
72. Kessler H. *Angew Chem Int Ed*. 1970;9:219–235.
73. Kolaskar AS, Sarathy KP. *Biopolymers*. 1980;19:1345–1355.
74. Khosla MC, Smeby RR, Bumpus FM. *J Am Chem Soc*. 1972;94:4721–4724.
75. Bayó-Puxan N, Tulla-Puche J, Albericio F. *Eur J Org Chem*. 2009;:2957–2974.
76. (a) Stewart DE, Sarkar A, Wampler JE. *J Mol Biol*. 1990;214:253–260. (b) Ramachandean GN, Mitra AK. *J Mol Biol*. 1976;107:85–92.
77. (a) Rydon HN, Smith PWG. *J Chem Soc*. 1956:3642–3650. (b) Hardy GW, Lowe LA, Sang PY, et al. *J Med Chem*. 1988;31:960–966.
78. (a) Bodanszky M, Martinez J. *Special Methods in Peptide Synthesis, Part B*. New York: Academic; 1983:5; p. 112. (b) Barany G, Merrifield B. *Methods of Peptide Synthesis, Part A*. New York: Academic; 1980:2; p. 1. (c) Schröder E, Lübke K. *Methods of Peptide Synthesis*. New York: Academic; 1965:p. 1.
79. Nutt RF, Holly FW, Homnick C, Hirschmann R, Veber DF, Arison BH. *J Med Chem*. 1981;24:692–698.
80. Goodman M, Stueben K. *J Am Chem Soc*. 1962;84:1279–1283.
81. Grathwohl C, Wüthrich K. *Biopolymers*. 1976;15:2025–2041.
82. Cappasso S, Mazzarella L. *J Chem Soc Perkin Trans*. 1999;2:329–332.
83. Gisin BF, Merrifield RB. *J Am Chem Soc*. 1972;94:3102–3106.
84. Pedroso E, Granadas A, de las Heras X, Eritja R, Giralt E. *Tetrahedron Lett*. 1986;27:743–746.
85. Capasso S, Vergara A, Mazzarella L. *J Am Chem Soc*. 1998;120:1990–1995.
86. (a) Meienhofer J. *J Am Chem Soc*. 1970;92:3771–3777. (b) Wenger RM. *Helv Chim Acta*. 1983;66:2672–2702.
87. Besser D, Greiner G, Reissmann S. *Lett Pept Sci*. 1998;5:299–303.
88. (a) Ueki M, Amemiya M. *Tetrahedron Lett*. 1987;28:6617–6620. (b) Coin I, Beerbaum M, Schmieder P, Bienert M, Beyermann M. *Org Lett*. 2008;10:3857–3860.
89. Alsina J, Giralt E, Albericio F. *Tetrahedron Lett*. 1996;37:4195–4198.
90. Isidro-Llobet A, Guasch-Camell J, Álvarez M, Albericio F. *Eur J Org Chem*. 2005:3031–3039.
91. Isidro-Llobet A, Álvarez M, Albericio. *Tetrahedron Lett*. 2005;46:7733–7736.
92. Guibé F. *Tetrahedron*. 1998;54:2967–3042.
93. Thieriet N, Alsina J, Giralt E, Guibé F, Albericio F. *Tetrahedron Lett*. 1997;38:7275–7278.
94. (a) Wenschuh H, Beyermann M, El-Faham A, Ghassemi S, Carpino LA, Bienert M. *J Chem Soc Chem Commun*. 1995:669–670. (b) Wenschuh H, Beyermann M, Rothemund S, Carpino LA, Bienert M. *Tetrahedron Lett*. 1995;36:1247–1250.

95. (a) Barlos K, Gatos D, Kapolos S, Papaphotiu G, Schäfer W, Yao W. *Tetrahedron Lett.* 1989;30:3947–3950. (b) Barlos K, Gatos D, Papaphotiu G, Schäfer W. *Liebigs Ann Chem.* 1993:215–220.
96. Borsuk K, van Delft FL, Eggen IF, ten Kortenaar PBW, Petersen A, Rutjes FPJT. *Tetrahedron Lett.* 2004;45:3585–3588.
97. (a) Celma C, Albericio F, Pedroso E, Giralt E. *Pept Res.* 1992;5:62–71. (b) Gillessen D, Felix AM, Lergier W, Studer RO. *Helv Chim Acta.* 1970;53:63–72. (c) Ainpour PA, Wickstrom E. *Int J Pept Prot Res.* 1980;15:225–235.

Chapter 2

β-Elimination Side Reactions

β-Elimination is a group of common side reactions that predominantly affect peptides bearing an electron-withdrawing substituent located on the side chain C^{β} position, e.g., Cys and phosphorylated Ser/Thr. These peptides could suffer from β-elimination mostly under base treatment. The consequence of this side reaction is the elimination of the substituent on C^{β} and the formation of dehydroalanine, and/or corresponding relevant adducts. Side chain protecting groups on Cys and base utilized for Fmoc deprotection have the most significant impacts on the tendency of the subjected derivative to undergo the β-elimination process.

2.1 β-ELIMINATION OF Cys SULFHYDRYL SIDE CHAIN

In nature there is a plethora of peptides with Cys residue located at their C-terminal position, e.g., somatostatin,[1] conotoxin[2] and defensin.[3] Chemical preparation of these natural compounds arouses considerable interests in the territory of peptide synthesis. During these investigations it has been figured out that one of the most destructive side reactions interrelated to Cys is its instinctively high tendency to suffer from racemization upon the activation of its carboxyl group.[4] This notorious challenge could be tackled from diverse aspects, e.g., Cys side chain protecting group, coupling reagent, solvent, organic base for the coupling reaction, and so on.[5] This side reaction will be discussed in a separate chapter.

Another side reaction frequently interfering with Cys residue is the β-elimination on its sulfhydryl side chain.[6] The mechanism of this process is depicted in Fig. 2.1. As is known, the Fmoc deblocking process in peptide synthesis is realized through an E1cB mechanism: initiated with the abstraction of the acidic hydrogen by a base such as piperidine. The formed carbamate derivative will subsequently undergo decarboxylation spontaneously to liberate the shielded amino group. In this process, however, H^{α} on Cys residue in the parental peptide is vulnerable to the base treatment, and the protected sulfhydryl derivative suffers from the degradation by means of splitting-off the substituent on the β-position on the Cys side chain, giving rise to the formation of a dehydroalanine intermediate. The generated dehydroalanine derivative would subsequently accommodate a nucleophilic attack from a piperidine molecule, and be transformed into 3-(1-piperidinyl)alanine adduct with a +51 amu molecular weight deviation from the target Cys-containing product. If the chosen base for Fmoc deblocking is

Side Reactions in Peptide Synthesis. http://dx.doi.org/10.1016/B978-0-12-801009-9.00002-1

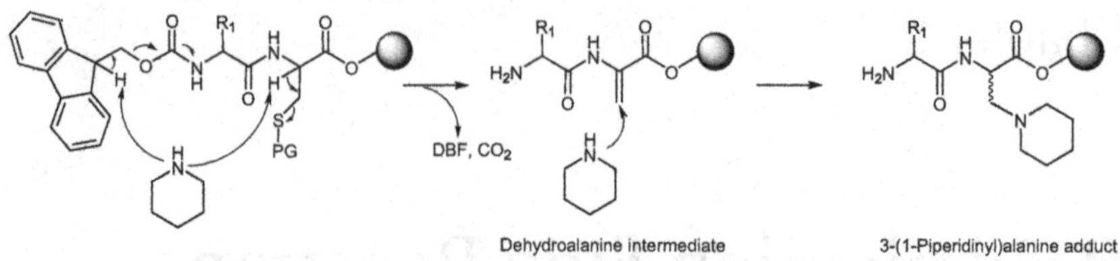

FIGURE 2.1 Piperidine-induced Cys β-elimination and subsequent piperidine addition.

4-methylpiperidine, the derived impurity originated from the above process will be increased by 65 amu in molecular weight,[6] in compliance with the proposed base-induced Cys β-elimination/Michael addition mechanism.

It has been discovered that various side-chain protected Cys derivatives, e.g., Cys(StBu),[7] Cys(Acm) and Cys(Trt)[6] could all be subject to the β-elimination side reaction, and the susceptibility to this process is communally controlled by the acidity of the subjected Cys-H^{α} and the properties of the protected sulfhydryl derivatives in terms of their leaving tendencies. Concretely speaking, the acidity of Cys-H^{α} relies on the following factors: (1) hydrogen on the parental Cys-N^{α}; (2) carboxyl group from Cys that is immobilized on the solid support; (3) protecting groups on the sulfhydryl side chain, normally in the form of Trt or Acm, whose intrinsic electron-withdrawing features endue the corresponding sulfhydryl derivatives with divergent leaving tendencies. Understandably, the superior leaving group could advantageously enhance the kinetics of the Cys β-elimination process.[8–10]

In a systematic investigation about the protecting group effects on Cys β-elimination[6] peptide H-Asp-Glu-Gln-Glu-Ala-Leu-Asn-Ser-Cys-OH/NH_2 was opted as the substrate containing C-terminal Cys structure. Under the same conditions with regard to resin, coupling reactions and the Fmoc deblocking reagent, striking differences were detected when Cys(Trt) or Cys(Acm) were utilized as the building block for the peptide assembly. No Cys β-elimination derived impurities were detected in the crude prepared from Cys(Trt), whereas almost half of the target product was transformed into the corresponding 3-(1-piperidinyl)alanine impurity when Cys(Acm) was adopted as the *C*-terminal Cys-building block. Other relevant studies have also reinforced the conclusion that Trt is superior to other sulfhydryl protecting groups in the context of suppressing the Cys β-elimination side reaction.[11] The inclination of Cys derivatives protected at its sulfhydryl side chain by diverse protecting groups to undergo β-elimination is aligned in descending order as Cys(StBu) ≥ Cys(Acm) > >Cys(Trt) (see also Fig. 2.2). Moreover, a new type of sulfhydryl protecting group Xan[12] has exhibited outstanding performances in terms of restraining the Cys β-elimination side reaction.[11]

The correlation between the susceptibility of Cys β-elimination and the types of employed solid supports in the corresponding SPPS have been investigated

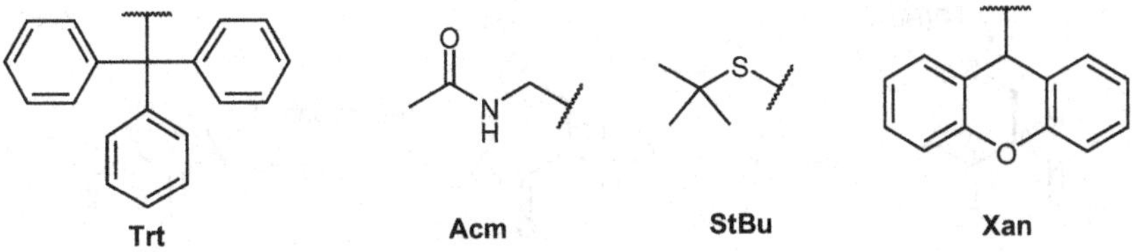

FIGURE 2.2 **Cys side chain sulfhydryl protecting groups.**

in the same study.[6] It was figured out that when Fmoc-Cys(Trt)-OH was selected as the building block for the assembly of the peptide H-Asp-Glu-Gln-Glu-Ala-Leu-Asn-Ser-Cys-OH on PAC-PEG/PS backbone resin no detectable Cys β-elimination impurities were generated, whereas Cys β-elimination was apparently induced when PS-backbone resin was employed as the solid support for the peptide assembly under otherwise identical conditions. This discrepancy indicates that PEG/PS hybrid resin disfavors Cys β-elimination compared with PS resin. However, when Fmoc-Cys(Acm)-OH was chosen as the starting material for the preparation of the same peptide, even the utilization of PAC-PEG/PS-based resin could not afford the target product while abundant 3-(1-piperidinyl) alanine impurity stemmed from Cys β-elimination was formed instead.

It is noted that Cys β-elimination takes place predominantly upon the piperidine treatment of the peptide derivative after the N^{α} from the concerned Cys is acylated by the upcoming amino acid, Fmoc-Cys(protecting group)-resin is nonetheless resistant to this side reaction.[6] This phenomenon might be attributed to the relatively weaker electron-withdrawing effect of unacylated Cys-N^{α}, while the enhanced electron-withdrawing impact of the acylated derivative would reasonably result in the elevated acidity of the Cys-H^{α} which in turn promotes the sulfhydryl β-elimination. Interestingly, when the position of the concerned Cys residing on the *C*-terminus of the original peptide is swapped with its *N*-terminal neighboring residue Ser, that is to say, when peptide H-Asp-Glu-Gln-Glu-Ala-Leu-Asn-Cys(Acm)-Ser-OH becomes the target via this intentional sequence modification, the haunting Cys β-elimination side reaction perplexing the original peptide synthesis will be tackled accordingly. This strikingly impressive comparison justifies the rule that the Cys β-elimination side reaction is predominantly affecting Cys residues located on the C-terminal positions.

Cys β-elimination could also take place in the $P(Bu)_3$-mediated Cys(StBu) side chain deblocking reaction.[13] The degraded *tert*-butyl thiolate released upon the reduction of Cys(StBu) by $P(Bu)_3$ could pull the H^{α} from the concerned Cys that induces the subsequent β-elimination of the sulfhydryl derivative functional group in the form of tributyl phosphine sulfide. The mechanism of this process is described in Fig. 2.3. Bases other than piperidine capable of inducing Cys β-elimination include Na/liq. NH_3 (reagent for Cys(Bzl) side chain deblocking),[14] NaOH,[15] and hydrazine.[16] Interestingly, a strong acid, e.g., HF, could invoke this side reaction as well.[17]

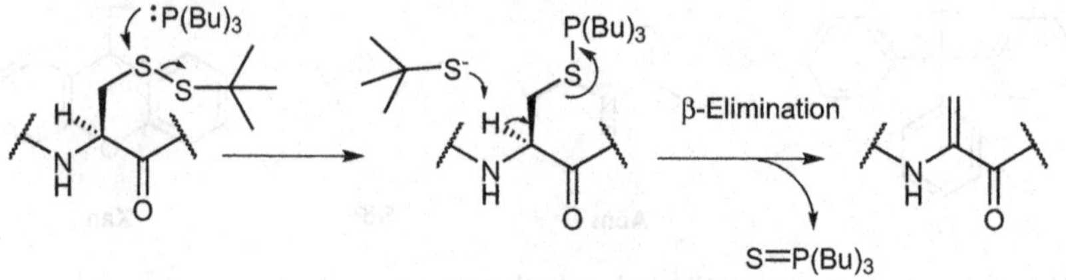

FIGURE 2.3 Proposed mechanism of the phosphine-induced Cys(StBu) side chain β-elimination.

Solutions addressing Cys β-elimination side reactions have been accordingly developed and tested. Methods such as alternation of Cys sulfhydryl protecting group and the utilization of more appropriate solid support are advisable when the Cys β-elimination side reactions wreak havoc on the target synthesis. Fine-tuning of the Fmoc deblocking solution is another strategy to tackle the Cys β-elimination. Beyermann and coworkers has ever-successfully reduced the extent of aspartimide formation applying 20% piperidine/DMF containing 0.01 M HOBt as a Fmoc deblocking solution.[18] Unfortunately, this method has proven ineffective for avoiding the occurrence of Cys β-elimination[6] (even the replacement of HOBt with the more acidic HOAt could not fully prevent this side reaction). Nonetheless, utilization of 2% DBU/DMF solution as Fmoc deblocking reagent in combination with a continuous-flow Fmoc cleavage pattern in SPPS can diminish the Cys β-elimination.[19] Lowering the concentration of bases for Fmoc deblocking is also of help in this connection.

Barany et al. has designed the strategy of immobilization of Cys derivative Fmoc-Cys(^{2}XAL$_4$)-OtBu* via its side chain on the solid support (see also Fig. 2.4), which is proved to diminish the Cys β-elimination which prevailed otherwise.[20]

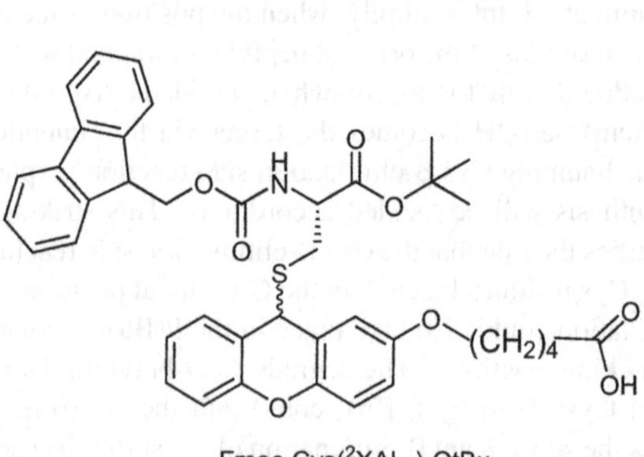

FIGURE 2.4 Chemical structure of Fmoc-Cys(^{2}XAL$_4$)-OtBu.

* ^{m}XAL$_n$: "m" symbolizes the substitution position of the linker, while "n" donates the number of the methylene groups in the linker.

One of the inherent advantages of this synthetic strategy lies in the fact that, if the referred Cys derivative suffers from side chain β-elimination during the SPPS process, the impurities induced by this means would be subsequently removed from the system simply by sufficient rinse of the affected peptidyl resin. The occurrence of this side reaction will unavoidably reduce the production yield. However, the formed side products will not be able to survive the following process, and will therefore not contaminate the crude. The potential risk that dehydroalanine and/or 3-(1-piperidinyl)alanine impurities, derived from the Cys β-elimination process might interfere with the purification of the target product is therefore minimized.

2.2 β-ELIMINATION OF PHOSPHORYLATED Ser/Thr

Phosphorylation–dephosphorylation equilibrium is regarded as one of the most important processes in cell regulation. Phosphorylation of peptide/protein could be realized either by means of enzyme catalysis or chemical synthesis. The most commonly applicable methods of peptide phosphorylation on solid phase could be basically classified into two categories: (1) utilization of prephosphorylated building blocks for the assembly of the target phosphopeptide; (2) global phosphorylation in a postsynthetic manner to phosphorylate the free hydroxyl functional groups on Ser, Thr, or Tyr substrate.[21] Both of the two strategies are associated with respective inherent advantages and disadvantages.

Phosphorylation of peptides with prephosphorylated amino acids as building blocks could utilize bis-protected derivatives Fmoc-Ser/Thr(PO_3R_2)-OH (illustrated in Fig. 2.5). Benzyl esters are normally opted as the protecting group in Fmoc phosphorylation chemistry that could be deblocked by TFA at peptide side chain global deprotection step.

Utilization of Fmoc-Ser/Thr(PO_3R_2)-OH (R = methyl, ethyl, *tert*-butyl, benzyl) building blocks for the preparation of phosphopeptides could potentially cause β-elimination side reaction[22,23] resembling Cys β-elimination. The mechanism of this process is illustrated in Fig. 2.6. The root cause of β-elimination of

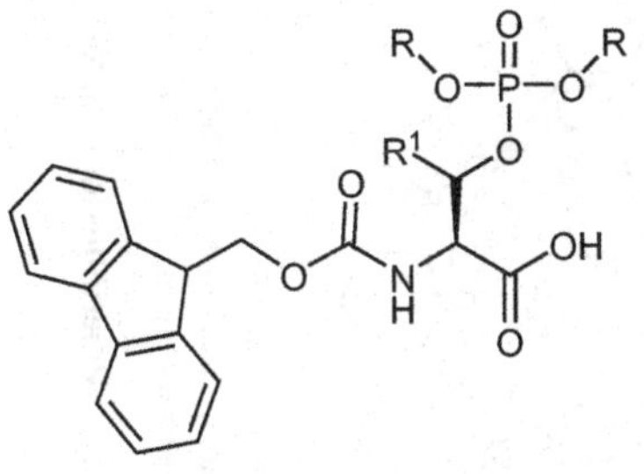

Fmoc-Ser/Thr(PO_3R_2)-OH
(Thr: R^1=CH_3; Ser: R^1=H)
R= methyl, ethyl, *tert*-butyl or benzyl

FIGURE 2.5 Chemical structure of Fmoc-Ser/Thr(PO_3R_2)-OH.

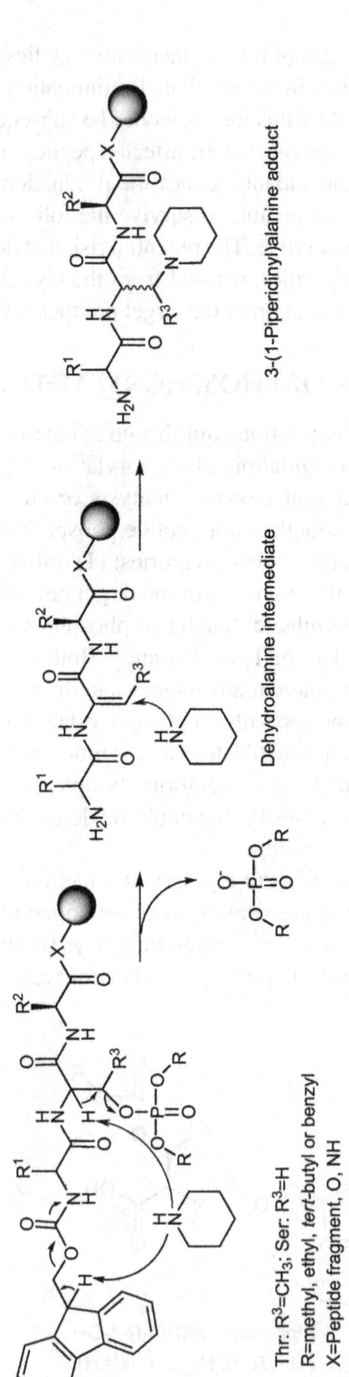

FIGURE 2.6 Piperidine-induced Ser/Thr(PO_3R_2) β-elimination.

phospho-serine/threonine is attributed to the piperidine treatment of N^{α}-Fmoc protected peptides. Acidic H^{α} on the subjected phospho-serine/threonine is abstracted by piperidine. The protected phosphoric acid is released in this process, and the degraded by-product dehydroalanine intermediate could be entrapped by piperidine to generate 3-(1-piperidine)alanine adduct.

In view of the high propensity of bis-protected phosphoamino acids, e.g., Ser/Thr(PO_3R_2)-OH (R = methyl, ethyl, *tert*-butyl or benzyl) to undergo β-elimination during peptide assembly, in combination with the outstanding utilities of their mono-protected counterparts Ser/Thr(PO_3R,H) (R = methyl, ethyl, *tert*-butyl, benzyl) (see also Fig. 2.7) with respect to phosphopeptide preparation,[24] these mono-protected phosphoamino acid derivatives have gained widespread applications in phosphopeptide synthesis in an effort to reduce the undesired β-elimination side reaction.[25,26] It has been verified by various investigations that mono-protected phosphoamino acids are more resistant to β-elimination relative to their bis-protected counterparts. This phenomenon might be attributed to the negative phosphate anion generated from the former during piperidine treatment which disfavors the release of the side chain phosphate derivative in a β-elimination manner.[25] Fmoc-Ser/Thr(PO_3R,H)-OH becomes nowadays the most frequently utilized phosphoamino acid building blocks in the Fmoc-mode SPPS of phosphopeptides.

Despite the fact that the mono-protected Fmoc-Ser/Thr(PO_3R,H)-OH building blocks have displayed improved performances in the context of β-elimination suppression compared with their bis-protected counterparts, this side reaction could not be thoroughly prevented in certain cases due to the extraordinarily high leaving tendencies of the subjected phosphate side chains. Under certain circumstances minimization of β-elimination in phosphopeptide synthesis process might be escalated to prioritized concerns. In light of the observations that phosphopeptide β-elimination takes place overwhelmingly at the base-mediated Fmoc deblocking step-optimization of this process will reasonably afford the solution to tackle this challenge.

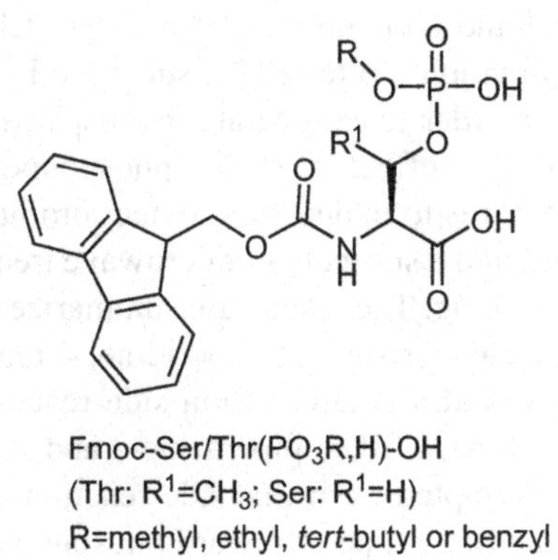

FIGURE 2.7 Chemical structure of Fmoc-Ser/Thr(PO_3R,H)-OH.

Factors like Fmoc deblocking reagents, location of the phosphoamino acids in the addressed phosphopeptides, the duration of the Fmoc deblocking treatment, and microwave irradiation have been opted and systematically investigated with respect to their influences on phosphopeptide β-elimination.[27] It has been revealed in the corresponding study that the addressed β-elimination took place majorly at the Fmoc cleavage step if Fmoc-Ser/Thr(PO_3R,H) resided at the *N*-terminus of the corresponding peptide chains. On the contrary, if the peptide assembly goes beyond the subjected Ser/Thr(PO_3R,H) residue, that is to say, when the phosphoamino acid is located in the *endo* position in the corresponding peptide sequence, the β-elimination side reaction on this residue upon base treatment will be attenuated. This finding contradicts nevertheless the conclusion drawn of *C*-terminal Cys sulfhydryl β-elimination elaborated in Section 2.1. However, this discrepancy alleges that utmost caution is supposed to be exerted to the step of Fmoc deprotection from Fmoc-Ser/Thr(PO_3R,H)-peptide derivative. It has also been figured out that Fmoc-Thr(PO_3Bzl,H)-peptide is much more resistant to β-elimination side reaction than its Fmoc-Ser(PO_3Bzl,H) counterpart.[27]

The extent of β-elimination on phosphopeptides is directly proportional to the duration of base treatment. This conclusion has been verified by the finding that elongated treatment of the phoshopeptidyl resin Fmoc-Ser(PO_3Bzl,H)-Val-Glu(OtBu)-resin by 20% piperidine/DMF solution will substantially increase the content of 3-(1-piperidyl)alanine adduct impurity detected in the crude product.[27] The energy input in the manner of microwave irradiation at the Fmoc deblocking step (condition: 20% piperidine/0.1 M HOBt/DMF, microwave 30 s, 40 W, 40°C) in SPPS of phosphopeptides would also drastically promote the occurrence of the β-elimination side reaction on the concerned phosphoamino acid residue.[27]

Routinely utilized Fmoc deblocking amines[28] have been addressed to systematic comparison in terms of their impacts on the stimulation of β-elimination on Fmoc-Ser/Thr(PO_3R,H)-peptides. Cyclohexylamine, DBU, piperazine, 4-methyl piperidine, morpholine, piperidine, and pyrrolidine are selected to this end and their impacts as a Fmoc deblocking reagent on phosphopeptide β-elimination are tested on substrate Fmoc-Ser(PO_3Bzl,H)-Val-Glu(OtBu)-resin. In order to exaggerate the disparities of these reagents with respect to their catalytic effects to induce phosphopeptide β-elimination, the concerned peptidyl resin undertakes twice ordinary base handlings (1×5 min, 1×20 min) and a subsequent microwave treatment (30 W, 40°C, 1.5 min) for Fmoc deblocking. The results are summarized in Table 2.1.[27] It is discernable through the comparison that piperidine, 4-methyl piperidine, and pyrrolidine have all invoked a β-elimination side reaction on the phosphoserine residue, while CHA-, DBU-, piperazine-, and morpholine-mediated Fmoc deprotection is exempted from this side reaction, among which CHA exhibits the most outstanding property toward phosphopeptide β-elimination suppression. This finding reinforces the conclusion that Fmoc deblocking

TABLE 2.1 Effects of Fmoc Deblocking Reagent on the Fmoc-Ser(PO_3Bzl,H)-Val-Glu(OtBu)-resin deFmoc Process

Amine	Structure	Concentration	Purity of the crude (%)
CHA	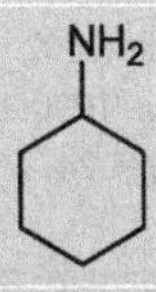	50% CHA/DCM	85.8
DBU	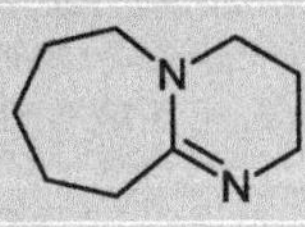	5% DBU/DMF	83.5
Piperazine	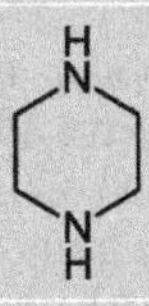	5% Piperazine/ DMF (w/v)	83.1
4-Methyl piperidine	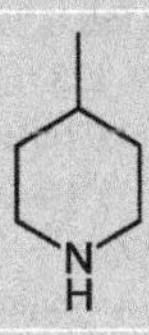	20% 4-Methyl piperidine/DMF	80.5
Morpholine	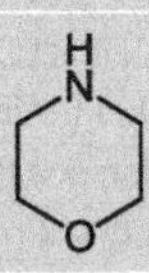	50% Morpholine/ DMF	79.3
Piperidine	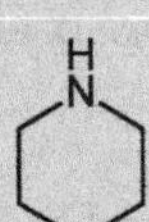	20% Piperidine/ DMF	73.6
Pyrrolidine	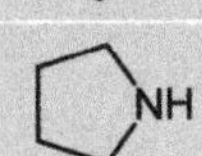	50% Pyrrolidine/ DMF	55.9

conditions for phosphopeptide are crucially important with respect to retaining the homogeneity of phosphopeptide, and minimize β-elimination during the assembly process.

REFERENCES

1. Alberti KG, Christensen NJ, Christensen SE, et al. *Lancet*. 1973;2:1299–1301.
2. McIntosh M, Cruz LJ, Hunkapiller MW, Gray WR, Olivera BM. *Arch Biochem Biophys*. 1982;218:329–334.
3. Ganz T. *Nat Rev Immunol*. 2003;3:710–720.

4. Kaiser T, Nicholson GJ, Kohlbau HJ, Voelter W. *Tetrahedron Lett.* 1996;37:1187–1190.
5. Han Y, Albericio F, Barany G. *J Org Chem.* 1997;62:4307–4312.
6. Lukszo J, Patterson D, Albericio F, Kates SA. *Lett Pept Sci.* 1996;3:157–166.
7. Erigja R, Ziehler-Martin JP, Walker PA, et al. *Tetrahedron.* 1987;43:2675–2680.
8. Photaki IJ. *J Am Chem Soc.* 1963;85:1123–1126.
9. Sokolovsky M, Wilchek M, Patchornik A. *J Am Chem Soc.* 1964;86:1202–1206.
10. Spande TF, Witkop B, Degani Y, Patchornik A. *Adv Protein Chem.* 1970;24:97–260.
11. Boulègue C, Musiol H-J, Prasad V, Moroder L. *Chim Oggi.* 2006;24:24–36.
12. Han Y, Barany G. *J Org Chem.* 1997;62:3841–3848.
13. Rijkers DTS, Kruijtzer JAW, Killian JA, Liskamp RM. *Tetrahedron Lett.* 2005;46:3341–3345.
14. Katsoyannis PG, Tometsko AM, Zalut C, Fukuda K. *J Am Chem Soc.* 1966;88:5625–5635.
15. Hiskey RG, Upham RA, Beverly GM, Jones WC. *J Org Chem.* 1970;35:513–515.
16. Maclaren JA, Savige WE, Swan JM. *Aust J Chem.* 1958;11:345–359.
17. Hallinan EA. *Int J Pept Protein Res.* 1991;38:601–602.
18. Dölling RM, Beyermann M, Haenal J, et al. *J Chem Soc Chem Commun.* 1994:853–854.
19. Wade JD, Bedford J, Sheppard RC, Tregear GW. *Pept Res.* 1991;4:194–199.
20. Barany G, Han Y, Hargittai B, Liu R-Q, Varkey JT. *Biopolymers.* 2003;71:652–666.
21. McMurray JS, Coleman IV DR, Wang W, Campbell ML. *Pept Sci.* 2001;60:3–31.
22. Lacombe JM, Andriamanampisoa F, Pavia AA. *Int J Pept Protein Res.* 1990;36:275–280.
23. Otvos L, Elekes I, Lee VM-Y. *Int J Pept Protein Res.* 1989;34:129–133.
24. Wakamiya T, Saruta K, Yasouka J, Kusumoto S. *Chem Lett.* 1993;22:1401–1404.
25. Wakamiya T, Saruta K, Yasouka J, Kusumoto S. *Chem Lett.* 1994;23:1099–1102.
26. Vorherr T, Bannwarth W. *Bioorg Med Chem Lett.* 1995;5:2661–2664.
27. Attard TJ, O'Brien-Simpson NM, Reynolds EC. *Int J Pept Res Ther.* 2009;15:69–79.
28. (a) Merrifield RB, Bach AE. *J Org Chem.* 1978;43:4808–4816. (b) Atherton E, Logan CJ, Sheppard RC. *J Chem Soc Perkin Trans.* 1981;1:538–546. (c) Chang C-D, Waki M, Ahmad M, Meienhofer J, Lundell EO, Haug JD. *Int J Pept Protein Res.* 1980;15:59–66. (d) Harrison JL, Petrie GM, Noble RL, Beilan HS, McCurdy SN, Culwell AR. In: Hugli TE, ed. *Techniques in Protein Chemistry.* San Diego: Academic Press; 1989:506–516. (e) Carpino LA, Sadat-Aalaee D, Beyermann M. *J Org Chem.* 1990;55:1673–1675.

Chapter 3

Peptide Global Deprotection/ Scavenger-Induced Side Reactions

Successful assembly of target peptide chains on solid supports does not necessarily indicate the final triumph of SPPS. The following detachment of immobilized peptides from solid supports (peptide cleavage) and the deblocking of the side chain protecting groups (global deprotection) are also extremely decisive for the success of the concerned peptide synthesis. HF[1] and TFA[2] are routinely utilized as acid for the purpose of peptide cleavage and global deprotection in the scope of Boc- and Fmoc-chemistry, respectively. The cleaved protecting groups are existing in the reaction systems in the form of carbocations,[3] sulfonyl cations,[4,5] or other derivatives[6,7] before being quenched by the corresponding scavenger. These reactive electrophilic species can nevertheless react with the liberated nucleophilic functional groups on the side chains of the peptides, e.g., thioether group on Met,[8,9] phenolic group on Tyr,[10] indolyl on Trp,[4,7,11] hydroxyl on Ser/Thr,[12] sulfhydryl on Cys,[13–15] and guanidino on Arg,[16] in either a reversible or irreversible manner, and give rise to diverse adduct side products. In order to prevent or minimize the occurrence of such side reactions, appropriate nucleophiles[3,17] are supposed to be concomitantly charged into TFA or HF cleavage solutions in order to protect peptide global deprotection reactions by *in situ* entrapping the degraded carbocations, sulfonyl cations, and other reactive species. By this means the homogeneities of the functional groups from the target peptides are assured of, in the process of peptide cleavage and global deprotection.

Even though the liberated functional groups from the target peptide products are protected by the shielding effect of the added scavengers at the step of peptide cleavage and global deprotection, diverse side reactions could still unavoidably take place and corresponding impurities are generated. The core content of this chapter will be focused on the behavior, mechanism and potential solution for these side reactions invoked in the process of peptide cleavage and global deprotection.

Side Reactions in Peptide Synthesis. http://dx.doi.org/10.1016/B978-0-12-801009-9.00003-3

3.1 *TERT*-BUTYLATION SIDE REACTION ON Trp DURING PEPTIDE GLOBAL DEPROTECTION

The vulnerable indolyl group from the Trp side chain could suffer from a plethora of side reactions during peptide global deprotection like *tert*-butylation;[18] alkylation by spacer cations degraded from resins;[19] sulfonation by Mtr,[7] Pmc,[20] or Pbf [21] ions derived from the corresponding Arg side chain protecting groups; oxidation;[22] and dimerization.[23,24] *Tert*-butylation side reactions affecting indolyl side chain on Trp will be elaborated in this section.

Basically, nucleophilic reagents utilized in the process of peptide assembly generally do not pose serious challenges on the Trp indolyl side chain (Trp-indolyl nitrozation could take place during the preparation of the acyl azide derivative[25]). Trp protected at its indolyl side chain is normally exempted from alkylation side reaction in an acid-free environment. However, upon the introduction of TFA that removes the protecting groups on Trp the originally shielded indolyl functional group is now exposed to the environment that contains abundant electrophilic species like *tert*-butyl cation, sulfonyl cation, acyl cation, benzyl cation, and so on. The irreversible reactions of the unprotected indolyl side chains on Trp with these reactive derivatives are, therefore, triggered in an acidic milieu.[21]

The indolyl side chain on Trp is considerably susceptible to electrophilic attacks under acidic conditions, and the modification is predominantly addressing its 2-position.[20,26] Other indolyl positions are also subject to the alkylation derivatization and multiple alkylation side reaction that could occur to Trp residue in some cases.[18] Trp N^{in} might be deprotonated upon the treatment with strong bases, and exposed to various side reactions in the form of indole anion. The predominant modification site under such circumstance is 1-position (see also Fig. 3.1).

Tert-butylation is one of the most frequently occurred alkylation side reactions that affect the nucleophilic Trp-indolyl side chain. In the process of Fmoc-mode peptide synthesis some reactive amino acid, e.g., Ser, Thr, Tyr, Asp, or Glu are protected at their side chains in the form of *tert*-butyl ethers, or *tert*-butyl esters. *Tert*-butyl cations are generated upon the TFA treatment of the protected peptide precursors consisting of these residues. The released *tert*-butyl cations might function further with the abundant TFA in the reaction system to form the reactive species *tert*-butyl trifluoroacetate,[6] which is regarded as one of the major culprits for *tert*-butylation side reaction induced in peptide global deprotection.[21] Other *tert*-butylation resource, such as Boc protecting group is decomposed upon TFA treatment. The possible mechanism of Boc acidolysis is as depicted in Fig. 3.2.[6,27] CO_2 and isobutene are released in this process, and the latter is protonated and transformed into *tert*-butyl cation in TFA solution. Or alternatively, isobutene might react with TFA to form *tert*-butyl trifluoroacetate. Meanwhile, it has also been reported that Boc is degraded directly into *tert*-butyl cation upon acidolysis.[28]

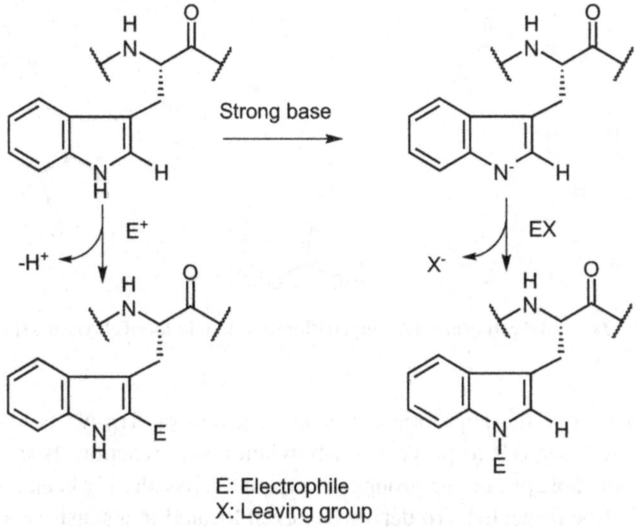

FIGURE 3.1 Electrophilic attack on Trp-indolyl side chain in acid/basic conditions.

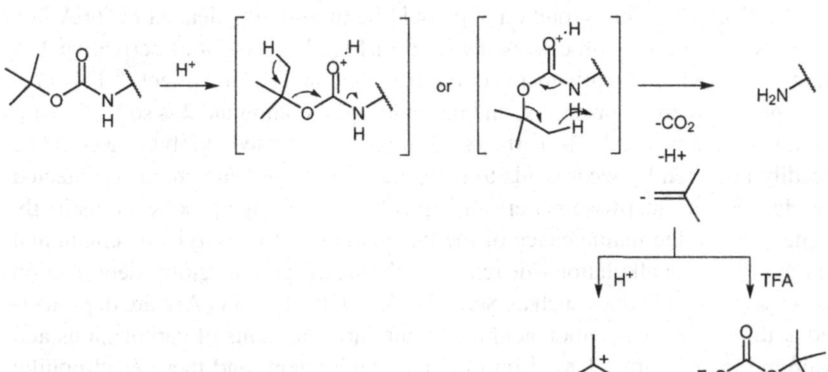

FIGURE 3.2 Proposed mechanism of Boc acidolysis process.

The abundant *tert*-butyl cations and *tert*-butyl trifluoroacetates generated in the peptide global deprotection could be potentially entrapped in the form of alkylating agents by various reactive nucleophilic functional groups on peptides, among which Trp-indolyl and Cys-sulfhydryl could be involved. The process of Trp *tert*-butylation is described in Fig. 3.3. The deprotected indolyl functional group on Trp could react with either *tert*-butyl cation or *tert*-butyl trifluoroacetate, and give rise to the corresponding *tert*-butylated Trp derivative with a molecular weight increase of +56 amu relative to the target product. This *tert*-butylation process could basically take place at all reactive positions on indolyl ring but overwhelmingly at 2-position.

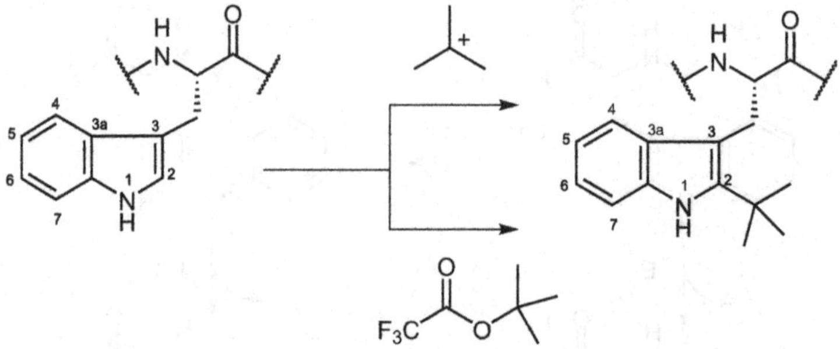

FIGURE 3.3 ***Tert*-butyl cation and *tert*-butyl trifluoroacetate-induced alkylation side reaction on Trp side chain.**

Introduction of the appropriate protecting group on Trp-N^{in} is one of the most effective methods to prevent *tert*-butylation side reaction. Boc is one of the most applicable protecting groups for Trp, and it has already been confirmed that the N^{in}-Boc protected Trp derivative, even located in a sensitive sequence toward Trp *tert*-butylation, is evidently unaffected by *tert*-butylation[21] and sulfonation[29] side reactions compared with its side chain unprotected counterpart.

N^{in}-Boc protecting group on Trp could be thoroughly cleaved by TFA in a stepwise mode. This process is depicted in Fig. 3.4. Trp(Boc) derivative **1** is first converted to its carbamate counterpart **2** upon TFA treatment.[30] This process proceeds pretty promptly, and the stability of compound **2** is so high that it could be even isolated. The carboxyl group on the affected indolyl ring could be readily removed by weak acids to regenerate the target side chain deprotected product **3**. This stepwise pattern of Trp N^{in}-Boc cleavage process is distinctly beneficial for the maintenance of the intactness of Trp indolyl side chain, and the avoidance of alkylation side reactions during the peptide global deprotection process. When residues such as Ser, Thr, Asp, Glu, Tyr, and Arg are deprotected at their side chains upon acid treatment, large amounts of carbocations and sulfonyl cations are released into the reaction system, and these electrophilic species might attack the liberated indolyl ring on the Trp side chain in case

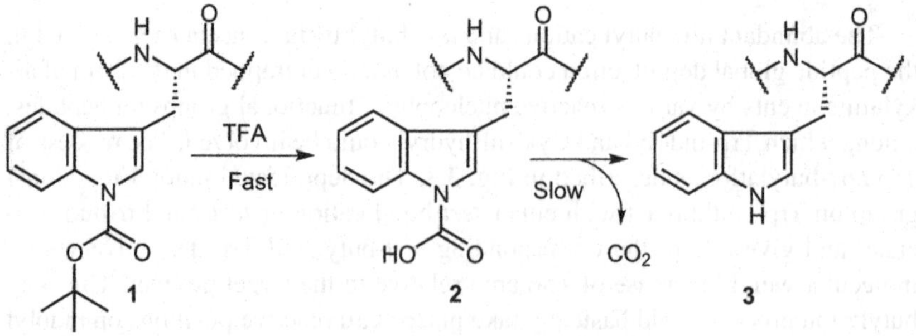

FIGURE 3.4 **Acid cleavage of N^{in}-Boc protecting group on Trp.**

they are not sufficiently scavenged. Meanwhile, N^{in}-Boc protecting groups on Trp undergo degradation simultaneously but the acidolysis process could be retained at the step of Trp-carbamate intermediate. The existence of carboxyl group on the Trp indolyl ring could not only protect the 1-position from potential electrophilic attacks by various reactive cation species, but also decrease the nucleophilicity on the other indolyl position which might be susceptible to divergent side reactions, e.g., alkylation or sulfonation, assuring by this means the integrity of the concerned Trp residue during TFA-mediated peptide global deprotection.[31]

Appropriate scavengers are routinely added into the peptide global deprotection reactions as to drive competitive reactions against TFA to entrap and deactivate the released *tert*-butyl cations. Moreover, the generated reactive *tert*-butyl trifluoroacetate would also be quenched by these added scavengers.[6] For example, reagent K (82.5% TFA/5% phenol/5% H_2O/5% thioanisole/2.5% EDT) could effectively suppress side reactions affecting the indolyl ring on Trp, during peptide global deprotection.[3] Among the constituents of reagent K EDT exhibits the most effective performance in terms of scavenging *tert*-butyl cations and *tert*-butyl trifluoroacetate.[6] On top of that water is another good reagent for quenching *tert*-butyl cations and *tert*-butyl trifluoroacetate.[3,32] The employment of these scavengers in peptide global deprotection in combination with the appropriate Trp side chain protection strategies could effectively suppress Trp-*tert*-butylation side reactions during peptide synthesis.

3.2 Trp ALKYLATION BY RESIN LINKER CATIONS DURING PEPTIDE CLEAVAGE/GLOBAL DEPROTECTION

Besides the *tert*-butylation side reaction that vulnerable Trp residue suffers at the peptide cleavage/global deprotection step, other undesired Trp modification could also be invoked in this process, such as degraded linker cations mediated Trp alkylation.[19] The major cause of this side reaction is the acid-induced acidolytic fragmentation of linker, anchored on the solid supports. The linker fragments are released into the solution phase in this process in the form of various cations, which could attack the indolyl ring on Trp by means of electrophilic addition. Or alternatively, if these reactive cationic derivatives still remain on the solid support unquenched in the peptide cleavage process, the reactions between these nucleophiles and the liberated Trp indolyl side chain would result in irreversible immobilization of the affected peptide chains on the solid supports[19,26] and drop of the production yield as a consequence. Taking the Wang resin and the HMPA resin for instance, both of these two acid-labile resins could afford corresponding peptide acid products upon the treatment by concentrated TFA solution. The desired fission between the C-terminus of the corresponding peptide chain, and the linker anchored on the solid supports releases the peptide product from the resin (see also Fig. 3.5, Site A). Under some circumstances, however, potentially acid-labile alkyl phenyl ether bond might also be subject to

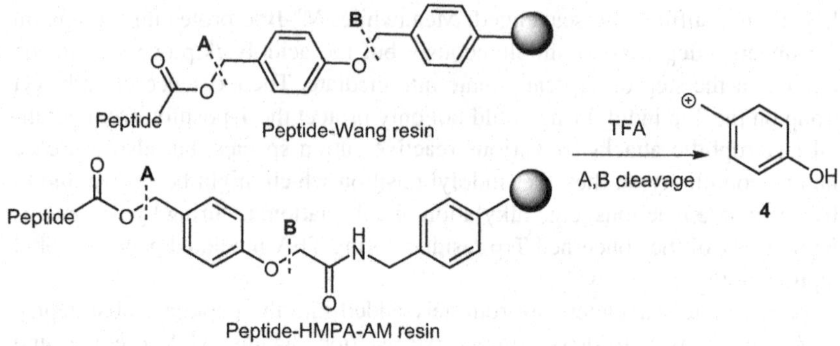

FIGURE 3.5 Acidolysis of Wang and HMPA-AM peptidyl resins.

acidolysis in an highly concentrated TFA solution[33] (see also Fig. 3.5, Site B). In case of simultaneous fragmentation having taken place at both A and B site on the Wang or HMPA-AM resins in the TFA-mediated peptide cleavage/global deprotection reaction, 4-hydroxybenzyl cation **4** will be generated, and released into the peptide solution as a consequence[34] (Fig. 3.5).

Besides hydroxymethyl phenoxy type linkers on Wang and HMPA, some other SPPS functional resins for peptide amides synthesis, e.g., Rink resin, Rink amide MBHA resin, and PAL resin [(5-(4-(9-fluorenylmethyloxycarbonyl) aminomethyl-3,5-dimethoxyphenoxy)valeric acid)] could also be subject to the undesired acidolytic cleavage between the spacer and their corresponding solid support backbone.[19,35] As illustrated in Fig. 3.6, if the acid treatment of the peptidyl PAL-MBHA resin results in regular fragmentation at site A, as well as the unexpected fission at site B, PAL-derived 5-(3,5-dimethoxy-4-methylphenoxy) pentanamide cation **5** will be generated, and released into the solution phase in this process; similarly, irregular acidolysis at both site B and C on peptidyl Rink resin gives rise to 4-hydroxybenzyl cation **4**, and the simultaneous astray cleavage at sites B and C on peptidyl Rink MBHA resin releases cationic 2-(4-methylphenoxy)acetamide **6**. All these cations released from the undesired cleavage between the corresponding spacers, and the resin backbone would potentially trigger a side reaction owing to their inherent electrophilic features in case they have not been sufficiently scavenged at the peptide cleavage step.

One of the most considerably affected amino acids that suffer from the alkylation modification by the linker cations during peptidyl resin cleavage is the indolyl ring on Trp side chain. As has been introduced in the previous section, Trp is susceptible to alkylation, e.g., *tert*-butylation driven by *tert*-butyl cation and/or *tert*-butyl trifluoroacetate released and formed in the process of peptide side chain global deprotection. The alkylation side reaction dictated by decomposed resin linker cations discussed in this section is nevertheless occurring at peptidyl resin cleavage/global deprotection "two-in-one" step. If these reactive cationic species could not be sufficiently and promptly quenched in the course

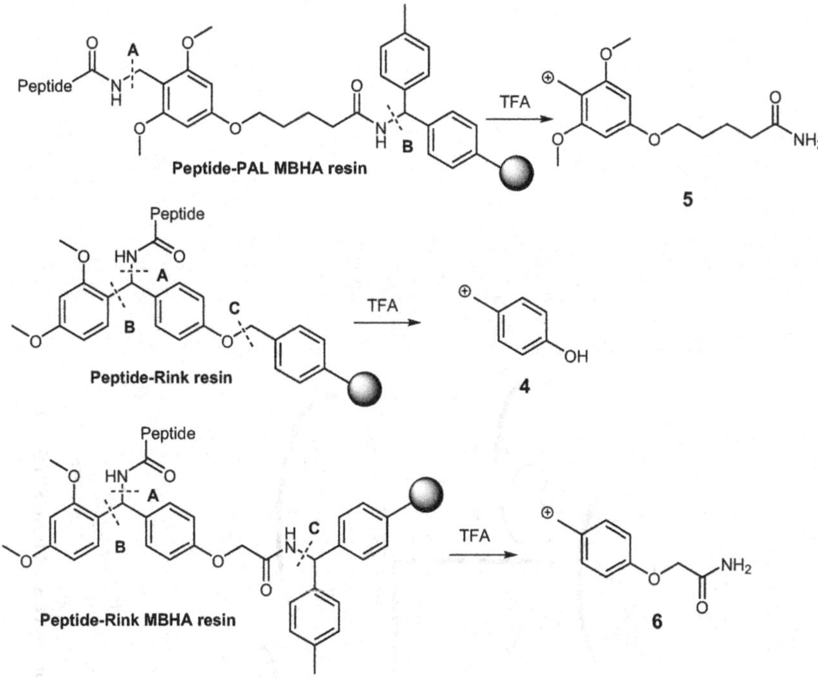

FIGURE 3.6 TFA acidolysis of PAL, Rink and Rink MBHA peptidyl resins.

of the addressed reaction, they might induce alkylation modifications on the liberated functional groups on the peptides, and form diverse side chain-modified peptide impurities. The irreversible reactions between Trp indolyl groups with these decomposed cationic linkers would unavoidably result in the occurrence of Trp alkylation side reactions. As illustrated in Fig. 3.7, if the side chain unprotected Trp is entrapped by diverse SPPS resin spacer cations it would be transformed into different alkylated side products, e.g., derivative **8** with a MW increase by 106 amu that is invoked by 4-hydroxybenzyl cation **4**. Similarly, alkylated Trp impurity **9** (ΔMW = +265 amu) is caused by the decomposed PAL-linker fragment 5-(3,5-dimethoxy-4-methylphenoxy)pentanamide cation **5**. While Trp impurity **10** (ΔMW = +163 amu) is derived through the indolyl side chain alkylation process by 2-(4-methylphenoxy)acetamide cation **6**. Likewise the released 4-hydroxybenzyl cation **4** could function with abundant TFA molecules in the reaction system, and give rise to the formation of cationic 4-trifluoroacetyoxybenzyl cation **7** which could consequently modify the free indolyl ring on the Trp residue, and result in the formation of a side product **11** with a MW increase of 202 amu.[21]

Similar to Trp *tert*-butylation, most above alkylation side reactions initiated by spacer cations take place on the 2-position of the indolyl ring, whereas under

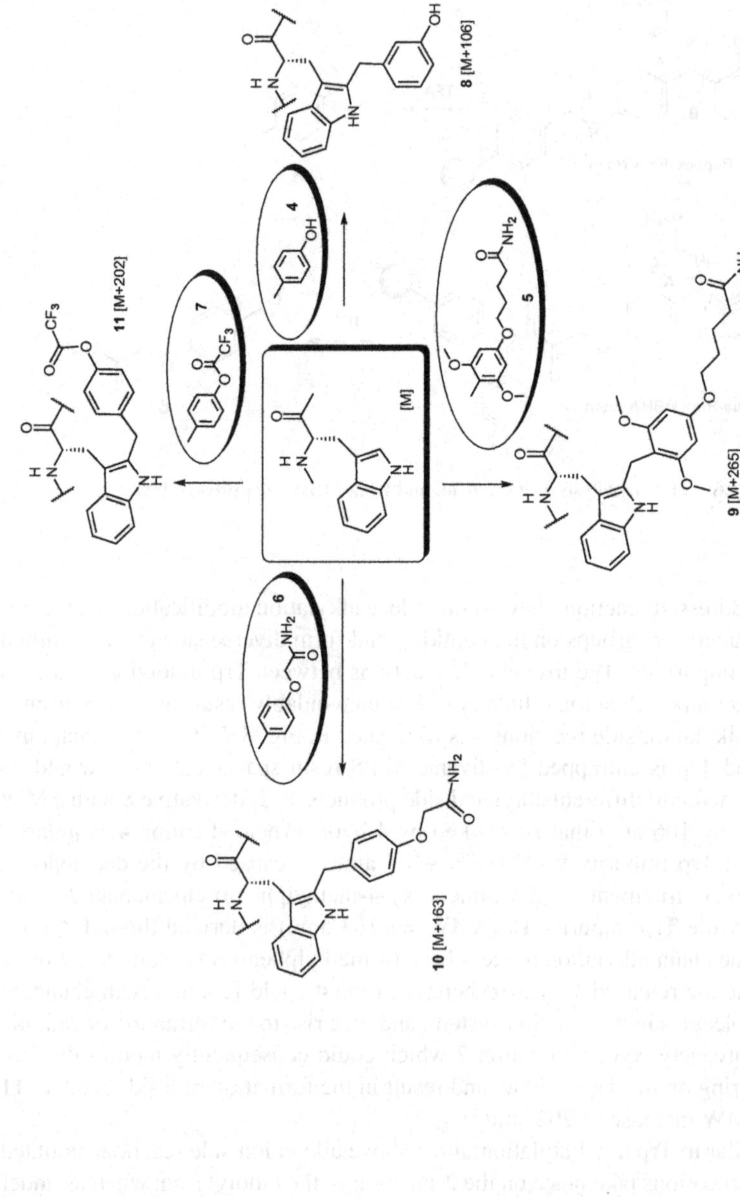

FIGURE 3.7 Trp benzylation side reactions by linker cations cleaved from SPPS resins.

certain circumstances multiple benzylation might occur.[26] The probability and extent of these side reactions might be sequence dependent, that is to say, whether a certain Trp residue could be affected by linker cation-dictated alkylation at the step of peptidyl resin cleavage/global deprotection partially depends on the position of the subjected Trp residue in the concerned peptide sequence. The unambiguous correlation between the Trp susceptibility to alkylation and its sequential location has nevertheless not been explicitly clarified, and contradictory conclusions to this end do exist.[26,36]

Targeted tactics have been developed addressing the decomposed SPPS resin spacer mediated Trp modification. Transection of the original one-step cleavage strategy into the stepwise cleavage/global deprotection process could principally improve the crude product integrity and reduce the susceptibility to linker decomposition by highly concentrated TFA solution imperative for one-step peptide cleavage procedure. A fully side chain protected peptide precursor could be obtained through the treatment of highly acid-labile peptidyl resin by a diluted TFA solution, bypassing in this manner the employment of highly concentrated TFA that would potentially exaggerate the irregular fragmentation of the resin spacer, and induce the Trp alkylation by released spacer cations as a consequence. The fully side chain protected peptide derivatives could be subsequently addressed to a dedicated separate side chain protecting group cleavage step, that removes simultaneously diverse protecting groups on the corresponding side chains. For the cases in which Trp benzylation side reactions is extraordinarily liable the utilization of the super acid-sensitive SPPS solid supports, e.g., CTC resin or Sieber resin for the preparation of peptide acids and amides, respectively, is deemed as a rational strategy to tackle the problem. The Boc protecting group on the indolyl side chain of Trp remains intact in the process of peptidyl resin treatment by diluted TFA solution, preventing by this means the potential Trp alkylation modification by various reactive cations in the reaction system. This stepwise peptidyl resin cleavage and global deprotection strategy could largely suppress Trp benzylation side reactions initiated by the decomposed SPPS resin cations.[37]

An alternative tactic to alleviate the referred side reaction is to utilize the SPPS resins possessing spacers with distinctive resistance toward strong acid treatment, minimizing the aberrant spacer decomposition in this process. Taking Wang resin, for example, it is a conventional resin adopting *p*-alkoxybenzyl alcohol as the spacer that is prone to suffer from the acidolysis side reaction. New generations of the Wang resin are, therefore, modified on its *o*- or *p*-position with hydroxymethyl group by introducing an electron-donating group that enhances the stability of the corresponding spacer to strong acids.[38] Many new SPPS resin spacers that are devoid of irregular fragmentation are designed and applied to SPPS[38,39] (see also Fig. 3.8). It has been reported that the functional spacers like PAL or Rink could be less susceptible for the acidolysis if they are linked to the aminomethyl group instead of MBHA group on solid support (Yang, Y., unpublished results).[35] The aminomethyl-linked SPPS resin spacer

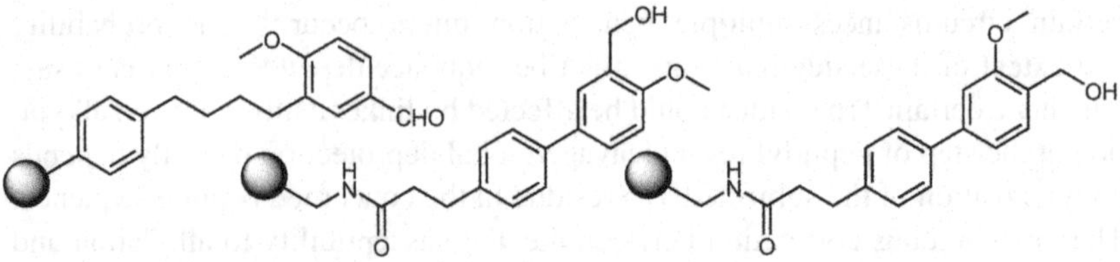

FIGURE 3.8 Acid-stable SPPS resin spacers.

remains stable even in the presence of HF, and is consequently exempted from the undesired acidolysis side reactions.[40]

The extent of the SPPS resin spacer acidolytic decomposition in the process of peptide cleavage is evidently proportional to acid concentration.[35] If the concentration of TFA solution for peptide cleavage is elevated the severity of Trp benzylation by spacer cations might be accordingly intensified. Suppression of this side reaction is, therefore, achievable through a meticulous fine-tuning of the concentration of the applied TFA solution in this process. Moreover, shortening of the duration of the peptidyl resin acid treatment is also helpful to reduce the extent of Trp modification by the degraded resin spacer cations.

Under circumstances where there is very limited room to modify, or opt SPPS resin spacers, the most effective strategy to curb the undesired resin spacer decomposition and subsequent Trp alkylation is to add appropriate nucleophilic scavengers in the TFA solution. These scavenging derivatives could sufficiently and effectively entrap the released spacer cations prior to the deleterious electrophilic attacks on the indolyl ring on Trp. It has been proved by variant studies that rational concoction of solution for peptidyl resin cleavage/global deprotection affords the most ideal tactics to suppress the Trp alkylation side reaction. Albericio et al.[19] has figured out that scavengers, e.g., phenol, thiophenol, benzyl mercaptan, dimethyl phosphite, and Tri-*n*-butylphosphine are relatively ineffective in quenching PAL-derived 5-(3,5-dimethoxy-4-methylphenoxy)pentanamide cation and preventing Trp residues in the target peptides from being alkylated; on the other hand, reagent R (anisole/thioanisole/EDT) could effectively protect Trp residues from the electrophilic attacks by PAL linker cation. It was disclosed in another investigation[26] that anisole posed to be an excellent additive to suppress the Trp alkylation by spacer cationic derivatives, whereas this side reaction was drastically stimulated when EDT/TIS in place of anisole was charged into the deprotection cocktail for the corresponding peptide global deprotection. Stathopoulos et al.[33] found that 1,3-dimethoxybenzne added into TFA/TIS system could considerably suppress the formation of 4-hydroxybenzylamide impurity on peptide C-terminus. The occurrence of this side reaction is irrelevant to Trp but due to the aberrant bond fission at site B and C depicted in Fig. 3.6 in terms of peptide-Rink resin. It is deduced by the author that the property of DMB (1,3-dimethoxybenzene) in the context of suppression of the irregular bond scissoring at site B is attributed to the structural similarity between DMB and the substance derived from spacer bond cleavage at site B. The addition of DFB into the peptide-Rink resin cleavage/peptide global deprotection could, therefore, affect the equilibrium of the potential spacer decomposition in a strong acidic milieu. The subjected irregular resin spacer degradation and the formation of reactive spacer cations are therefore minimized, and the integrity of Trp is thus guaranteed. This improvement is nevertheless not achieved through the effects of scavenging the release benzyl cation derivatives by DMB.

3.3 FORMATION OF Trp-EDT AND Trp-EDT-TFA ADDUCT IN PEPTIDE GLOBAL DEPROTECTION

Trp-containing peptides are susceptible to many potential side reactions at the TFA-mediated peptide side chain global deprotection step, among which *tert*-butylation and benzylation induced by *tert*-butyl cations/*tert*-butyl trifluoroacetate, and resin spacer cations, respectively, have been introduced in the previous sections. Furthermore, some vulnerable Trp residues are also affected by TFA/EDT dictated modifications at this step, and are transformed into cyclic dithioacetal adducts as a consequence.[7] The side product formed in this process is the Trp derivative **12** that is modified at the 2-position on the indolyl ring (see also Fig. 3.9) with a molecular weight increase by 172 amu relative to the parental molecule.

This side reaction was first detected in a peptide side chain global deprotection reaction[7] in which 15 equiv. EDT was charged as scavenger into a TFA/H_2O (95:5) solution. When the temperature of the concerned peptide side chain global deprotection reaction was increased to 50°C in an effort to quantitatively cleave the Mtr protecting group off the Arg side chain, a side product with a molecular weight increase by 172 amu relative to the target product was detected. Its chemical structure was confirmed by NMR as the Trp derivative modified at the 2-position on the indolyl side chain by TFA/EDT in the form of five-member ring dithioacetal **12** (Fig. 3.9). Moreover, dedicated experiments have also been conducted to investigate the Trp modification by the TFA/EDT

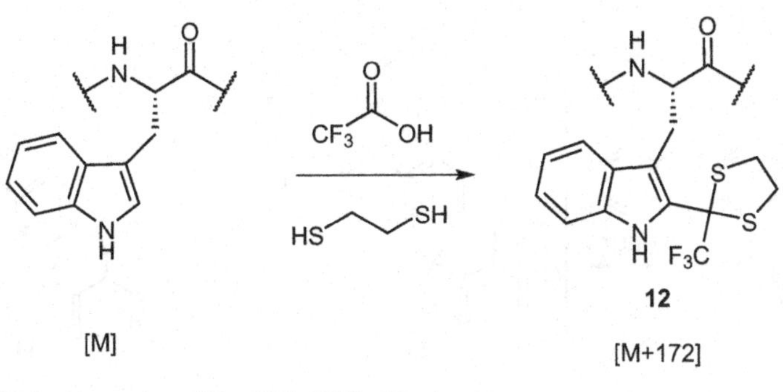

FIGURE 3.9 Formation of Trp-TFA-EDT adduct.

condensation (Yang, unpublished results). Fmoc-Trp(Boc)-OH was subject to TFA/EDT a mediated side chain deblocking reaction, and the peptide solution was stressed at high temperature overnight, abundant derivatives with MW 598.7 amu were found in the reaction system (MW of Fmoc-Trp-OH is 426.5 amu, ΔMW = +172 amu) which is in line with the TFA/EDT dithioacetal adduct of Fmoc-Trp-OH.

The formation of cyclic dithioacetal is an ordinary acid-catalyzed organic reaction between dithiol-derivatives and carbonyl compounds, normally exploited for the temporary protection of carbonyl functional groups.[41] For example, 1,3-dithiolane related compounds could be derived through acid-catalyzed condensation between 1,2-ethandithiol and the corresponding carbonyl reagents.[42] It has been discovered in a study[43] that abundant impurities with ΔMW = +76 amu were emerged during the chemical preparation of Bpa-containing peptide when EDT was charged into the peptide global deprotection reaction as a scavenger. The referred side product was identified as dithioacetal adduct **14** formed via the hemimercaptol intermediate **13**[44] (Fig. 3.10). It is noted that DTT in place of EDT in the global deprotection cocktail would not induce similar dithioacetal condensation, indicating the inherent correlation of the occurrence of this side reaction with steric effects such as the stability of the corresponding dithioacetal derivative.

It is not intensively investigated with regard to peptide synthesis entangled in the above Trp modification by EDT/TFA condensation, but impurities with +172 amu MW are, nevertheless, haunting a substantial number of peptide productions. Some affected peptides are devoid of componential Trp residues (Yang, unpublished results), this phenomenon might hint that TFA/EDT-condensed dithioacetal cations[45] was formed prior to its attack on the corresponding substrates, e.g., indolyl ring on Trp, or other aromatic residues such as Tyr or Phe under acidic conditions, giving rise to side products with MW increase by 172 amu.

This kind of side reaction will basically be intensified in peptide global deprotection at a high temperature which is deemed as imperative for the quantitative removal of some less acid-labile protecting groups, e.g., Mtr or Pmc on Arg

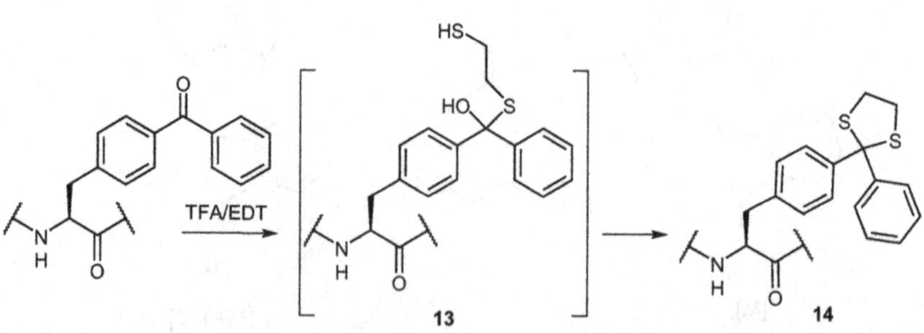

FIGURE 3.10 EDT/TFA-induced dithioacetal condensation on Bpa-containing peptide.

side chain.[12,16] A higher temperature might facilitate the formation of TFA-EDT dithioacetal cations, and promote its electrophilic attack on the corresponding substrate residues, e.g., Trp, Phe, or Tyr. New types of protecting groups on Arg, e.g., MIS (1,2-dimethylindole-3-sulfonyl)[46] are therefore, designed and utilized in an effort to address the inherently correlated side reactions in the process of peptide side chain global deprotection reactions. The MIS protecting group could be cleaved off the Arg side chain under milder conditions relative to those for the removal of routine Arg protecting groups, e.g., Pmc, Mtr, or even Pbf, bypassing in this manner the adverse high temperature condition, and suppressing the potential dithioacetal condensation on Trp.

Another tactic in an effort to minimize the undesired dithioacetal modification on Trp is to address the root cause of this side reaction by blocking the formation of TFA-EDT dithioacetal cationic species. In view of the mentioned observation that Bpa-containing peptide is almost unaffected by TFA-DTT mediated global deprotection, it is advisable to accordingly prepare the peptide global deprotection solution in which EDT is replaced by DTT or other mercaptan derivatives such as 1,8-octanedithiol. The employment of such a fine-tuned deprotection cocktail in the corresponding peptide side chain global deprotection could effectively suppress the occurrence of dithioacetal modification on the susceptible residues. Moreover, rational reduction of the global deprotection duration might diminish the extent of this side reaction as well. The N^{in}-Boc protection on the Trp indolyl side chain, even in the form of transient carbamate protection during acidolysis, will substantially decrease the density of electron cloud on its 2-position, and render it less susceptible for modification by diverse electrophiles.

3.4 Trp DIMERIZATION SIDE REACTION DURING PEPTIDE GLOBAL DEPROTECTION

The behaviors of Trp dimerization exert extraordinarily significant impacts on the physiological properties of the subjected proteins.[47] Under the circumstances of peptide chemical synthesis, the indole rings on Trp side chain could be subject to a dimerization process as well, which might be inherently different from the *in vivo* protein dimerization in the context of the corresponding bond connections. Dimerized peptides via the Trp indolyl side chain exhibit different behaviors as to UV and fluorescence spectra relative to their monomer counterparts and their spatial conformation and the stability will be altered as well.[48]

Peptide dimerization via Trp takes place dominantly at the step of the TFA- or the HF-mediated peptide side chain global deprotection,[23,48] which is catalyzed by the presence of strong acids. The mechanism of this side reaction is to some extent similar to that of Mannich-type dimerization (see also Fig. 3.11).[24,49] Two Trp residues from the respective peptides are connected at their 2-position on the indolyl ring as consequences.[50] The whole process is initiated by the formation of cationic Trp derivative **15** through protonation, that

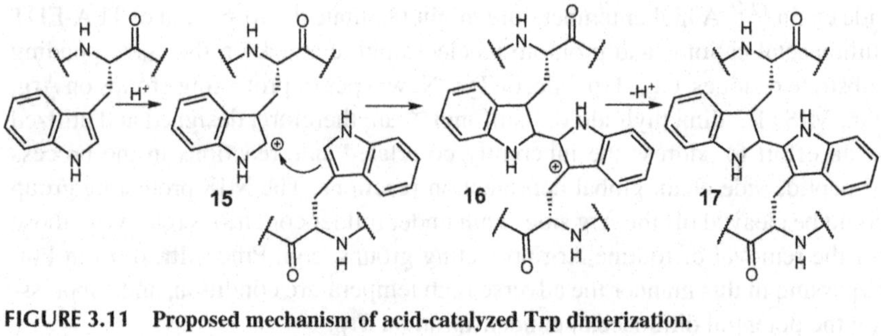

FIGURE 3.11 Proposed mechanism of acid-catalyzed Trp dimerization.

subsequently undertakes the nucleophilic attack from another neutral Trp, and generates cationic intermediate **16**. This dimeric derivative undergoes spontaneous deprotonation process to give rise to the stable indole–indoline product **17**. Similar dimerization process might also take place on other indole derivatives, for example, indole-3-acetic acid could be transformed into the 2,2'-indole–indoline dimer upon TFA treatment.[51]

Indole–indoline derivative **17** could be further oxidized in the presence of acid to the corresponding indole–indole dimer counterpart **18** (Fig. 3.12). For example, the successive treatment of Trp-containing peptide by TFA and DDQ could give rise to the formation of intramolecular or intermolecular indole–indole dimeric peptide products.[52] Indole–indoline dimer **17** could also be oxidized under milder conditions, e.g., air oxidation or light irradiation[53] to the corresponding indole–indole counterpart **18**.

The extent of Trp dimerization is evidently proportionate to the duration of the subjected peptide treatment by TFA.[50,54] It is hence recommended to rationally shorten the peptide global deprotection reaction so as to minimize the potential Trp dimerization. Boc-protected Trp has also been verified to be less susceptible to this side reaction.[48]

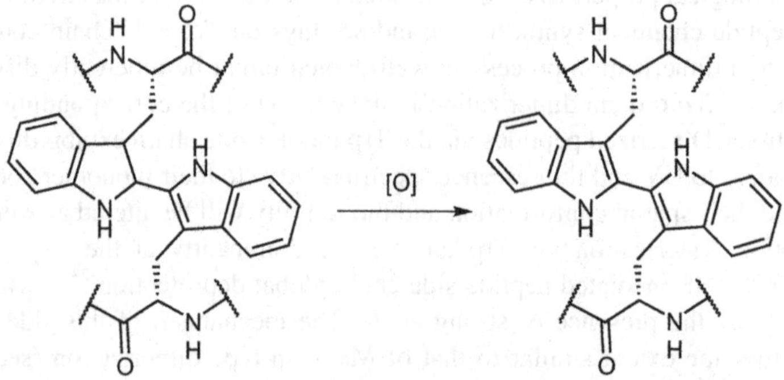

17 18

FIGURE 3.12 Oxidation of indole–indoline derivative to indole–indole dimer.

3.5 Trp REDUCTION DURING PEPTIDE GLOBAL DEPROTECTION

The trialkylsilane family could serve as mild reducing agents under acidic conditions. Substituent species connected to the Si atom are decisively crucial for the properties of the Si—H bond, and is hence posed to be a critical factor for the reduction potential of the concerned parental trialkylsilane compound. This family could also be applied as scavengers in peptide side chain global deprotection reactions as to quench reactive cationic species, particularly the relatively stable Trt cations. Trialkylsilanes manifest superior performances to EDT in the context of Trt cation scavenging.[55] In most cases trialkylsilanes in acidic milieu could be deemed as a hydrogen donor in the process of ionic hydrogenation.[56] Stable cations released from the corresponding protected amino acid precursors at the step of peptide global deprotection, e.g., Trt, Acm, and Tmob cations, could be irreversibly entrapped by trimethylsilane, and converted to the inert triphenylmethane (see also Fig. 3.13) facilitating by this means the effective cleavage of Trt protecting groups, and suppression of the potential retritylation side reactions provided that released Trt cations were not sufficiently quenched.

The most frequently utilized trialkylsilane compounds with respect to peptide synthesis include TIS and TES. The bulky isopropyl substituents on TIS entrust the molecule evident steric effects that could be superbly manifested by the features with respect to reaction selectivity. On the other hand, the price of TES is significantly lower than that of TIS, which is evidently advantageous for the industrial production of peptide substances, and its utilization in place of TIS is frequently adopted in an effort to lower the costs of the corresponding peptide manufacturing.

TES is intrinsically equipped with a stronger reduction potential relative to its bulkier counterpart TIS. This particular feature could, nevertheless, lead to the undesired reduction of indole derivative under acidic condition to the corresponding indoline.[57] In line with this process, conversion of Trp to Trp-indoline counterpart[55] with a molecular weight increase by 2 amu could be induced at the step of peptide side chain global deprotection in the presence of TES (see also Fig. 3.14). This reduction side reaction might be initiated by the formation of a relatively stable cationic Trp intermediate upon TFA treatment (see

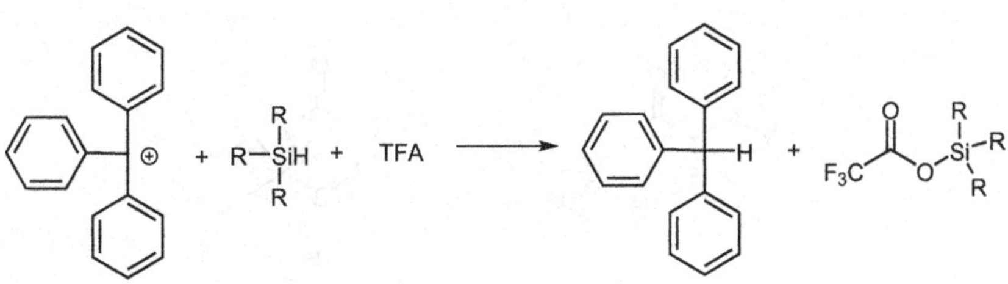

FIGURE 3.13 Scavenging of Trt cation by trialkylsilane.

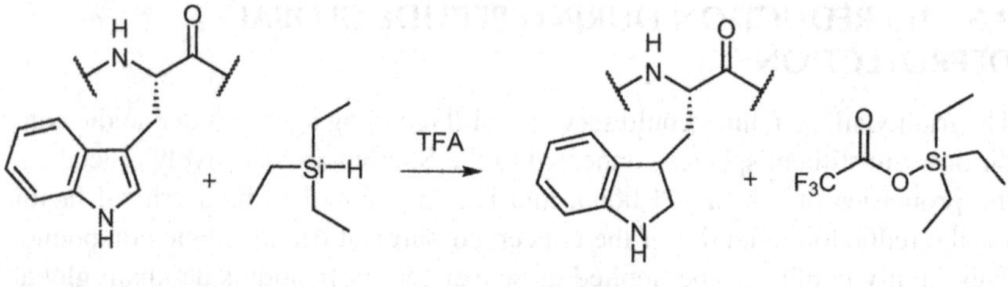

FIGURE 3.14 TES-induced Trp reduction.

also Fig. 3.11), which is subsequently reduced by TES to the corresponding indoline derivative. The mechanism of this process is basically akin to that of the reduction of olefin to commensurate alkane under acidic condition by trialkylsilane.[58,59]

Similar to the most side reaction addressing Trp in peptide synthesis, Trp reduction by trialkylsilane would be effectively suppressed upon the protection of the indolyl side chain by Boc.[55] TIS in place of TES as the scavengers for peptide global deprotection could also substantially restrain the occurrence of the Trp reduction in acidic milieu.

3.6 Cys ALKYLATION DURING PEPTIDE GLOBAL DEPROTECTION

The sulfhydryl functional group on the Cys side chain could suffer from irreversible alkylation side reactions in the peptide side chain global deprotection reaction. *Tert*-butyl cations derived from Boc or tBu protecting groups and *tert*-butyl trifluoroacetate, if not duly and sufficiently quenched, would probably function with the sulfhydryl groups on Cys side chains under acidic condition,[60] and give rise to the formation of *tert*-butyl thioether impurity **19** (see also Fig. 3.15).

Similar to this side reaction, Cys could also be reversibly modified by Trt cations in the process of peptide side chain global deprotection, and converted back to Cys(Trt) derivative (see also Fig. 3.16).

The most effective solution to address various alkylation side reactions on the Cys sulfhydryl side chain is the employment of the appropriate scavengers that are capable of sufficiently entrapping the released cations during peptide

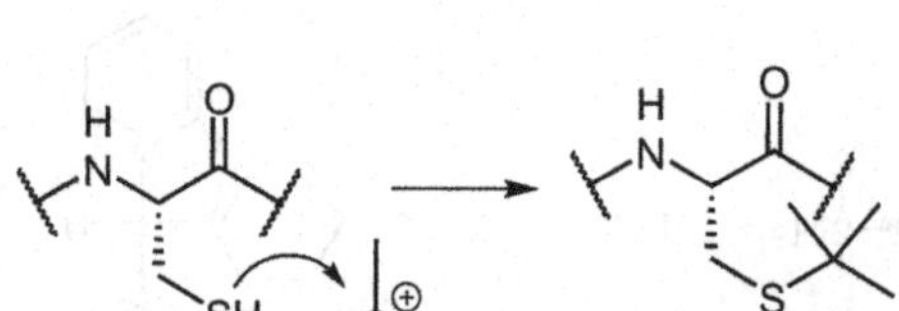

19

FIGURE 3.15 Cys *tert*-butylation.

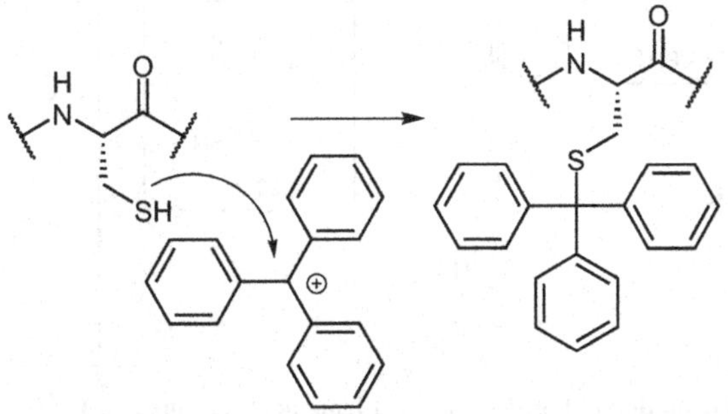

FIGURE 3.16 Tritylation of Cys.

side chain global deprotection reactions, and protecting the liberated sulfhydryl groups from potential electrophilic attacks by these cationic alkylating agents. EDT has been verified to be one of the most potent scavengers for *tert*-butyl cation while trialkylsilanes such as TES or TIS could effectively quench Trt cation. Additionally, these scavengers in the TFA solution for peptide side chain global deprotection reaction could efficaciously suppress the potentially hazardous alkylation side reactions on the Cys side chain, and thus assure the integrity of the Cys sulfhydryl groups in the process of peptide side chain global deprotection reactions.

3.7 FORMATION OF Cys-EDT ADDUCTS IN PEPTIDE GLOBAL DEPROTECTION REACTION

The free sulfhydryl side chain on Cys could be covalently connected to another sulfhydryl group in the form of an intramolecular or intermolecular disulfide bridge via air-mediated oxidation. This reaction normally proceeds under weak basic conditions, whereas the newly formed disulfide bond could participate in the process of thiol-disulfide exchange with other thiolate derivatives,[61] leading to scrambling of the disulfide bridge complex. The undesired formation of disulfide bond, and thiol-disulfide exchange during TFA-mediated peptide global deprotection received little attention, and relatively scarce pertinent information is therefore available. It has been reported[62] that a Cys(Acm)-containing peptide **20** treated by TFA/EDT in the process of a peptidyl resin cleavage and/or global deprotection was partially converted to Cys-EDT adduct **21**, whereas the employment of other scavengers in place of EDT would restrain the occurrence of the similar side reactions. Nevertheless, Cys derivative **22** generated from Cys(Acm) premature cleavage, and its disulfide dimer **23** were formed instead (Fig. 3.17).

The explicit mechanism of the previous side reaction was not clarified in the investigation. Actually it is not an unusual phenomenon in the process of peptide synthesis that an impurity with $\Delta MW = +92$ amu is emerged (Yang, unpublished results), all these concerned cases share some striking similarities:

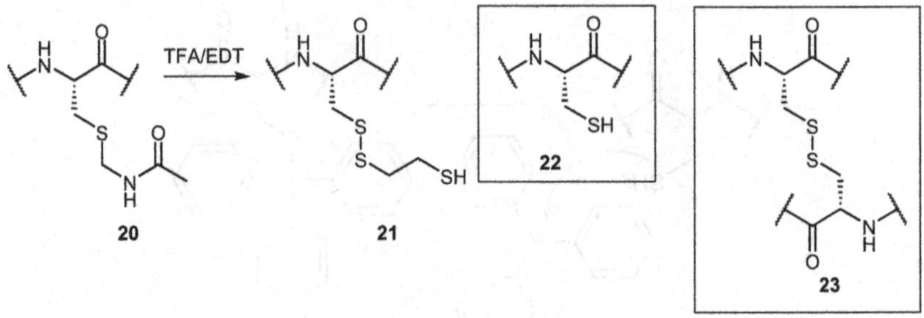

FIGURE 3.17 Formation of Cys-EDT adduct from acid treatment of Cys(Acm).

(1) the affected target peptides contain Cys residues with free sulfhydryl side chains (Cys is not involved in disulfide bridge formation); (2) most of the susceptible Cys residues are originally protected at their sulfhydryl side chains by Trt; (3) TFA is normally applied for the final global deprotection treatment; and (4) EDT functions as one of the componential scavengers in the global deprotection cocktail. In order to elucidate the features of the previous side reaction from a macroaspect, systematic comparison with respect to peptide global deprotection reactions under different conditions has been designed, and conducted on certain susceptible peptide substrates in which DTT and EDT served as an individual scavenger in the corresponding TFA solution, respectively. In the DTT-participated experiment no Cys-DTT adducts were detected even after stressing the subjected reaction. On the contrary, an impurity with a molecular weight increase by 92 amu relative to the target product was formed in the EDT-directed peptide global deprotection reactions. Moreover, treatment of this side product with TCEP regenerated the target peptide product. It could be deduced from these observations that the root cause of the formation of the side product with $\Delta MW = +92$ amu should be attributed to the disulfide bridge formation between the liberated sulfhydryl functional group on Cys and the scavenger EDT even under acidic conditions, which gives rise to the occurrence of the Cys-EDT adduct **24** (see also Fig. 3.18).

The propensity of free Cys residue to undergo disulfide bond formation with EDT (under acid condition) might be subjected to the specific microenvironment in which the concerned Cys is located. The environmental attribute might

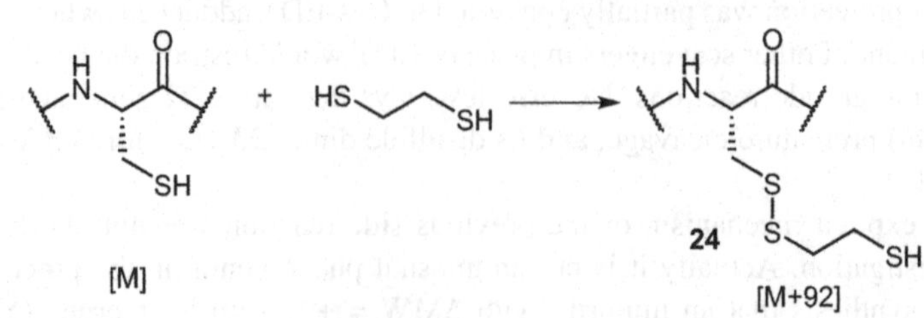

FIGURE 3.18 Formation of Cys-EDT adduct.

substantially affect the pK_a value of the sulfhydryl substituent on Cys.[63] Cys residues with relatively lower pK_a value could be regarded as more active derivatives since the ratio of thiolate/thiol on these compounds in acidic milieu are higher than those from less acidic counterparts. As a consequence thiolate would be able to participate in the process of thiol/thiolate exchange as a nucleophile, and trigger various undesired side reactions.[64,65] For instance, if a neighboring residue to the subjected Cys bears charge on its side chain, the pK_a value of the sulfhydryl substituent on this Cys could be evidently affected by its side-chain charged neighboring residue.[66] Positively charged residues, e.g., Lys and Arg could lower the pK_a value of the neighboring Cys-sulfhydryl, and increase its nucleophilicity accordingly, whereas negatively charged Asp or Glu functions conversely. This phenomenon partially explains the apparent diversity of the inclination of Cys residues toward EDT addition during peptide side chain global deprotection reactions.

There are some controversial opinions toward the mechanism of this side reaction in which the disulfide bridge is formed between the Cys residue and the various exterior mercaptan compounds. The affected Cys might be firstly converted to the corresponding sulfenic acid Cys-OH **25** by oxidant.[67] The resultant hydroxyl group on the sulfenic acid moicty will increase the susceptibility of the subjected sulfur atom to the nucleophilic attack by various thiol or thiolate derivatives[68–70] and will hence facilitate the formation of a disulfide derivative **26** (Fig. 3.19). The disulfide bond formation via the Cys sulfenic acid intermediate is regarded by some researchers as a radical process.[71]

Another proposed mechanism for the formation of the Cys-EDT adduct is related to the process of the thiol-disulfide exchange. EDT utilized in the concerned peptide side chain global deprotection reaction could be partially converted to its disulfide dimeric counterpart **27** (Fig. 3.20) by uncontrolled oxidation upon inappropriate storage,[61] whereas this concomitant byproduct is charged together with EDT as scavenger component to the TFA-mediated peptide side chain global deprotection reaction. Dimeric EDT derivative **27** could stimulate thiol-disulfide exchange with the liberated sulfhydryl functional group on Cys that results in the formation of a Cys-EDT adduct **28**.

Similar to the process of Cys-EDT adduct formation, another Cys-related side product with a molecular weight increase of 148 amu relative to the target product is also frequently detected in the peptide production (Yang, unpublished

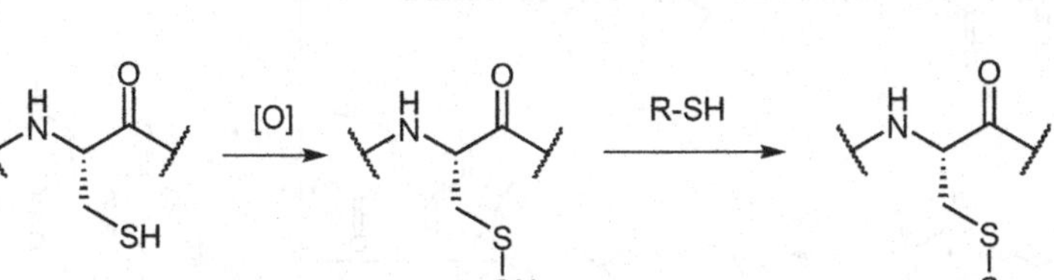

25 **26**

FIGURE 3.19 Formation of Cys-SOH intermediate and disulfide derivative.

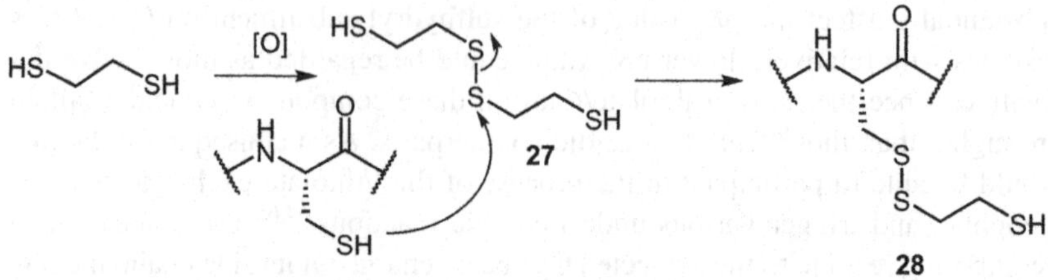

FIGURE 3.20 **Thiol-disulfide exchange between Cys and dimeric EDT.**

results) despite the fact that this side reaction is not intensively reported in the literature. Akin to the Cys-EDT adduct, the referred side reaction also addresses the Cys-containing peptides that suffer from the undesired derivatization at the step of peptide side chain global deprotection. All affected peptide syntheses share a similarity in that EDT is utilized as a scavenger ingredient in the concerned reaction. Similarly, the isolated side product with ΔMW = +148 amu relative to the target product could be regenerated to the original Cys derivative with TCEP treatment, and analogous side products are not aroused once DTT in place of EDT is utilized as the scavenger component in the subjected peptide side chain global deprotection reaction.

The possible mechanism of the side reaction is proposed in Fig. 3.21. As is illustrated previously, the liberated sulfhydryl substituent on the Cys side chain could be modified reversibly by the scavenger EDT to the Cys-EDT adduct impurity **29** with a molecular weight increase by 92 amu. Since mercaptan derivatives could basically serve as scavenging substances for *tert*-butyl cations and *tert*-butyl trifluoroacetate released in the process of the peptide side chain global deprotection, derivative **29** bearing a free sulfhydryl functional group could also participate in the cation quenching-process in the form of a scavenger. When compound **29** entraps a surrounding *tert*-butyl cation, or *tert*-butyl

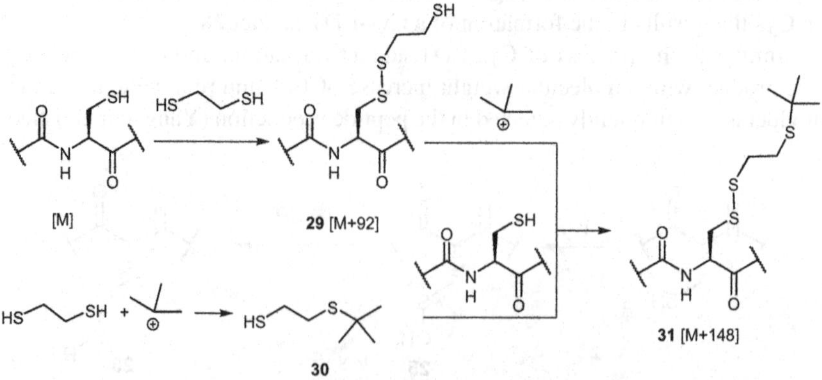

FIGURE 3.21 **Possible routes of Cys-EDT-tBu adduct formation.**

trifluoroacetate it will be irreversibly converted to a corresponding Cys-EDT-tBu adduct **31** whose molecular weight is 148 amu higher than the target Cys product. Or alternatively, the free EDT molecule is first transformed into 2-(*tert*-butylthio)ethanethiol **30** in the peptide side chain global deprotection reaction by *tert*-butyl cation or *tert*-butyl trifluoroacetate. Reactive compound **30** could subsequently function with the liberated Cys sulfhydryl side chain, and give rise to the formation of the Cys-EDT-tBu adduct **31**.

Although no in-depth investigations on the exact mechanism of the Cys-EDT adduct derivative formation have hitherto been systematically conducted, it could nonetheless be summarized that the solutions addressing this kind of side reaction could be reflected as: (1) replacement of EDT by nonmercaptan scavengers in peptide side chain global deprotection reactions. Under many circumstances, H_2O in place of EDT could serve as a qualified scavenger for *tert*-butyl cations.[32] (2) In case of peptide production, that is, in unconditional need of sulfhydryl-type scavengers alternative mercaptan derivatives, e.g., DTT, DODT, or 1,8-octanedithiol could be tested.[72] Caution should be exerted on the storage of these compounds to avoid the uncontrolled contact with air and formation of active dimeric derivatives. (3) Regeneration of the target Cys-peptides could be realized via meticulous reduction of the impurities by reducing agents like TCEP.

3.8 PEPTIDE SULFONATION IN SIDE CHAIN GLOBAL DEPROTECTION REACTION

Sulfonyl protecting groups such as Pmc, Mtr, or Pbf on guanidino side chains of Arg residues exhibit relatively enhanced stabilities toward TFA treatment compared with other ordinary protecting groups in peptide synthesis, introducing by this means unnegligible challenges in the context of quantitative removal of sulfonyl protecting groups, and the regeneration of Arg residues, particularly for those Arg-abundant peptides.[3,73] Sulfonyl cations cleaved off the Arg side chains might trigger undesired modifications on susceptible substrates, e.g., Trp,[7] Ser/Thr,[12] Tyr,[74] or even Arg itself[16] during peptide side chain global deprotection reactions that give rise to the formation of diverse corresponding sulfonated side products.

The occurrence of sulfonation side reactions at the step of peptide global deprotection is synergistically-dependent on the existence of nucleophilic sulfonation substrates on the target peptide molecule, and the survival of the sulfonyl cation species that escape the quenching process during the subjected reaction. If no scavengers are charged into the peptide global deprotection system, Pmc cations cleaved off Arg(Pmc) residues might be converted into a sulfonic acid derivative **32**, as well as dimeric sulfone **33**[74] that are potentially reactive toward diverse nucleophilic functional groups on peptides. For instance, Trp, Tyr, and Fmoc moieties in the target peptides might be modified by the unquenched sulfonyl cations, and transformed into the Pmc-adduct **34** or corresponding sulfonic acid derivative **35** (see also Fig. 3.22).

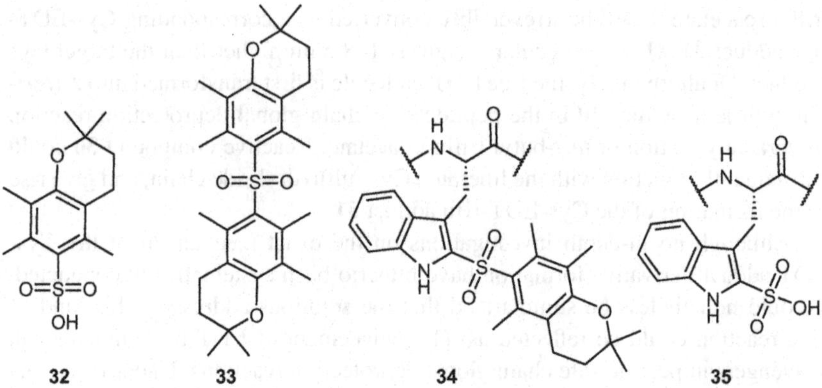

FIGURE 3.22 **Sulfonated peptide side products-induced by Pmc derivatized compounds.**

Posttranslational *O*-sulfonation of serine and threonine residues in natural proteins is a known process in eukaryote evolution.[75] Nevertheless, undesired sulfonation of Trp, Tyr, and other susceptible aromatic residues, as well the hydroxyl-containing amino acids, e.g., Ser and Thr could take place in the process of chemical peptide synthesis. This process is mostly invoked by the sulfonyl cations cleaved off Arg guanidino side chains at the step of the peptide side chain global deprotection.[12] Sulfonated amino acid side products Ser(SO_3H) **36** and Thr(SO_3H) **37** are formed in this process upon the derivatization of side chain unprotected Ser and Thr precursors, respectively (see also Fig. 3.23). Sulfonated side products with molecular weight increase of multiples of 80 amu are frequently detected in the crude peptide products whose side chain protected precursors contain abundant Arg(Pbf) residues. The occurrence of this side reaction might be attributed to the formation of mixed anhydride derivative CF_3COOSO_2Ar generated from the function between benzenesulfonic acid, e.g., compound **32**, and abundant TFA molecules in the reaction system.[74] This reactive mixed anhydride could derivatize the hydroxyl substituents on Ser or Thr,

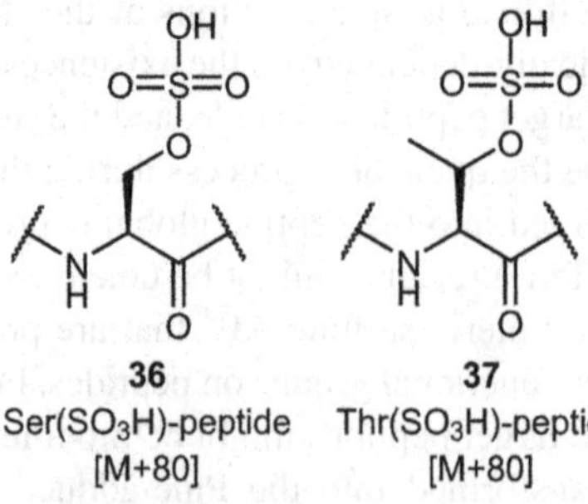

FIGURE 3.23 **Sulfonated Ser and Thr side products.**

and give rise to the formation of corresponding sulfonated side products with a molecular weight increase by 80 amu.

Arg itself might also be subjected to a sulfonation side reaction in peptide side chain global deprotection reaction. Unlike other susceptible amino acid residues that are modified by the cleaved sulfonyl cations, sulfonated Arg side product Arg(SO_3H) could be induced by the irregular cleavage of the side chain sulfonyl protecting group.[16] As is illustrated in Fig. 3.24, if the guanidino side chain protecting group Pbf undergoes a sulfonamide bond cleavage at site b upon TFA treatment, Pbf protecting group would be deblocked from the host Arg residue in the form of sulfonyl cation, and the target Arg product will be hence regenerated. On the other hand, if an irregular bond scission at site a between the 2,2,4,6,7-pentamethyl-2,3-dihydrobenzofuran and the sulfonamide moiety, Arg(SO_3H) derivative **38** will be formed instead in the manner of a sulfonated Arg impurity with a molecular weight increase by 80 amu.

Meanwhile, it has been discovered that the location of the subjected Arg residues could evidently impact on the inclination and extent of Arg sulfonation side reactions.[20] For example, if Trp and Arg(Pmc) are separated by one amino acid residue, that is to say, if the concerned peptide contains -Trp-Xaa-Arg(Pmc)- unit, the susceptibility of the Trp- to Pmc-mediated sulfonation side reaction would be substantially enhanced compared with other primary structural patterns. Moreover, the type of the interval amino acid could also exert an influence on the occurrence of the sulfonation side reaction during peptide side chain global deprotection. These phenomena could to some extent sustain the hypothesis that the referred sulfonation side reaction might be initiated by an intramolecular rearrangement of the concerned sulfonyl protecting group between the sourced Arg residue and the corresponding substrate, e.g., the indolyl ring on Trp.

The solutions for the suppression of undesirable sulfonation side reactions in the process of peptide side chain global deprotection might be developed from the aspects of the inherent features of the subjected sulfonyl protecting

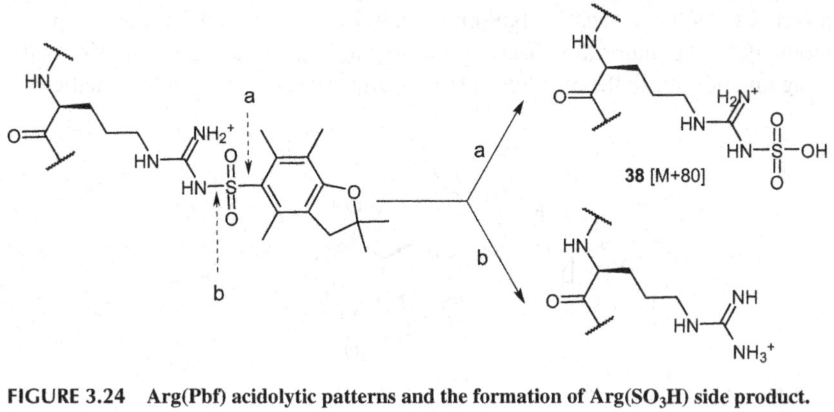

FIGURE 3.24 Arg(Pbf) acidolytic patterns and the formation of Arg(SO_3H) side product.

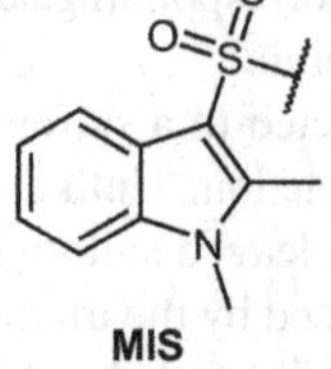

FIGURE 3.25 Arg guanidino side chain protecting group MIS.

groups. Pbf could considerably decrease, even though incapable of thoroughly suppressing, the occurrence of Trp sulfonation side reaction compared with other sulfonyl protecting groups, e.g., Pmc. This advantageous effect is particularly evident when the addressed Trp is protected by Boc on its indolyl side chain.[76] New generations of the Arg side chain protecting groups have been accordingly designed, and utilized to address the undesired features of the ordinary guanidino protecting groups in terms of their inferior acid lability in peptide side chain global deprotection reaction, as well as the resultant implicative side reactions such as sulfonation. MIS protecting group (Fig. 3.25) for the Arg guanidino side chain represents a promising candidate to this end.[46] N^{ω}-MIS exhibits drastically elevated acid lability relative to the ordinary Arg protecting groups, e.g., Pbf and Pmc. The advantages of MIS could also be manifested by its decomposition pattern. Pbf- and Pmc- sulfonyl cations could undergo irregular acidolysis (see also Fig. 3.24, site a) to give rise to 2,2,4,6,7-pentamethylfuran and 2,2,5,7,8-pentamethylchroman,[5] respectively, which would facilitate resultant sulfonation side reactions. On the contrary, MIS-OH byproduct derived from the acidolysis of the MIS protecting group exhibits a substantially enhanced stability, and is devoid of further fragmentation and initiation of a sulfonation side reaction.

In case of extreme susceptibilities to sulfonation side reactions, $N^{\omega,\omega'}$-Bis-Boc protected Arg derivative Fmoc-Arg(Boc)$_2$-OH (see also Fig. 3.26) might be resorted to as the alternative building block for the assembly of target peptide products.[77] The major drawback of this Arg derivative lies in its inferior coupling kinetics due to the disadvantageous steric effect,[78] but it could nonetheless

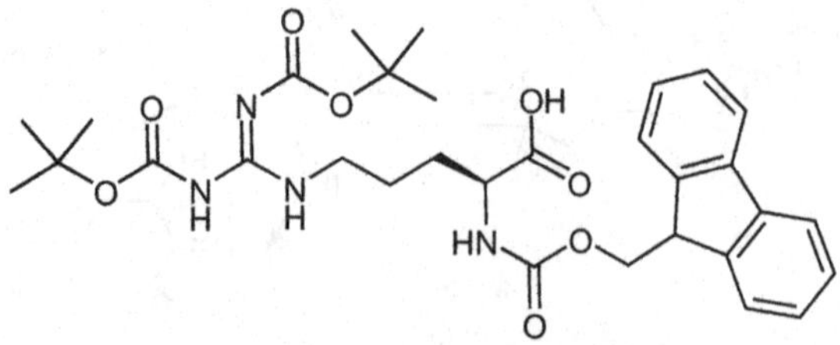

FIGURE 3.26 Fmoc-Arg(Boc)$_2$-OH.

totally avoid the sulfonation side reactions on Trp, Tyr, and other susceptible amino acids.

On the other hand, scavengers play critically important roles with respect to the extents of sulfonation side reactions. Highly-efficient scavengers for sulfonyl cations could effectively quench the released sulfonyl cations in the process of peptide side chain global deprotection reaction, and block the pathway of sulfonyl cations to modify the sensitive amino acid residues. Phenol stands out in the ordinary scavengers in this connection (Yang, unpublished results) and water could also be employed as a quenching agent for sulfonyl cations.[12,16]

3.9 PREMATURE Acm CLEAVAGE OFF Cys(Acm) AND Acm *S*→*O* MIGRATION DURING PEPTIDE GLOBAL DEPROTECTION

Application of Acm as a protecting group for the sulfhydryl substituent on the Cys side chain drastically facilitates the development of the synthesis of disulfide-bridge-containing peptides.[79] Acm as a sulfhydryl protecting group introduces a new degree of orthogonality that substantially enriches the methodologies for site-selective preparation of disulfide compounds, particularly the one containing multiple disulfide bonds.

Under most circumstances Acm on Cys side chains remains stable either in acid or basic solutions, and could hence be applied in Boc/Bzl[80] and Fmoc/tBu[81] peptide chemistry. A reciprocal Cys(Acm) pair could release their Acm protecting groups and be covalently connected via an intra- or intermolecular disulfide bond upon I_2[82] or $Tl(tfa)_3$[83] treatment. Alternatively, Hg(II)[84] or silver triflate[85] could be utilized to selectively deblock Acm, and liberate the sulfhydryl groups on the corresponding Cys residues. Contradictory opinions exist toward the stability of Cys(Acm) under strong acidic conditions. In spite of the assertions that the Acm protecting groups on the Cys side chains remain intact in the process of acid- or base-mediated cleavage of N^{α}-protecting groups, and the peptide side chain global deprotection in TFA, or HF solution, opposite conclusions are drawn from the observations that HF-[86–89] or TFA-directed[62,90] peptide side chain global deprotection could lead to the premature cleavage of Acm from the corresponding Cys(Acm) precursors, and the occurrence of this undesired process is predominantly determined by the composition of scavengers utilized in the concerned global deprotection reaction.[62]

Thioanisole that routinely serves as a scavenger for peptide side chain global deprotection could induce premature Acm cleavage off the Cys(Acm) precursor and consequent irreversible Acm modification on susceptible substrates, e.g., Tyr, Ser, and Thr residues. It is already known that thioanisole in TFA could facilitate the deblocking of N^{α}-Z protecting groups on peptides via "push–pull mechanism";[91] whereas N^{α}-Z remains extremely stable toward TFA in the absence of thioanisole. Similarly, a relevant study[89] indicates that Cys(Acm) does not undergo side chain Acm cleavage during 3-h treatment by anhydrous TFA, while the addition of 5% thioanisole in the reaction system provokes evident

Acm cleavage off the corresponding Cys(Acm) residues. Analogous to thioanisole, anisole could also induce premature Acm cleavage during peptide side chain global deprotection.[86] In view of these observations, it is highly advisable to utilize deprotection cocktails devoid of thioanisole and mercaptan scavengers providing Cys(Acm), Cys(StBu), or Cys(tBu) residues are contained in the subjected protected peptide precursor in order to prevent the undesired premature cleavage of the sulfhydryl protecting groups in this process.[89,92]

Intramolecular or intermolecular migration of Acm groups from Cys(Acm) to the substrate Ser/Thr hydroxyl substituents could take place upon $Tl(tfa)_3$ or I_2 treatment of the subjected peptides.[93] This side reaction might become particularly severe for Ser- and/or Thr-abundant peptides, giving rise to the formation of Ser(Acm) or Thr(Acm) impurities (see also Fig. 3.27). The root cause of this side reaction could be attributed to the insufficient quenching of the Acm cations release in this process, and the resultant irreversible function between the free Acm cations and the hydroxyl groups on Ser/Thr side chains. Not only Ser/Thr, but also Asn/Gln could be subjected to Acm modification, and transformed to Asn(Acm) and Gln(Acm) side products.[94] Taking into account this side reaction, glycerol could be charged as Acm cation substrates into the above processes in order to effectively suppress Acm *S→O* migration.

Similar to the aforementioned side reaction peptide containing -Tyr-Cys(Acm)- sequence could accommodate the Acm migration upon acid treatment. Acm shifts from the Cys sulfhydryl side chain to the *ortho*-position on the neighboring Tyr aromatic side chain[89] (Fig. 3.28). This apparent Acm *S→C* migration could nevertheless proceed via an intervening Acm *S→O* shift step, and the derived hydroxyl-Acm intermediate is transformed to the final C^{Acm}-Tyr isomer through an acid-catalyzed rearrangement.[95] Employment of appropriate

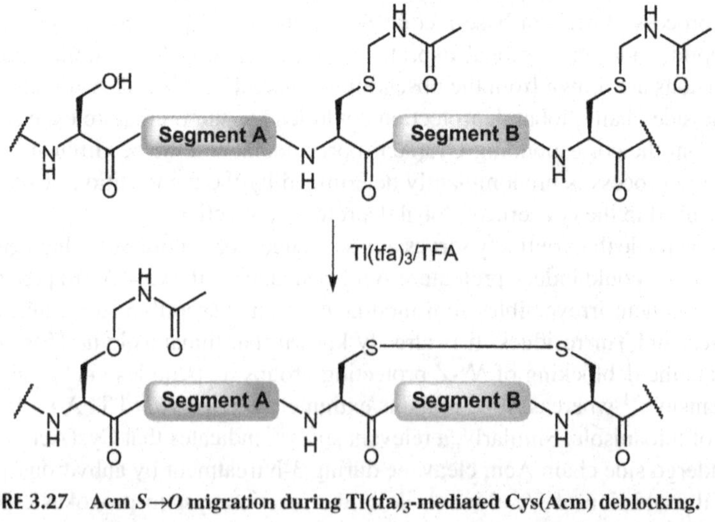

FIGURE 3.27 Acm *S→O* migration during $Tl(tfa)_3$-mediated Cys(Acm) deblocking.

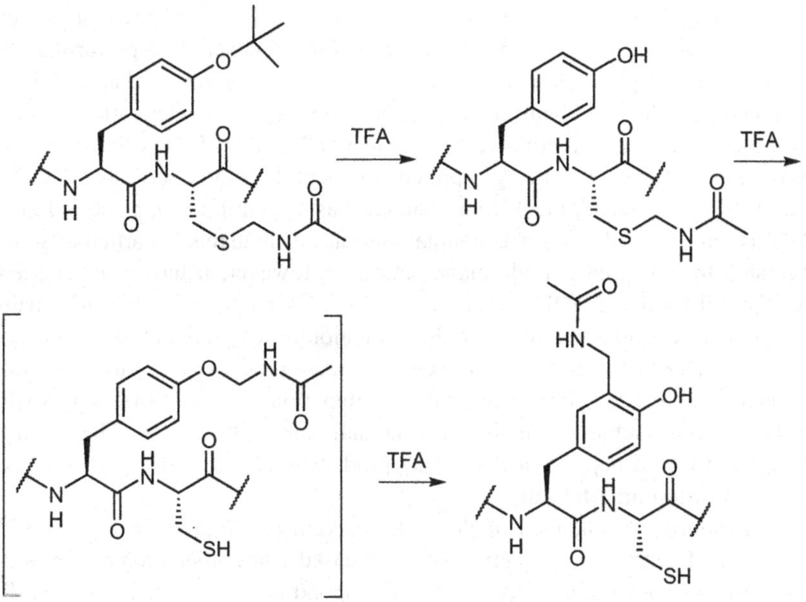

FIGURE 3.28 Possible mechanism of Acm *S*→*C* migration from Cys(Acm) to the neighboring Tyr.

scavengers might be incompetent to thoroughly prevent the premature cleavage of Acm from the precursor Cys(Acm) residue but it could nonetheless attenuate the electrophilic attack of the released Acm cations on the substrate Tyr. It has been discovered that the addition of phenol to TFA solution could diminish the formation of Tyr(Acm) side product generated in the process of peptide side chain global deprotection.[90]

3.10 METHIONINE ALKYLATION DURING PEPTIDE SIDE CHAIN GLOBAL DEPROTECTION WITH DODT AS SCAVENGER

One of the major concerns for the chemical synthesis of Met-containing peptides is the potential alkylation side reaction affecting the sulfur atom on the thioether side chain. The carbocations generated at the step of peptide side chain global deprotection could provoke alkylation of the thioether moiety on Met, and give rise to various sulfonium salt side products that undergo subsequently diverse degradation.[96] Unlike other nucleophilic amino acid residues Met could rapidly function with a variety of electrophiles within a pH spectrum as broad as pH > 2.[97] This distinctive property renders Met more vulnerable to the modifications by carbocations and other electrophiles released in the process of the TFA-mediated peptide side chain global deprotection. Undesired Met alkylation could be provoked if the generated electrophiles during peptide global deprotection are not sufficiently and duly quenched by the corresponding scavengers.

A variety of mercaptan-family scavengers, e.g., EDT and DTT are regularly employed in peptide synthesis thanks to their outstanding performances to quench electrophilic derivatives like *tert*-butyl cations, released at the TFA-mediated peptide side chain global deprotection step. New generations of the mercaptan scavenger alternative to classic EDT and DTT have been tested, and have commenced being applied in peptide manufacturing. DODT (3,6-dioxa-1,8-octanedithiol),[72] for instance, has been utilized in place of EDT/DTT as scavenger, thanks to its nonmalodorous attribute that is particularly appreciated in industrial peptide manufacturing. However, it has been reported that the utilization of DODT as scavenger for TFA-mediate peptide side chain global deprotection could result in the formation of a significant extent of impurity, with a +117 amu molecular weight increase relative to the target peptide product.[98] It has been confirmed by the combination of LC-MS-MS and NMR analysis that the subjected modification actually takes place on the Met residue from the affected peptide, and the side product is identified as *tert*-butylthio ethylated sulfonium Met salt.

The putative mechanism of this side reaction is proposed in Fig. 3.29.[98] First, DODT entraps a *tert*-butyl cation, released from other protected amino acid residue, and is converted to the corresponding *S-tert*-butylated DODT, which undergoes intramolecular cyclization, and gives rise to the formation of either *tert*-butyl thiiranium **39** or dioxanium derivative **40**. The function of the active compound **39** or **40** with the thioether side chain on Met will result in the formation of *tert*-butylthio ethylated Met derivative **41**. Regardless of the pertinence of the proposed mechanisms the existence of ether moiety in the DODT structure seems to be responsible for the occurrence of this side reaction. This

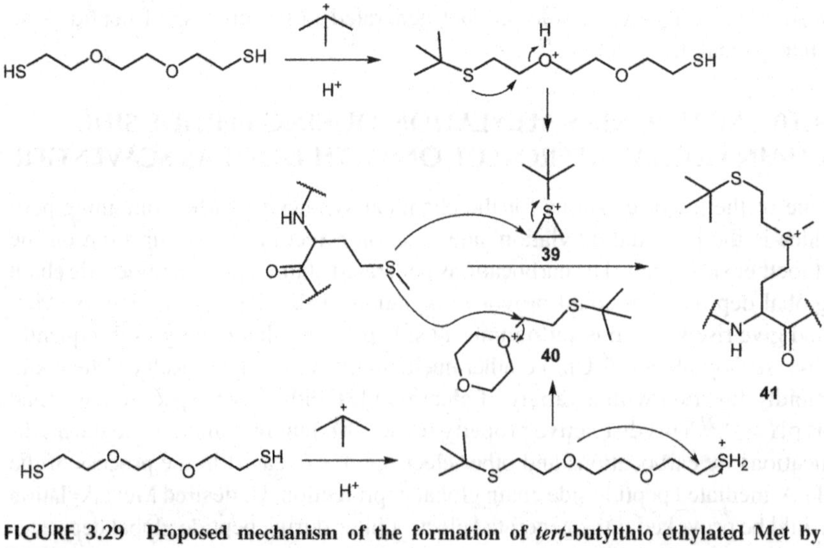

FIGURE 3.29 Proposed mechanism of the formation of *tert*-butylthio ethylated Met by DODT.

assertion has been verified in the same investigation through the replacement of DODT by ODT (1,8-octanedithiol) that is void of the ether functionality. The impurity with +117 amu molecular weight increase that is substantially formed in the presence of DODT is thoroughly suppressed when ODT is utilized as scavenger under otherwise identical peptide global deprotection conditions.

The undesired formation of *tert*-butylthio ethylated Met impurity could be circumvented by the judicious choice of mercaptan scavenger, or the adoption of protected Met(O) as the building block that could regenerate the Met at a separate step. Moreover, this impurity is not stable, and could be quantitatively dealkylated to regenerate the original Met compound by gentle warming in dilute acidic solutions.[99]

3.11 THIOANISOLE-INDUCED SIDE REACTIONS IN PEPTIDE SIDE CHAIN GLOBAL DEPROTECTION

Thioanisole is a routinely utilized scavenger in peptide synthesis to entrap the released nucleophilic species at the step of peptide side chain global deprotection. The employment of thioanisole as a scavenger is recommended for the process of the cleavage of many protecting groups, e.g., benzyloxycarbonyl,[91] Tosyl,[100] Mts,[101] and Mtr.[102] Moreover, the properties of thioanisole with respect to circumventing the undesired Met oxidation in the process of peptide side chain global deprotection are also intensively exploited.[103]

In spite of its versatility thioanisole could indeed induce various uncontrolled modifications on sensitive amino acids, e.g., Met, Trp, Glu, or Asp during acid-mediated peptide side chain global deprotection, particularly in the presence of highly concentrated strong acids such as TFMSA or HF.[54] It is, therefore, advisable to avoid the employment of thioanisole for the side chain global deprotection of Trp- and Met-containing peptides directed by HF, or TFMSA in an effort to bypass the alkylation side reactions addressing these residues by thioanisole.

It has been elucidated in Section 3.9 that thioanisole could provoke premature cleavage of the sulfhydryl protecting groups from Cys(Acm) and Cys(StBu) during the peptide side chain global deprotection process which might be capable of triggering subsequent tandem side reactions.

Another side reaction initiated by thioanisole in the process of peptide side chain global deprotection is the undesired C—O bond fission. For instance, *O*-methyl tyrosine could be transformed to the corresponding tyrosine by the synergistic effects of strong acid and thioanisole via a demethylation pathway. This undesired process is initiated by the protonation of aryl ether oxygen atom (hard base) by hard acid H^+. The subsequent nucleophilic attack of the sulfur atom from thioanisole molecule (soft base) on the electron-deficient methyl carbon (soft acid) leads to the cleavage of the subjected C—O bond and the simultaneous formation of a new C—S bond (Fig. 3.30). The process of *O*-methyltyrosine demethylation is thus induced by thioanisole and a strong acid, e.g., TFMSA in a concerted manner. This process is regarded as a

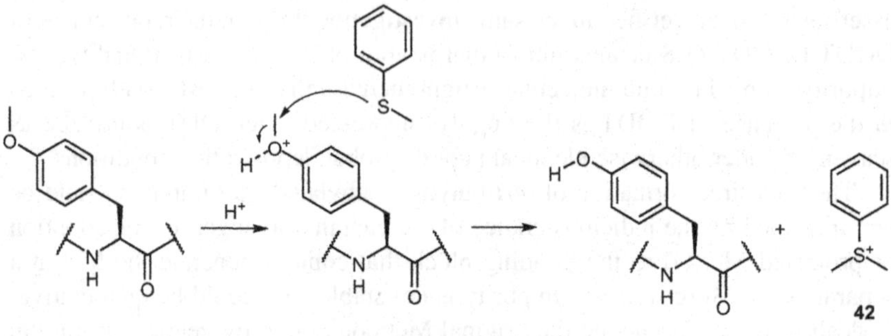

FIGURE 3.30 Thioanisole-induced demethylation of Tyr(Me).

synergic action of a soft nucleophile (thioanisole) and a hard electrophile (H^+) on the substrate (*O*-methyltyrosine) via a push–pull mechanism.[104] Lewis acid, such as TMSOTf, could also provoke the demethylation side reaction in the presence of thioanisole. The kinetics of this reaction is apparently correlated with the property of the attacking sulfur nucleophiles. Thioanisole is more effective in facilitating demethylation than dimethyl sulfide, thanks to the stabilizing effect of the conjugated phenyl group on the generated sulfonium ion **42**. It is noted that thioanisole is regarded as a reversible scavenger with respect to the peptide side chain global deprotection since the formed cationic thioanisole-adduct could realkylate the nucleophilic functional groups on the peptide substrates. It is, therefore, not advisable to use this kind of reversible scavenger alone in the subjected peptide side chain global deprotection reaction. The concomitant addition of irreversible scavengers, e.g., cresol should be performed in order to quench the potentially reactive thioanisole sulfonium cations, and prevent the realkylation of susceptible peptide functional groups in this process.

It will be separately elaborated in Section 6.5 that bromoacetylated peptide could suffer from decomposition invoked by the thioether side chain on Met residue. Similarly, the process of global deprotection of bromoacetylated peptides could be affected by the presence of thioanisole out of the same reason.[105] The nucleophilic attack of the sulfur atom from the thioanisole molecule on the peptide bromoacetyl moiety will lead to the formation of the corresponding sulfonium peptide **42** that subsequently undergoes various degradations (Fig. 3.31).

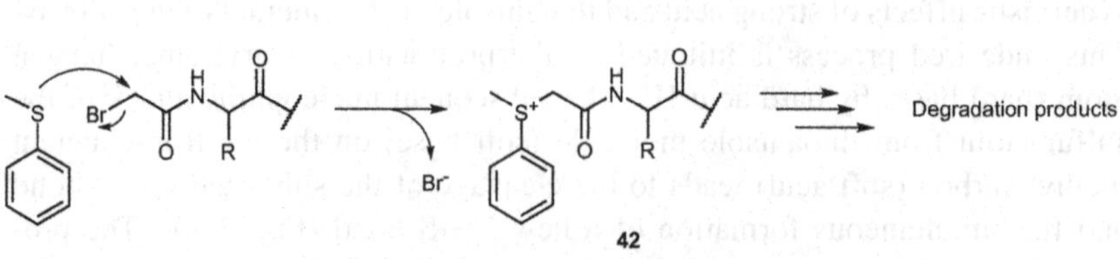

FIGURE 3.31 Thioanisole-induced *N*-bromoacetylated peptide degradation.

REFERENCES

1. Sakakibara S, Shimonishi Y, Kishida Y, Okada M, Sugihara H. *Bull Chem Soc Jpn.* 1967;40:2164–2167.
2. Schwyzer R, Rittel W. *Helv Chim Acta.* 1961;44:159–169.
3. King DS, Fields CG, Fields GB. *Int J Pept Protein Res.* 1990;36:255–266.
4. Riniker B, Floersheimer A, Fretz H, Sieber P, Kamber B. *Tetrahedron.* 1993;49:9307–9320.
5. Ramage R, Green J, Blake AJ. *Tetrahedron.* 1991;47:6353–6370.
6. Lundt BF, Johansen NL, Vølund A, Markussen J. *Int J Pept Protein Res.* 1978;12:258–268.
7. Sieber P. *Tetrahedron Lett.* 1987;28:1637–1640.
8. Noble RL, Yamashiro D, Li CH. *J Am Chem Soc.* 1976;98:2324–2328.
9. Sieber P, Riniker B, Brugger M, Kamber B, Rittel W. *Helv Chim Acta.* 1970;53:2135–2150.
10. Lundt BF, Johansen NL, Markussen J. *Int J Pept Protein Res.* 1979;14:344–346.
11. Johnson T, Sheppard RC. *J Chem Soc Chem Commun.* 1991:1653–1656.
12. Jaeger E, Remmer HA, Jung G, et al . *Biol Chem Hoppe-Seyler.* 1993;374:349–362.
13. Barany G, Merrifield RB. In: Gross E, Meienhofer J, eds. *The Peptides Analysis, Synthesis, Biology. Vol. 2, Special Methods in Peptide Synthesis, Part A.* New York: Academic Press; 1979:1–298.
14. Photaki I, Taylor-Papadimitriou J, Sakarellos C, Mazarakis P, Zervas L. *J Chem Soc C.* 1970:2683–2687.
15. Buellesbach EE, Danho W, Heilbig H-J, Zahn H. In: Siemion LZ, Kupryszweski G, eds. *Peptides 1978.* Wroclaw: Wroclaw University Press; 1979:643–646.
16. Beck-Sickinger AG, Schnorrenberg G, Metzger J, Jung G. *Int J Pept Protein Res.* 1991;38:25–31.
17. Fields GB, Noble RL. *Int J Pept Protein Res.* 1990;35:161–214.
18. Wuensch E, Jaeger E, Kisfaludy L, Loew M. *Angew Chem Int.* 1977;16:317–318.
19. Albericio F, Kneib-Cordonier N, Biancalana S, et al. *J Org Chem.* 1990;55:3730–3743.
20. Stierandová A, Sepetov NF, Nikiforovich GV, Lebl M. *Int J Pept Protein Res.* 1994;43:31–38.
21. Guy CA, Fields G. *Methods Enzymol.* 1997;289:67–83.
22. Fontana A, Toniolo C. *Fortschr Chem Org Naturst.* 1976;33:309–449.
23. Omori Y, Matsuda Y, Aimoto S, Shimonishi Y, Yamamoto M. *Chem Lett.* 1976;5:805–808.
24. Andreu D, García FJ. *Lett Pept Sci.* 1997;4:41–48.
25. Agarwal KL, Kenner GW, Sheppard RC. *J Chem Soc C.* 1969:954–958.
26. Giraud M, Cavelier F, Martinez J. *J Pept Sci.* 1999;5:457–461.
27. Agami C, Couty F. *Tetrahedron.* 2002;58:2701–2724.
28. Ashworth IW, Cox BG, Meyrick B. *J Org Chem.* 2010;75:8117–8125.
29. Choi H, Aldrich JV. *Int J Pept Protein Res.* 1993;42:58–63.
30. Franzén H, Grehn L, Ragnarsson U. *J Chem Soc Chem Commun.* 1984:1699–1700.
31. White P. Peptides, chemistry and biology. In: Smith JA, Rivier JE, eds. *Proceedings of the Twelfth American Peptide Symposium.* Leiden, The Netherlands: ESCOM; 1992:537–538.
32. Sieber P. In: Bricas E, ed. *Peptide 1968.* Amsterdam: North-Holland; 1968:236.
33. Stathopoulos P, Papas S, Tsikaris V. *J Pept Sci.* 2006;12:227–232.
34. Cironi P, Tulla-Puche J, Barany G, Albericio F, Álvarez M. *Org Lett.* 2004;6:1405–1408.
35. Yraola F, Ventura R, Vendrell M, et al. *QSAR Comb Sci.* 2004;23:145–152.
36. Atherton E, Cameron LR, Sheppard RC. *Tetrahedron.* 1988;44:843–857.
37. Han Y, Botems SL, Hegyes P, et al. *J Org Chem.* 1996;61:6326–6339.
38. Gu W, Silverman RB. *Org Lett.* 2003;5:415–418.
39. Colombo A, de la Figuera N, Fernàndez JC, Fernàndez-Forner D, Albericio F, Forns P. *Org Lett.* 2007;9:4319–4322.

40. Mitchell AR, Kent SBH, Erickson BW, Merrifeld RB. *Tetrahedron Lett.* 1976;42:3795–3798.
41. Green TW, Wuts PGM. *Protective Groups in Organic Synthesis.* New York: John Wiley and Sons, Inc; 1991:198–207.
42. Haroutounian SA. *Synthesis.* 1995;1:39–40.
43. Breslav M, Becker J, Naider F. *Tetrahedron Lett.* 1997;38:2219–2222.
44. March J. *Advanced Organic Chemistry.* New York: John Wiley & Sons, Inc; 1992:894.
45. Fields CG, Fields GB. In: Epton R, ed. *Innovation and Perspectives in Solid Phase Synthesis: Peptides, Proteins and Nucleic Acids.* Birmingham, UK: Mayflower Worldwide Limited; 1994:251.
46. Isidro A, Latassa D, Giraud M, Álvarez M, Albericio F. *Org Biomol Chem.* 2009;7:2565–2569.
47. Medinas DB, Gozzo FC, Santos LFA, Iglesias AH, Augusto O. *Free Radic Biol Med.* 2010;49:1046–1053.
48. Ösapay K, Tran D, Ladokhin AS, White SH, Henschen AH, Selsted ME. *J Biol Chem.* 2000;275:12017–12022.
49. Biggs B, Presley AL, v.Vranken DL. *Bioorg Med Chem.* 1998;6:975–981.
50. Hashizume K, Shimonishi Y. *Bull Chem Soc Jpn.* 1981;54:3806–3810.
51. Bergman J, Koch E, Pelcman B. *Tetrahedron Lett.* 1995;36:3945–3948.
52. Stachel SJ, Habeeb RL, van Vranken DL. *J Am Chem Soc.* 1996;118:1225–1226.
53. Carter DS, van Vrankan DL. *Tetrahedron Lett.* 1996;37:5629–5632.
54. Isidro-Llobet A, Álvarez M, Albericio F. *Chem Rev.* 2009;109:2455–2504.
55. Pearson DA, Blanchette M, Baker ML, Guindon C. *Tetrahedron Lett.* 1989;30:2739–2742.
56. Doyle MP, McOsker CC. *J Org Chem.* 1978;43:693–696.
57. Lanzilotti AE, Littell R, Fanshawe WJ, McKenzie TC, Lovell MF. *J Org Chem.* 1979;44:4809–4813.
58. Kursanov DN, Parnes ZN, Loim NM. *Synthesis.* 1974:633–651.
59. Carey FA, Tremper HS. *J Org Chem.* 1971;36:758–761.
60. Nacagawa Y, Nishiuchi Y, Emura Y, Sakakibra S. In: Okawa K, ed. *Peptide Chemistry 1980.* Osaka, Japan: Protein Research Foundation; 1981:41.
61. Houk J, Whitesides GM. *J Am Chem Soc.* 1987;109:6825–6836.
62. Singh PR, Rajopadhye M, Clark SL, Williams NE. *Tetrahedron Lett.* 1996;37:4117–4120.
63. Ying J, Clavreul N, Sethuraman M, Adachi T, Cohen RA. *Free Radic Biol Med.* 2007;43:1099–1108.
64. Tan JT, Bardwell JCA. *Chem Bio Chem.* 2004;5:1479–1487.
65. Kim JR, Yoon HW, Kwon KS, Lee SR, Rhee SG. *Anal Biochem.* 2000;283:214–221.
66. Bulaj G, De La Cruz R, Azimi-Zonooz A, et al. *Biochemistry.* 2001;40:13201–13208.
67. Rehder DS, Borges CR. *Biochemistry.* 2010;49:7748–7755.
68. Allison WS. *Acc Chem Res.* 1976;9:293–299.
69. Toennies G. *J Biol Chem.* 1937;122:27–47.
70. Ullrich V, Kissner R. *J Inorg Chem.* 2006;100:2079–2086.
71. Clavreul N, Adachi T, Pimental DR, Ido Y, Schoneich C, Cohen RA. *FASEB J.* 2006;20:518–520.
72. Teixeira A, Benckhuijsen WE, de Koning PE, Valentijn ARPM, Drijfhout JW. *Protein Pept Lett.* 2002;9:379–385.
73. Green J, Ogunjobi OM, Ramage R, Stewart ASJ, McCurdy S, Noble R. *Tetrahedron Lett.* 1988;29:4341–4344.
74. Riniker B, Hartmann H. Peptides: chemistry, structure and biology. In: Rivier JE, Marshall GR, eds. *Proceedings of the Eleventh American Peptide Symposium.* Leiden: ESCOM; 1990:950–952.

75. Medzihradszky KF, Darula Z, Perlson E, et al. *Mol Cell Proteom.* 2004;3:429–443.
76. Cynthia GF, Fields GB. *Tetrahedron Lett.* 1993;34:6661–6664.
77. Verdini AS, Lucietto P, Fossati G, Giordani C. *Tetrahedron Lett.* 1992;33:6541–6542.
78. Verdini AS, Lucietto P, Fossati G, Giordani C. In: Smith JA, Rivier JE, eds. *Peptides: Chemistry and Biology.* Leiden: ESCOM; 1992:562.
79. (a) Veber DF, Milkowski JD, Varga SL, Denkewalter RG, Hirschmann R. *J Am Chem Soc.* 1972;94:5456–5461. (b) Kamber B. *Helv Chim Acta.* 1971;54:927–930.
80. Sakakibara S. *Biopolymers.* 1995;37:17–28.
81. Atherton E, Pinori M, Sheppard RC. *J Chem Soc Perkin Trans.* 1985;1:2057–2064.
82. Shinohara K, Kilpatrick M. *J Am Chem Soc.* 1934;56:1466–1472.
83. Fujii N, Otaka A, Funakoshi S, Bessho K, Yajima H. *J Chem Soc Chem Commun.* 1987:163–164.
84. Tam JP, Shen ZY. *Int J Pept Protein Res.* 1992;39:464–471.
85. Nomizu M, Utani A, Shiraishi N, Yamada Y, Roller PP. *Int J Pept Protein Res.* 1992;40:72–79.
86. van Rietschoten J, Pedroso Muller E, Granier C. Peptides. In: Goodman M, Meienhofer J, eds. *Proceedings of the Fifth American Peptide Symposium.* New York: Wiley; 1977:522.
87. Lyle TA, Brady SF, Ciccaarone TM, et al. *J Org Chem.* 1987;52:3752–3759.
88. Wade JD, Fitzgerald SP, McDonald MR, McDougall JG, Tregear GW. *Biopolymers.* 1986;25(suppl):S21–S37.
89. Atherton E, Sheppard RC, Ward P. *J Chem Soc Perkin Trans.* 1985;1:2065–2073.
90. Engebretsen M, Agner E, Sandoshan J, Fischer PM. *J Pept Res.* 1997;49:341–346.
91. Kiso Y, Ukawa K, Akita T. *J Chem Soc Chem Commun.* 1980:101–102.
92. Eritja R, Ziehler-Martin JP, Walker PA, et al. *Tetrahedron.* 1987;43:2675–2680.
93. Lamthanh H, Roumestand C, Deprun C, Ménez A. *Int J Pept Protein Res.* 1993;41:85–95.
94. Mendelson WL, Tickner AM, Holmes MM, Lantos I. *Int J Pept Protein Res.* 1990;35:249–257.
95. Spanninger PA, Rosenberg JL. *J Am Chem Soc.* 1972;94:1973–1978.
96. Cooper AJL. In: Keller MD, Kiene RP, Kirst GO, Visscher PT, eds. *Biological and Environmental Chemistry of DMSP and Related Sulfonium Compounds.* New York: Plenum Press; 1996:13–27.
97. Gurd FRN. *Methods Enzymol.* 1967;11:532–541.
98. Harris PWR, Kowalczyk R, Yang S-H, Williams GM, Brimble MA. *J Pept Sci.* 2014;20:186–190.
99. Gundlach HG, Moore S, Stein WH. *J Biol Chem.* 1959;234:1761–1764.
100. Kiso Y, Satomi M, Ukawa K, Akita T. *J Chem Soc Chem Commun.* 1980:1063–1064.
101. Yajima M, Akaji K, Mitani N, et al. *Int J Pept Protein Res.* 1979;14:169–176.
102. Antherton E, Sheppard RC, Wade JD. *J Chem Soc Chem Commun.* 1983:1060–1062.
103. (a) Huang H, Rabenstein DL. *J Pept Res.* 1999;53:548–553. (b) Yajima H, Kanaki J, Kitajima M, Funakoshi S. *Chem Pharm Bull.* 1980;28:1214–1218.
104. Kiso Y, Nakamura S, Ito K, et al. . *J Chem Soc Chem Commun.* 1979:971–972.
105. Robey FA. In: Pennington MW, Dunn BM, eds. *Peptide Synthesis Protocols, Methods in Molecular Biology. Vol. 35.* Totowa, New Jersey: Humana Press; 1994:79.

Chapter 4

Peptide Rearrangement Side Reactions

Undesirable peptide rearrangement represents a category of common side reactions occurring in the process of peptide manufacture as well as storage. Due to the existence of a plethora of functional groups and divergent conditions, e.g. different pH values that the peptide has to be subjected to in the process of manufacture, in combination with somewhat unpredictable synergistic steric effects, peptide rearrangement is prone to take place on some labile individuals under certain circumstances. One of the challenges inherent to these types of side reactions is that the derived side products are frequently isomer to the target peptides, and the development of pertinent analytical methods with respect to the re-arranged peptide impurities hence poses a critical task. Moreover, the rearrangement process might also result in the formation of impurities with a redundantly incorporated residue/segment, which is difficult to be separated from the target product. In order to alleviate the somewhat overwhelming stresses these impurities exert on the downstream process (purification) of peptide production, they should be preferably addressed in the upstream process (synthesis) so as to minimize the occurrences by efficient synthetic strategies. In this chapter, representative rearrangement processes that daunt peptide synthesis are elucidated, and the corresponding potential solutions are discussed.

4.1 ACID CATALYZED ACYL *N→O* MIGRATION AND THE SUBSEQUENT PEPTIDE ACIDOLYSIS

Acyl migration between nitrogen and oxygen atoms is a common process in organic chemistry that has been utilized in many compound syntheses, among which the *O*-acyl isopeptide synthetic tactic (Fig. 1.2) as an outstanding exemplar has gained widespread applications in the domains of peptide preparation, particularly for those challenging syntheses complicated by an aggravated peptide chain aggregation along the subjected peptide assembly.[1] The focus of this chapter will, nevertheless, deviate from the beneficial aspects of the O-acyl isopeptide methodology, and focus on the side reactions originating from the acyl *N→O* migration interconnected with the utilization of the O-acyl isopeptide building block in the process of the affected peptide synthesis.

Side Reactions in Peptide Synthesis. http://dx.doi.org/10.1016/B978-0-12-801009-9.00004-5

Protein autocatalytic degradation refers to the phenomenon that some natural proteins containing specific sequences are subject to fragmenting at certain sites, and being decomposed by this means into diverse segments.Many protein autocatalytic degradation processes are provoked on an intramolecular *N→O* acyl migration chemical basis, and lead to the formation of an isomeric ester counterpart as an intermediate.[2] The Acyl *N→O* migration process was initially detected under the circumstances of substrate treatment by strong acids, e.g., H_2SO_4, HF, or HCl.[3] Along with the widespread application of Fmoc-chemistry, TFA-catalyzed acyl *N→O* migration is becoming one of the most frequently detected side reactions in peptide synthesis.[4] This side reaction might be induced at the step of an acid-mediated peptide side chain global deprotection, and its severity is evidently correlated to the specific sequences from the parental peptide. For instance, the substitution of a residue simply by its D-amino acid counterpart in the subjected peptide sequence might substantially affect the tendency of this side reaction.[4] Besides, the existence of a phosphoamino acid in the peptide sequence might also significantly facilitate acyl *N→O* migration, and the extent thereof could be subject to the specific location of the referred phosphoamino acid residue in the peptide sequence.[5] The acceptors of the acid-catalyzed acyl shift in peptide synthesis are normally those residues that bear nucleophilic substituents, e.g., Ser, Thr, or Cys, whereas this sector will be focused on *N→O* acyl migration exclusively.

The mechanism of acid-catalyzed acyl *N→O* migration is proposed in Fig. 4.1. The hydroxyl group on the Ser or Thr from peptide **1** initiates an acid-catalyzed nucleophilic attack on its preceding backbone amide bond, and generates a 5-member-ring hydroxyoxazolidine intermediate **2**,[6] and the latter will be subsequently rearranged to the corresponding ester isomer **3** in acidic condition. It has also been asserted that a concentrated H_2SO_4-catalyzed acyl *N→O* migration proceeds via an oxazoline intermediate.[7] An acid-catalyzed acyl *N→O* migration is a reversible process that could be redirected to an amide formation in neutral or basic conditions. The protonation of the released amino group on peptide **3** in an acidic milieu is a critical impetus to drive the equilibrium in the direction of ester formation. Under neutral or basic circumstances ester **3** will undergo a prompt acyl *O→N* shift to regenerate the original amide **1**

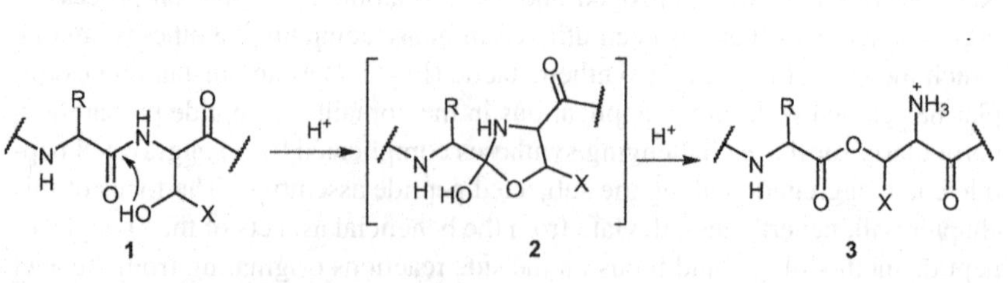

FIGURE 4.1 Proposed mechanism of acid-catalyzed peptide acyl *N→O* migration.

compound.[8] This equilibrium is regarded as the theoretical basis for the application of O-acyl isopeptide building blocks in peptide synthesis.

Acyl *N→O* migration could take place at the step of a peptide side chain global deprotection. As illustrated in Fig. 4.2 an ordinary peptide **4** that is subject to the process of a side chain protecting group cleavage might be affected by a concomitant acyl *N→O* migration at the site of a -Xaa-Ser/Thr- sequence, and be rearranged to the corresponding isomeric ester **5**. The latter could undergo hydrolytic degradation in an aqueous condition during synthesis, workup, isolation, purification or even storage, and be decomposed to the resultant peptide fragments **6** and **7**.[9]

Basically, the acyl *N→O* migration process affects the *N*-terminal Ser substrate more frequently than an endo-Ser residue.[10] Taking peptide **8** with *N*-Ac-Ser moiety for instance (Fig. 4.3), the N^{α}-group regenerated from the N^{α}-Fmoc protected precursor undertakes acetic acid-mediated acetylation process and the derived N^{α}-Ac-Ser peptide is subsequently subject to a TFA-dictated peptide side chain global deprotection. The N-terminal N^{α}-Ac-Ser moiety on the concerned peptide **8** will partially undergo TFA-catalyzed acyl *N→O* migration under this circumstance to give rise to an isomeric H-Ser(Ac) impurity **9**. If the

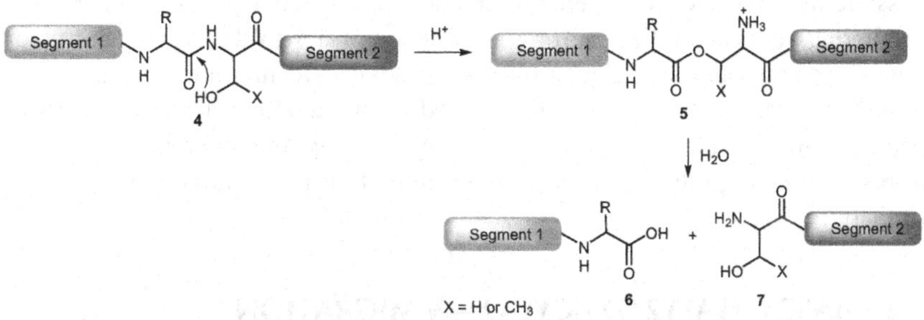

FIGURE 4.2 **Acidolysis of peptide containing -Xaa-Ser/Thr- unit.**

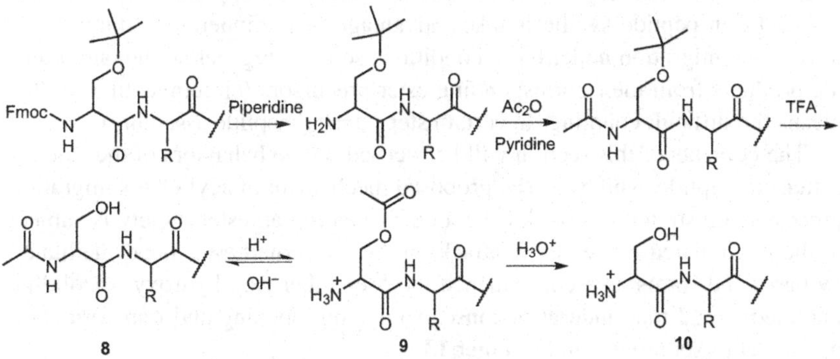

FIGURE 4.3 ***N→O*** **acyl migration on N^{α}-Ac-Ser peptide and subsequent de-acetylation process.**

ester derivative **9** in acidic milieu is subject to further hydrolysis degradation, the side chain acetylated intermediate could be converted to the corresponding deacetylated side product **10**. Overall the degradative outcome could be deemed as a consequence of N^{α}-deacetylation. Nevertheless, this decomposition process might be facilitated by the described acyl *N*→*O* migration effect.

Attention is supposed to be paid to the syntheses of these acyl migration-prone peptide products: (1) the employment of strong acids, e.g., HF, H_2SO_4, HCl, and TFMSA should be avoided as much as possible; (2) if strong acids could not be spared in the peptide global deprotection, the addition of water as a scavenger or a cosolvent are supposed to be prevented; (3) a lowering of the acid concentration for the removal of side chain protecting groups; and (4) the extent of acyl *N*→*O* migration is directly interconnected to the duration of the acid treatment of the subjected peptide.[5] Rational shortening of acid stress on susceptible peptides could hence effectively diminish the undesirable peptide ester formation induced by an acyl *N*→*O* migration. The potential subsequent peptide ester hydrolysis, and resultant backbone fragmentation could be accordingly alleviated; and (5) caution should be exerted in the process of product purification and storage. Product fractions collected from RP-HPLC containing acidic additives from eluents should be subject to the lyophilization as soon as possible in order to decrease the possible acid-catalyzed acyl *N*→*O* migration; or alternatively, neutralize the product fractions prior to the subsequent handlings; (6) in the case of the occurrence of an acyl *N*→*O* migration regeneration of the target products that could be realized via the treatment of the concerned peptide with a readily removable base, e.g., a dilute ammonia solution so as to reverse the equilibrium of acyl *N*—*O* migration to the direction of amide formation.

4.2 BASE CATALYZED ACYL *O*→*N* MIGRATION

The base catalyzed acyl *O*→*N* shift could be regarded as the reverse reaction of an acid catalyzed acyl *N*→*O* migration. The widely applied O-acyl isopeptide tactic in peptide synthesis takes advantage of the inherent features of an acyl *O*→*N* migration under basic conditions so as to regenerate the target amide products from their corresponding ester precursors (circumventing by this means the difficult couplings at certain steps during peptide assembly).

The contents of this section will be focused on the behaviors of the side reactions in peptide synthesis. The proposed mechanism of acyl *O*→*N* migration process is illustrated in Fig. 4.4. Peptide **11** bearing an ester moiety is subject to the N^{α}-initiated nucleophilic attack on the concerned ester bond facilitated by base, and forms as a consequence the 5-member-ring hydroxyoxazolidine intermediate **12** that undergoes consequently ring opening and transformation into the corresponding amide isomer **13**.

Base catalyzed acyl *O*→*N* migration in peptide synthesis could function as the source of the formation of trifluoroacetate side products. In the Boc-mode

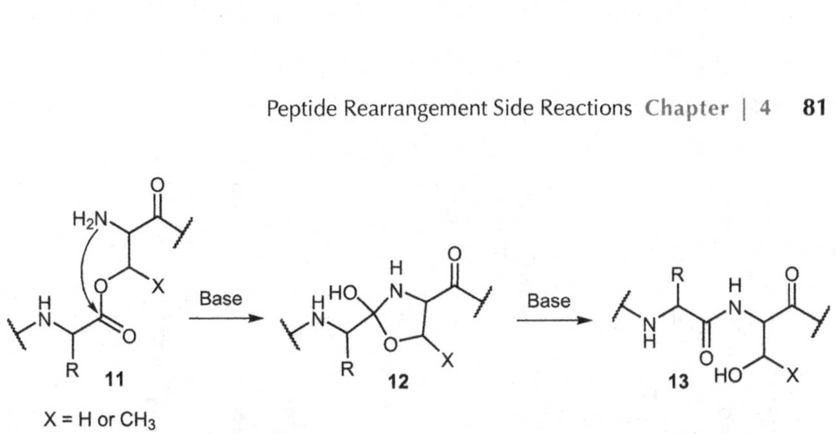

FIGURE 4.4 Proposed mechanism of base-catalyzed peptide acyl *O*→*N* migration.

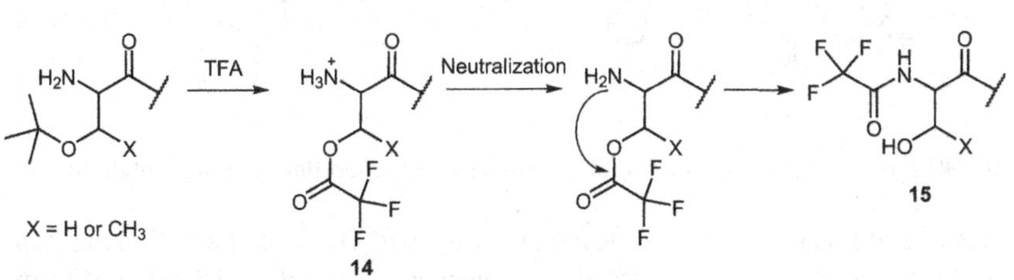

FIGURE 4.5 N^{α}-trifluoroacetylation side reaction on N-terminal Ser/Thr peptide.

SPPS TFA utilized for N^{α}-Boc removal might function with the residual unacylated hydroxyl group on the resin, and the formed trifluoroacetyl derivatives could provoke the acyl *O*→*N* migration upon the subsequent base treatment at the neutralization step. The subjected trifluoroacetyl group will shift under such circumstances from the hydroxyl to the N^{α} on the peptide backbone, and give rise to the formation of a N^{α}-trifluoroacetyl peptide side product that consequently causes the termination of a peptide chain elongation.[11] Analogously, during the synthesis of N-terminal Ser or Thr peptides the liberated hydroxyl side chain on the subjected Ser/Thr could be functionalized by TFA (or the possible residual trifluoroacetic acid anhydride) at the step of the peptide side chain global deprotection that gives rise to the formation of a corresponding H-Ser/Thr(TFA)-derivative **14**. This intermediate might undergo trifluoroacetyl *O*→*N* shift subsequently during workup or purification to generate the N^{α}-trifluoroacetylated side product **15**.[12] (Fig. 4.5).

Owing to the undesired base-catalyzed acyl *O*→*N* migration abundant isomeric amide impurities **19** were formed during the preparation of the cyclic derivative **18** as per the original synthetic strategy (Fig. 4.6).[13] The on-resin linear precursor peptide **16** was first treated by $Pd(PPh_3)_4$ to remove the Ally and Alloc protecting groups from the originally attached D-Asp-C^{α} and D-Ala-N^{α}, respectively. The liberated carboxyl and amino group was subsequently covalently spliced by a newly formed amide bond through a PyBOP-mediated coupling, giving rise to the formation of the cyclic intermediate **17**. The subsequent piperidine treatment of **17** was conducted in an effort to deblock the Fmoc protecting group from Thr-N^{α} to afford the protected cyclic peptide **18**. The original

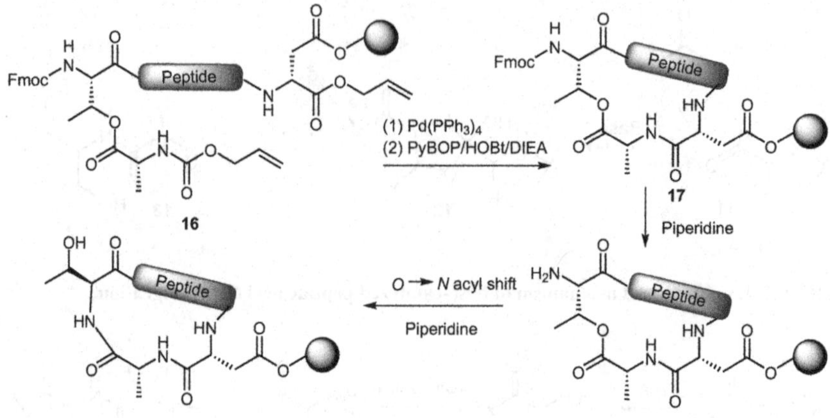

FIGURE 4.6 Acyl *O*→*N* migration side reaction in the preparation of cyclic peptide 18.

objective of this treatment is to acylate the liberated Thr-N$^{\alpha}$ at the following step for the assembly of the target product. However, intermediate **18** suffered from the undesired acyl *O*→*N* migration upon piperidine treatment. The original labile ester bond between the Thr hydroxyl side chain and D-Ala carboxylate backbone undertook the nucleophilic attack from Thr-N$^{\alpha}$ and was replaced by a newly formed amide bond between the referred Thr and D-Ala. Side product **19** was hence provoked on the basis of a base-catalyzed acyl *O*→*N* migration. A solution eliminating this side reaction is accordingly developed in which Trt instead of Fmoc was adopted as the N$^{\alpha}$-protecting group for the concerned Thr. Since N$^{\alpha}$-Trt could be deblocked by dilute acid solution the root cause of the acyl *O*→*N* migration with respect to base-catalyst is specifically addressed and the undesired isomerization from amide to ester in the final product is effectively prevented.

Peptide synthesis adopting a "minimal-protecting" strategy in which Ser/Thr residues are assembled into the target peptide chains in the form of side chain unprotected derivatives could inherently be entangled in base-catalyzed acyl *O*→*N* migration side reactions. For example, the chemical synthesis of phospho-Ser/Thr peptide with "global phosphorylation" tactic (Fig. 4.7) might potentially bring about base-catalyzed acyl *O*→*N* migration and a series of resultant side reactions. Ser/Thr residues are incorporated into the target precursor peptide **20** in the form of side chain unprotected species. When the assembly of the linear peptide is completed "global phosphorylation" is conducted in which the free β-hydroxyl groups on Ser/Thr are converted to their corresponding phospho-Ser/Thr counterparts to give rise to the target phosphopeptide **21**.[14]

Ser/Thr residues with free β-hydroxyl group are supposed to be incorporated into the to-be-phosphorylated peptide precursor as per the "global phosphorylation" strategy. However, side reactions, e.g., the redundant incorporation of the subjected Ser/Thr residues, and the resultant over-phosphorylation are prone to occur during this process. The root cause of the above side reactions are

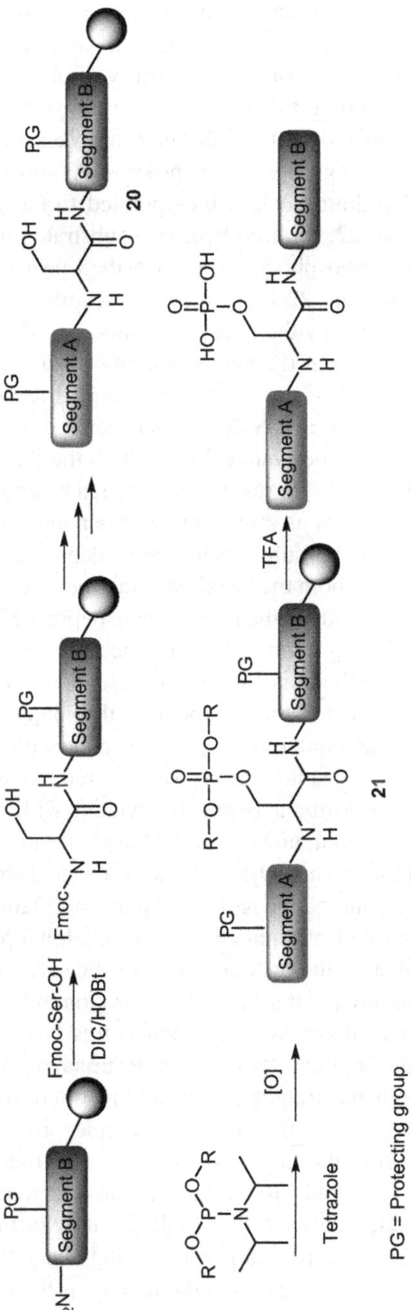

FIGURE 4.7 Global phosphorylation strategy for phosphopeptide preparation via assembly of peptide with side chain unprotected Ser/Thr.

attributed to the base-catalyzed acyl *O→N* migration occurring during side chain unprotected Ser/Thr incorporation and subsequent handlings. In spite of the widespread applications of peptide assembly via "minimal protection strategy" [15] and its implicative utilization in the territory of phosphopeptide preparation,[16] side reactions are prone to be provoked at the step of the Fmoc-Ser/Thr-OH coupling predominantly due to the undesirable acylation of the unprotected Ser/Thr-β-hydroxyl groups in this process. As illustrated in Fig. 4.8 the Fmoc-Ser/Thr-OH building block is incorporated to the growing peptide chain to form intermediate **22**, the free hydroxyl substrate on **22** is supposed to be subject to the global phosphorylation when the linear peptide construction is completed. Intermediate **22** is, however, susceptible to over-acylation on the free β-hydroxyl groups by the excessive Fmoc-Ser/Thr-OH in the reaction system. Despite the fact that the nucleophilicity of the hydroxyl group is substantially weaker than that of the N^{α}-group, it is still reactive towards the activated Fmoc-Ser/Thr-OH species. Side products derived from the mentioned over-acylation process include derivative **24** in which the free hydroxyl group on the to-be-phosphorylated Ser/Thr residue is modified by another molecule of Fmoc-Ser/Thr-OH in the form of an ester moiety. When the side chain-acylated intermediate **24** undertakes the subsequent Fmoc deblocking by piperidine treatment the N^{α} groups from both the backbone, and the side chain Ser/Thr are simultaneously liberated to produce the intermediate peptide **25**, that undergoes base-catalyzed acyl *O→N* migration in the presence of piperidine. This rearrangement process results in the relocation of the redundantly incorporated Ser/Thr residue from the side chain to the backbone of the subjected peptide. As a consequence side product **26** containing repetitive sequence -Ser/Thr-Ser/Thr- is generated. At the "global phosphorylation" step the redundantly incorporated Ser/Thr residue will be indiscriminately phosphorylated, without taking the potential steric effects into account, and the over-phosphorylated side product **27** bearing -Ser/Thr(PO_3H_2)-Ser/Thr(PO_3H_2)- unit is hence formed (Yang, Y., unpublished results). In summary, the redundant phosphorylation side reactions occurring in the preparation of phosphopeptide via a "global phosphorylation" tactic are basically ascribed to the over-acylation of the side chain-unprotected Ser/Thr residue and the consequent acyl *O→N* migration in basic milieu.

No general solutions could universally address this type of side reactions induced by base-catalyzed acyl *O→N* migration. Rational modification or revision of the subjected synthetic strategy is normally resorted to so as to circumvent or alleviate the adverse acyl migration. For example, prior to the potentially detrimental base treatment of the susceptible peptide substrate it will be effective to temporarily block the inductive amino group that readily initiates the undesirable acyl *O→N* shift. This solution could partially or thoroughly shield the nucleophilicity of the concerned amino group and hence the resultant acyl migration. The meticulous design of the synthetic route is therefore particularly essential for the peptide individuals bearing susceptible moieties to undergo acyl *N↔O* shift upon pH variation.

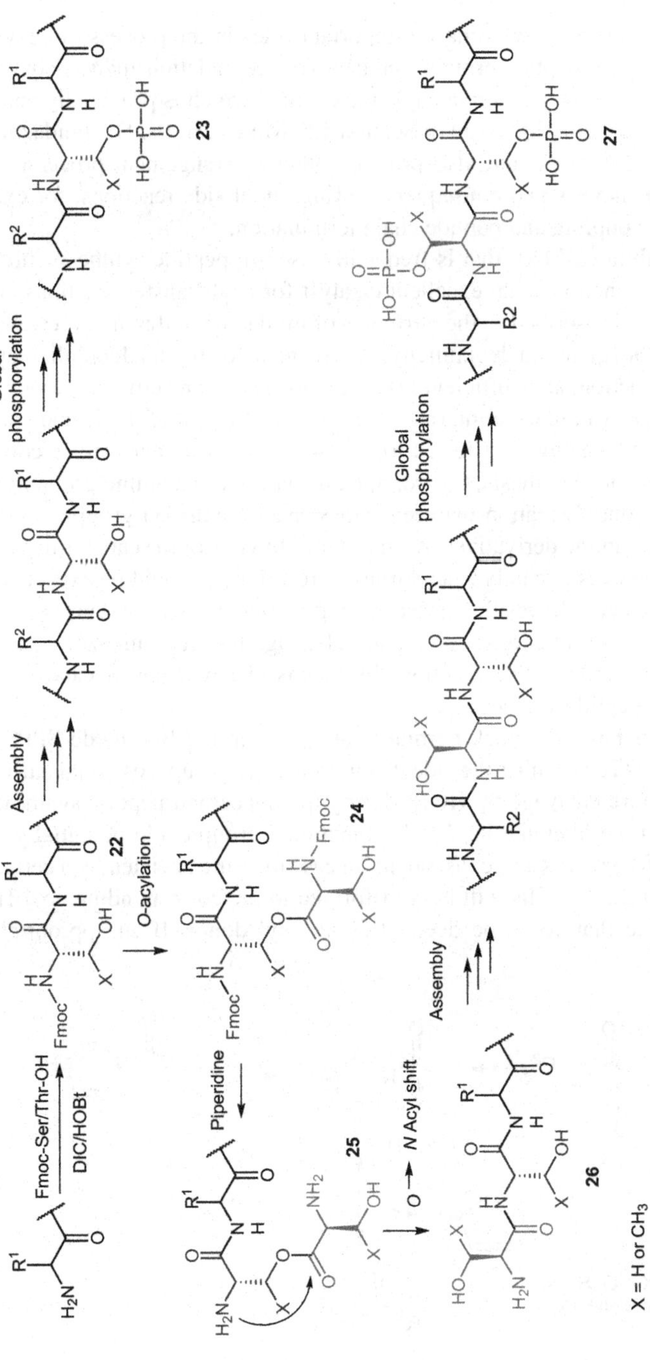

FIGURE 4.8 Side chain-unprotected Ser/Thr induced over-phosphorylation side reaction.

4.3 His-N^{im}- INDUCED ACYL MIGRATION

His residue plays extraordinarily important roles in the process of enzyme catalysis, such as peptide/protein autocatalytic degradation induced by the nucleophilicity of the His imidazolyl side chain,[17] which is principally analogous to the side reaction discussed in Section 1.5. Moreover, the His imidazolyl side chain as a catalyst could also provoke other rearrangement processes in the chemical synthesis and consequently bring about side reactions, for example, redundant coupling and peptide chain termination.

Resembling DMAP that is frequently used in peptide synthesis, imidazole could also function as an excellent catalyst for acyl transfer reactions. The intrinsic merit is ascribed to the attributes of imidazole in that it possesses simultaneously adequate nucleophilicity as a reagent for the nucleophilic acyl substitution reaction, and sufficient leaving tendency as an activating group on the generated acyl-imidazole intermediate. The mechanism of imidazole-catalyzed acyl transfer reaction is referred to Fig. 4.9. Imidazole derivatives could also catalyze peptide synthesis, for example, acylation of an amino group by active ester,[18] lactone formation between thioester and hydroxyl group,[19] as well as acylation of amine derivatives by thioester.[20] In view of the capabilities of imidazole derivatives to catalyze acyl transfer reactions, it could be expected that if the side chain of His residue in certain peptide molecules is not protected, or its protecting group undergoes premature cleavage the free imidazolyl functional group might induce side reactions by means of acyl transfer catalysis in the process of peptide synthesis.

The His building block normally takes part in the Boc-mode SPPS in the form of His(Tos), whilst the side chain protecting group Tos could suffer from the premature cleavage by HOBt during the concerned peptide synthesis process,[21] and the liberated imidazolyl functional group could potentially initiate acyl transfer process as a consequent. In case of being acylated by a certain acyl substituent the free His will be transformed to the corresponding acyl-His-N^{im} intermediate that could be deemed as an acyl donor. If an appropriate acyl

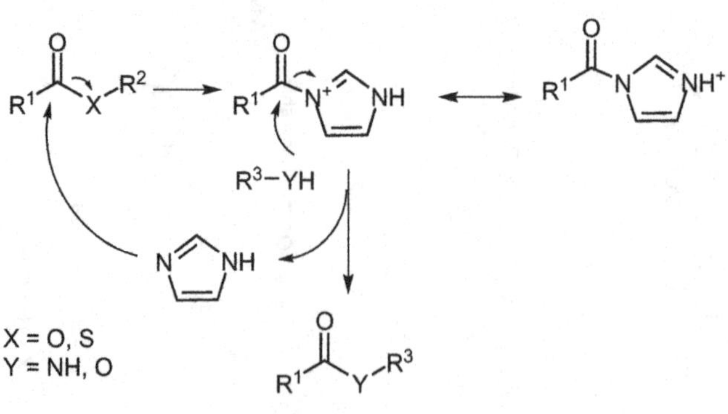

FIGURE 4.9 Mechanism of imidazole catalyzed acyl transfer reaction.

acceptor, e.g. a free amino or a hydroxyl group lies in proximity, the acyl substituent bearing on the His side chain could be transferred in proper conditions, and give rise to the corresponding amide or ester side products. As is revealed in Fig. 4.10, peptide **28** containing His(Tos) residue could undergo Tos premature cleavage at repetitive amino acid coupling steps facilitated by HOBt, and is consequently converted to an intermediate **29** bearing side chain-unprotected His residue. The liberated imidazolyl substituent on the affected His could remain intact in the following peptide assembly process partially due to the steric effects that shield the exposed His-N^{im} from the acylation by the forthcoming amino acids, in particular those sterically hindered amino acids. Alternatively, if the affected histidine indeed undertakes acylation modification on its liberated imidazolyl side chain, the generated His-N^{im}-acylated intermediate will be readily subjected to hydrolysis, and regenerate the free imidazolyl functional group as a consequence. The overall impact on a free His side chain is therefore not manifested by this means.

In spite of the ability to maintain the integrity in most peptide synthetic conditions His-N^{im}-acylation is relatively more prone to occur in the process of Boc-Gly-OH coupling thanks to its advantageous steric effect compared with other bulkier residues. The resultant His-N^{im}-acylation by Boc-Gly-OH will engender derivative **30** containing Boc-Gly-$(Xaa)_n$-His(Boc-Gly-)-moiety. In light of the Boc-SPPS process intermediate **30** is subsequently subjected to successive N^{α}-Boc deblocking, and N^{α}-neutralization to give derivative **31**. Treatment of peptide chains immobilized on solid support by tertiary amines, e.g., DIEA is conducted in an effort to neutralize the N^{α} group protonated at the preceding TFA-mediate N^{α}-Boc cleavage step. However, based-catalyzed $N^{im} \rightarrow N^{\alpha}$ acyl migration could be induced at the neutralization step owing to the synergistic effects of the reduced Gly steric hindrance and an advantageous peptide conformation.[21] The redundantly incorporated Gly residue that is originally linked to the histidine imidazolyl side chain is thus relocated and squeezed into the peptide backbone by virtue of $N^{im} \rightarrow N^{\alpha}$ acyl migration. As a consequence impurity **32** with an endo-Gly residue and Δ MW = +57 amu is formed.

Besides the susceptible N^{α} functional group the free hydroxyl group in a peptide might serve as the receptor of His-induced acyl migration, resulting in the formation of *O*-acylated side products.[22] Reactions between hydroxyl groups and active esters are normally sluggish. This phenomenon serves as the theoretical basis for the adoption of the "minimal protection" peptide synthetic strategy in which Ser/Thr could be incorporated into the growing peptide chains in the form of side chain unprotected species. However, if a His residue with free imidazolyl side chain happens to be simultaneously present in the concerned peptide intermediate, the hydroxyl substituent on Ser/Thr might become the target of the His-catalyzed acyl $N^{im} \rightarrow O$ migration process, rendering the originally inert hydroxyl groups to be susceptible to acylation. The scheme of this kind of side reaction is exemplified in Fig. 4.11. The convergent segment condensation is conducted between *N*-terminal component peptide active

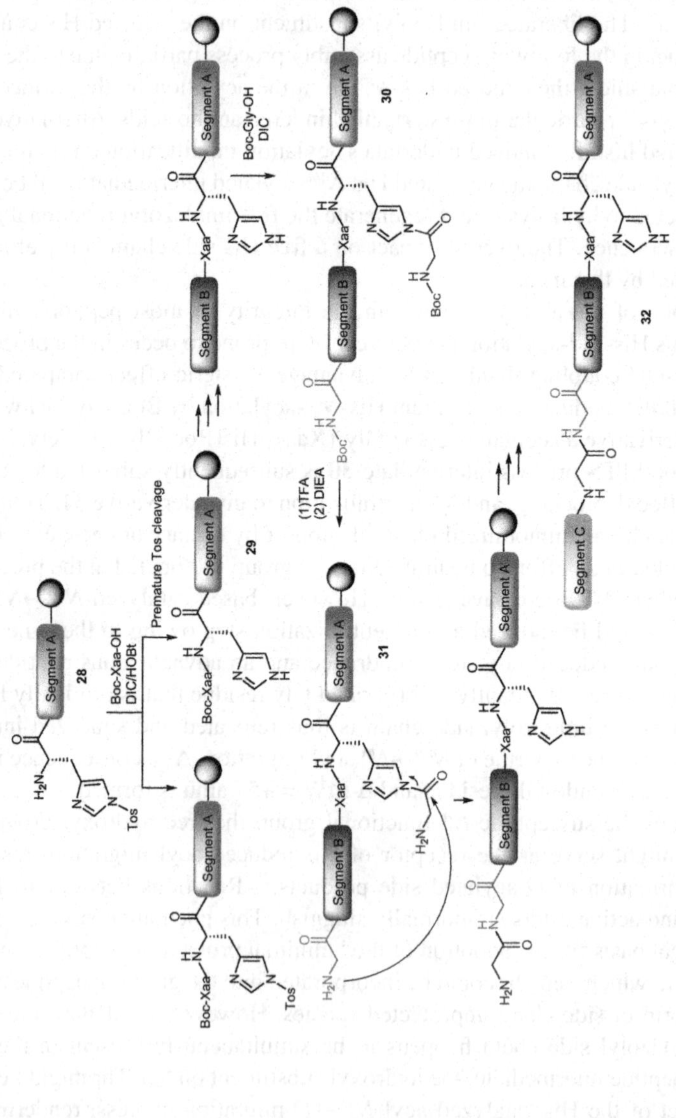

FIGURE 4.10 Redundant Gly incorporation induced by His-mediated acyl $N^{im}\rightarrow N^{\alpha}$ transfer.

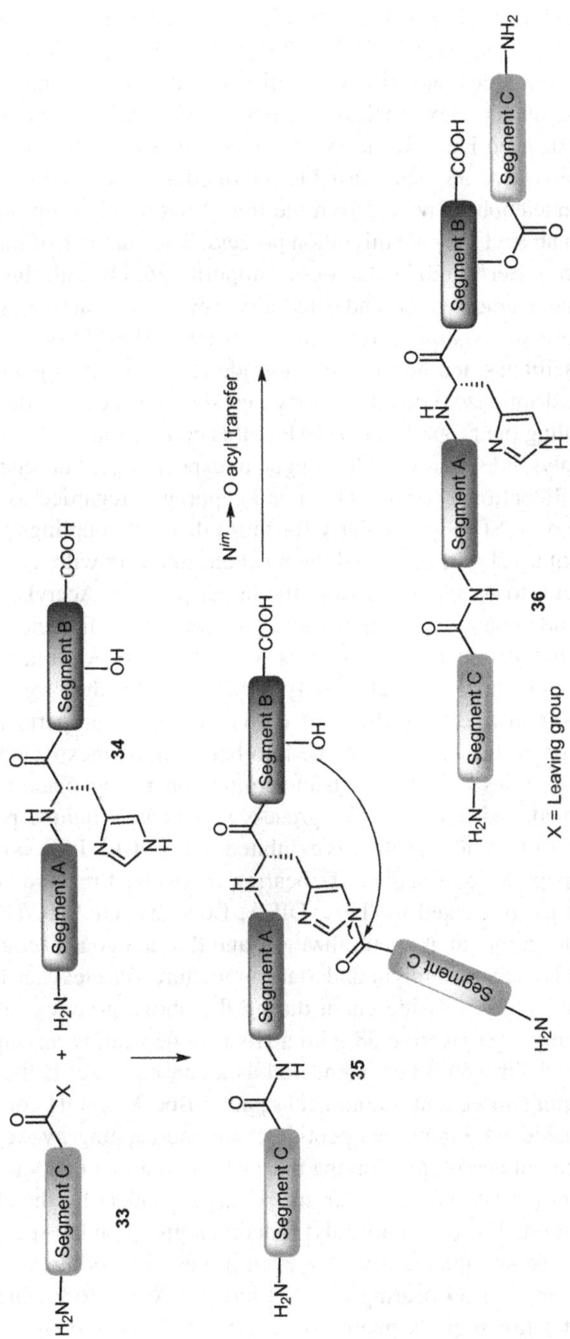

FIGURE 4.11 **His mediated acyl $N^{im}\rightarrow O$ migration.**

ester **33** and *C*-terminal component peptide fragment **34** bearing side chain unprotected His and Ser/Thr. The desired conjugation reaction is supposed to take place between the active ester moiety on **33** and backbone N^{α} functional group on **34**. A resultant amide bond will be established in this process that covalently splices the 2 segments. Nevertheless, the undesired reaction between the imidazolyl substituent on His and the excessive active ester **33** might generate a N^{im}-acylated derivative **35**. This unstable intermediate could subsequently be subjected to a nucleophilic attack from the free β-hydroxyl group on Ser/Thr, and involved in an acyl $N^{im}\rightarrow O$ migration process. The outcome of this process is the formation of Ser/Thr side chain-ester impurity **36**. Overall, this complex process could be deemed as the undesired alcoholysis of an active ester **33** by the hydroxyl derivatives; however, the involvement of His-N^{im} as a catalyst for acyl transfer facilitates and accelerates this side reaction. Analogously, it has been found[23] that imidazole could serve as a catalyst for the Ser side chain acylation, resembling the property of DMAP in this connection.

His-imidazolyl side chain might trigger unexpected peptide chain termination as well. Intentional peptide chain end-capping is regarded as a regular synthetic strategy in SPPS particularly for those difficult couplings that could hardly be quantitatively realized, and the resultant impurity with a deletion sequence is difficult to be separated from the target product. Acetylation of the unacylated peptide chains after amino acid coupling by acetic acid anhydride could be applied under such circumstances so that the derived truncated acetylated peptide chains could be sufficiently separated from the target products by virtue of their instinctively different chromatographic properties. Peptide chain termination could, however, come into being in an unexpected manner. Among various causes of peptide chain termination taking place in peptide synthesis His-mediated acyl $N^{im}\rightarrow N^{\alpha}$ process represents a unique pathway.[24] The mechanism of this side reaction is exhibited in Fig. 4.12. In Boc-chemistry SPPS process peptide intermediate **37** bearing a His(Tos) residue is sequentially and repetitively treated by TFA, DIEA, Boc-Xaa$_n$-OH/DIC/HOBt with the aim of N^{α}-Boc removal, N^{α} neutralization, and Boc-amino acid coupling, respectively. His(Tos) residue might undertake premature Tos cleavage by HOBt, and liberate the imidazolyl side chain during the above processes due to its inherent instability.[21] Derivative **38** with a His residue bearing an unprotected imidazolyl side chain would be engendered as a consequence. If the incorporation of a certain subsequent amino acid residue Boc-Xaa$_n$-OH could not be quantitatively achieved, intentional peptide chain end-capping by Ac_2O/DIEA would be carried out accordingly taking into the account that the potential des-Xaa$_n$ impurity might be inseparable from the target product by the chromatographic purification. The free imidazolyl functional group on the subjected His might, however, be simultaneously acetylated in this process by Ac_2O, giving rise to the side product **39** bearing His(Ac) moiety. At the following N^{α}-Boc removal/neutralization steps the peptidyl resin is treated successively by TFA and DIEA. The acetyl group immobilized on His-N^{im} might migrate to the peptide

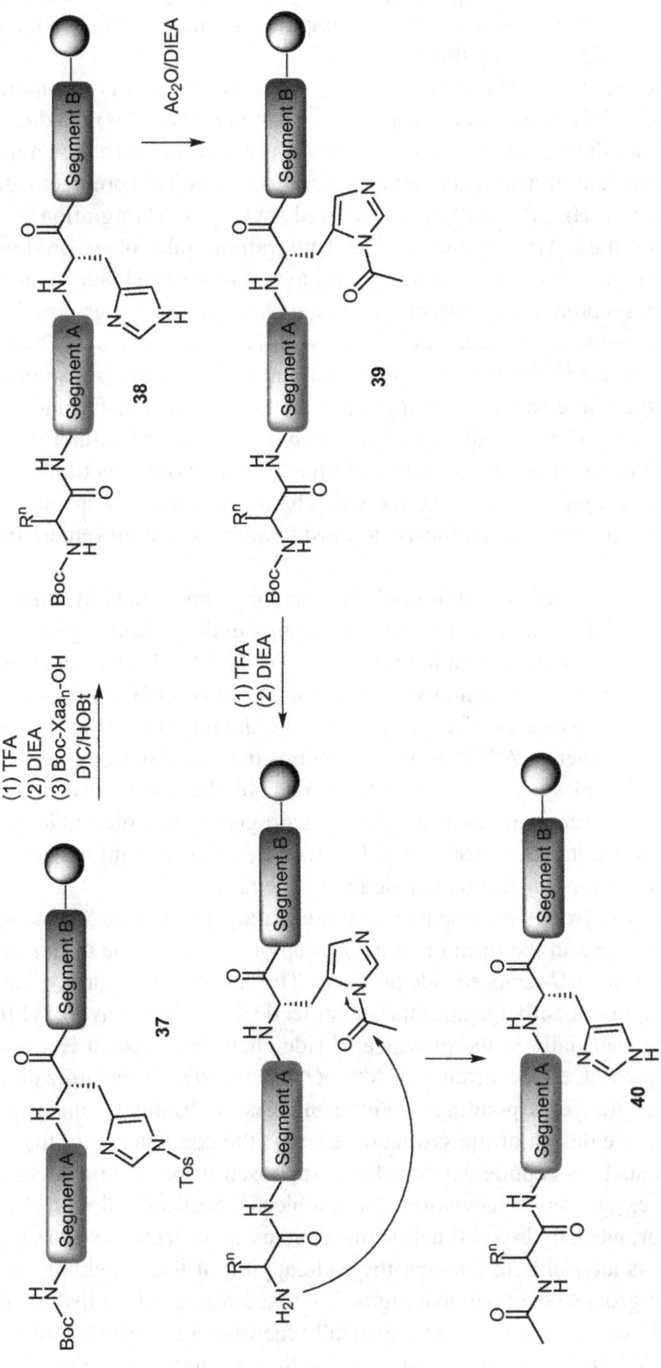

FIGURE 4.12 His/Ac$_2$O-mediated unintentional peptide chain termination.

backbone N^{α} group during this process facilitated by DIEA, leading to the formation of backbone N^{α}-acetylated impurity **40** that terminates the affected growing peptide chain by this means.

In view of the mechanism of the above side reaction, it is recommended to avoid Ac_2O/DIEA treatment on peptidyl resin once His(Tos) residue is incorporated into the peptide chain in order to suppress or diminish the unintentional peptide chain termination concertedly induced by His(Tos) premature cleavage, acetylation of His-N^{im} and DIEA catalyzed acyl $N^{im} \rightarrow N^{\alpha}$ migration.

In summary, His-N^{im}-induced acyl migrations take place predominantly via His derivatives, which bear an unprotected imidazolyl side chain or those that undergo premature deblocking of side chain protecting groups. The histidine imidazolyl substituent could react with DIC by virtue of its basicity and nucleophilicity.[25] Furthermore, side chain unprotected His is also prone to be subjected to diverse side reactions, for example, acyl transfer and racemization. In spite of the feasibility of the employment of side chain unprotected His derivatives in peptide synthesis with a "minimal deprotection" strategy, it is highly recommended to use the side chain protected His species as building blocks for peptide synthesis in consideration of the aforementioned side reactions.

His-induced side reactions could be somehow diminished by enhancing the stabilities of the imidazolyl protecting groups under peptide synthesis conditions. In view of the inherent lability of N^{im}-Tos and N^{im}-Boc the employment of these derivatives as the imidazolyl protecting groups for His should be avoided as much as possible. N^{im}-Trt has become the default His protecting groups in Fmoc-SPPS whereas N^{im}-Bom group exhibits increased stability relative to Tos in Boc-SPPS. Despite the fact that N^{im}-Bom could be decomposed to formaldehyde upon its removal via acidolysis or hydrogenation, which induces various side reactions, this drawback could be offset by rational employment of appropriate scavengers for formaldehyde in the due process.[26]

Hydroxyl groups are capable of participating in the His-N^{im}-induced acyl transfer process in the form of an acyl receptor. The outcome of this process is the formation of *O*-acylated side products. This attribute of hydroxyl derivatives requests consequently the appropriate protection on Ser/Thr hydroxyl functional groups especially in the presence of side chain unprotected His residues in order to prevent the occurrence of $N^{im} \rightarrow O$ acyl transfer. If the utilization of free Ser/Thr in the corresponding peptide synthesis is absolutely unavoidable due to the unique design of the synthetic strategy, the equivalency of the acylating reagents such as peptide active ester is supposed to be fine-tuned so as to reduce the extent of over-acylation on the subjected Ser/Thr hydroxyl side chains. Moreover, when hydroxyl functionalized resins are utilized as solid supports for SPPS, it is advisable to conduct the end-capping of the remaining unacylated hydroxyl groups post resin loading by the first amino acid, so that the potential acyl transfer side reactions synergistically facilitated by resin hydroxyl groups and His-N^{im} could be minimized.[27] As for the preparation of sensitive peptides containing His residue it is not recommended to end-cap the unconverted peptide chains by Ac_2O/DIEA in order to prevent the undesired peptide termination caused by His-N^{im}-induced $N^{im} \rightarrow N^{\alpha}$ acetyl migration.

REFERENCES

1. (a) Yoshiya T, Sohma Y, Kimura T, Hayashi Y, Kiso Y. *Tetrahedron Lett.* 2006;47:7905–7909. (b) Taniguchi A, Yoshiya T, Abe N, et al. *J Pept Sci.* 2007;13:868–874. (c) Coin I. *J Pept Sci.* 2010;16:223–230.
2. Paulus H. *Chem Soc Rev.* 1998;27:375–386.
3. (a) Lenard J, Hess GP. *J Biol Chem.* 1964;239:3275–3281. (b) Elliott DF. *Biochem J.* 1952;50:542–550. (c) Levy D, Carpenter FH. *Biochemistry.* 1970;9:3215–3222.
4. Carpino LA, Krause E, Sferdean CD, Bienert M, Beyermann M. *Tetrahedron Lett.* 2005;46:1361–1364.
5. Eberhard H, Seitz O. *Org Biomol Chem.* 2008;6:1349–1355.
6. (a) Bergmann M, Brand E, Weimann F. *Z Physiol Chem.* 1923;131:1–17. (b) Phillips AP, Baltzly R. *J Am Chem Soc.* 1947;69:200–204.
7. Wakamiya T, Tarumi Y, Shiba T. *Chem Lett.* 1973;233–236.
8. Shao Y, Paulus H. *J Peptide Res.* 1997;50:193–198.
9. Bodanszky M, Martinez J. *Synthesis.* 1981;333–356.
10. Mouls L, Subra G, Enjalbal C, Martinez J, Aubagnac J-L. *Tetrahedron Lett.* 2004;45:1173–1178.
11. Kent SBH, Mitchell AR, Engelhard M, Merrifield RB. *Proc Natl Acad Sci.* 1979;76:2180–2184.
12. Hübener G, Göhring W, Musiol H-J, Moroder L. *Pept Res.* 1992;5:287–292.
13. Stawikowski M, Cudic P. *Tetrahedron Lett.* 2006;47:8587–8590.
14. (a) Perich JW, Johns RB. *Aus J Chem.* 1990;43:1623–1632. (b) Perich, JW., Johns, RB. 1988;29:2369–2372.
15. (a) Gagnon P, Huang X, Therrien E, Keillor JW. *Tetrahedron Lett.* 2002;43:7717–7719. (b) Pearson AJ, Chelliah MV, Bigna GC. *Synthesis.* 1997;536–540. (c) Sin N, Meng L, Auth H, Crews CM. *Bioorg Med Chem.* 1998;6:1209–1217.
16. (a) Attard TJ, O'Brien-Simpson N, Reynolds EC. *Int J Pept Res Ther.* 2007;13:447–468. (b) McMurray JS, Coleman IV DR, Wang W, Campbell WL. *Biopolymers.* 2001;60:3–31.
17. Mazur RH, Schlatter JM. *J Org Chem.* 1963;28:1025–1029.
18. Wieland T, Vogeler K. *Angew Chem Int Ed Engl.* 1963;2:42.
19. Li Y, Giulinatti M, Houghten RA. *Org Lett.* 2010;12:2250–2253.
20. Li Y, Yongye A, Giulianotti M, Martinez-Mayorga K, Yu Y, Houghten RA. *J Comb Chem.* 2009;11:1066–1072.
21. Kusunoki M, Nakagawa S, Seo K, Hamana T, Fukuda T. *Int J Peptide Protein Res.* 1990;36:381–386.
22. Bodanszky M, Fink ML, Klausner YS, Natarajan S, Tatemoto K, Yiotakis AE, et al. *J Org Chem.* 1977;42:149–152.
23. Girin SK, Shvachkin YP. *Int J Peptide Protein Res.* 1985;25:200–205.
24. Ishiguro T, Eguchi C. *Chem Pharm Bull.* 1989;37:506–508.
25. Rink H, Riniker B. *Helv Chim Acta.* 1974;57:831–835.
26. Mitchell MA, Runge TA, Mathews WR, Ichhpurani AK, Harn NK, Dobrowolski PJ, et al. *Int J Peptide Protein Res.* 1990;36:350–355.
27. Pessi A, Mancini V, Filtri P, Chiappinelli L. *Int J Peptide Protein Res.* 1991;39:58–62.

Chapter 5

Side Reactions Upon Amino Acid/Peptide Carboxyl Activation

Activation of the carboxyl group on either subjected amino acid building block or peptide segment that is supposed to be spliced to the reciprocal peptide fragment bearing to-be-acylated amino group or other relevant substituent constitutes one of the key steps in the peptide synthesis process. Various side reactions, nevertheless, could be triggered at this step and lead to the formation of abundant side products such as that with side chain modification, redundant amino acid incorporation, or end-capped truncated sequence that terminates the peptide chain elongation or conjugation. Alternatively, some types of side reactions occurred upon carboxyl activation could transfer the affected reaction components into the inert or less reactive species that consequently modify the reaction stoichiometry, leading to the incomplete conversion or change of reaction kinetics which might trigger further side reactions, e.g., racemization.

5.1 FORMATION OF *N*-ACYLUREA UPON PEPTIDE/AMINO ACID-CARBOXYL ACTIVATION BY DIC

Carbodiimide-type compounds are one of the most intensively utilized coupling reagents for the activation of carboxyl groups on amino acids/peptides. Abundant investigations have been carried out to explore the properties and applications of these reagents for peptide synthesis.[1] DIC and DCC represents the most popular carbodiimide coupling reagents and the mechanism of carboxyl activation by these reagents is illustrated in Fig. 5.1. The first step involves the proton abstraction of the to-be-activated carboxyl derivative **1** by carbodiimide **2** and the consequent formation of protonated carbodiimide **3**. The deprotonated carboxyl derivative **4** initiates subsequently a nucleophilic attack on **3** and gives rise to the formation of active *O*-acylisourea intermediate **5**, which is supposed to undertake the nucleophilic addition by N^{α} groups on the incoming amino acids or peptides **7** to form the target product **8** and urea **9** as by-product.

Diverse side reactions might be aroused in the above carbodiimide-mediated amino acid activations. Conversion of active *O*-acylisourea intermediate to the corresponding inert *N*-acylurea[2] represents one of the most frequently occurred

Side Reactions in Peptide Synthesis. http://dx.doi.org/10.1016/B978-0-12-801009-9.00005-7

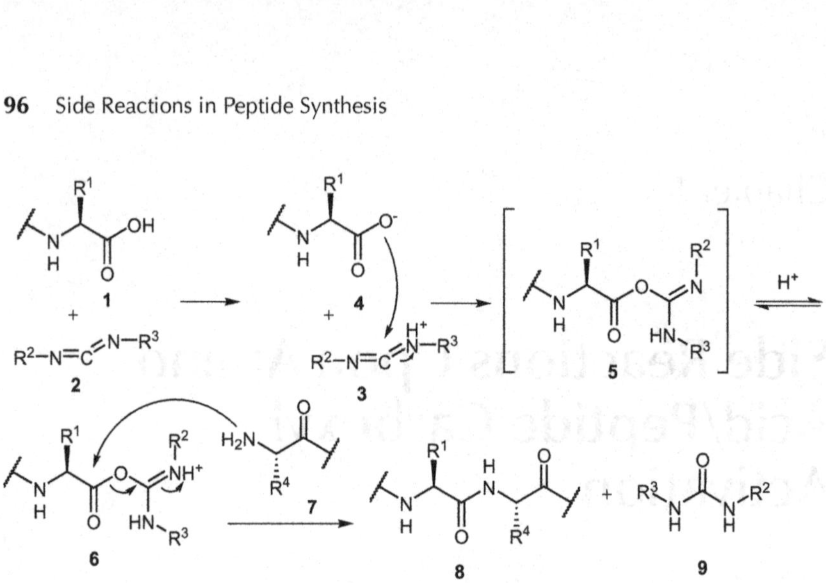

FIGURE 5.1 Mechanism of carbodiimide-mediated carboxyl activation and amide bond formation.

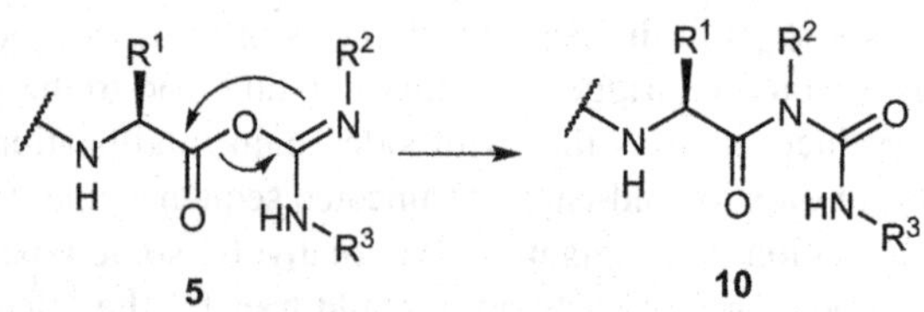

FIGURE 5.2 Formation of *N*-acylurea from *O*-acylisourea via *O*→*N* acyl migration.

side reactions in this process. The mechanism of the conversion is illustrated in Fig. 5.2. The impetus of this side reaction lies in the spontaneous conversion of *O*-acylisourea intermediate **5** to the corresponding isomeric *N*-acylurea **10** via an intramolecular acyl *O*→*N* migration.

The referred acyl *O*→*N* migration on *O*-acylisourea active ester species would not only adversely impact on the yields of the target coupling reactions but also introduce challenges on product purification provided the formed *N*-acylurea by-products survive the subsequent workup processes. The occurrence of this side reaction is significantly dictated by the reaction conditions since decrease of the reaction temperature could effectively suppress *N*-acylurea formation[3] whereas increase of the polarity of the reaction solvent would considerably stimulate this side reaction. This conclusion has been verified by the observations that the extents of *N*-acylurea formation side reactions are reduced in the reactions conducted in DCM, CCl_4 and benzene while drastically increased in DMF, ACN, DMSO or water.[4]

Since the inherent chemical structure of *O*-acylisourea is the root cause of the undesired acyl *O*→*N* migration and the resultant formation of *N*-acylurea impurity, employment of coupling additives, for example, HOSu or HOBt during the carboxyl activation to transform *in situ* the *O*-acylisourea intermediate into the corresponding active esters is advised. This strategy could substantially lessen the severity of acyl *O*→*N* migration side reactions. As is elucidated in Fig. 5.1, the addition of these weak acidic compounds could partially protonate the *O*-acylisourea intermediate **5** to give rise to its cationic counterpart **6.** This conversion decreases the nucleophilicity of the affected nitrogen that dictates the potential acyl shift and consequently reduces the possibility of the undesirable acyl *O*→*N* migration. More importantly, the addition of HOBt or HOSu could rapidly convert the unstable *O*-acylisourea to the corresponding benzotriazyl or succinimidyl active ester, respectively. This synthetic strategy could not only diminish the formation of *N*-acylurea, but also decrease the racemization of amino acids or peptides at the step of carboxyl activation.

5.2 URONIUM/GUANIDINIUM SALT COUPLING REAGENTS-INDUCED AMINO GROUP GUANIDINATION SIDE REACTIONS

Uronium/Guanidinium salt coupling reagents such as TBTU, HBTU, and HATU have already been widely applied in peptide synthesis territories (reviews about uronium/guanidinium salt coupling reagents are referred to Ref. 5). The mechanism of HATU-mediated amino acid/peptide activation is exhibited in Fig. 5.3. Amino acid/peptide **11** is firstly converted to its carboxylate counterpart **12** by DIEA treatment. Compound **12** is subsequently derivatized by HATU and converted to the highly active *O*-acylurea salt intermediate **13**. This intermediate could be entrapped *in situ* by additive HOAt and give rise to the formation of active ester **15**. The latter might be subjected to isomerization with respect to

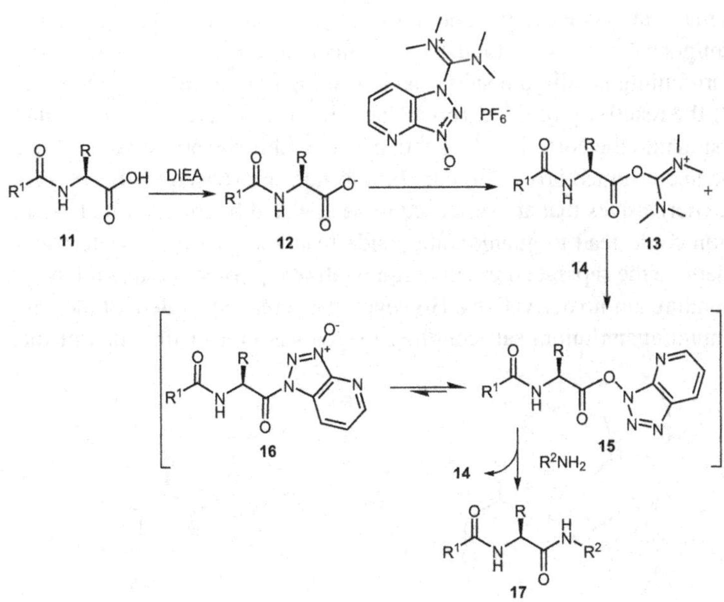

FIGURE 5.3 HATU-mediated carboxyl activation and peptide bond formation.

acyl $O \rightarrow N$ migration to generate the less reactive species *N*-acyl derivative **16**. The activated amino acid/peptide will subsequently undergo aminolysis to provide the target amide compound **17**. It has not been unambiguously identified whether the activated amino acid/peptide is partially involved in the amide bond formation in the form of *O*-acylurea salt **13**. The majority of the active species is, however, participating in the reaction as active ester **15/16**.[6]

Side reactions might be induced by uronium/guanidinium salt coupling reagents in the process of amino acid/peptide coupling reactions, among which guanidination of amino groups represents one of the most frequently occurred side reactions. Guanidination is principally originated from the undesired functions between uronium/guanidinium salts and amino derivatives.[7] The mechanism of this side reaction is illustrated in Fig. 5.4. The amino functional group that is supposed to accommodate the acylation by the reciprocal carboxylate derivative could be simultaneously involved in the competitive reaction with the uronium/guanidinium salts, giving rise to the side products **18** bearing guanidino moieties. As the formed guanidino derivatives remain chemically stable at the subsequent peptide synthesis steps even in highly concentrated TFA solution,[8] the elongation of peptide chains will be inevitably terminated once the guanidination formation takes place on the peptide N^{α} functional groups. As a consequence, truncated peptide impurities with *N*-terminal guanidino group will be generated.

In consideration of the ordinary processes of peptide manufacture, the causes of guanidination formation on the amino functional group could be attributed to the following conditions: (1) carboxyl compounds are charged simultaneously with uronium/guanidinium salts, and DIEA to the amino derivatives without a separate pre-activation treatment. The transient co-existence of amino species and uronium/guanidinium salts facilitates the nucleophilic attack of the amino groups on uronium/guanidinium salts, and hence the formation of guanido impurities; (2) the reactivity of the carboxylate is too low to assure the quantitative conversion into the corresponding active ester within the preactivation time slot, and the unconsumed uronium/guanidinium salts are subsequently charged to the amino derivatives that are supposed to be acylated by the activated ester. This situation could lead to guanidination side reaction; (3) carboxylate compounds undertake the separate pre-activation treatment prior to its function with the corresponding amino derivatives. However, the actual equivalent of the employed uronium/guanidinium salt coupling reagents is greater than that of the

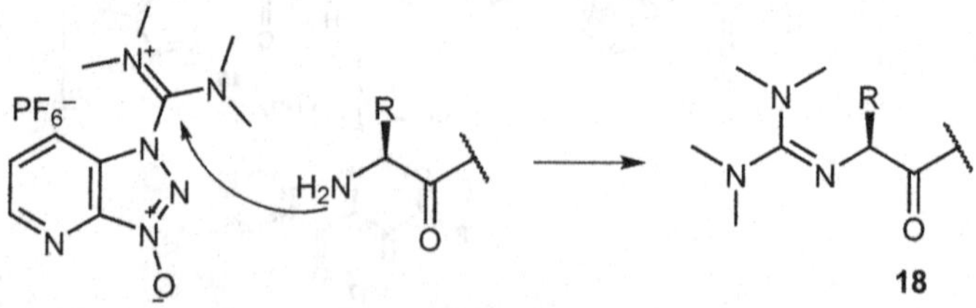

18

FIGURE 5.4 HATU-directed guanidination of amino group.

carboxylate substrate. This improper stoichiometry, either originated from an incautious process development such as the ignorance of the actual content of carboxyl derivative in the starting materials or imprudent practical operation, would inevitably lead to the variation of the actual stoichiometry of the reactants. Guanidination formation could hence be provoked in the case of uronium/guanidinium salts that are employed in excess, relative to a carboxyl reactant. This adverse effect will be particularly magnified when the target coupling reaction runs sluggishly; (4) under the circumstances, e.g., head-to-tail intramolecular peptide cyclization, the reactant carboxylate, and amino moiety will have to simultaneously exist in one parental peptide molecule. The activation of carboxylate will be inevitably carried out in the presence of an amino group. This situation might therefore inherently trigger a guanidination side reaction.

Rational modification of the synthetic routes or pertinent improvement on the production processes could be resorted to in consideration of the above root causes of the guanidination formation side reactions. Pre-activation of the carboxylate compounds is normally advisable so as to avoid the undesired functions between the unconsumed coupling reagents and the amino components in the coupling reactions. In case of the sluggish carboxyl activation process a modest elongation of the pre-activation would consume the uronium/guanidinium salt coupling reagents quantitatively. It is to note, however, that this kind of process modification could result in intensified racemization of carboxylate components of amino acids/peptides.[9] The optimum in terms of relevant process parameters could be explored and obtained via rational and systematic investigation, for example, Design of Experiments (DoE).

The stoichiometry of reactants is another crucial parameter that affects the occurrence of guanidination side reaction. Special care is to be exerted to this end in case the undesired function between uronium/guanidinium salt and amino group is prompt, and the competitive guanidination side reaction significantly interferes with the target acylation of the amino derivative. A slight decrease of the equivalent of the coupling reagents cf. carboxylate components might alleviate the potential guanidination side reaction at the sacrifice of small quantities of the carboxylate compounds that could not be converted to the corresponding active esters. For example, 0.95 equiv. coupling reagents cf. carboxylate could be employed in the pre-activation process. As a consequence, this stoichiometry will cause the loss of 5% carboxylate components. However, this trivial impact could be readily offset by the fact that excessive carboxyl derivatives are normally utilized in the process of peptide assembly reactions. A rational decrease of the coupling reagent equivalent in peptide head-to-tail cyclization reaction might be helpful to diminish the extent of potential guanidination side reaction.[10] Nevertheless, this side reaction might be particularly intensified in the process of peptide cyclization on a solid support[11] which could be ascribed to the sluggish cyclization kinetics intrinsically correlated with the enhanced steric hindrance compared with that in solution-phase peptide cyclization. Phosphonium salts like PyBOP or PyAOP are the choice of coupling reagents for the extreme cases thanks to their inherent features that these phosphonium salts are inert to amino functional groups.[7] This distinctive merit is frequently taken advantage of when severe guanidination side reactions haunt in the due peptide synthesis such as peptide head-to-tail cyclization and convergent peptide segment condensation.

5.3 δ-LACTAM FORMATION UPON Arg ACTIVATION REACTION

The guanidino side chain of Arg might induce diverse side reactions in peptide synthesis due to its strong nucleophilicity. Ideally all the $N^{\delta,\omega,\omega'}$ functionalities on the Arg-guanidino side chain should be properly protected so as to avoid potential side reactions. In spite of the inventions, for example, Arg(Trt)$_3$ derivatives,[12] their practical applications in peptide synthesis have not been actually realized. δ-lactam formation on Arg during its activation represents one of the common side reactions inherently occurring to this residue in peptide synthesis owing to the insufficient protection on the guanidino side chain.[13] The scheme of this side reaction is described in Fig. 5.5. The unprotected N^{δ} on Arg guanidino side chain might initiate a nucleophilic attack on the activated carboxylate and consequently result in the intramolecular cyclization on Arg to give rise to a six-member ring δ-lactam derivative.

In actual peptide synthesis, the N^{ω} or $N^{\omega'}$ on Arg guanidino side chain is normally shielded by a protecting group which limits the nucleophilicity of N^{δ} as well via its electron-withdrawing effect or steric hindrance. Potential δ-lactam formation upon Arg activation is hence limited by this means. If the subjected Arg is involved in the corresponding SPPS an individual amino acid building block, its conversion into the δ-lactam derivative in carboxyl-activation process would basically not introduce new impurities into the target product as the generated δ-lactam by-product is much less reactive toward aminolysis than its active ester counterpart and will be removed from the reaction system simply by peptidyl-resin rinse after the coupling reaction is completed. However, the

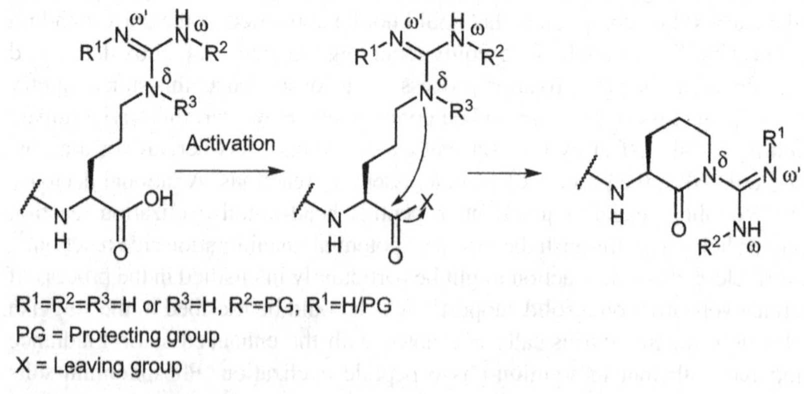

FIGURE 5.5 δ-lactam formation during Arg carboxyl-activation.

transformation of Arg into δ-lactam upon carboxyl-activation will inevitably result in the loss of the starting material and even the substoichiometric Arg derivatives relative to the amino components. This extremity would bring about the formation of peptide impurities with des-Arg deletion sequences. Incomplete Arg coupling is not a seldom phenomenon in SPPS which necessitates Arg recoupling process. This problem might be partially attributed to Arg δ-lactam formation during its carboxyl-activation. If this side reaction addresses the *C*-terminal Arg residue on a peptide segment that is supposed to be activated at its carboxyl group and involved in the subsequent convergent condensation with the reciprocal peptide fragment, the δ-lactam formation will inactivate the *N*-terminal peptide components and thus terminate the target conjugation. The extent of Arg δ-lactam formation is subject to the carboxyl-activation conditions. An investigation has figured out that Arg δ-lactam formation side reaction could be drastically stimulated in microwave-mediated peptide synthesis.[14]

Since Arg δ-lactam formation is directly correlated to its intrinsic side chain protection pattern, rational protection on Arg guanidino side chain could be counted on as to diminish the possible intramolecular cyclization on Arg during its carboxyl-activation. Nitro is deemed as one of the most applicable protecting groups on Arg-$N^{\omega'}$ owing to its outstanding electron-withdrawing effect.[15] The nucleophilicity of Arg-N^{δ} will be substantially reduced upon the introduction of nitro protecting group which will in turn decrease the risk of Arg δ-lactam formation. Bis-Boc protection on $N^{\omega,\omega'}$ has been confirmed to be invalid as to thoroughly suppress the occurrence of Arg δ-lactam formation upon carboxyl-activation.[16] Adoc that possesses a distinctive steric hindrance compared with that of Boc could be utilized as a protecting group on Arg guanidino side chain, whereas bis-Adoc protected Arg derivative $N^{\delta,\omega}$-bis(Adoc)-Arg could entirely inhibit the Arg δ-lactam formation. Nevertheless, this derivative turns out to be incapable of sufficiently shielding the $N^{\omega'}$ moiety from the undesired acylation side reactions.[17] An inherent problem correlated with the employment of $N^{\delta,\omega}$-bis(Adoc)-Arg lies in its instability once the Arg-N^{α} is liberated.[18] Consequently *N*-terminal $N^{\delta,\omega}$-bis(Adoc)-Arg will be susceptible to cyclization in MeOH and converted to the cyclic derivative **19** (Fig. 5.6). This transamination side reaction is particularly evident in weak acidic condition.

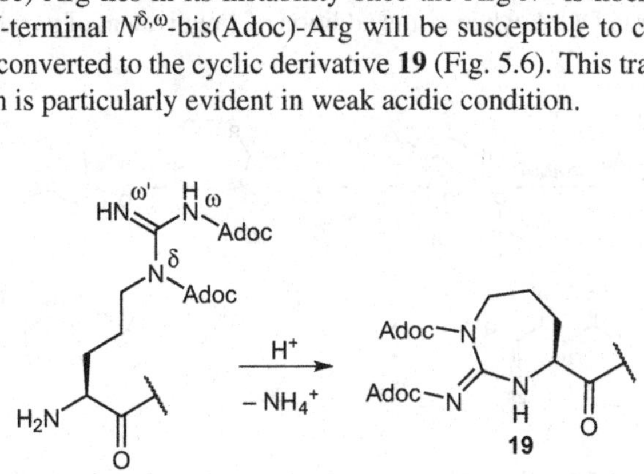

FIGURE 5.6 Cyclization of $N^{\delta,\omega}$-bis(Adoc)-Arg derivative.

In consideration of the susceptibility of Arg to δ-lactam formation in the process of carboxyl-activation, it is advisable to fine-tune the subjected Arg preactivation in order to diminish this side reaction, since the Arg active ester formed at the preactivation step will be unavoidably subjected to the cyclization in the absence of the incoming amino component. Preactivation duration and temperature are the most critical factors in this connection.

5.4 NCA FORMATION UPON Boc/Z-AMINO ACID ACTIVATION

Boc-amino acids could be suffered from fragmentation side reaction in the process of carboxyl-activation under certain circumstances and be transformed into the corresponding NCA derivatives.[19] The extent of this side reaction is basically correlated to the conditions under which the subjected Boc-amino acid is activated. The process of NCA formation upon Boc-amino acid carboxyl-activation and the resultant dipeptide formation is illustrated in Fig. 5.7. Boc-AA-OH **20** is firstly converted to its active ester **21** by DCC, and the latter might undergo intramolecular cyclization in the absence of the incoming nucleophile, and give rise to 2-*tert*-butoxy-5(4*H*)-oxazolone derivative **22**. **22** is subsequently protonated by the acidic coupling additive, e.g., HOBt or 4-nitrophenol, and generates the protonated oxazolone derivative **23**, which in turn undergoes acidolytic degradation by means of de-*tert*-butylation, and is converted into a *N*-carboxyanhydride derivative **24**. Active compound **24** could further function with 4-nitrophenol or HOBt to form carbamate **25** that undergoes decarboxylation to give amino acid ester **26**. By virtue of its nucleophilicity **26** could react with **21** or **22** in the system to engender Boc-dipeptide ester **27**. The process from amino acid **20** to oxazolone **22** coincides partially with N^{α}-urethane-protected amino acid racemization upon carboxyl-activation. The Boc-protected oxazolone **22** could be subjected to either racemization or de-*tert*-butylation/NCA formation as described in Fig. 5.7. It is to note that the NCA formation is particularly prone to occur in DCM, whereas the addition of 1 equivalent of pyridine into the

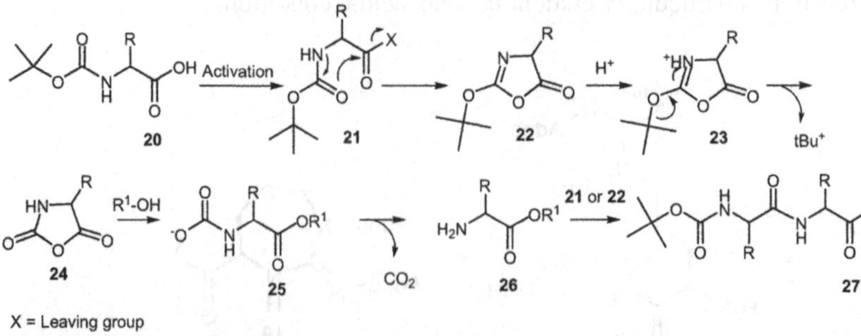

FIGURE 5.7 NCA formation and subsequent dipeptide formation during Boc-amino acid carboxyl-activation.

DCC-mediated Boc-amino acid carboxyl-activation reaction could to some extent suppress this side reaction.[20]

The NCA formation upon Boc-amino acid activation is most significantly manifested in the preparation of Boc-amino acid chloride. The propensity of amino acid chlorides to undergo NCA formation could address both N^{α}-Boc-[21] and N^{α}-Z-protected amino acids,[22] whereas the former is more susceptible to this side reaction. The intrinsic instabilities of Boc- and Z-amino acid chlorides doom their inapplicability in the territories of peptide production. In a few cases in which Boc-amino acid chlorides are utilized as starting materials for peptide synthesis, they are normally handled under very low temperature (−30 to −20°C)[21] or applied to non-α-amino acid substrates.[23] An attempt to prepare Z-Aib-Cl ended up with the formation of Aib-oxazolone derivative as product.[24]

Unlike Boc/Z-amino acids, Fmoc-amino acid chlorides with acceptable stability could be prepared and are commercially available.[25] In spite of the incompetence of N^{α}-Boc/Z-amino acid chlorides as starting materials for peptide synthesis, their acid fluoride counterparts N^{α}-Boc-AA-F[26] and N^{α}-Z-AA-F[27] could be employed for some difficult couplings. This attribute is due to the reduced activities of Boc/Z-AA-F compared with their acid chloride counterparts and the attenuated tendencies to undergo oxazolone formation and the subsequent de*tert*-butylation or debenzylation. In view of this property, N^{α}-Boc/Z-AA-F could be resorted to when a difficult amino acid coupling is encountered. The formation of NCA derivative from N^{α}-Boc/Z-AA-F via oxazolone intermediate as well as the consequent dipeptide formation will be considerably suppressed compared with the corresponding amino acid chlorides.

5.5 DEHYDRATION OF SIDE CHAIN-UNPROTECTED Asn/Gln DURING CARBOXYL-ACTIVATION

"Minimal protection" strategy has been successfully applied both in stepwise and convergent segment condensation types of peptide synthesis.[28] The employment of the side chain-unprotected amino acid residues or peptide fragments as starting materials for the corresponding peptides synthesis could partially alleviate the inherent solubility problems brought forward by the introduction of the side chain protecting groups. However, as some of the side chain functional groups are exposed to diverse reactants at the step of amino acid/peptide activation as well as other treatment various side reactions might be provoked in this process.

Asn residue with free amide side chain could suffer from side reaction during its activation in which Asn is converted to β-cyanoalanine side product.[29,30] Gln is subjected to the similar side reaction as well but the extent thereof is lower than that of Asn. The mechanism of this "dehydration" process is depicted in Fig. 5.8. Asn **28** with free amide side chain is firstly converted to its active ester **29** upon carboxyl activation, and the latter initiates subsequently a base-catalyzed intramolecular cyclization process in which the amide

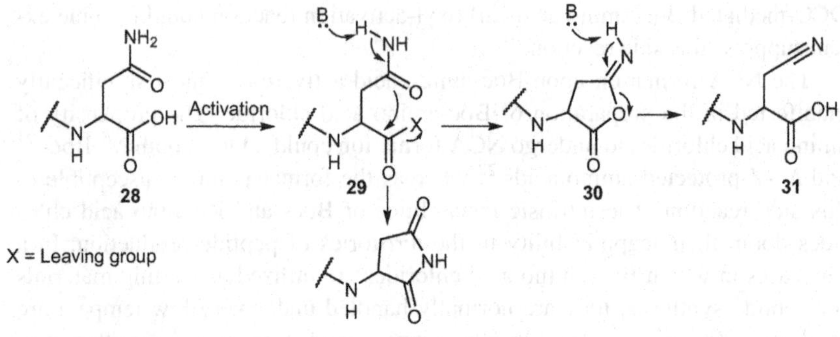

FIGURE 5.8 Mechanism of the dehydration of side chain-unprotected Asn during carboxyl-activation.

substituent mediates a nucleophilic attack on the backbone ester moiety and gives rise to the isoimide derivative **30**.[31] Ring opening could address compound **30** through base treatment to give β-cyanoalanine derivative **31**, which might be incorporated as a building block into the peptide chain by coupling reagents in the reaction system. Alternatively, compound **29** could be directly transformed into aspartimide derivative **32** that might undergo aminolysis subsequently to give an Asn-related derivative.

The extent of side chain-unprotected Asn/Gln to suffer from dehydration is largely dependent on the amino acid activation conditions.[32] This side reaction could not only be provoked in carbodiimide-mediated couplings, but is also susceptible to uronium/guanidinium salt- and phosphonium salt-directed carboxyl-activation.[33] Since Asn/Gln dehydration is catalyzed by base, the choice of carbodiimide coupling reagents and the addition of acidic *N*-hydroxylamine additives for the concerned Asn/Gln coupling could alleviate this side reaction and the consequent formation of cyano-impurities.[30,34] Employment of preactivated ester Fmoc-Asn/Gln-OPfp as the starting material for the concerned peptide synthesis has been verified effective ininhibiting the formation of cyano side products.[30] The side chain-unprotected Asn/Gln building blocks, once assembled into the target peptide chain, will be no more subjected to the dehydration side reaction at the following synthetic steps, which is in line with the mechanism of this process described in Fig. 5.8. The incorporated β-cyanoalanine residue could be regenerated to the original Asn by HF or even TFA treatment, although the latter is correlated with a much inferior kinetics.[34]

In view of the potential dehydration side reactions that might affect side chain-unprotected Asn/Gln residues, as well as the low solubility of Fmoc-Asn/Gln-OH in ordinary organic solvents and their aggregation tendencies[35] which will consequently lead to decreased reaction kinetics, it is recommended to adopt the side chain-protected Asn/Gln species as building blocks for peptide synthesis, except otherwise specially requested. The most frequently utilized amide protecting groups include Trt and Xan.

5.6 FORMATION OF H-β-Ala-OSu FROM HOSu-CARBODIIMIDE REACTION DURING AMINO ACID CARBOXYL-ACTIVATION

HOSu is used as an *N*-hydroxylamine-type of coupling additive in peptide synthesis, and DCC/HOSu is widely utilized as condensation agents for solution-phase peptide segment conjugations.[36] HOSu exhibits somewhat superior properties relative to HOBt with respect to racemization-prone coupling reactions, although the latter is capable of enhancing the kinetics of the concerned reaction compared with HOSu.[37] One of the most appreciable attributes of succinimide esters lies in their isolability and superior stability under low temperature. Moreover, some amino acid succinimide esters are commercially available which might considerably facilitate various protein modifications.[38] Amino acid succinimide ester is particularly appropriate for the coupling reaction with amino component that bear simultaneous a free α-carboxyl group,[39] avoiding by this means the potential indiscriminate carboxyl-activation under *in situ* activation conditions. Another advantage of HOSu over other *N*-hydroxylamine coupling additives is its relatively weaker acidity (pKa_{HOAt} = 3.47, pKa_{HOBt} = 4.60, pKa_{HOOBt} = 3.97, pKa_{HOSu} = 7.19).[40] This feature could be exploited for the preparation of some extraordinarily acid-sensitive peptides. For example, Nps (2-nitrophenylsulfanyl) protecting group which is partially labile in the presence of HOBt or HOAt[41] is nonetheless compatible with HOSu and no acidolytic degradation could be detected in HOSu-mediated reaction.[42] Moreover, N^{im}-tosyl protected His could suffer from premature side chain deprotection due to the inherent its incompatibility with HOBt.[43] The premature His(Tos) side chain deblocking process could consequently initiate a series of side reactions (referred to Section 4.3) whereas His(Tos) remains stable in the presence of HOSu.[44]

In spite of the mentioned advantages over other *N*-hydroxylamine derivatives HOSu could nevertheless engender some side reactions in carbodiimide-mediated amino acid coupling processes, among which the incorporation of β-Ala residue into the target peptide chains represents the most noteworthy one.[45] This side reaction is triggered in the process of amino acid carboxyl-activation by synergistic function between one molecule of DCC and three HOSu molecules. DCC can react with HOSu in THF, forming by this means succinimidoxycarbonyl-β-alanine-hydroxysuccinimide ester[46] that is reactive for further derivatization. If this process takes place during the preparation of amino acid succinimidyl ester and the above by-product is not sufficiently removed from the product at isolation and purification steps, it will be inevitably involved concomitantly in the subsequent peptide synthesis and incorporated into the peptide chain in the form of β-Ala.

The mechanism of the above side reaction is depicted in Fig. 5.9. First, DCC reacts with a molecule of HOSu to form *O*-succinimidyl isourea **33**, which

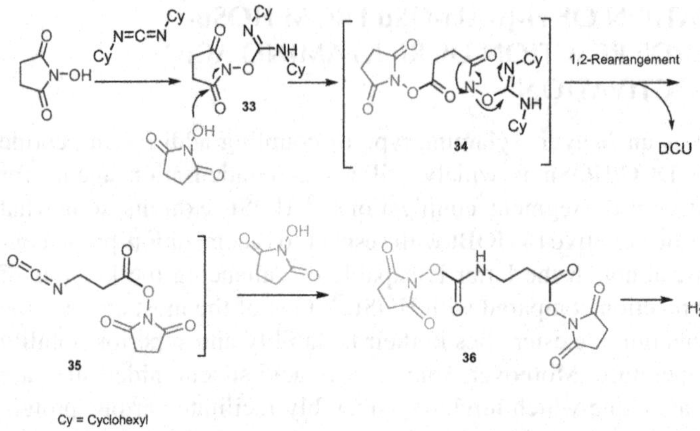

FIGURE 5.9 Proposed mechanism of succinimidoxycarbonyl-β-alanine-hydroxysuccinimide ester formation via the function between DCC and HOSu.

could then undergo ring opening by the nucleophilic attack from another HOSu molecule and be converted to the corresponding adduct **34**. Derivative **34** might be a virtual intermediate in this process since it undergoes 1,2-rearrangement to give the isocyanate derivative **35**. This process is regarded as the key step shared by the Lossen, Curtius, and Hoffmann Rearrangement. Isocyanate **35** could principally react with a third HOSu molecule to give rise to the formation of the very unstable intermediate succinimidoxycarbonyl-β-alanine-hydroxysuccinimide ester **36**, which is subsequently degraded into β-Ala-OSu **37** owing to the extraordinary instability of the *O*-succinimidyl carbamate moiety.[47] The formed by-product **37** could further function with the activated amino acid Fmoc-Xaa$_1$-OSu and the derived adduct Fmoc-Xaa$_1$-β-Ala-OSu can be incorporated into the to-be-acylated peptide sequence Xaa$_2$-peptide, giving rise to endo-β-Ala side product Fmoc-Xaa$_1$-β-Ala-Xaa$_2$-peptide. Alternatively, hydroxysuccinimide ester **37** could function as an acylating agent that firstly reacts with the N^{α} group on Xaa$_2$-peptide to engender an intermediate containing H-β-Ala-Xaa$_2$-peptide unit, and this intermediate could be subsequently acylated by Fmoc-Xaa$_1$-OSu in the reaction system to give rise to the same endo-β-Ala side product Fmoc-Xaa$_1$-β-Ala-Xaa$_2$-peptide. H-β-Ala-Xaa$_2$-peptide might even function with another molecule of **37** to generate double endo-β-Ala side product H-β-Ala-β-Ala-Xaa$_2$-peptide. The overall consequence of this side reaction will be the introduction of multiple β-Ala residues into the target peptide sequence.

This side reaction takes place predominantly in the sluggish amino acid coupling with inferior kinetics. HOSu is supposed to be replaced under such circumstances by other *N*-hydroxylamine as coupling additive. Or alternatively, HOSu could be charged into the reaction mixture after the addition of carbodiimide coupling reagent in order to minimize the undesired functions between carbodiimide and HOSu and the consequent formation of side products.

5.7 BENZOTRIAZINONE RING OPENING AND PEPTIDE CHAIN TERMINATION DURING CARBODIIMIDE/HOOBt MEDIATED COUPLING REACTIONS

HOOBt (also named as HODhbt, 3-hydroxy-1,2,3-benzotriazin-4(3*H*)-one, see also Fig. 5.10) is an ordinary *N*-hydroxylamine coupling additive frequently utilized in peptide synthesis in an effort to increase the coupling efficiency and/or decrease the extent of racemization.[48–50]

HOOBt is normally utilized together with carbodiimide-type coupling reagents like DIC or DCC to mediate the condensation reactions of amino acid or peptide.[51] This reagent has exhibited excellent properties in terms of enhancing the coupling efficiency and minimizing racemization side reactions. Furthermore, amino acid HOOBt esters are also equipped with improved resistance against undesirable alcoholysis in solvents, for example, TFE, relative to the active esters derived from other coupling additives.[50,52,53] These distinctive attributes endue HOOBt the applicability in some special organic solvents for the peptide fragments which are sparsely soluble in other ordinary solvents. DCC/HOOBt has been widely utilized as a qualified coupling reagent for solution-phase peptide segment condensation reactions.[33,54]

In spite of the above advantages of HOOBt as a coupling additive, its inherent chemical structure has to some extent limited its applicability in peptide synthesis due to the distinctive HOOBt-provoked side reactions. The triazinone moiety in HOOBt is susceptible to an undesirable ring opening during DCC mediated carboxyl-activation. This process results in the formation of an active intermediate, which would consequently lead to peptide chain termination. The scheme of this side reaction is illustrated in Fig. 5.11. HOOBt reacts firstly with DCC to generate derivative **38**. This compound can accommodate the nucleophilic attack from another HOOBt molecule. The consequence of this process is the triazinone ring opening and the formation of 2-azido-4-oxo-1,2,3-benzotriazin-3-yl benzoate **39** upon releasing a molecule of DCU (*N,N'*-dicyclohexylurea). Compound **39** with an active ester moiety, once added to the to-be-acylated amino component, could react with the N^{α} functional group and terminate the growing peptide chain by means of the generation of N^{α} -2-azidobenzoyl-amino acid/peptide **40**.[51]

In view of the above side reaction caused by the aberrant function of HOOBt with carbodiimide coupling reagent, corresponding process modification could

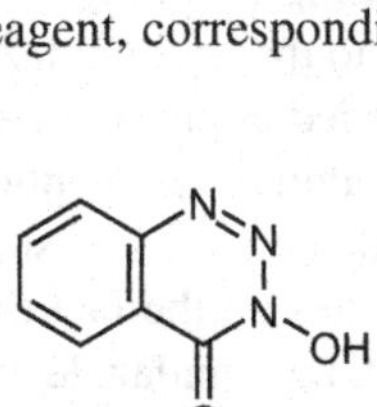

HOOBt (HODhbt)

FIGURE 5.10 Chemical structure of HOOBt (HODhbt).

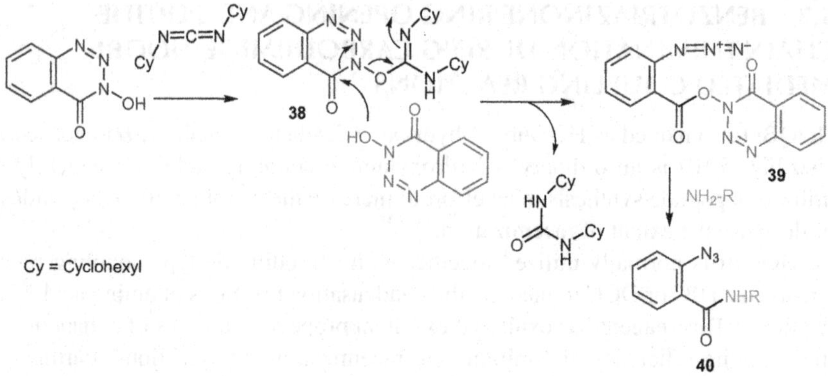

FIGURE 5.11 **Formation of 2-azido-4-oxo-1,2,3-benzotriazin-3-yl benzoate via DCC and HOOBt.**

be tested. For instance, DCC could be firstly added to the carboxyl-component amino acid/peptide to lead a 5 min preactivation process, HOOBt is then charged to the reaction system as to limit the occurrence of the mentioned side reaction by preventing the function between free DCC and HOOBt.[51,55] This strategy is principally analogous to the one designed to address the side reaction of β-alanine formation in DCC/HOSu-mediated amino acid/peptide activation that is introduced in Section 5.6.

5.8 PEPTIDE CHAIN TERMINATION THROUGH THE FORMATION OF PEPTIDE *N*-TERMINAL UREA IN CDI-MEDIATED COUPLING REACTION

CDI is a commonly utilized coupling reagent in peptide synthesis that is capable of converting the carboxyl group to the corresponding acyl imidazolide. This active derivative could subsequently undertake aminolysis by N^{α} on the incoming amino acid/peptide. The mechanism of CDI-mediated peptide coupling reaction is described in Fig. 5.12. Carboxyl group that is supposed to be activated firstly transfers a proton to CDI, forming the corresponding carboxylate and CDI anion. Reaction between these two derivatives gives rise to the formation of mixed anhydride **41** by releasing a molecule of imidazole. Intermediate **41** will be spontaneously degraded to the acyl imidazolide **42** upon splitting a CO_2 molecule. Derivative **42** is regarded as an excellent acylating agent that is consumed by N^{α} on the incoming amino acid/peptide by releasing an imidazole molecule to form the target amide bond. One distinctive advantage of CDI over carbodiimide coupling reagents lies on the fact that the by-products formed in CDI-mediated coupling reaction, i.e., imidazole and CO_2, which could be readily removed from the reaction system, alleviating by this means the potentially tedious workup process. Furthermore, CDI also finds its applications in territories other than peptide synthesis such as the preparation of urea[56] or carbamate

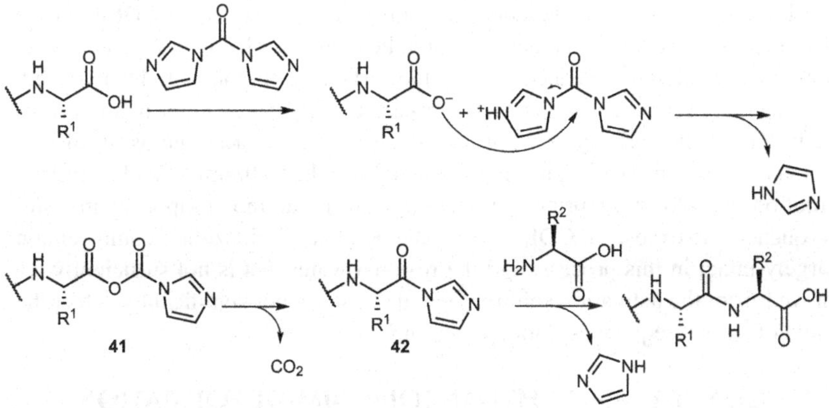

FIGURE 5.12 Mechanism of CDI-mediated amino acid coupling.

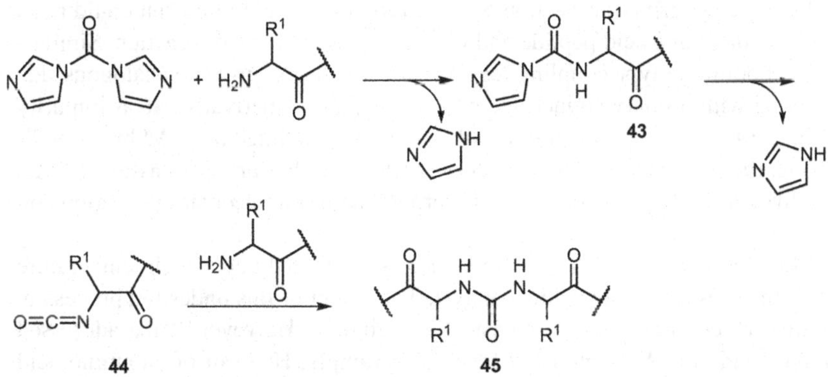

FIGURE 5.13 Urea formation and peptide chain termination by reaction between CDI and N$^{\alpha}$ on amino acid/peptide.

materials[57] from the corresponding amines or alcohols. It could appropriately replace the highly toxic phosgene as the condensation agent in these cases.

In spite of the role of CDI as an appropriate replacement for carbodiimide coupling reagents for amino acid/peptide carboxyl-activation, its widespread application in peptide production has not been actually realized. This phenomenon is partially ascribed to the high cost and hygroscopicity of CDI compared with carbodiimide coupling reagents. Moreover, excessive CDI in the condensation reaction system could provoke undesirable function with N^{α} functional groups and generate imidazolecarboxamide derivative **43,** which is subsequently converted to isocyanate **44** by eliminating an imidazole molecule. Active isocyanate compound **44** could further react with N^{α} functional group on another molecule of amino acid/peptide to give rise to urea derivative **45** and terminate in this manner the growing peptide chain[58] (Fig. 5.13).

Taking the above side reaction into consideration, excessive CDI in amino acid/peptide condensation reaction should be avoided. That is to say, the equivalent of CDI should not exceed that of the carboxyl component to be activated. In case of any uncontrollable factors like the unknown content of impurities or waters in the carboxyl substances which makes the accurate assay of carboxyl material impractical, it is advisable to rinse the CDI/amino acid activation solution by cold water prior to its addition to the amino component in order to quench the excessive CDI, if any. The activated imidazolide could remain largely intact in this process, and the residual water that is not sufficiently removed from the activation solution would normally not significantly affect the yield of the subsequent acylation reaction.[59]

5.9 GUANIDINO OR HYDANTOIN-2-IMIDE FORMATION FROM CARBODIIMIDE AND N^{α} GROUP ON AMINO ACID/PEPTIDE

It has been described in Section 5.2 that uronium/guanidinium salt could react with N^{α} on amino acid/peptide and induce guanidination side reaction. Similarly, carbodiimide-type coupling reagents are also able to provoke analogous side reaction with primary amines and generate guanidino derivative **46** as impurity (Fig. 5.14). In case of an amino acid ester that is modified at its N^{α} by carbodiimide, the generated guanidino intermediate could be further transformed into the hydantoin-2-imide derivative **47** through an intramolecular cyclization and elimination of an alcohol.[60]

In peptide synthesis this side reaction is facilitated under high temperature or in the presence of an acidic catalyst. The extent of this undesired process in routine reaction conditions is basically negligible. However, if the addressed amino acid is in N^{α}-protonated form, for example, HCl salt of an amino acid ester, the susceptibility to carbodiimide-induced guanidination and/or resultant hydatoin-2-imide formation will be accordingly enhanced.[61]

5.10 SIDE REACTIONS-INDUCED BY CURTIUS REARRANGEMENT ON PEPTIDE ACYL AZIDE

At the very beginning era of peptide chemistry T. Curtius invented the methodology for amino acid condensations by means of acyl azides as an active acylating agent.[62] This synthetic strategy has obtained widespread applications

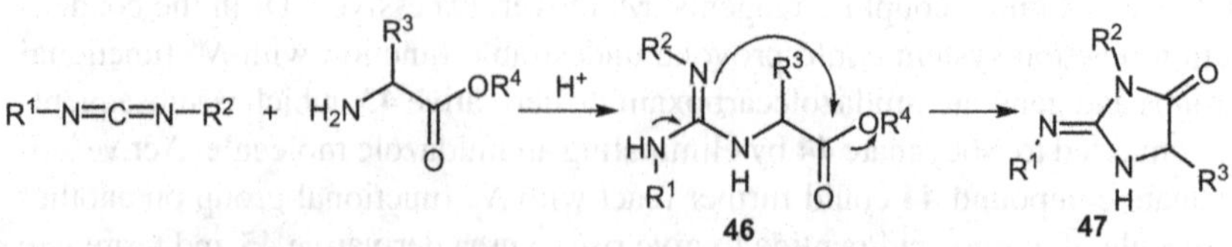

FIGURE 5.14 Reaction between amino acid ester and carbodiimide.

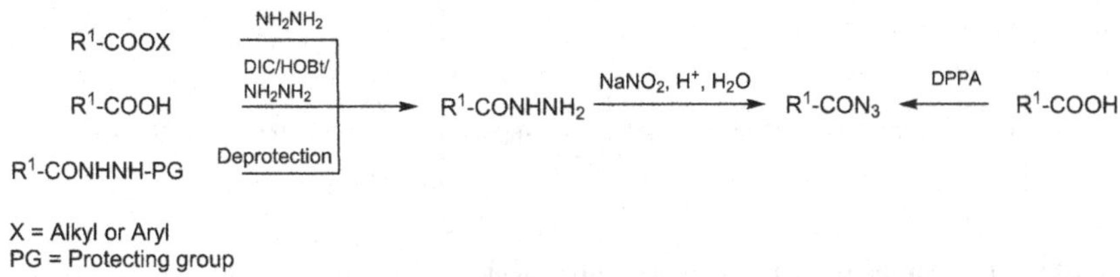

FIGURE 5.15 Scheme of acyl azide preparation.

both in ordinary stepwise peptide synthesis and "maximal or minimal protection" mode segment condensations.[63] One of the distinctive advantages of acyl azide peptide synthetic strategy lies in its low tendency to suffer from racemization via oxazol-5(4*H*)-one intermediate formation[64] as well as its compatibility with side chain-unprotected Ser, Thr, His and Trp in ordinary peptide synthetic conditions.[65]

The scheme of acyl azide preparation is illustrated in Fig. 5.15. Amino acid/peptide hydrazide is firstly obtained from the corresponding alkyl/aryl ester or active ester via hydrazinolysis, or alternatively through the deblocking of the protected amino acid/peptide hydrazide precursor. The derived hydrazide intermediate is subsequently subjected to HNO_2 treatment to generate the corresponding amino acid/peptide azide. The more applicable strategy for acyl azide preparation is the employment of an azide agent, for example, DPPA,[64] which converts carboxyl derivative directly to its acyl azide counterpart.

The generated amino acid/peptide azide could be directly charged to the following acylation reaction without a separate isolation step. The pH value of the acyl azide-involved condensation reaction could be fine-tuned by bases like DIEA or triethylamine. The reaction between acyl azide and the amino component is principally conducted under weak basic condition.

Various side reactions could be induced in acyl azide-guided peptide synthesis among which Curtius rearrangement represents the most significant.[66] Curtius rearrangement is regarded as an important synthetic strategy in organic synthesis which enables the conversion of carboxylic acid to the corresponding amine derivative.[67] However, this process, once triggered in peptide synthesis, might induce undesirable formation of urea and carbamate side products. The mechanism of Curtius rearrangement is referred to Fig. 5.16. Acyl azide could release N_2 upon being heated and undergo a concomitant 1,2-rearrangement process to engender the corresponding isocyanate derivative.[68] It is to note that the configuration of R remains intact during the earlier process.

In case acyl azide is converted to the corresponding isocyanate via the Curtius rearrangement process in peptide synthesis, it could react with a plethora of nucleophiles like amines and alcohols to give rise to various side products.[69] As exhibited in Fig. 5.17, peptide azide **48**, in case not sufficiently derivatized

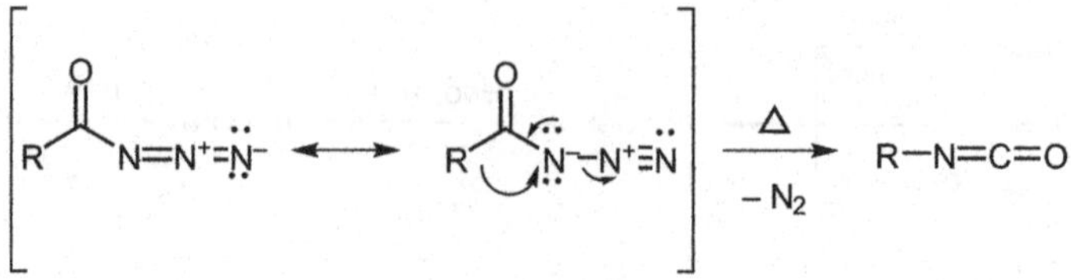

FIGURE 5.16 **Mechanism of Curtius rearrangement.**

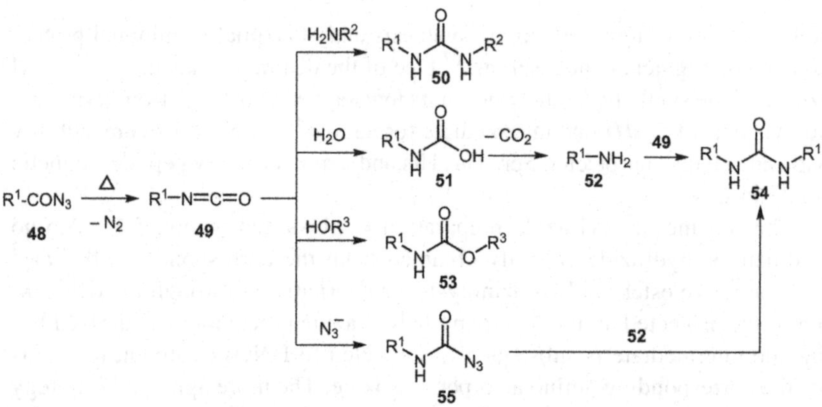

FIGURE 5.17 **Side reactions induced by isocyanate intermediate derived from the Curtius rearrangement of acyl azide.**

by the incoming amino component, could be transformed to the corresponding isocyanate **49** through Curtius rearrangement upon heating, which will in turn provoke diverse side reactions. The reaction between isocyanate **49** and the N^{α} on the incoming amino acid or peptide gives rise to the unsymmetric urea **50** with a 15 amu molecular weight increase compared with the target amide product. If residual water exists in the reaction system, it will hydrolyze isocyanate **49** to give the corresponding carbamic acid **51**, which undergoes spontaneous decarboxylation to form amine derivative **52**. The generated amine **52** by this means is deprived of one carbon relative to the starting material acyl azide **48**. This process serves as the underlying principle of amine preparation from carboxylic acid via Curtius rearrangement. It is regarded, nevertheless, as an undesired side reaction in the process of acyl azide-mediated peptide synthesis. The generated amine derivative **52** is able to further react with another molecule of isocyanate **49** to form the symmetric urea **54**. The overall effect of this process is the formation of one molecule of inert urea impurity at the sacrifice of two molecules of carboxyl derivatives. Although the formation of symmetric urea side product from the corresponding amino acid azide will not affect the integrity of target peptide product as unsymmetric urea **50** does, this side reaction consumes nonetheless two equivalents of acylating agent **48** and is hence disadvantageous from the perspective of production mass balance. The uncontrollable exaggeration of this side reaction might make the acylating agent insufficient for the target coupling reaction. Since acyl azide material prepared from the due process could be directly added to amino component without isolation, the possibly excess azide agent in the system, N_3^- anion, might function with isocyanate **49** to form carbamoyl azide derivative **55**. Although the electrophilicity of this compound is weaker than that of its acyl azide counterpart **48**, it is still able to consume a molecule of isocyanate **49** to form the symmetric urea derivative **54**.

On top of standard amino acids/peptide azides, which are affected by Curtius rearrangement, some special amino acids, such as side chain-unprotected Ser and His might be affected by further side reactions dictated by Curtius rearrangement. As illustrated in Fig. 5.18, the free β-hydroxyl substituent on the Ser side chain could initiate a nucleophilic attack on the backbone isocyanate moiety to form a five-member-ring derivative **56**.[70] Analogously, the imidazole functional group on side chain-unprotected His is also capable of inducing similar process via its N^{π} substituent to give the side product **57**.[71] Besides, if the starting material hydrazide **59** is not quantitatively converted to the target acyl azide product **58**, a reaction between these two compounds might take place to generate the symmetric diacyl hydrazide **60**.[72]

Since all the above side reactions proceed with isocyanate intermediate formation dictated by Curtius rearrangement, it is highly recommended to shorten the existence period of the labile acyl azide derivatives as much as possible and the incoming amino components are supposed to be charged promptly to the subjected acyl azides as to minimize the undesirable Curtius rearrangement. Meticulous control of reaction temperature is also deemed as an effective solution to suppress Curtius rearrangement in light of the concern that this process is evidently temperature dependent.

Another common side reaction occurring readily in the process of peptide/amino acid azide preparation via nitrosylation of acyl hydrazine is the formation of amide. This phenomenon was detected for the first time during the synthesis of Z-Lys(Z)-N_3 through $NaNO_2$/HCl procedure which resulted in the formation of Z-Lys(Z)-NH_2 as major side product.[73] Since analogous impurity was not detected in the process of amino acid azide preparation from acid chloride and sodium azide, it is proposed that the occurrence of this side reaction is due to the degradation of the nitrosylated acyl hydrazine as indicated in Fig. 5.19.[74] The side reaction with respect to amide formation in this process could be alleviated by the meticulous fine-tuning of the reaction conditions, such as homogenous solutions, for example, in DMF, DMSO, high acidity and low temperature (−30 to −5°C). Utilization of organic nitrite such as *tert*-butyl or butyl nitrite could also suppress amide formation during peptide/amino acid azide preparation.

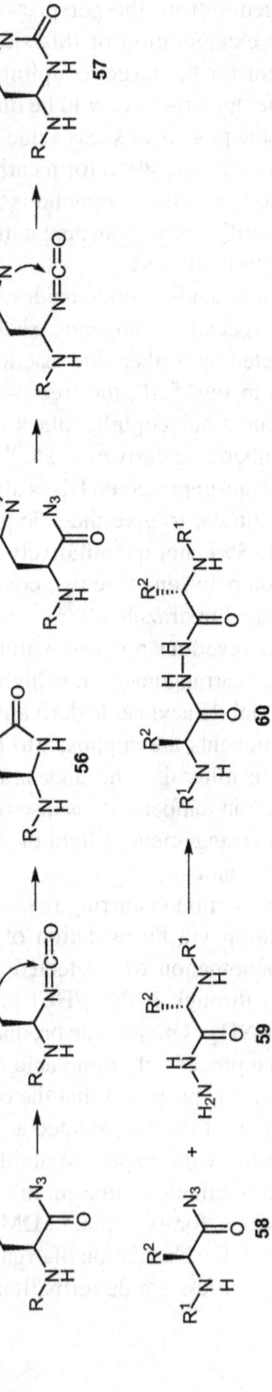

FIGURE 5.18 Other acyl azide-mediated side reactions.

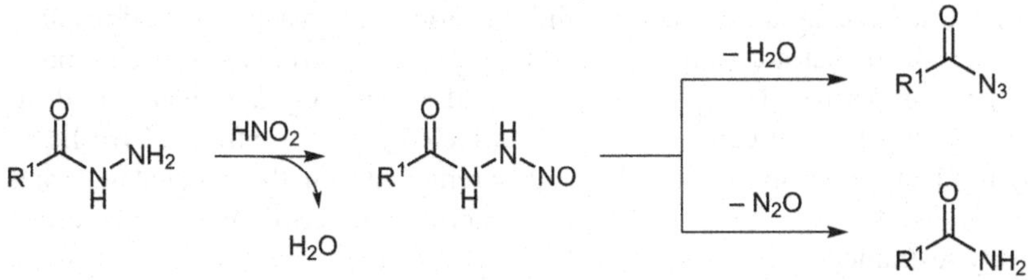

FIGURE 5.19 Formation of amide from the preparation azide via nitrosylation of acyl hydrazine.

5.11 FORMATION OF PYRROLIDINAMIDE-INDUCED BY PYRROLIDINE IMPURITIES IN PHOSPHONIUM SALT

Phosphonium salt coupling reagents have acquired broad applications in peptide synthesis by virtue of their inertness to amino functional group and the consequent suppression of the potential guanidination side reaction at the step of amino acid activation and coupling. This distinctive attribute of phosphonium salt is particularly beneficial for the couplings between residues with significant steric hindrances as well as peptide cyclization. Under these circumstances phosphonium salt is preferred over carbodiimide or uronium/guanidinium salt as condensation agent.

Nonetheless it is noteworthy that the commercial phosphonium coupling reagents such as PyBOP, PyAOP, and PyBroP might contain residual pyrrolidine derivatives (0.5% w/w). These compounds are capable of inducing side reactions that irreversibly modify the carboxyl groups. Pyrrolidinamide impurities instead of the target amide products will be generated from this side reaction[75] (Fig. 5.20). The residual pyrrolidine impurity in the phosphonium salts will principally not be incorporated into the final product in stepwise peptide synthesis in spite of the loss of the staring carboxyl material. On the contrary, if this process takes place in the solution-phase peptide segment condensation

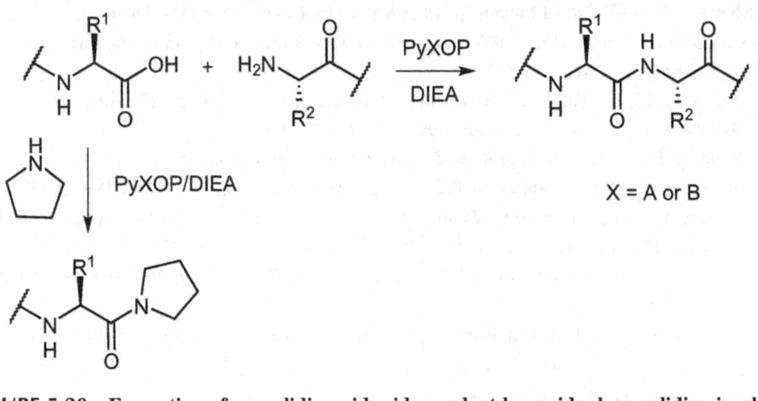

FIGURE 5.20 Formation of pyrrolidinamide side product by residual pyrrolidine in phosphonium salt.

or peptide head-to-tail cyclization, the concerned side reaction would result in side products that are similar to the target product with respect to chromatographic properties. This phenomenon could accordingly aggravate the difficulty for the down-stream process of the affected peptide synthesis. Pyrrolidine by-products in phosphonium salt could be removed from the affected materials by means of recrystallization so that the above side reactions could be effectively restrained.

REFERENCES

1. (a) Wendlberger G. *Houben-Weyl.* 1974;15:101. (b) Rich DH, Singh J. In: Gross E, Meienhofer J, eds. *The Peptides Analysis, Synthesis, Biology*. New York: Academic; 1979:241. (c) Albert JS, Hamilton AD. In: Paquette LA, ed. *Encyclopedia of Reagents for Organic Synthesis*. Chichester, UK: Wiley; 1995:1751. (d) Pottorf RS, Szeto P. In: Paquette LA, ed. *Encyclopedia of Reagents for Organic Synthesis*. Chichester, UK: Wiley; 1995:2430. (e) Griffin AM. In: Paquette LA, ed. *Encyclopedia of Reagents for Organic Synthesis*. Chichester, UK: Wiley; 1995:1438.
2. DeTar DF, Silverstein R. *J Am Chem Soc*. 1966;88:1013–1019.
3. (a) Helferich B, Böschagen H. *Chem Ber*. 1959;92:2813–2827. (b) Schneider F. *Hoppe-Seyler's Z Physiol Chem*. 1960;302:82.
4. (a) Sheehan JC, Hess GP. *J Am Chem Soc*. 1955;77:1067–1068. (b) Anderson GW, Callahan FM. *J Am Chem Soc*. 1958;80:2902–2903.
5. (a) Lloyd-Williams P, Albericio F, Giralt E. *Chemical Approaches to the Synthesis of Peptide and Proteins*. Boca Raton, FL: CRC Press; 1997: pp. 53–55. (b) Albericio F, Carpino LA. Fields GB, eds. *Methods in Enzymology*, vol. 289. New York: Academic; 1997:104. (c) Mergler M. Pennington MW, Dunn BM, eds. *Methods in Molecular Biology*, vol. 35. Totowa, New Jersey: Humana; 1994:1. (d) Kates SA, Triolo SA, Griffin GW, et al. In: Epton R, ed. *Innovations and Perspectives in Solid Phase Synthesis 1998*. Birmingham, UK: Mayflower; 1996:41. (e) Fields CG, Lloyd DH, MacDonald DL, Otteson KM, Noble RL. *Peptide Res*. 1991;4:95. (f) Schnölzer M, Alewood PE, Jones A, Alewood D, Kent SBH. *Int J Peptide Protein Res*. 1992;40:180–193. (g) Humphrey JM, Chamberlin AR. *Chem Rev*. 1997;97:2243–2266. (h) El-Faham A, Albericio F. *Chem Rev*. 2011;111:6557–6602.
6. Dourtoglou V, Gross B, Lambropoulou V, Zioudrou C. *Synthesis*. 1984;572–574.
7. Albericio F, Bofill JM, El-Faham A, Kates SA. *J Org Chem*. 1998;63:9678–9683.
8. Gausepohl H, Pieles U, Frank RW. In: Smith AJ, Rivier JE, eds. *Peptides: Chemistry and Biology*. Leiden: ESCOM; 1998:523.
9. (a) Carpino LA, El-Faham A, Albericio F. *Tetrahdron Lett*. 1994;35:2279–2282. (b) Han Y, Albericio F, Barany G. *J Org Chem*. 1997;62:4307–4312.
10. Ehrlich A, Heyne H-U, Winter R, et al. *J Org Chem*. 1996;61:8831–8838.
11. (a) Arttamangkul S, Arbogast B, Barofsky D, Aldrich JV. *Lett Pept Sci*. 1996;3:357–370. (b) Story SC, Aldrich JV. *Int J Peptide Protein Sci*. 1994;43:292–296. (c) Arttamangkul S, Muury TF, DeLander GE, Aldrich JV. *J Med Chem*. 1995;38:2410–2417.
12. Gazis E, Bezas B, Stelakatos GC, Zervas L. In: Young GT, ed. *Peptides 1962*. Oxford: Pergamon; 1963:17.
13. (a) Paul R, Anderson GW, Callahan FM. *J Org Chem*. 1961;26:3347–3350. (b) Rittel WR. *Helv Chim Acta*. 1962;45:2465–2473.
14. Novabiochem® public technical report.
15. Bergmann M, Zervas L, Rinke H. *Hoppe-Seyler's Z Physiol Chem*. 1934;224:40–44.

16. Cezari MH, Juliano L. *Pept Res*. 1996;9:88–91.
17. (a) Atherton E, Sheppard RC. Udenfriend S, Meienhofer J, eds. *The Peptides: Analysis, Synthesis, Biology*, vol. 9C. New York: Academic; 1987:1. (b) Rink H, Sieber P, Raschdorf F. *Tetrahedron Lett*. 1984;25:621–624.
18. (a) Jäger G, Geiger R. *Chem Ber*. 1970;103:1727–1747. (b) Moroder L, Marchiori F, Borin G, Scoffone E. *Biopolymers*. 1973;12:729–750.
19. Bodanszky M, Klausner YS, Bodanszky A. *J Org Chem*. 1975;40:1507–1508.
20. Benoiton NL, Lee YC, Chen FMF. *Int J Peptide Protein Res*. 1993;41:587–597.
21. Losse G, Wehrstedt K-D. *Z Chem*. 1981;21:148–149.
22. Bergmann M, Zervas L. *Ber Dtsch Chem Ges*. 1932;65:1192–1201.
23. (a) Howard MH, Sardina FJ, Rapoport H. *J Org Chem*. 1990;55:2829–2838. (b) Bold G, Steiner H, Moesch L, Walliser B. *Helv Chim Acta*. 1990;73:405–410.
24. Leplawy MT, Jones DS, Kenner GW, Sheppard RC. *Tetrahedron*. 1960;11:39–51.
25. (a) Carpino LA, Cohen BJ, Stephens Jr KE, Sadat-Aalaee SY, Tien J-H, Langridge DC. *J Org Chem*. 1986;51:3732–3734. (b) Carpino LA. *J Org Chem*. 1988;53:875–878.
26. (a) Carpino LA, Mansour EME, Sadat-Aelaee D. *J Org Chem*. 1991;56:2611–2614. (b) Savrda J, Chertanova L, Wakselman M. *Tetrahedron*. 1994;50:5309–5322.
27. (a) Bertho J-N, Loffet A, Pinel C, Chao HG, DeSelms RH. *J Am Chem Soc*. 1990;112:9651–9652. (b) Polese A, Formaggio F, Crisma M, et al. *Chem Eur J*. 1996;2:1104–1111. (c) Gratias R, Konat R, Kessler H, et al. *J Am Chem Soc*. 1998;120:4763–4770.
28. (a) Hirschmann R, Nutt RF, Veber DF. et al. *J Am Chem Soc*. 1969;91:507–508. (b) Beacham J, Dupuis G, Finn FM, et al. *J Am Chem Soc*. 1971;93:5526–5539.
29. Stroup AN, Cole LB, Dhingra MM, Gierasch LM. *Int J Peptide Protein Res*. 1990;36:531–537.
30. Gausepohl H, Kraft M, Frank RW. *Int J Peptide Protein Res*. 1989;34:287–294.
31. Stammer C. *J Org Chem*. 1961;26:2556–2560.
32. Rovero P, Pegoraro S, Bonelli F, Triolo A. *Tetrahedron Lett*. 1993;34:2199–2200.
33. Nishiuchi Y, Nishio H, Inui T, Bódi J, Kimura T. *J Pept Sci*. 2000;6:84–93.
34. Mojsov S, Mitchell AR, Merrifield RB. *J Org Chem*. 1980;45:555–560.
35. Bedford J, Hyde C, Johnson T, et al. *Int J Peptide Protein Res*. 1992;40:300–307.
36. (a) Wünsch E, Drees F. *Chem Ber*. 1966;99:110–120. (b) Weygand F, Hoffmann D, Wünsch E. *Z Naturforsch B*. 1966;21:426–428.
37. Izdebski J, Kunce D. *J Pept Sci*. 1997;3:141–144.
38. (a) Nielsen O, Buchardt O. *Synthesis*. 1991;819–821. (b) Waugh SM, DiBella EE, Pilch PF. *Biochemistry*. 1989;28:3448–3455. (c) Koplin E, Niemeyer CM, Simon U. *J Mater Chem*. 2006;16:1338–1344.
39. (a) Anderson J, Zimmerman JE, Callahan FM. *J Am Chem Soc*. 1964;84:1839–1842. (b) Moroder L, Göhring W, Lucietto P, et al. *Hoppe-Seylers Z Physiol Chem*. 1983;364:1563–1584.
40. Koppel I, Koppel J, Leito I, Pihl V, Grehn L, Ragnarsson U. *J Chem Res Synop*. 1993;446: *J Chem Res, Miniprint*, (1993) 3008.
41. Geiger R, König W. Gross E, Meienhofer J, eds. *The Peptides*, vol. 3. New York: Academic; 1981:3.
42. Wünsch E, Wendlberger G. *Wien Tieraerztl Monatsschr*. 1986;73:164.
43. Geiger R, Treuth G, Burow F. *Chem Ber*. 1973;106:2339–2346.
44. Fujii T, Kimura T, Sakakibara S. *Bull Chem Soc Jpn*. 1976;49:1595–1601.
45. Sieber P, Iselin B. *Helv Chim Acta*. 1968;51:622–632.
46. Gross H, Bilk L. *Tetrahedron*. 1968;24:6935–6939.
47. Isidro-Llobet A, Just-Baringo X, Ewenson A, Álvarez M, Albericio F. *Pept Sci*. 2007;88: 733–737.

48. König W, Geiger R. *Chem Ber*. 1970;103:2024–2033.
49. Carpino LA, El-Faham A, Albericio F. *J Org Chem*. 1995;60:3561–3564.
50. Kuroda H, Chen Y-N, Kimura T, Sakakibara S. *Int J Peptide Protein Res*. 1992;40:294–299.
51. König W, Geiger R. *Chem Ber*. 1970;103:2034–2040.
52. Nozaki S. *Chem Lett*. 1997;26:1–2.
53. Nozaki S. *J Peptide Res*. 1999;54:162–167.
54. Nishiuchi Y, Inui T, Nishio H, et al. *Proc Natl Acad Sci USA*. 1998;95:13549–13554.
55. Atherton E, Holder JL, Meldal M, Sheppard RC, Valerio RM. *J Am Chem Soc Perkin Trans*. 1988;1:2887–2894.
56. Staab HA. *Justus Liebigs Ann Chem*. 1957;609:75–83.
57. Staab HA. *Justus Liebigs Ann Chem*. 1957;609:83–88.
58. Staab HA, Benz W. *Angew Chem*. 1961;73:66.
59. Paul R, Anderson GW. *J Org Chem*. 1962;27:2094–2099.
60. DeTar DF, Silverstein R, Rogers Jr FF. *J Am Chem Soc*. 1966;88:1024–1030.
61. Merrifield RB, Gisin BF, Bach AN. *J Org Chem*. 1977;42:1291–1295.
62. (a) Curtius T. *Ber Dtsch Chem Ges*. 1902;35:3226–3228. (b) Curtius T, Levy L. *J Prakt Chem*. 1904;70:89–108.
63. Meienhofer J. Gross E, Meienhofer J, eds. *The Peptides: Analysis, Synthesis, Biology*, vol. 1. New York: Academic; 1979:198.
64. Shioiri T, Ninomiya K, Yamada S. *J Am Chem Soc*. 1972;94:6203–6205.
65. (a) Determann H, Wieland T. *Ann Chem*. 1963;670:136–140. (b) Shioiri T, Yamada S. *Chem Pharm Bull*. 1974;22:855–858. (c) Shioiri T, Yamada S. *Chem Pharm Bull*. 1974;22:859–863.
66. Chorev M, MacDonald SA, Goodman M. *J Org Chem*. 1984;49:821–827.
67. Banthorpe DV. In: Patai S, ed. *The Chemistry of the Azido Group*. New York: Wiley; 1971:397.
68. Brückner, R, *Reaktionsmechanismen, 3. Auflage*, Elsevier GmbH, Spektrum Akademischer Verlag: München, 2004; 623–626.
69. (a) Schnabel E. *Justus Liebigs Ann Chem*. 1962;659:168–184. (b) Chelli M, Ginanneschi M, Papini AM, Pinzani D, Rapi G. In: Schneider CH, Eberle AN, eds. *Peptides 1992*. Leiden: ESCOM; 1993:255.
70. Fruton JS. *J Biol Chem*. 1942;146:463–470.
71. Chelli M, Ginanneschi M, Papini AM, Pinzani D, Rapi G. *J Chem Res Synop*. 1993;118.
72. Harris JI, Fruton JS. *J Biol Chem*. 1951;191:143–151.
73. Prelog V, Wieland P. *Helv Chim Acta*. 1946;29:1128–1132.
74. Honzl J, Rudinger J. *Collect Czech Chem Commun*. 1961;26:2333–2344.
75. Alsina J, Barany G, Albericio F, Kates SA. *Lett Pept Sci*. 1999;6:243–254.

Chapter 6

Intramolecular Cyclization Side Reactions

Due to the concomitant existence of abundant functional groups within the same individual peptide molecule intramolecular functions between the reciprocal substituents will become unavoidable when the circumstances are in favor of such processes. Under certain circumstances peptide intramolecular reactions will result in the formation of cyclic impurities as appropriate. For instance, aspartimide formation represents one of the most notorious side reactions in peptide synthesis. On top of that pyroglutamate and hydantoin formation are also frequently addressing chemical preparation of peptide materials. Common intramolecular cyclization side reactions are discussed in this dedicated chapter: with respect to the mechanism, influencing parameters, and the corresponding solutions.

6.1 ASPARTIMIDE FORMATION

In spite of the profound applications of aspartimide derivatives in various territories[1] its formation is regarded as a serious side reaction in peptide synthesis. Aspartimide formation could be both acid[2] and base-catalyzed[3] and could affect solid phase[4] as well as solution phase-peptide synthesis.[5] It is noted that glutamate residue could suffer from a similar side reaction that results in a six-member ring glutarimide derivative.[6] Nonetheless, aspartate serves as a more appropriate and representative template for this category of side reaction and the content of this chapter will be narrowed down to aspartimide formation whereas glutarimide side reaction will not be elaborated separately.

The mechanism of aspartimide formation is shown in Fig. 6.1. The aspartate residue participating in the peptide synthesis is normally protected on its carboxyl side chain by an appropriate ester that is supposed to be removed at the peptide side chain global deprotection step. Nevertheless, this ester moiety might be attacked by various nucleophiles in the process of peptide synthesis. In light of this concern peptide **1** containing -Asp(X)-Xaa- sequence could be transformed into a five-member ring aspartimide **2** through a nucleophilic attack of the backbone amide on the subjected ester substituent from Asp(X) side chain. It is noted that the proton exchange of on aspartimide moiety is

Side Reactions in Peptide Synthesis. http://dx.doi.org/10.1016/B978-0-12-801009-9.00006-9

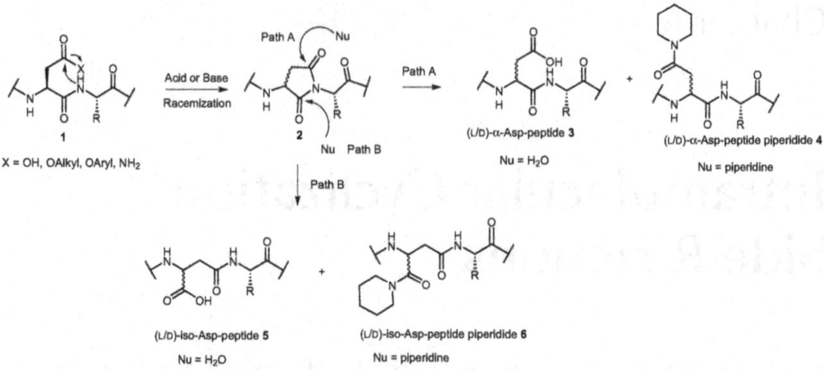

FIGURE 6.1 Mechanism of the formation of aspartimide and relevant impurities.

accelerated accordingly due to its increased acidity compared with that of H^{α} on the Asp precursor,[7] consequently leading to the aggravated C^{α} racemization and the formation of an enantiomeric aspartimide derivative **2**. Compound **2** might be able to survive the following reactions and be presented in the final product as an impurity with −18 amu molecular weight difference relative to the target product. Or alternatively, it might undergo the ring-opening process through nucleophilic addition. The nucleophilic attack on aspartimide moiety could proceed via Path A or B as indicated in Fig. 6.1. Aspartimide derivative **2** might undergo hydrolysis at the following reaction steps and be converted to the corresponding (L/D)-α-Asp **3** (Path A) and/or (L/D)-iso-Asp **5** (Path B). Formation of derivative **3** or **5** could hardly be detected by means of ordinary mass spectrometry and it is normally quite difficult to separate these isomers from the target product by means of ordinary reverse-phase liquid chromatography.[8] If the nucleophile mediating aspartimide ring-opening is piperidine that is utilized at the Fmoc cleavage steps, aspartimide derivative **2** would be accordingly converted to the corresponding (L/D)-α-Asp piperidide **4** (Path A) or (L/D)-iso-Asp piperidide **6** (Path B). The occurrence of these impurities could be readily detected by mass spectrometry since the molecular weights thereof are increased by 67 amu relative to that of the target product. It is noteworthy that the precursor of aspartimide could either be aspartate derivatives in the form of side chain-protected ester (X = OR), Asn derivatives (X = NH_2 or NHTrt),[9] or side chain-unprotected aspartate. All of these derivatives could be transformed into aspartimide upon treatment by strong acid,[10,11] base such as hydrazine,[12] or even during the storage in the solution.[11]

Aspartimide formation was originally detected in Boc/Bzl peptide chemistry. It could take place at the step of HF-mediated peptidyl resin cleavage/peptide side chain global deprotection, TFA-directed repetitive N^{α}-Boc acidolytic removal, neutralization of the liberated N^{α} after Boc cleavage, and even amino acid coupling in the presence of auxiliary tertiary amine.[13] Ever since Fmoc/tBu chemistry found its widespread applications in peptide synthesis the severity of aspartimide formation and its impact on peptide synthesis have been substantially intensified owing to the fact that this side reaction could occur readily at the repetitive piperidine-mediated N^{α}-Fmoc deprotection steps. The effect of aspartimide formation is accumulative as each piperidine treatment of peptidyl resin during SPPS could enhance the extent of aspartimide side product. This distinctive attribute of aspartimide formation is inherently different from DKP and pyroglutamate formation in peptide synthesis. The latter two side reactions address predominantly certain sensitive peptide sequences and their occurrence would be suppressed once the corresponding syntheses go beyond the subjected "susceptible" sequences. The difference between aspartimide formation and DKP/pyroglutamate formation in this respect is attributed to the fact that the aspartimide formation is induced by the nucleophilic attack of the backbone amide instead of N^{α} on peptide *N*-terminus, which is the underlying source for DKP and pyroglutamate formation. The intrinsic features of aspartimide formation determine the accumulative property of this side reaction, which could basically take place during the whole process of peptide synthesis after the incorporation of the concerned Asp. Generally, the more circles of piperidine treatments thc peptide has to undertake after the coupling of the subjected Asp residue, the more intensified aspartimide formation it will provoke.

Aspartimide formation could take place, as already been indicated, not only at the step of piperidine-mediated N^{α}-Fmoc deprotection and amino acid coupling, but also in the process of TFA-directed peptide side chain global deprotection, and even during peptide purification and storage. It has been figured out that aspartimide formation is more prone to take place under basic environment than in acidic milieu[14,15] and it becomes consequently one of the most challenging side reactions in Fmoc/tBu peptide synthesis.

6.1.1 Factors That Influence Aspartimide Formation

It could be deduced from the mechanism of aspartimide formation that many factors could potentially impact on its occurrence. These factors are elaborated systematically in the following sections.

6.1.1.1 Base

Since aspartimide formation could be catalyzed by base, all base-participated steps in peptide synthesis could theoretically facilitate this side reaction. Auxiliary tertiary amines are normally imperative for amino acid activations and couplings and aspartimide formation could therefore be triggered at this step in the presence of tertiary amines.[2] The type of tertiary amines utilized at the step of amino acid coupling is verified to be influential on the propensity of aspartimide formations. DIEA is less prone to induce aspartimide formation relative to trimethylamine thanks to the marked steric hindrance of the former.[10,16]

The concentration of the employed tertiary amine is critical for the tendency of aspartimide formation.[17]

Now that it is imperative to treat N^{α}-Fmoc protected peptidyl resin with secondary base in order to liberate the N^{α}-group prior to the amino acid coupling, it is understandable that aspartimide could be induced at this step and the extent thereof is much more significant than that generated during the tertiary amine-mediated amino acid couplings.[10] The type and concentration of the secondary base adopted for N^{α}-Fmoc deprotection would exert decisive effects on the inclination and extent of aspartimide formation.[18]

6.1.1.2 Acid

It has been verified that aspartimide formation could be catalyzed by acids as well. HF is utilized in Boc/Bzl peptide chemistry as the acid to release the peptide chains from the solid support and remove the protecting groups from the corresponding shielded amino acids. Aspartimide formation could be triggered at the step of HF-mediated peptide cleavage and side chain global deprotection.[10] Other acids, such as TFMSA, 6N HCl and concentrated TFA are also capable of inducing aspartimide formation.[15,19–21]

6.1.1.3 Protecting Groups on Asp Side Chain Carboxyl Group

The type of protecting group on the Asp side chain is one of the most decisive factors that regulate the tendency of aspartimide formation in the process of peptide synthesis. In view of the mechanism of aspartimide formation that is elucidated in Fig. 6.1, this process is initiated by the nucleophilic attack of the amide backbone from the -Asp(X)-Xaa- unit on the carboxylate side chain of the concerned Asp, followed by the release of a hydroxyl derivative as the leaving group, and the simultaneous formation of a 5-member ring intermediate. This process implies that the features of the protecting group on the β-carboxylate side chain of the concerned Asp residue would be one of the most influential factors with respect to the regulation of the likelihood of aspartimide formation. The steric effect and electron-withdrawing property of the adopted protecting group on Asp side chain stand out as the most predominant factors in this connection. Basically, protecting groups with less electron-withdrawing effect and/or increased bulkiness could lower the tendency of the potential aspartimide formation.

The carboxyl-side chain of Asp is normally protected in the form of benzyl ester in Boc/Bzl peptide chemistry. However, Asp(OBzl) residue tends to undergo readily aspartimide formation in an acidic environment,[13] whereas some extraordinarily sensitive -Asp(OBzl)-Xaa- sequences could even be quantitatively transformed into its aspartimide counterparts upon acid treatment.[10,22] From this aspect benzyl ester could not be regarded as an ideal protecting group for the Asp carboxyl-side chain. In some special cases chemists even intentionally utilized Asp(OBzl) as a synthon of aspartimide. Cyclohexyl ester Asp(OChx) disfavors aspartimide formation compared with Asp(OBzl), which might be attributed to the increased steric hindrance of the former.[20,23] Under certain circumstances, Asp(OChx) could substantially inhibit aspartimide formation,[24] even though it is incapable of suppressing its occurrence thoroughly. The advantages of cyclohexyl ester over benzyl ester as Asp carboxyl-side chain protecting group could also be reflected by its capability to restrain DKP formation in which the concerned Asp(X) residue is entangled.[24] Aside from Asp(OBzl) other Asp derivatives, for example, Asp(OAll) also incline to facilitate aspartimide formation.[25] Asp(OtBu) building block, which has already found widespread applications in Fmoc-mode peptide synthesis, exhibits significant superiority in terms of the suppression of aspartimide formation in the process of peptide synthesis, thanks to its considerably enhanced bulkiness.[4,21,26]

6.1.1.4 Solid Support for Peptide Synthesis

Under certain circumstances, the choice of resin for the solid phase peptide synthesis could also exert evident influences on the inclination of aspartimide formation. For example, the employment of CTC resin as the solid support for SPPS makes it possible to detach the peptides in the form of side chain-protected precursors from the solid supports by dilute acid solution, attenuating by this means the stress of strong acid-facilitated protonation/activation of ester group on Asp side chain and consequently minimizing the aspartimide formation compared with the occasion if concentrated strong acids were applied instead.[19,21] A study [27,28] indicates that when less hydrophobic material such as PEGA or CLEAR® (Cross-Linked Ethoxylate Acrylate) is used as the backbone of the concerned SPPS resin in place of polystyrene, the extent of aspartimide formation induced in the process of due SPPS, if any, could be lowered. The improvement is attributed to the disruption of the noncovalent aggregations between the individual peptide chains immobilized on the solid supports once less hydrophobic resin is utilized. This distinctive attribute leads to the increased N^{α}-Fmoc deblocking kinetics, which could in turn shorten the duration of piperidine treatment of the addressed peptidyl resin and consequently lower the extent of aspartimide formation by attenuated base stress.[27]

6.1.1.5 Temperature

Similar to most other side reactions, the extent of aspartimide formation could be effectively reduced at lower temperature. The stringent control over the reaction temperature is highly desirable in particular when the peptide bearing sensitive sequence is addressed to base or acid treatment since under such circumstances temperature will substantially impact on the extent of aspartimide formation. It has been verified that by lowering the temperature the aspartimide formation will be effectively suppressed and the purity of the corresponding product is evidently increased.[16,29] On the contrary, a temperature increase will unavoidably lead to the aggravated aspartimide formation side reaction.[30]

6.1.1.6 Solvent

The attributes of solvent, polarity in particular, are directly correlated to the tendency of aspartimide formation in the process of peptide synthesis. In light of the mechanism of aspartimide formation, the increase of solvent polarity would contribute to the stabilization of the polar intermediates formed in this process, consequently driving the reaction equilibrium to the direction of aspartimide formation. Basically, the tendency of aspartimide formation in different solvents is aligned in a descending order DMSO > DMF > >THF > DCM.[18] DCM is the choice of solvent or component of solvent mixture for the amino acid couplings in which aspartimide formation poses a serious problem.[31] THF could replace DMF as the solvent for piperidine solution if aspartimide formation is severely proceeding at the piperidine-mediated N^{α}-Fmoc deblocking step.[32] The overall performance of DCM with regard to aspartimide suppression is superior to that of THF.[7] Aprotic polar solvents such as DMF outperform protic solvents methanol, ethanol, and butanol in terms of alleviating aspartimide formation side reaction.[33]

6.1.1.7 Peptide Sequence

The chemical and physical microenvironments in which the susceptible Asp is located play extraordinarily important role with respect to the inclination of aspartimide formation. Since this process is initiated by the nucleophilic attack of the backbone amide on the carboxyl-side chain of Asp, it is reasonable to indicate that the inherent properties of Xaa in sequence -Asp(X)-Xaa- could significantly dictate aspartimide formation via its electronic and steric effect.[34,35] There are indeed controversial opinions towards the dependence of aspartimide formation on peptide sequence. The seemingly conflicting assertions in the context of sequence dependence could be somehow comprehended from the aspect that no individual peptide could be selected as an absolute rational template for the investigation on the sequence dependence of aspartimide formation. A certain -Asp-Xaa- unit in one peptide molecule, which suffers from a severe aspartimide formation does not necessarily have to cause the same extent of aspartimide side product in another peptide. Now that the aspartimide formation is synergistically regulated by many factors whereas the explicit interactions between these parameters could hardly be accurately and unequivocally extrapolated in the context of investigations on sequence dependence of aspartimide formation.

In light of a plethora of studies, with regard to the sequence dependence of aspartimide formation it could be concluded that when the neighboring Gly is resided on *C*-terminal side of the concerned Asp, that is to say when the subject peptide contains -Asp(X)-Gly- unit, aspartimide formation will be extremely prone to take place both in acidic and basic environment.[36,37] This phenomenon could be attributed to the advantageous steric effect of Gly which favors the nucleophilic attack of the backbone amide on Asp carboxyl side chain. Aside from aspartimide formation Gly is also liable to induce other side reactions such as hydantoin[38] and DKP formation.[39] The distinctively high tendency of peptide containing -Asp(X)-Gly- unit to undergo aspartimide formation is largely ascribed to the remarkably reduced bulkiness of Gly compared with other amino acids. In view of the evident inclination to undergo aspartimide formation peptides bearing -Asp(X)-Gly- sequence are frequently opted as appropriate substrates for the purpose of investigations on this side reaction.[40–42]

Another amino acid that is acknowledged to readily induce aspartimide formation in basic condition is asparagine regardless of its amide side chain protecting group such as Trt or more acid labile Mtt. A peptide containing -Asp(X)-Asn(Trt/Mtt)- unit is liable to aspartimide formation.[4,18,32] Asn is also readily subjected to aspartimide formation in acidic milieu.[3,10] In addition, Asp(OtBu) itself frequently promotes the formation of an aspartimide side product[4,43] and the synthesis of the peptide accommodating repetitive Asp residue, for example, -Asp(OtBu)-Asp(OtBu)- hence renders a challenging task due to its marked high tendency to undergo aspartimide formation. There are controversial opinions toward the Ala and Gln(Trt) with regard to their propensities to induce aspartimide formation. Some investigations indicate that -Asp(OtBu)-Ala/Gln(Trt)- sequence in certain peptides leads to aspartimide formation,[3,26,43] whereas others assert that no aspartimide impurities are detected in their peptide syntheses when Ala/Gln(Trt) is resided to *C*-terminal of Asp.[4] These contradictory results could originate from the fact that the tendency of aspartimide formation is not independently decided by sequence alone. Other parameters such as peptide conformation might outweigh the influence of the type of amino acid neighboring to Asp and could hence intensify or attenuate the aspartimide formation under diverse circumstances. The influences of the presence of Ala/Gln(Trt) neighboring Asp might not be dominant enough to regulate the inclination of aspartimide formation and their effects could be counterbalanced by other factors that might be ignored unintentionally in the corresponding studies. The propensity of aspartimide formation from the peptide bearing -Asp(X)-His(Y)- unit is seemingly pH dependent in that it is relatively stable under basic condition[3] whereas it tends to cyclize in acidic environment.[44] This discrepancy might be due to the inherent differences between the mechanisms of base- and acid-catalyzed aspartimide formation. Since imidazolyl side chain on His is normally protected on its N^{τ} group, the unshielded N^{π} group could basically be involved in the process of aspartimide formation.

It has been widely accepted that unprotected Ser and Thr could markedly stimulate aspartimide formation in both basic[4] and acidic conditions.[3] However, their side chain-protected counterparts are relatively resistant to aspartimide formation.[4] This phenomenon could imply the neighboring-group effects from the free β-hydroxyl group on Ser/Thr. They might be entangled in the process of aspartimide formation.[13]

The type of side chain-protecting groups on the amino acid that is *C*-terminal neighboring to Asp could impact on the propensities of their parental peptides towards aspartimide formation. For example, Cys(Acm) and Arg(Pbf) are more prone to provoke aspartimide formation compared with Cys(Trt) and Arg(Pmc), respectively.[4,30,32,43]

6.1.1.8 Peptide Conformation

The influences of peptide conformation on the occurrence of aspartimide formation have not yet been explicitly revealed. It has, however, been figured out that the mutation of a critical amino acid in certain peptide sequences could markedly affect the inclination and extent of aspartimide formation. For example, in the synthesis of the analogs of corticotrophin-releasing hormone (CRH) via Fmoc/tBu peptide chemistry[34] the major components in the crude product are identified as the target peptides. However, corresponding D-amino acid scan on the subjected peptides indicates that the aspartimide-sensitive sequence -Asp^{25}(OtBu)-Gln^{26}(Trt)- is influenced by its C-terminal sequence -Leu^{27}-Ala^{28}- in that replacement of -Leu^{27}-Ala^{28}- unit in the process of D-amino acid scan by -D-Leu^{27}-D-Ala^{28}- would trigger aspartimide formation on Asp^{25}. The major components in the crude product obtained from the D-amino acid scan are aspartimide and the corresponding derivatized impurities. Although this study does not elucidate the unambiguous correlations between the propensity of aspartimide formation and the peptide conformation, it implies at least that this process could be influenced to some extent by the spatial alignment of the affected domain from the concerned peptide.

6.1.1.9 Microwave

In spite of the fact that the introduction of microwave technology has exerted a powerful impetus to peptide synthesis, particularly for those hardly achievable by conventional synthetic means, it could, on the other hand, induce some side reactions like racemization and aspartimide formation.

It has been revealed in a study[45] that the synthesis of a peptide with a sequence DFDFDFEpGEFDFDFD did not give rise to aspartimide formation followed by the conventional peptide synthetic process, whereas the introduction of microwave irradiation drastically increased the ratio of aspartimide relevant impurities to 35.4% in the protected intermediary peptide product. The addition of 0.1 M HOBt to 20% piperidine/DMF for Fmoc cleavage while sustaining microwave input during the synthesis process lowered the aspartimide relevant derivatives to 15.4%, which remained nevertheless unacceptable. Aspartimide formation induced by microwave could be alleviated by means like temperature lowering, replacement of piperidine by piperazine or addition of HOBt into Fmoc cleavage solution.[46] In spite of these improvements, the propensity of aspartimide formation upon the introduction of microwave irradiation should not be ignored or underestimated, especially for those peptide individuals containing sensitive sequences.

6.1.2 Solutions for Aspartimide Formation

In light of the earlier systematic analysis of the potential factors, which could impact on the occurrence of aspartimide formation solutions with the aim to suppress or alleviate this side reaction have been exploited and applied to the corresponding peptide synthesis.

6.1.2.1 Protecting Groups on β-Carboxyl Group of Asp

In spite of the fact that *tert*-butyl ester and cyclohexyl ester protecting groups shielding Asp carboxyl-side chain could markedly reduce the extent of aspartimide formation compared with benzyl ester, they are nevertheless incompetent to this end in certain cases in which the subjected Asp is extremely susceptible to aspartimide formation. Development of more effective Asp side chain protecting groups to suppress aspartimide formation is therefore mandatory in this connection. Qualified protecting groups for Asp are supposed to satisfy the following criteria: (1) it should be orthogonal to N^{α}-protecting group, that is to say, the protecting group on Asp carboxyl-side chain should remain intact and sufficiently stable during the repetitive N^{α}-protecting group deblocking treatment along the whole peptide synthesis; (2) it could be removed readily and sufficiently at the step of peptide side chain global deprotection to fully regenerate the shielded Asp side chain; (3) it should possess a rational electron-withdrawing property, making it well balanced between the desired protection on the Asp side chain-carboxyl group and reasonable tendency to be removed once the concerned peptide assembly is completed; (4) it should be provided with advantageous steric effect in order to resist the potential nucleophilic attack of backbone amide on the protected Asp carboxyl-side chain; (5) it should bestow the addressed Asp derivatives reasonable solubility in ordinary organic solvents for peptide synthesis; and (6) its introduction should not adversely affect the efficient coupling of the subjected Asp building block.

Some Asp protecting groups with various extents of bulkiness have been designed from the perspective of steric effects in terms of aspartimide formation inhibition. These bulky protecting groups are expected to restrain the attack of backbone amide on Asp side chain by virtue of their evident steric hindrance. The chemical structures of these derivatives are non-exhaustively depicted in Fig. 6.2.

Trt is generally not adopted for the protection on the Asp/Glu-carboxyl side chain due to its extraordinary sensitivity to acid.[47] Its utilization could nonetheless be facilitated through rational chemical structure modifications by increasing its stability in order to take advantage of its inherent bulkiness attribute to minimize the aspartimide formation. Some derivatives of Trt-protected Asp such as Fmoc-Asp(OPyBzh)-OH **8**, Fmoc-Asp(OPhFl)-OH **9**, and Fmoc-Asp(OPp)-OH **11**[30] have unexpectedly augmented the extent of aspartimide formation compared with Fmoc-Asp(OtBu)-OH. The pyridine

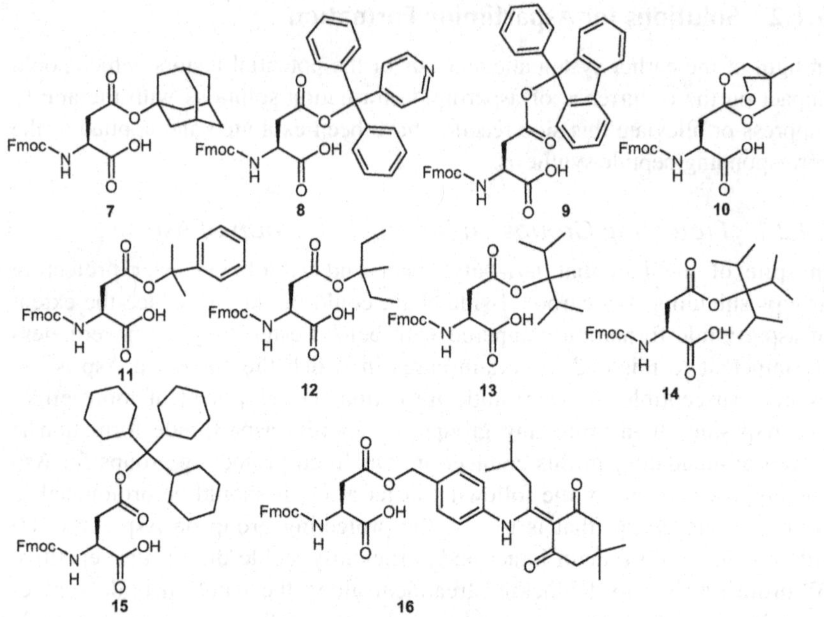

FIGURE 6.2 **Fmoc-Asp-OH with various side chain protecting groups.**

moiety on OPyBzh substituent could participate in the process of proton abstaction from the subjected backbone amide and catalyze by this means the subsequent nucleophilic attack of amide anion on the Asp side chain, and the resultant aspartimide formation. The aggravation of aspartimide formation in case of OPhFl protection has not yet been reasonably elucidated. It might be inherently attributed to its inferior stability in basic environment. A separate study indicated that peptide with an *N*-terminal Glu(OPhFl) residue could be converted into the corresponding pyroglutamate derivative in the presence of morpholine.[47] In summary, most aforementioned Trt-derived protecting groups are not eligible for Asp side chain protection in spite of their apparent steric hindrances.

Fmoc-Asp(OBO)-OH **10** represents another protection strategy for Asp carboxyl-side chain through the transformation of Asp into its ortho-ester Asp(OBO)[48] and the recovery of shielded carboxyl group via a two-step hydrolysis once the target peptide assembly is completed. Unfortunately, utilization of **10** as Asp precursor is still incapable of preventing aspartimide formation. This incompetence might be due to the fact that ortho-ester of OBO is gradually hydrolyzed in the process of peptide synthesis to the corresponding 2,2-dimethoxylpropylester derivative that facilitates the aspartimide formation at the following steps. This inherent deficiency of Asp(OBO) has limited its widespread application as a eligible substitution for Asp(OtBu).

Bulky cyclic aliphatic esters might be utilized as protecting groups for the Asp carboxyl-side chain as reflected by the utilization of the Asp(OChx) building block as the replacement for Asp(OBzl) in an effort to suppress the aspartimide side reaction. Neither the enlargement of cyclohexyl ester on Asp(OChx) to the corresponding cycloheptyl/cyclooctyl esters nor its shrink to cyclobutyl/cyclopentyl ester counterparts could remarkably affect the tendency of aspartimide formation.[49] The evidently bulkier derivative Fmoc-Asp[O(1-Ada)]-OH **7** exhibits limited stability and its performance as to suppress piperidine-induced aspartimide formation is inferior to that of Asp(OtBu).[4]

Dmab ester substituent on Fmoc-Asp(ODmab)-OH **16** could be sufficiently removed in 2% hydrazine/DMF solution, it introduces a new degree of orthogonality with respect to peptide side chain protection strategy. However, it has been verified that Asp(ODmab) could intensify the aspartimide formation in spite of its distinctive bulkiness.[19] Its application in peptide synthesis, if necessary, is therefore advisable to be accompanied by other complementary methodologies such as backbone amide protection so as to restrain aspartimide formation. Another Asp derivative Asp(OAll) equipped with appreciable protection orthogonality is notoriously prone to trigger aspartimide formation and in certain extreme conditions, the peptide containing Asp(OAll) unit could even be quantitatively converted to its aspartimide counterpart.[50]

Thanks to the prominent performances of Asp(OtBu) to suppress aspartimide formation it has been intensively exploited in terms of Asp side chain *tert*-butyl ester derivatization in an effort to improve the suppression of aspartimide formation, among which Fmoc-Asp(OMpe)-OH **12**[30,51] and Fmoc-Asp(ODie)-OH **13**[36] have exhibited superior properties to Asp(OtBu) in terms of aspartimide inhibition. In view of the discrepant performances of more or less equally bulky OMpe/ODie protecting groups on one side and O(1-Ada)/OPyBzh/OPhFl on the other, it is implied that the increased rigidity of the latter protecting group category is adverse to the suppression of aspartimide formation.[30] Fmoc-Asp(OTim)-OH **14** and Fmoc-Asp(OTcm)-OH **15** could be derived from further modifications on the basis of OMpe and ODie protecting groups. Their distinctive bulky side chains exert nevertheless evident negative impacts on the coupling kinetics of the corresponding Asp derivatives. Furthermore, they could not exhibit improved performances as to suppress aspartimide formation.[36] These attributes have limited their applications in peptide production.

6.1.2.2 Base

Since base-catalyzed aspartimide formation is initiated by the proton abstraction from the peptide amide backbone, the rational choice of base for peptide synthesis is understandably regarded as one of the most effective strategies to suppress aspartimide formation. This methodology is in line with the effective solutions for many other base-induced side reactions such as amino acid racemization, DKP formation, pyroglutamate formation, and Cys/Ser/Thr β-elimination in that elaborate screening of applicable bases is advised to minimize these kinds of base-catalyzed side reactions. Generally, lowering the pKa or enlarging the bulkiness of the concerned bases could be advantageous to restrain aspartimide formation.

Utilization of certain bulky tertiary amine, for example, *N,N*-dimethylaniline and triphenylamine as auxiliary tertiary base in SPPS would drastically slow down the affected amino acid coupling kinetics due to their excessively low basicity. This adverse property could result in insufficient amino acid couplings in peptide synthesis in spite of their merits to suppress aspartimide formation.[2]

Considering that piperidine-induced aspartimide formation is the predominant cause of this side reaction, the focus of aspartimide inhibition is understandably centered on the screening of alternatives to piperidine. The routinely utilized Fmoc cleavage reagents such as solutions of 20% (v/v) piperidine, 6% (w/v) piperazine, 20 % (v/v) 1-hydroxypiperidine, 0.02 M TBAF and 2% (v/v) DBU have been systematically compared in a dedicated investigation.[52] It was figured out that piperazine and 1-hydroxypiperidine could remarkably lower the extent of aspartimide formation compared with piperidine. A similar conclusion has also been drawn from other investigations,[19,35,46] whereas TBAF and DBU are more prone to induce aspartimide formation than piperidine.[43,52] Owing to their trivial nucleophilicity DBU and TBAF would not generate adducts such as Asp-piperidide. It has been reported that the mixture of hexamethyleneimine/HOBt/NMP/DMSO 4:50:4:71:71 (v/v/w/v/v) could be employed as Fmoc cleavage solution. It has exhibited superior performance to piperidine in terms of aspartimide suppression.[43] This mixture was originally envisaged as reagent to remove Fmoc protecting groups from peptide thioester precursor[53] by virtue of its compatibility with a thioester functional group. This fine-tuned base solution could also find the application so as to alleviate base-catalyzed aspartimide formation.

6.1.2.3 *Protection on Backbone Amide and Application of Pseudoproline*

No matter whether the subjected aspartimide formation is catalyzed by acid or base it involves the nucleophilic attack of backbone amide on the carboxyl-side chain of Asp, the fundamental strategy of aspartimide formation inhibition is therefore to mask the nucleophilicity of the concerned amide. Peptide backbone amide protecting groups such as Dmb have found appreciable applications for the preparation of aspartimide formation-susceptible peptides. Addition of electron-withdrawing substituents to the scaffold of benzyl protecting group would increase its acidic lability,[54] which facilitates its removal upon acid treatment. Hmb is another routinely utilized backbone amide protecting groups in peptide synthesis.[30,55] The incorporation of -Asp(OtBu)-*N*-Hmb-Xaa- unit into the peptide of interest could thoroughly suppress the occurrence of aspartimide formation and the backbone amide protection strategy is regarded as the most effective solution against this side reaction. Aside from Dmb and Hmb, many other derivatives (Fig. 6.3) could be utilized as peptide backbone amide

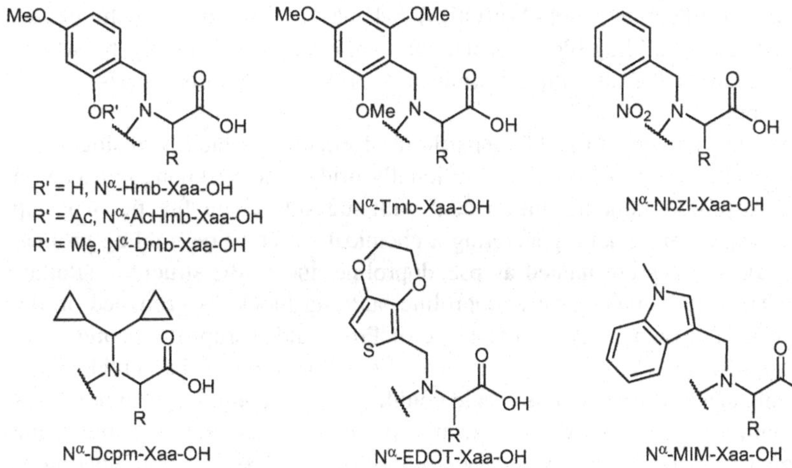

FIGURE 6.3 Peptide amide backbone protecting groups.

protecting groups, among which Dcpm (dicyclopropylmethyl), MIM (1-methyl-3-indolylmethyl) and EDOT (3,4-ethylenedioxy-2-thienyl) outperform Dmb with regard to the coupling kinetics of the corresponding parental amino acid and acid liability.[37,41] Furthermore, these moieties will not be acylated as Hmb during peptide synthesis due to the absence of the reactive hydroxyl substituent.

Backbone amide protection has also exhibited advantages in the territories other than aspartimide suppression. For example, -Xaa-Gly- moiety in certain peptides could suffer from backbone amide acylation by highly activated carboxylate derivatives, giving rise to the formation of imide impurities such as hydantoin,[56] or alternatively it could be acylated during peptide chain elongation by activated amino acids.[57] These side reactions frequently address susceptible Gly residue that bears the least steric hindrance, legitimating the development of backbone-protected Gly derivatives. Another advantage of amide backbone protection is reflected by the improvement of difficult amino acid couplings in SPPS. The intensified inter-chain aggregation on the solid support during SPPS would generally result in the rigidifying of the affected peptide chains and impede the efficient couplings of the incoming amino acids. Certain peptide secondary structures, most notably β-sheet,[58] could drastically hinder the smooth peptide chain elongation once formed. The incorporation of backbone amide protecting group could block the formation of peptide interchain hydrogen bonds and therefore disrupt the secondary structures, enhancing by this means the flexibility of the peptide chain and promoting the coupling with amino acids.

Since the incorporation of Fmoc-Asp(OtBu)-OH to a peptide chain on which H-*N*-Hmb-Xaa is located on the *N*-terminus is normally difficult especially for the bulky amino acids such as Val, it is advisable to utilize the preformed dipeptide Fmoc-Asp(OtBu)-*N*-Hmb-Xaa-OH under such circumstances as the building block for the concerning peptide assembly, bypassing in this manner the difficult coupling of Asp(OtBu) derivative to H-*N*-Hmb-Xaa peptide.

Another strategy to tackle aspartimide formation, which is analogous to amide backbone protection, is to rationally bridge the backbone nitrogen on the aspartimide-inducing amino acid with the side chain functional group on the same amino acid, generating a chemical structure resembling proline. These derivatives are named as pseudoproline due to the structural similarity.[59] The employment of pseudoproline building blocks is grounded on the following reasoning: (1) the existence of Pro could disrupt the peptide secondary structure, leading to the increased solubilization of the peptide chain and enhanced kinetics of amino acid coupling;[60] (2) peptides containing -Asp-Pro- unit are basically exempted from aspartimide formation by virtue of the inherent property of the secondary amino group on Pro.[29] In combination of the aforementioned merits of Pro it is advisable to transform Ser, Thr, and Cys which bear a β-hydroxyl or sulfhydryl substituent on its side chain into the corresponding pseudoproline derivatives and incorporate these derivatized building blocks into the target peptide as a synthon of the corresponding Ser/Thr/Cys. The affected peptide synthesis could thus be improved by virtue of the aforementioned merits bestowed by pseudoproline building blocks. Ser/Thr/Cys pseudoproline derivatives in the form of -Ser($\psi^{Me, Me}$Pro)-, -Thr($\psi^{Me, Me}$Pro)-, or -Cys($\psi^{Me, Me}$Pro)- **17** could be reversed back to their original structures **18** upon the treatment by concentrated TFA solution (Fig. 6.4).[60] This strategy could be adopted in an effort to suppress aspartimide formation or promote amino acid coupling. The limitation of the pseudoproline strategy is that it necessitates Ser, Thr, or Cys that bears a reactive β-functional group, which could be bridged to their α-amino groups to form the five-member ring structure of pseudoproline. The reactivity of N^{α} on the pseudoproline is normally low due to the steric hindrance of dimethyl group on its 2-position, which might impede the target acylation process. In view of this limitation, pseudoproline derivatives are normally pretransformed into the corresponding

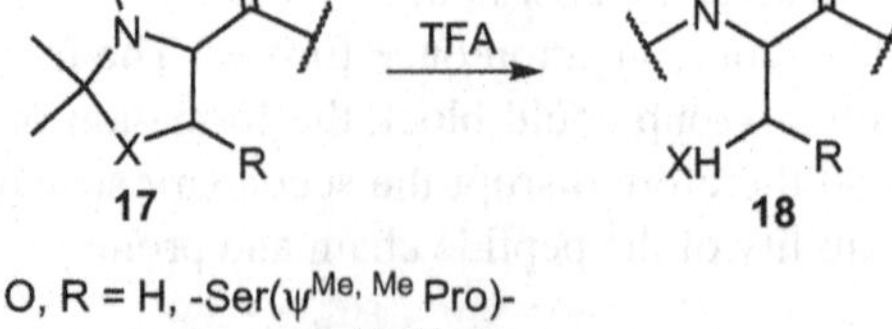

X = O, R = H, -Ser($\psi^{Me, Me}$Pro)-
X = O, R = CH_3, -Thr($\psi^{Me, Me}$Pro)-
X = S, R = H, -Cys($\psi^{Me, Me}$Pro)-

FIGURE 6.4 Acidolysis of pseudoproline derivatives.

dipeptide Fmoc-Xaa-Ser/Thr/Cys($\psi^{Me, Me}$Pro)-OH in solution as a building block [Xaa = Asp(OtBu) when suppression of aspartimide formation is concerned]. This underlying strategy is essentially resembling the employment of Fmoc-Asp(OtBu)-*N*-Hmb-Xaa-OH dipeptide building block in SPPS.

6.1.2.4 N-Hydroxylamine and Phenol Derivatives

N-hydroxylamine and phenol derivatives have already been employed in peptide synthesis to alleviate amino acid racemization in the process of amino acid activation and coupling[61] or to inhibit guanidinium formation on peptide *N*-terminus by certain coupling reagents.[62] It has also been confirmed by a plethora of studies that *N*-hydroxylamine and phenol derivatives could also be utilized as additives to N^{α}-Fmoc cleavage solution in an effort to diminish the aspartimide formation taking place at the peptide N^{α}-Fmoc deblocking step.[4,46,52] In consideration of the mechanism that base-induced aspartimide information is initiated via the deprotonation of peptide backbone amide the addition of acidic *N*-hydroxylamine or phenol derivatives such as HOAt, HOBt, Oxyma, HOPfp, and DNP could interfere with the equilibrium of the detrimental amide deprotonation and reduce consequently the extent of aspartimide formation.[13] Aspartimide inhibition by acidic *N*-hydroxylamine and phenol derivatives could be inherently analogous to the mechanism of aspartimide suppression by the acidic side-chain unprotected Glu, Asp or Tyr neighboring on *C*-terminal of Asp(OtBu) in peptide sequence -Asp(OtBu)-Glu/Asp/Tyr-,[13,63] the existence of these acidic derivatives could interfere with the process of backbone amide proton abstraction and consequently inhibit the following nucleophilic reaction that results in aspartimide formation.

The contribution of *N*-hydroxylamine and phenol derivatives in terms of aspartimide inhibition could be reflected at the synthetic steps other than N^{α}-Fmoc removal. They are routinely charged as additives to coupling reagents into the amino acid activation reactions. It has been demonstrated that the addition of *N*-hydroxylamine could drastically reduce the content of aspartimide side product that is formed at amino acid coupling steps.[13] Acidic nonhydroxylamine derivatives like pentachlorophenol or DNP could evidently diminish aspartimide formation as well.[13] However, no direct correlation between the acidity of the additive and its effect to reduce aspartimide formation could be aligned, which might be partially due to the assertion that the steric effects of these additives could contribute to the suppression of aspartimide formation as well. Furthermore, concentration of the subjected *N*-hydroxylamine and phenol derivatives is also regarded as a significant parameter with regard to aspartimide reduction. This concern could be comprehended from the dual features of these acidic derivatives as to simultaneously inhibit aspartimide formation and impede amino acid couplings through protonation of the N^{α} to be acylated. A rational balance is therefore supposed to be reached in order to minimize aspartimide formation while keeping the kinetics of amino acid coupling unaffected.[13]

6.1.2.5 *N*$^{\alpha}$-Protecting Groups

The alternative *N*$^{\alpha}$-protecting groups to Fmoc could be resorted to in case aspartimide formation is markedly susceptible at the step of *N*$^{\alpha}$-Fmoc deblocking. The removal of these *N*$^{\alpha}$-protecting groups could be exempted from base treatment, which is the underlying root cause for aspartimide formation. This strategy is analogous to that of the prevention of DKP formation elucidated in Section 1.10 since both DKP and aspartimide formation are induced predominantly at the step of piperidine-mediated *N*$^{\alpha}$-Fmoc deprotection and alternative *N*$^{\alpha}$-protecting groups that could be removed under neutral or acidic conditions are regarded as the solution addressing the root cause of these side reactions. The most frequently employed alternative *N*$^{\alpha}$-protecting groups, which are in compliance with SPPS requirements and tBu-category of side chain protecting groups, include *N*$^{\alpha}$-Trt, *N*$^{\alpha}$-Alloc, *N*$^{\alpha}$-pNZ etc. Removal of *N*$^{\alpha}$-Alloc could be carried out under neutral environment in the presence of Pd(0) catalyst and allyl scavengers.[64] *N*$^{\alpha}$-pNZ protecting groups could be effectively reduced by $SnCl_2$ or $Na_2S_2O_4$ and cleaved spontaneously afterwards[40] or removed via hydrogenolysis.[65] The mechanism of this process could be referred to Fig. 1.24. Dilute TFA solution is capable of efficiently removing *N*$^{\alpha}$-Trt from the addressed peptide chains immobilized on compatible solid supports. All of the aforementioned *N*$^{\alpha}$-protecting groups could be quantitatively deblocked with ease in acidic or neutral conditions and the base-induced aspartimide formation will be hence suppressed by this means.

6.1.2.6 Fine-Tuning of Asp β-Carboxyl Activation

It is to note that aspartimide formation could take place not only at the steps of *N*$^{\alpha}$-Fmoc removal and amino acid coupling, but also in the process of Asp β-carboxylate activation. The Asp β-carboxylate side chain is frequently subjected to activation process in peptide synthesis in an effort to accommodate targeted modifications such Asn *N*-glycosylation[66] or side-chain to backbone, side-chain to side-chain cyclization.[31] The process of aspartimide formation induced in peptide *N*-glycosylation and side-chain to side-chain cyclization is depicted in Figure 6.5 and Figure 6.6, respectively. The underlying similarity between these two cases lies in the activation of Asp β-carboxylate. The prerequisite for the convergent synthesis of *N*-glycosylated peptide **22** is the liberation of carboxylate substituent on the substrate Asp side chain to afford derivative **19**. It is necessary to active the concerned carboxylate into its active species **20**, which is supposed to be subsequently trapped by the incoming appropriate glycosylamine and transformed into the target *N*-glycopeptide **22**. The intervening activation of the Asp carboxyl-side chain, however, entrusts the affected carboxylate with an ideal leaving group for a potential nucleophilic attack from the backbone amide, which results in the formation of aspartimide derivative **21**. The severity of aspartimide formation could be even aggravated when the convergent condensation reaction is slowed down due to

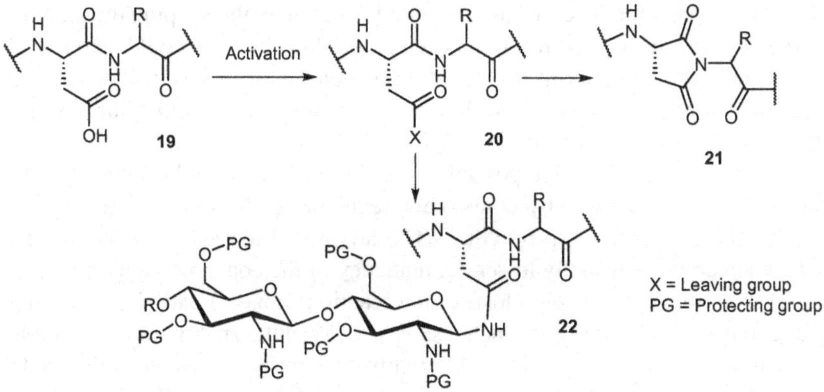

FIGURE 6.5 Aspartimide formation during peptide *N*-glycosylation.

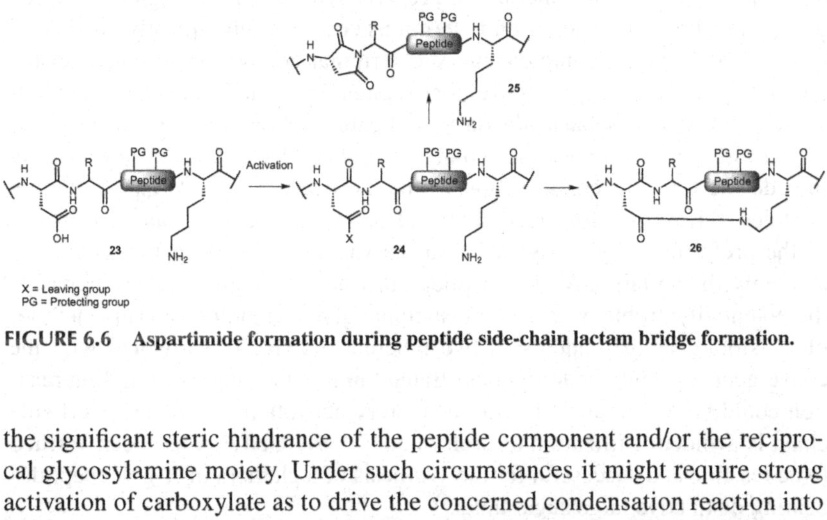

FIGURE 6.6 Aspartimide formation during peptide side-chain lactam bridge formation.

the significant steric hindrance of the peptide component and/or the reciprocal glycosylamine moiety. Under such circumstances it might require strong activation of carboxylate as to drive the concerned condensation reaction into completion. However, intensified activation on the Asp carboxylate side chain might exaggerate the severity of potential aspartimide formation as a consequence. Similarly, activation of Asp carboxyl-side chain on the linear peptide precursor **23** in an effort to form a lactam bridge between reciprocal Asp and Lys and produce cyclic peptide **26** could also be adversely affected by the undesirable formation of aspartimide impurity **25** derived from the Asp side chain-activated intermediate **24**.

The underlying drive of aspartimide formation in the above two cases of peptide derivatization is the activation of the Asp carboxyl-side chain which substantially enhances the electrophilicity of the subjected carboxylate and facilitates as a consequence the nucleophilic attack from the backbone amide. It is therefore advisable to fine-tune the process of the Asp side chain activation in an effort to suppress the potential aspartimide formation while keeping the kinetics of the target condensation reaction acceptable. This objective could be realized via the utilization of appropriate coupling reagents/additives as well as rational adjustments of the reaction conditions in terms of solvent, temperature, orders of reactant addition, and so on.

A solution to address the possible aspartimide formation during the preparation of *N*-glycopeptide by convergent segment condensation strategy is to replace the concerned Asp by Glu.[67] The additional methylene group on the side chain could drastically lower the tendency of the concerned amino acid to suffer from this kind of ring-closure side reaction. It has already been verified that glutarimide formation is much less favored than the analogous aspartimide formation in identical conditions. Alternatively, peptide *N*-glycosylation could be conducted on a peptide segment in which Asp is located on the *C*-terminus. Aspartimide formation would not affect the concerned peptide segment since no -Asp-Xaa- amide bond that is imperative for this side reaction is available under such circumstances. The preglycosylated peptide segment bearing glycosylated Asn derivatives on *C*-terminus could be subsequently condensed to the corresponding complementary *C*-terminal component to construct the target *N*-glycopeptide. The convergent segment condensation scheme demonstrated in Fig. 6.5 is basically only applicable for the sugar units with less significant steric hindrances since bulkier oligosaccharides could considerably impede the efficient condensation to the sterically hindered Asp-containing peptides.[68] It is therefore recommended under such circumstances to resort to the preformed *N*-glycosylated Asn derivatives as building blocks for the assembly of the target *N*-glycopeptide since this synthetic strategy bypasses the potentially problematic posttranslational glycosylation step. This methodology simultaneously addresses two inherent drawbacks correlated with the convergent segment condensations strategy in that the sluggish coupling reaction could be accelerated and the mandatory activation on Asp carboxyl-side chain is avoided. Utilization of preformed glycosylated Asn has been verified to be feasible and successful for the synthesis of a plethora of *N*-glycopeptides bearing distinctive large sugar units.[69]

6.1.2.7 Methanolysis of Aspartimide

It has been elucidated that aspartimide intermediate generate in the course of peptide synthesis could result in the formation of impurities such as D-α-Asp- and/or L/D-iso-Asp peptides as depicted in Fig. 6.1. The chromatographic properties of these derivatives could be very similar to that of the target product, which inevitably imposes serious challenges to the process of product purification. This intrinsic problem caused by aspartimide formation could be potentially alleviated via methanolysis treatment of the aspartimide-affected product, transforming by this means the cyclic aspartimide derivatives into the corresponding Asp methyl esters. Titration of peptide product contaminated with aspartimide impurities by 2% (v/v) DIEA methanol solution could quantitatively

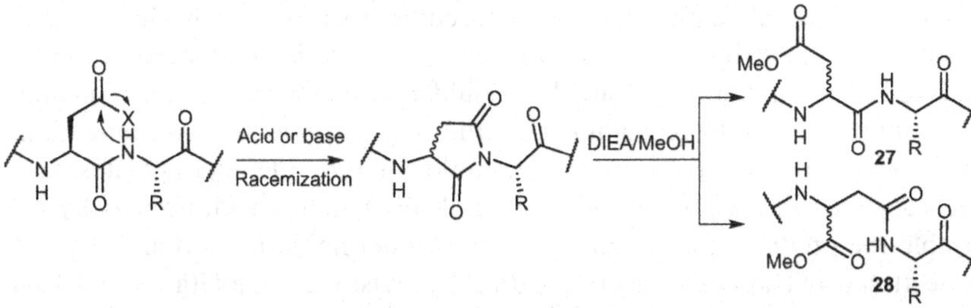

FIGURE 6.7 Methanolysis of aspartimide derivative.

convert the cyclic aspartimide into the corresponding L/D-α-Asp(OMe) **27** and L/D-iso-Asp(OMe) derivative **28** as depicted in Fig. 6.7.[15,22] This methanolysis process could magnify the differences in terms of chromatographic attributes between the aspartimide-derived impurities and the target peptide which would facilitate the subsequent product purification. In spite of the incompetence of this method to suppress aspartimide formation its employment could improve the affected production process since aspartimide-relevant impurities could be removed more effectively and sufficiently in this manner.

The strategy of peptide methanolysis could also find its application in the reaction in-process control. For example, in the synthesis of side-chain to side-chain cyclic peptide **26** depicted in Fig. 6.7, possible aspartimide side products **25** could not be differentiated from target product **26** by mass spectrometry since these two isomers have the same molecular weight. In this case inappropriate conclusion might be drawn out of the misleading MS results. Methanolysis treatment of the product mixture and the detection of possible -Asp(OMe)- derivative is thus helpful to analyze the occurrence of aspartimide formation by mass spectrometry.

6.2 Asn/Gln DEAMIDATION AND OTHER RELEVANT SIDE REACTIONS

6.2.1 Mechanism of Asn/Gln Deamidation

Asn/Gln-containing peptides are frequently involved in deamidation side reactions in which their amide side chains are converted into the corresponding carboxylates, reflected by a molecular weight increase by 1 amu. The deamidation process transforms the neutral Asn/Gln amide side chains into negatively charged carboxylates in physiological conditions and affects potentially the conformations of the addressed peptides/proteins and their stabilities.[70,71] It was originally assumed that Asn/Gln deamidation was simply a consequence of Asn/Gln direct acid- or base-catalyzed hydrolysis that releases a molecule of NH_3. However, in view of the observation that this "deamidation" process was frequently accompanied by the configuration change of C^{α} on the affected

Asn/Gln the actual deamidation is supposed to proceed via a mechanism differing from direct hydrolysis. Moreover, isoAsp derivatives were normally detected in the mixture of Asn/Gln deamidation reaction. This phenomenon motivated chemists to reexamine the true origins of Asn/Gln deamidation. In abundant investigations cyclic aspartimide/glutarimide intermediates have been detected in the process of Asn/Gln deamidation, which principally resemble the products formed in aspartimide/glutarimide formation described in Section 6.1. These findings rationalize the revision of Asn/Gln deamidation mechanism.

Since kinetics of Asn deamidation is normally faster than that of Gln, most investigations to this end have been focused on Asn derivatives. The proposed mechanism of deamidation of the Asn-containing peptide is described in Fig. 6.8.[72,73] The amide backbone on the -Asn-Xaa- unit could undergo deprotonation in basic condition, rendering the backbone amide into an anionic nucleophile which subsequently attacks the amide substituent on Asn side chain. This process gives rise to the formation of the tetrahedral intermediate **29** which is subsequently converted to aspartimide derivative **30** upon releasing a NH_3 molecule. Similar to the aspartimide intermediate derived from Asp ring closure process, Asn-induced aspartimide intermediate **30** is also susceptible to racemization on its C^{α} due to the accelerated proton exchange of H^{α}, giving rise to the formation of racemate **31**. This imide derivative is evidently vulnerable to nucleophilic attack by water. Ring opening adopting Path A generates L/D-Asp derivative **32** whereas L/D-iso-Asp **33** is

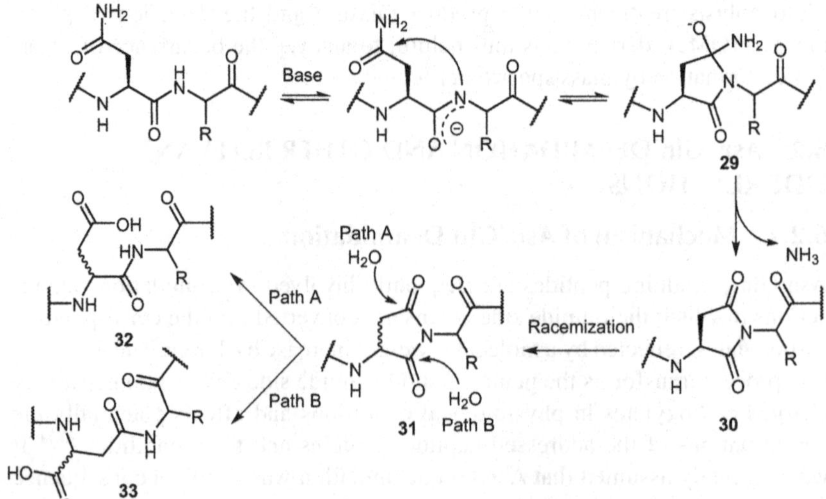

FIGURE 6.8 Proposed mechanism of Asn deamidation and subsequent transformation into Asp/iso-Asp.

formed when Path B is followed. This mechanism is principally in line with the observation that Asn/Gln deamidation process is always accompanied by the concurrent racemization and isoAsp formation. Generally, isoAsp derivatives **33** are more prone to be formed than its Asp counterparts **32** in spite of some exceptions.[74] The process of aspartimide ring opening has been verified to be pH dependent.[75]

Although most Asn/Gln deamidation processes follow the route via aspartimide intermediate, possibility of direct hydrolysis of Asn/Gln to their Asp/Glu counterparts should not be excluded. For instance aspartimide formation from -Asn-Pro-sequence is rare due to the secondary amine property of Pro. Under such circumstances direct hydrolysis of the Asn amide side chain could be realized. Alternatively, -Asn-Pro- might undergo fragmentation process resembling -Asp-Pro- acidolytic degradation (Section 1.4). This phenomenon could be attributed to the possible formation of aspartimide intermediate from -Asn-Pro- in the manner of quaternary amide cation, which is extremely unstable and readily susceptible to hydrolysis, leading to the fragmentation of -Asn-Pro- and the consequent formation of two segments with Asn as the *C*-terminus on one fragment and Pro as the *N*-terminus from the other.[75] It is to note that direct hydrolysis of Gln to Glu could take place more frequently than that of Asn[76] duc to the fact the formation of six-member ring glutarimide is less favored than aspartimide, facilitating by this means the competitive direct hydrolysis of Gln side chain. A sound proof of direct Asn hydrolysis is that the content of isoAsp derivatives formed in this manner is drastically lower than that derived via the formation of aspartimide intermediate and the subsequent aspartimide hydrolysis.[71]

Another relevant side reaction affecting Asn/Gln amide side chain is basically analogous to the Asn/Gln deamidation process. This side reaction does not result in the transformation of Asn/Gln to the corresponding Asp/Glu, while it results in the cleavage of the -Asn-Xaa- backbone amide bond. It has been detected that peptides bearing -Asn-Pro- or -Asn-Leu- unit[75] are susceptible to intramolecular cyclization and subsequent fragmentation will be incubated at 100°C and under pH 7.4. The proposed mechanism of this process is illustrated in Fig. 6.9. The amide substituent on Asn side chain initiates a nucleophilic attack on the backbone amide that results in the peptide fragmentation and the formation of segment **34** with aspartimide located on the *C*-terminus and segment **35** with Pro/Leu as *N*-terminal residue.

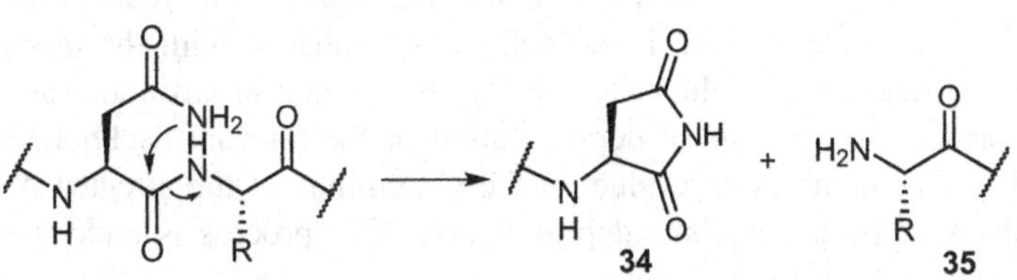

FIGURE 6.9 **Proposed mechanism of fragmentation of peptide containing -Asn-Xaa- unit.**

The aspartimide moiety might undergo further hydrolysis and a deamidation process. This side reaction is evidently favored when Xaa in the addressed sequence -Asn-Xaa- is a secondary amino acid, for example, Pro or the one with a bulky side chain. On the contrary, peptides containing -Asn-Gly- unit are basically exempted from this side reaction but facilitate the aspartimide formation by virtue of the steric advantage of Gly. This phenomenon accounts for the fact Asn residue might be subjected to diverse side reactions with different intrinsic mechanisms in accordance with the corresponding peptide sequences.[75]

6.2.2 Factors Impacting on Asn/Gln Deamidation

6.2.2.1 pH Value

It has been verified that pH value is critical for the patterns of Asn/Gln deamidation[77] in that most Asn/Gln undergoes deamidation via aspartimide/glutarimide intermediate under neutral or basic conditions whereas direct hydrolysis of Asn/Gln amide side chain is more favored in acidic milieu. It has been indicated from relevant studies that the kinetics minimum of the investigated Asn/Gln-peptide deamidation process is basically located in the range of pH 4–6 and the concerned deamidation would be accelerated no matter whether the pH is increased or decreased from this region.[78] The deprotonation process of amide backbone that initiates the aspartimide formation is understandably subjected to pH value. Moreover, the fluctuation of pH value could also impact on the amino acids in proximity which might be involved in Asn deamidation process in the manner of neighboring effects.[73] Generally, high pH value promotes Asn deamidation by intensified deprotonation of the backbone amide which in turn favors the resultant aspartimide formation, whereas low pH induces the deamidation of Asn predominantly by means of direct hydrolysis. It is rational to assert, from the perspective of diverse mechanisms of Asn deamidation, that the rate-determining step under diverse pH ranges could also be different.[72]

6.2.2.2 Peptide Sequence

The inclination of Asn/Gln-peptide to undergo deamidation process could be considerably affected by the corresponding peptide sequence. The amino acid neighboring Asn on its *C*-terminus could impact much more profoundly on its tendency to be subjected to deamidation than the one resided on its N-terminus.[79] This observation is basically in compliance with the mechanism of Asn/Gln deamidation elucidated in Fig. 6.8 in that aspartimide intermediate is formed on the basis of deprotonation of the relevant backbone amide, while it is the amino acid residue on the *C*-terminus of the affected Asn that is involved in the prerequisite deprotonation. This process is understandably interconnected with the electronic attribute and steric effect of the concerned amino acid residue. Concretely, it has been verified by many studies that peptide containing -Asn-Gly- unit is dramatically susceptible to Asn deamidation.[71,75,80] This intrinsic attribute of -Asn-Gly- with regard to Asn deamidation has been rationally employed in protein chemistry for the specific fragmentation of the amide bond between Asn and Gly by hydroxylamine treatment.[81,82] The impact of the residue on Asn *C*-terminus of in terms of Asn/Gln deamidation could also be reflected by its steric effect on shielding the nucleophilic attack of the backbone anionic amide on the subjected Asn/Gln side chain,[79] which is basically analogous to the steric effect of the neighboring residue on Asp-mediated aspartimide formation. The impacts of neighboring groups on Asn deamidation have been investigated by dedicated chemical synthesis of target peptides with various sequences.[83] Different propensities of Asn/Gln deamidation are detected when the neighboring residues to the addressed Asn/Gln are systematically screened. Last but not least, neighboring residues to Asn such as Ser/Thr, His, Lys, Trp, and Asp/Glu might be involved in the process of Asn deamidation by virtue of catalytic effects of hydroxyl, imidazolyl, amino, indolyl, and carboxyl side chains, respectively.[74,79]

6.2.2.3 Peptide Conformation and Other Factors

The secondary peptide structures might interfere with the process of Asn/Gln deamidation. For instance, if the concerned Asn residue in resided in α-helix[84] or β-sheet[85] motif, it could be locked in a microenvironment which is somehow shielded from the potential deamidation mutation, manifested by the drastically decreased deamidation kinetics.

The configuration of the subjected Asn might influence its susceptibility to deamidation as well. It was discovered in a study[75] that peptide H-Val-Tyr-Pro-Asn-Gly-Ala-OH, a segment from adrenocorticotropic hormone (ACTH), would be less susceptible to Asn deamidation when the Asn residue is replaced by D-Asn. This phenomenon could be interpreted from the mechanism of Asn deamidation in that the side chain of Asn is involved in the ring closure process of aspartimide formation as electrophile which might be regarded in most cases as the rate determining step.[72] The conversion of the addressed Asn configuration could therefore reasonably affect the process of Asn deamidation by interfering with the intermediary aspartimide formation step.

Aside from pH value and peptide sequence factors which could affect the Asn/Gln deamidation process, there are further factors including temperature, ionic strength, solvent polarity, viscosity and so forth.[77,86,87] An increase of temperature could potentially accelerate Asn deamidation.[75,88] The addition of organic solvents with low dielectric strength into the peptide aqueous solution could retard the process of Asn deamidation,[89] probably owing to their destabilization effects on the deprotonated backbone amide anion that attacks the side chain of Asn to form an aspartimide intermediate.[90]

Finally, it is noted that deamidation of Asn/Gln could take place not only in the process of peptide synthesis, but also during product purification, storage, and utilization.[88]

6.2.3 Influences of Asn/Gln Deamidation on Peptide Chemical Synthesis

As has been elucidated in the previous chapter Asn and Gln residues could participate in peptide synthesis in the form of side chain-unprotected derivatives. The purpose of such "minimal protection" strategy is to solve the inherent problems associated with the fully protected peptide species in that the latter is frequently sparsely soluble in ordinary organic solvents, which impedes efficient peptide synthesis such as convergent segment condensation. Another advantage of the utilization of side chain-unprotected Asn/Gln is correlated with the common phenomenon in peptide synthesis that the removal of the Trt protecting group from the *N*-terminal Asn/Gln is sometimes stubbornly difficult which might require extreme deprotection conditions.[91] Employment of side chain-unprotected Asn/Gln derivatives as the building block for the relevant peptide synthesis could therefore bypass this intrinsic challenge in terms of sluggish Asn/Gln side chain deprotection.

The free amide side chain on Asn/Gln could, however, provoke many side reactions such as dehydration of Asn/Gln side chain during carboxyl-activation and the resultant formation of β-cyano alanine impurity, which has been described in Section 5.5. This side reaction takes place exclusively on side chain-unprotected Asn/Gln and its occurrence would be suppressed once the referred Asn/Gln is incorporated into the target peptide chain. Asn/Gln residues with free amide side chain could nevertheless be subjected to a deamidation side reaction, which could potentially be haunted in the whole process of peptide synthesis. Contrary to the common consensus, peptide with side chain-unprotected Asn/Gln is normally less soluble but more liable to chain aggregation.[92] This phenomenon could be attributed to the reinforced formation of intermolecular hydrogen bonds facilitated by the amide side substituents on Asn/Gln side chain.[93] In view of the above concerns it is recommended to utilize Asn/Gln building blocks in the form of side chain-protected species for the target peptide synthesis thanks to the improved attributes with respect to the enhanced resistance to deamidation and aggregation.

Asn is normally protected on its amide side chain by Trt in the standard Fmoc/tBu peptide chemistry. Protecting groups on the amide substituent could shield Asn from aspartimide formation to some extent, and this effect is much more profound compared with the preventive effects of Asp side chain-protecting group on aspartimide formation that protected Asp derivatives have to suffer from. Nevertheless, *N*-glycosylated Asn is known to be susceptible to aspartimide formation following the route depicted in Fig. 6.8.[66] One of the most effective strategies to suppress this side reaction, as has been elaborated in Section 6.1.2, is the combination of Asp(OAll) orthogonality and backbone amide protection in the form of -Asp(OAll)-*N*-Hmb-Xaa- precursor. Allyl ester on Asp side chain could be removed selectively to liberate the carboxylate substituent, which could be subsequently derivatized by a glycosylamine compound to

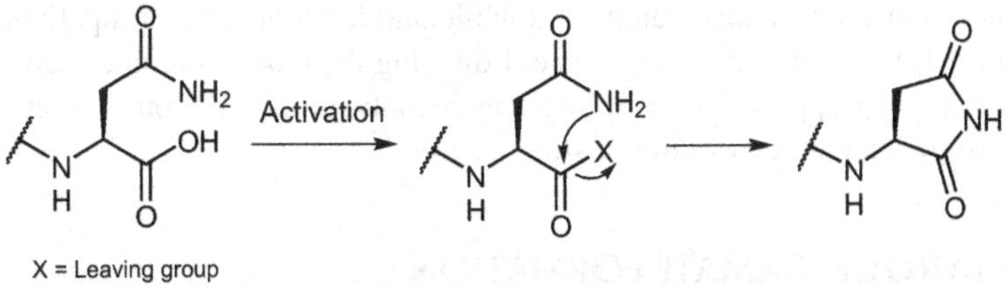

FIGURE 6.10 Aspartimide formation during activation of C-terminal side chain unprotected Asn.

form the corresponding *N*-glycosylated Asn product. The employment of backbone amide protecting groups is adopted in view of the concern that Asp(OAll) is much more prone to suffer aspartimide formation compared with Asp(OtBu). The amide backbone-protecting group could be sufficiently removed at the peptide side chain global deprotection step. Synthesis of *N*-glycosylated peptide following this strategy could principally effectively suppress the potential Asn deamidation side reaction.

Another possible pathway of aspartimide formation is supposed to be described on this occasion. The mechanism is depicted in Fig. 6.10, which is inherently similar to the process of -Asn-Xaa- fragmentation displayed in Fig. 6.9. This side reaction takes place during the activation of side chain-unprotected Asn or peptide bearing unprotected Asn at the *C*-terminus. The unshielded amide side chain on Asn could induce a nucleophilic attack on the activated α-carboxyl group and form a *C*-terminal aspartimide residue as a consequence which could be subject to a further hydrolysis–deamidation process.

Aspartimide formation by this means could take place not only during the activation of side chain-unprotected Asn,[94] but also upon the carboxyl-activation of *N*-glycosylated Asn building block.[17] No impurities would be assembled into the target peptide product when the corresponding Asn derivative is subjected to this kind of side reaction at the step of amino acid activation in SPPS (the virtual function between peptide N^{α}-group that is supposed to be acylated and the formed Aspartimide derivative is not taken into account herein), as the formed unconverted aspartimide derivatives will be removed from the reaction system simply by resin rinsing. However, if this side reaction takes place during the convergent segment condensation in solution phase and the unprotected Asn happens to be located on the *C*-terminus of the *N*-component, its transformation into aspartimide upon carboxyl-activation could terminate the desired segment condensation since aspartimide is much less reactive toward N^{α}-group compared with the corresponding activated carboxylate. The occurrence of such side reaction would lower the effective equivalent of the Asn-bearing component and might lead to the incomplete segment condensation. A stimulative factor in favor of this side reaction will be the sluggish target reaction between the activated Asn and the reciprocal amino components that inevitably leaves the activated Asn species unacylated for a while and facilitates the competitive aspartimide formation as a consequence. Lowering the reaction temperature and/or utilization of appropriate coupling reagent/solvent system could potentially decrease the extent of this kind of side reaction.[17]

6.3 PYROGLUTAMATE FORMATION

Pyroglutamate derivatives play important roles as chiral building blocks in organic synthesis.[95] They could also be found on the *N*-terminus of some naturally occurring proteins.[96] Pyroglutamate originates from the lactam formation of *N*-terminal Gln or Glu in proteins (Fig. 6.11). Gln is more susceptible than Glu to accommodate pyroglutamate formation,[97] which impose serious side reactions on peptide synthesis under certain circumstances, terminating the growing peptide chain as a consequence.[98,99]

This side reaction was first detected in Boc/Bzl peptide chemistry.[100] When the subjected peptide chain elongation reaches the Gln residue and its N^{α}-Boc protecting group is removed by acidolysis,[101,102] the liberated N^{α}-group could initiate a nucleophilic attack on the amide substituent on Gln side chain, regardless of whether the affected Gln residue is protected on its side chain or not. The nucleophilic addition will result in the formation of a five-member ring pyroglutamate derivative. It is noted that this process could take place in acidic environment despite the fact that pyroglutamate formation is initiated by N^{α}-group on peptide.

The focus of the investigation with regard to suppress pyroglutamate formation was initially centered on the step of N^{α}-Boc acidolysis[103,104] as well as the subsequent N^{α}-neutralization after the Boc deblocking.[105] It was reported that DMF-TFA adduct,[106] which was formed during resin washing by DMF after TFA-mediated N^{α}-Boc acidolysis, is one of the major incentives for pyroglutamate formation.[107] It was therefore advised to wash the resin with DCM both before and after N^{α}-Boc acidolysis so as to replace the residual DMF remaining in the peptidyl resin to prevent the formation of DMF-TFA adduct that induces pyroglutamate formation. The following in-depth study has, however, identified that pyroglutamate formation occurs predominantly at the step of a coupling

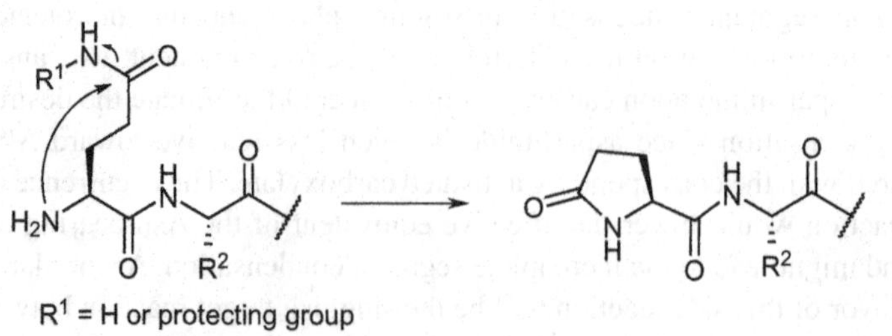

FIGURE 6.11 Pyroglutamate formation.

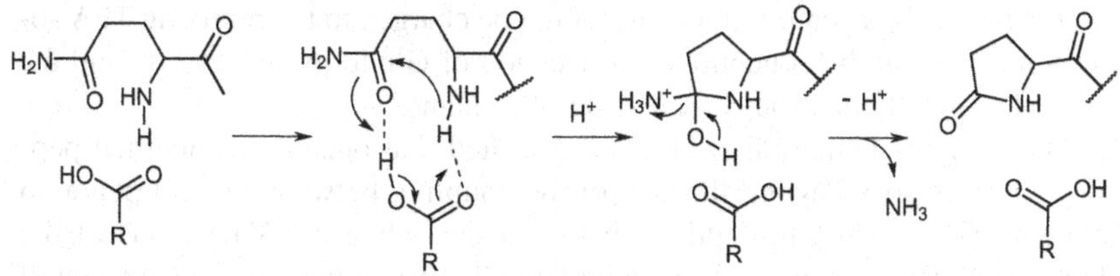

FIGURE 6.12 Proposed mechanism of acid-catalyzed pyroglutamate formation.

reaction between the incoming amino acid and the peptide bearing Gln residue on the *N*-terminus. The extent of pyroglutamate formation is apparently correlated with the kinetics of the concerned coupling reaction.[108] Pyroglutamate formation is known to be catalyzed by weak acids.[109–111] The possible mechanism of pyroglutamate formation is proposed in Fig. 6.12. This finding is actually in accordance with the observation that the acidolytic deblocking of the Boc group from N^{α}-Boc-Gln(X)-peptide could induce pyroglutamate as a side product. It has also been reported that amide derivatives could be subjected to the acid-catalyzed aminolysis process,[112] which shares the underlying principle with pyroglutamate formation.

In light of the findings from Dicmarchi et al. peptide bearing a side chain unprotected Gln at its *N*-terminus could be readily converted to the pyroglutamate counterpart in a short time upon treatment by weakly acidic HOBt or Boc-Ile-OH solution, whereas a stronger acid like HCl/Dioxane failed to induce pyroglutamate formation in otherwise identical conditions.[109] TFA/DCM solution is capable of transforming *N*-terminal Gln into pyroglutamate, but highly concentrated TFA solution (50%) will result in less extent of pyroglutamate formation relative to diluted TFA solution (0.5%). *N*-terminal Gln is seemingly more stable in basic milieu since neither DIEA- nor piperidine-treatment of the *N*-terminal-Gln peptide resulted in evident pyroglutamate formation. Moreover, addition of 0.02 M DIEA into the Boc-Ile-OH stressing solution decreased the pyroglutamate formation compared with the blank Boc-Ile-OH solution devoid of DIEA. These results basically sustain the conclusion that pyroglutamate formation is catalyzed by weak acid and are in compliance with the observation that pyroglutamate side product is predominantly generated during the coupling of the incoming amino acid with the *N*-terminal-Gln peptide. It has been reported that reverse phase chromatographic purification of *N*-terminal-Gln peptide with TFA as organic modifier in the eluents might be detrimental to the integrity of the product due to the occurrence of the pyroglutamate formation.[110] This observation verifies once more the conclusion that pyroglutamate formation is catalyzed by acid.

Interestingly, the intentional preparation of a peptide with pyroglutamate residue as its *N*-terminus could be realized by the incorporation of a Gln as its pyroglutamate synthon.[113] The precursor peptide bearing *N*-terminal-Gln is subsequently subject to the treatment by 4 M acetic acid/Dioxane solution at 40°C, converting the subjected Gln by this means quantitatively into pyroglutamate.

Alternatively, *N*-terminal-Gln peptide could be charged to the refluxing TFA solution to facilitate the concomitant conversion of Gln to pyroglutamate and the removal of protecting groups on peptide side chains.

The study of Dimarchi et al. also identified that once the concerned peptide synthesis goes beyond the step of the coupling between incoming amino acid and *N*-terminal-Gln peptide, that is to say, when the *N*-terminus of the elongating peptide is no more occupied by Gln, pyroglutamate formation will principally be quenched.[110] The nonaccumulative property of pyroglutamate formation complies essentially with the proposed mechanism elucidated in Fig. 6.11 in that it is the transient N^{α}-group on Gln instead of the backbone amide that initiates the ring closure of the addressed *N*-terminal Gln, whereas aspartimide/glutarimide formation is induced by the nucleophilic attack of the backbone amide that endows these side reactions with the accumulative features. Other factors affecting pyroglutamate formation include solvent,[109] temperature, pressure[111] as well as peptide sequences.[114]

Since pyroglutamate formation takes place predominantly during the process of acylation of the *N*-terminal-Gln peptide, it is reasonable to focus on this critical step as to minimize this side reaction. Acceleration of the subjected acylation of the *N*-terminal Gln could basically diminish pyroglutamate formation by converting the nucleophilic N^{α}-group on Gln to the target amide derivative that is inert to pyroglutamate formation. Utilization of highly effective coupling reagents and/or employment of large excess of the incoming amino acids could be adopted in an effort to accelerate the concerned Gln acylation. Alternatively, transformation of the incoming amino acid into its symmetric anhydride derivative as acylating agent could minimize the function between the *N*-terminal Asn and weak acid like Fmoc-Xaa-OH itself. Acid-catalyzed pyroglutamate formation could therefore be restrained. Preactivation of the addressed amino acid to its active ester might suppress this side reaction in a similar manner. It is noted that acidic *N*-hydroxylamine derivatives such as HOBt are frequently employed as additives for the amino acid coupling reactions, it is therefore imperative to fine-tune their concentrations if pyroglutamate formation posed a significant concern in such conditions. The polarity of the solvent in which pyroglutamate might occur plays a crucial role in the process of ring closure of the *N*-terminal Gln. It is advisable to adopt DMF instead of DCM as the solvent for Gln acylation reaction in an effort to minimize pyroglutamate formation.[109] This recommendation is made due to the increased kinetics of Gln acylation in DMF compared with that in DCM, shortening by this means the transient existence of the vulnerable unacylated *N*-terminal Gln in the reaction system. In addition, treatment of *N*-terminal-Gln peptides in the relevant processes such as chromatographic purification or lyophilization is supposed to be conducted with utmost caution. Strict control of pH value is recommended if pyroglutamate formation poses a serious challenge. The protection of Gln side chain could significantly reduce pyroglutamate formation[115] in spite of its incompetence to thoroughly suppress this side reaction.

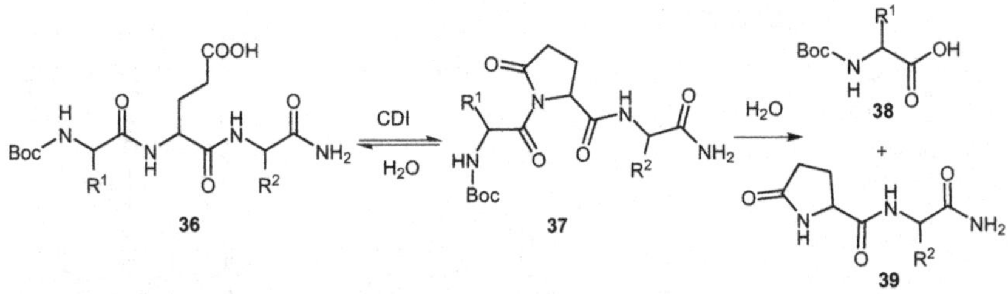

FIGURE 6.13 Formation of endo-pyroglutamate and its subsequent hydrolysis.

It is to note that although it has been indicated that pyroglutamate formation will discontinue once the affected Gln is no more located on the *N*-terminus of peptide, it does not necessarily mean that pyroglutamate reside could not reside in the central position of a peptide sequence. As indicated in Fig. 6.13 when an endo-Glu residue in peptide **36** is activated by CDI, its γ-carboxylate substituent in the form of an active ester could function with the backbone amide to form the corresponding endo-pyroglutamate derivative **37**,[116] which differs from the process of glutarimide formation as the latter is formed by the nucleophilic attack of the amide from the *C*-terminal residue neighboring the affected Gln. This structure is generally instable and might undergo fragmentation to give rise to a peptide fragment **39** bearing pyroglutamate residue as *N*-terminus and the reciprocal degradative fragment **38**. Under certain circumstances endopyroglutamate derivative **37** could also be reversed to its original endoGlu derivative **36** by hydrolysis, but the possibility of this conversion is much trivial compared with the mentioned fragmentation process. Furthermore, treatment of endoGlu-peptides with HF could also induce the cyclization, pyroglutamate formation[117] as well as the subsequent hydrolysis. Fragmentation of peptides consisting endoGlu residue might take place during storage, which is likewise attributed to the formation of endopyroglutamate intermediate.[118]

6.4 HYDANTOIN FORMATION

By virtue of the profound applications for structure-activity relationship (SAR) hydantoin derivatives (Fig. 6.14) have been intensively studied in the territories like organic synthesis and pharmaceuticals development.[119] Nevertheless, the formation of hydantoin might adversely impact on the peptide

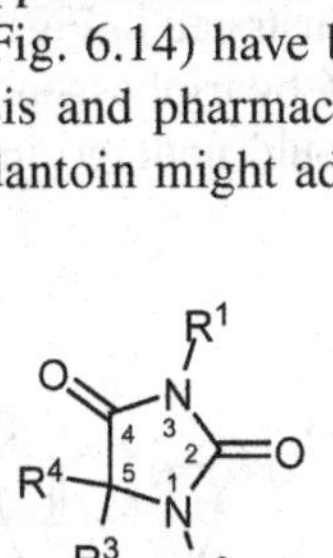

FIGURE 6.14 Structure of hydantoin derivative.

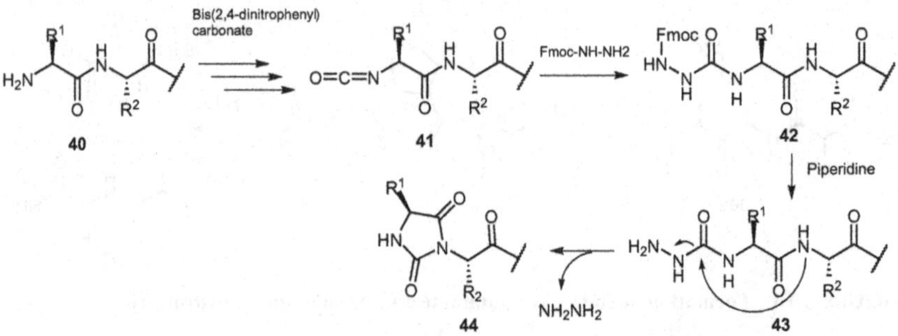

FIGURE 6.15 **Hydantoin formation during the preparation of Aza-peptide.**

syntheses as a common side reaction.[120,121] The underlying origin of hydantoin formation in peptide synthesis is somewhat similar to that of the base-catalyzed aspartimide formation. Hydantoin formation is principally triggered by the deprotonation of amide or urea functionality by base and the resultant anion could initiate an intramolecular nucleophilic attack on the carbonyl functionalities from ester, amide, urea, carbamate or hydrazide moieties that bear sufficient electrophilicity. This process could result in the formation of a five-member ring hydantoin derivative.

Hydantoin formation could readily take place in the preparative process of a plethora of peptidomimetics. The formation of hydantoin side product during solid phase Aza-peptide synthesis is schematized in Fig. 6.15. Peptide **40** is firstly transformed into the corresponding isocyanate derivative **41** by bis(2,4-dinitrophenyl) carbonate. Isocyanate **41** reacts subsequently with Fmoc-NH-NH_2 to give rise to Fmoc-Aza-peptide **42**, which is supposed to undergo N^α-Fmoc removal by piperidine and be converted to the target Aza-peptide **43**. The stability of this Aza-peptide is nevertheless low since its *N*-terminal hydrazinecarboxamide moiety could be subjected to the nucleophilic attack from the backbone amide and be transformed into hydantoin derivative **44** upon releasing a hydrazine molecule.

Hydantoin formation could also be formed at the intermediary step of active *N*-isocyanate peptide[122,123] as illustrated in Fig. 6.16. Peptide **45** is converted to the corresponding isocyanate **46** by triphosgene/DIEA treatment. Highly reactive isocyanate derivative **46** could undergo intramolecular cyclization under

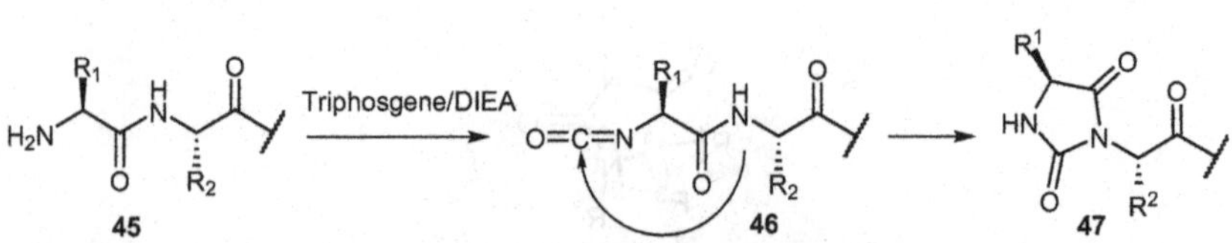

FIGURE 6.16 **Hydantoin formation during the preparation of N-isocyanate peptide.**

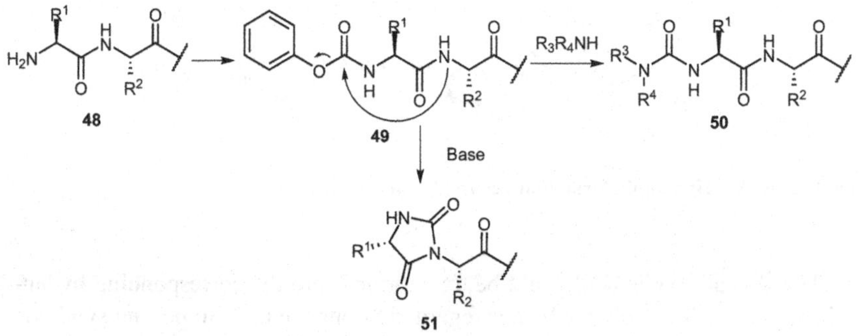

FIGURE 6.17 Hydantoin formation in the process of ureido peptide synthesis.

basic environment and transform into hydantoin derivative **47** even prior to its function with Fmoc-NH-NH_2 in the previous case.

Preparation of ureido peptide might also be accompanied with hydantoin formation as depicted in Fig. 6.17.[122] Ureido peptide is generally obtained from ordinary peptide **48** whose N^{α}-group is firstly activated and converted to carbamate **49** (or alternatively be transformed into isocyanate by phosgene or triphosgene, referring to Fig. 6.16). Carbamate **49** reacts subsequently with amine derivative to give rise to the target ureido peptide **50**. The formation of active intermediate *N*-carbamate **49** might, however, lead the reaction astray to the corresponding hydantoin derivative **51** via the intramolecular nucleophilic attack of backbone amide on the carbamate moiety and the release of a hydroxyl compound.

It is indicated by Fig. 6.17 that the preparation of ureido peptide with the strategy of aminolysis of *N*-carbamate peptide might be accompanied with the formation of hydantoin side product. The origin of this side reaction lies in the reactivity of *N*-terminal carbamate towards the backbone amide and the drive to form a sterically favored five-member ring derivative. Factors affecting this process include solvent, base, steric effect and so on. It has been indicated that DMF facilitates hydantoin formation more than THF.[124] Target ureido peptide could be obtained with a considerably enhanced purity in THF than in DMF. From the aspect of the amine derivative as an aminolysis agent it is predicted that the increase of the nucleophilicity and decrease of the steric hindrance will be advantageous to ureido peptide synthesis. On the contrary, hydantoin formation will be favored if the incoming amine is with exceptionally high basicity and/or significant steric hindrance. In another word, the role of the amine in the latter case is more like a base than a nucleophile so that the target aminolysis of *N*-carbamate will be impeded whereas the competitive hydantoin formation is facilitated. Some extreme cases indicate that *N*-carbamate peptide could be almost quantitatively converted to the corresponding hydantoin catalyzed by tertiary amines such as DIEA or DBU in DMF.[124]

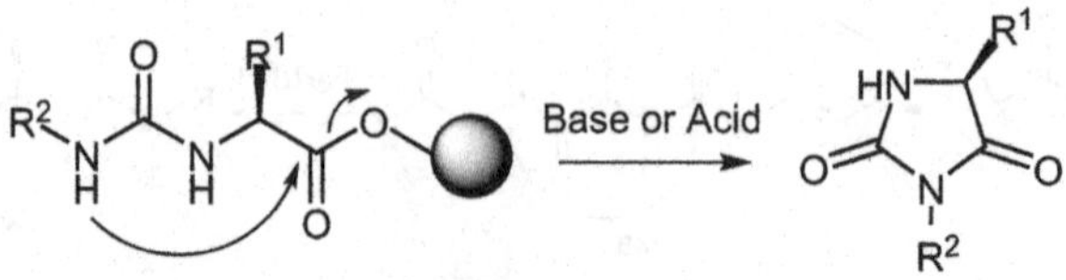

FIGURE 6.18 Hydantoin formation via ureido carboxylic acid.

Ureido carboxylic acid could be transformed into the corresponding hydantoin derivative (Fig. 6.18), which is regarded as one the most important synthetic strategies to produce hydantoin compounds.[125] This process could be realized in both basic[126] and acidic[125,127,128] environment. Similarly, solid phase synthesis of ureido peptide is also susceptible to hydantoin formation. Moreover, protected peptide precursor is supposed to be subject to TFA-mediated side chain global deprotection in which hydantoin formation might be provoked as well.[127]

In some peptide syntheses that adopt N^{α}-Boc/Bzl synthetic strategy the backbone carboxyl groups of the peptide intermediates are temporarily protected in the form of esters, which are supposed to be removed by base hydrolysis at the end of the synthesis.[129] This procedure could, nevertheless, give rise to the formation of hydantoin side product. It is to note that hydantoin formed in this manner could take place to both N^{α}-Z protected[130] and N^{α}-Boc protected peptides.[131] This process is reflected in the synthesis of Methionine-Enkephalin as depicted in Fig. 6.19.[131] The fully protected Methionine-Enkephalin **52** whose N^{α} and backbone carboxyl group is masked by Boc and methyl ester, respectively, is subjected to hydrolysis in NaOH solution in an effort to regenerate the *C*-terminal carboxyl group. However, the *N*-terminal Boc-Tyr(OBzl)-Gly- moiety might be suffered from a ring closure at the saponification step, forming a five-member ring hydantoin derivative **53** which could be further degraded to urea derivative **54** via hydrolysis. Hydantoin formation in this case is apparently pH dependent since the decrease of the pH value could restrain the hydantoin

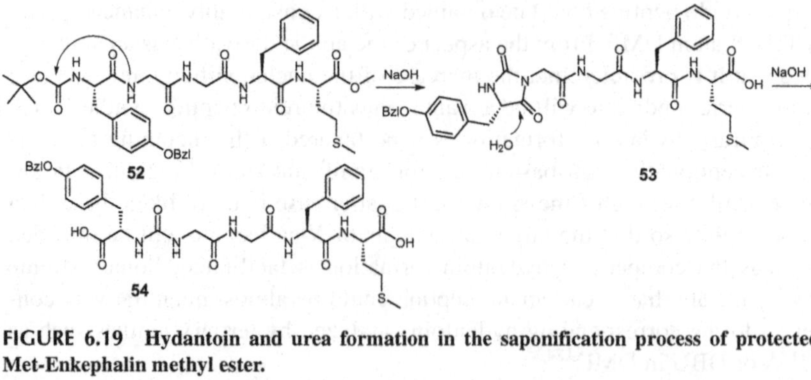

FIGURE 6.19 Hydantoin and urea formation in the saponification process of protected Met-Enkephalin methyl ester.

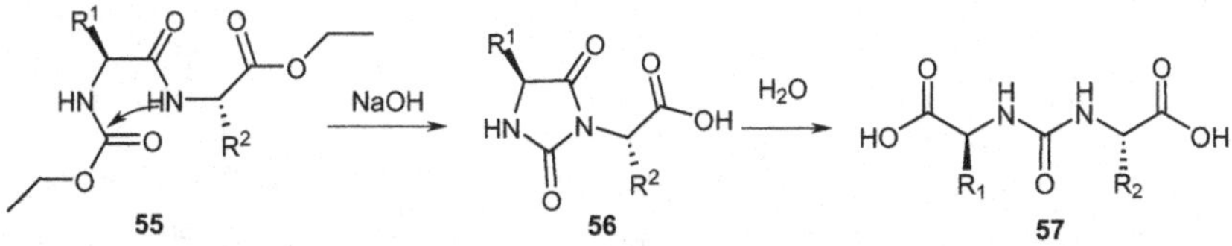

55 56 57

FIGURE 6.20 **Formation of hydantoin and urea from N^{α}-ethoxycarbonyl dipeptide ethyl ester.**

formation and impede the methyl ester saponification. The most effective solution to suppress hydantoin formation under such circumstances is to remove the N^{α}-Boc protecting group by acidolysis prior to NaOH-mediated saponification since no appropriate substrate could be available to hydantoin formation on peptide *N*-terminus in the absence of N^{α}-Boc protecting group.

The structures of the concerned peptides are crucial for their inclinations to undergo hydantoin formation. For instance, *N*-ethoxycarbonyl protected peptide ethyl ester **55** could be converted to hydantoin **56** upon base treatment, which subsequently undergoes hydrolysis to form urea derivative **57** (Fig. 6.20).[56,132] Although this process could occur to ordinary peptides such as N^{α}-Z protected peptide esters at the saponification step, hydantoin formation in such conditions is generally trivial. However, the extent of this side reaction could be strikingly augmented when $R^2 = H$. In another word, when the second amino acid from the *N*-terminal of the addressed peptide is Gly.[38,133] Hydantoin formation presents another example of a Gly-favored side reaction. The apparent readiness of hydantoin formation on Gly is majorly due to its advantageous steric effect, which considerably facilitates the ring closure process. This is in principle analogous to the evident susceptibility of -Asp-Gly- sequence to aspartimide formation elucidated in Section 6.1 and -Asn-Gly- to deamidation described in Section 6.2.

Preparation of polypeptide with NCA as building block might induce the formation of hydantoin derivative as well. The scheme of this process is displayed in Fig. 6.21. Arg(Pbf)-NCA **58** is employed as starting material for the synthesis of H-Arg(Pbf)$_n$-OH **62**. It firstly undergoes hydrolysis in basic condition to form *N*-carboxyl-Arg(Pbf)-OH **59** which is subjected to decarboxylation in acidic environment and neutralization to give H-Arg(Pbf)-OH **60**. Compound **60** could attack another Arg(Pbf)-NCA **58** under the basic condition to generate a *N*-carboxyl dipeptide **61** which sequentially initiates the polymerization process by this means to generate polypeptide H-Arg(Pbf)$_n$-OH **62** mixtures with diverse degree of polymerisation (DP). This seemingly smooth polymerization process could nevertheless be terminated by the occurrence of hydantoin formation when *N*-carboxyl dipeptide **61** deviates from the default "decarboxylation-nucleophilic addition" pathway at high pH value whereas an intramolecular cyclization of **61** takes place instead to form the hydantoin derivative **63** and terminates the target polymerization

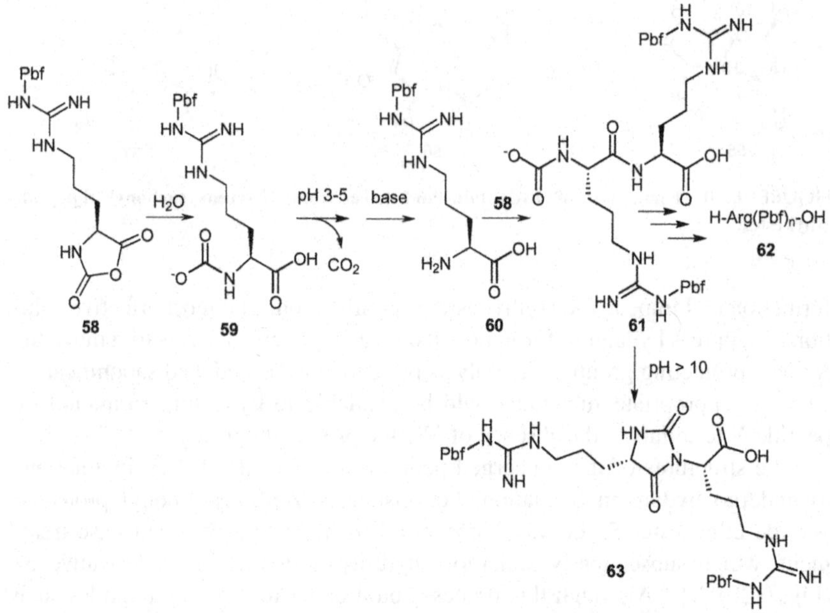

FIGURE 6.21 Hydantoin formation during polymerization of Arg(Pbf)-NCA.

process (Yang, Y., unpublished results). The occurrence of hydantoin formation by this means is evidently pH dependent in that an increase of the pH value would exaggerate the hydantoin side reaction.

Besides the above situations in which hydantoin derivatives are generated at different synthetic steps, some labile substituents could be spontaneously degraded to hydantoin derivative via an intramolecular rearrangement process, which could not be detected by means of mass spectrometry. Residue -Aph^5(L-Hor)- in degarelix, for instance, might be transformed into hydantoin isomer through the rearrangement on its dihydroorotamide substituent under basic condition,[134] the process thereof is displayed in Fig. 6.22.

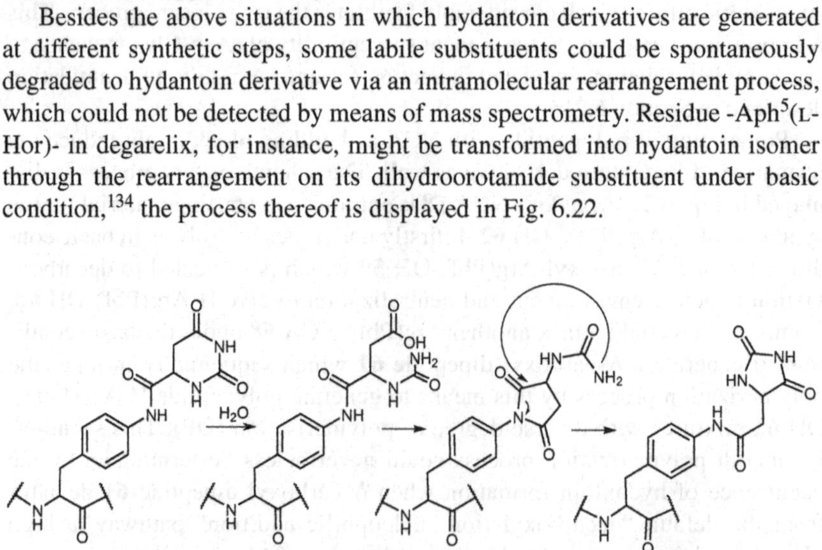

FIGURE 6.22 Rearrangement of Aph(L-Hor) to hydantoin derivative.

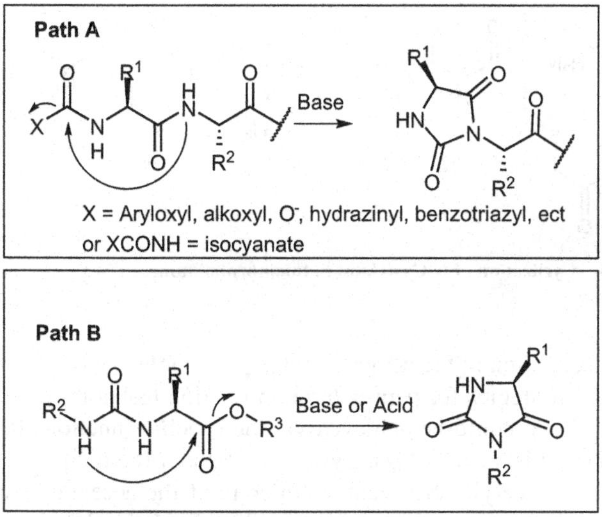

FIGURE 6.23 General routes of hydantoin formation in peptide synthesis.

In summary, most hydantoin side reactions in peptide synthesis could be roughly categorized into two classes from the aspects of reaction mechanism depicted in Fig. 6.23: (Path A) through nucleophilic attack of the backbone amide nitrogen (generally the nitrogen from the second amino acid from *N*-terminus) on the active *N*-terminal isocyanate or carbamoyl functionality such as benzyl carbamate, phenyl carbamate, ethyl carbamate, hydrazinecarboxamide, benzotriazole-1-carboxamide[135] under basic conditions; (Path B) through nucleophilic attack of nitrogen from the *N*-terminal urea unit on the backbone ester bond, or sometimes on an imide bond like the case of the hydantoin formation from Asp(L-Hor). This route is adopted meanwhile as one of the major strategies for the solid phase synthesis of hydantoin derivatives.

6.5 SIDE REACTIONS ON *N*-TERMINAL Cys(Cam) AND *N*-BROMOACETYLATED PEPTIDE

Certain peptides with an *N*-terminal Cys derivative might be susceptible to intramolecular cyclization side reactions in suitable conditions. The cyclization process that affects the *N*-terminal Cys(Cam) residue is described in Fig. 6.24. Cys(Cam) is normally derived from Cys by iodoacetamide treatment. It could undergo an intramolecular cyclization in 0.1 M NH_4HCO_3 solution or other weak base solution, giving rise to the thiomorpholinone derivative by releasing ammonia.[136] The molecular weight of this side product is decreased by 17 amu relative to that of the target Cys(Cam) product. This process takes place frequently in proteomic analysis.

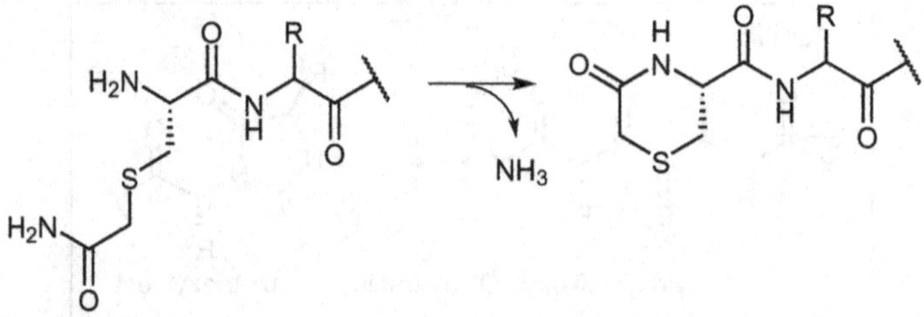

FIGURE 6.24 Cyclization of N-Cys(Cam) to thiomorpholinone.

Bromoacetylation of the N^{α}-group on peptide molecule is one of the regularly adopted strategies for peptide/protein ligation technology. Conjugation of peptide/protein could be achieved via the specific functions between reactive amino acids such as Cys, Lys, Met, His and the reciprocal fragment bearing a bromoacetyl substituent.[137] In spite of the acceptable stability of *N*-bromoacetyl moiety at the step of peptide side chain global deprotection both in Fmoc[138] and Boc[139] chemistry, it could react in neutral condition with the sulfhydryl group on cysteine side chain.[140] This process is displayed in Fig. 6.25 that the Cys side chain sulfhydryl substituent could initiate a nucleophilic attack at around neutral pH (6–8) on the *N*-terminal bromoacetyl group to form a ring structure via a thioether bond. Peptide dimerization would also be induced provided that this process occurs in an intermolecular manner. Although the nucleophilicity of sulfhydryl is drastically reduced in an acidic milieu, it is notwithstanding noted that mercaptan-type scavengers should be excluded from the reaction system for a side chain global deprotection reaction of bromoacetylated peptide. In spite of this concern, it has been reported that the addition of EDT to TFA solution for the side chain global deprotection of *N*-bromoacetylated peptide would not affect the integrity of the target product.[141]

The target ligation of bromoacetyl group with Lys-N^{ε} or N^{α} could be realized at pH > 9,[142] giving rise to the formation of a resultant secondary amine (monocarbamoylmethyl lysine derivative) or tertiary amine (dicarbamoylmethyl lysine derivative) as appropriate.[143] The imidazolyl substituent on His side chain

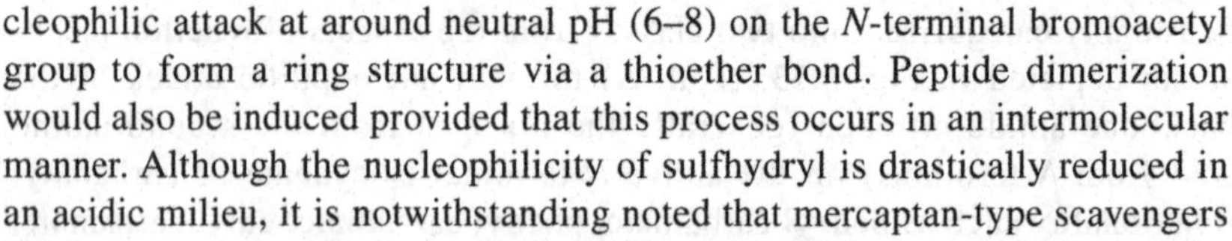

FIGURE 6.25 Cyclization side reaction between N-terminal bromoacetyl and sulfhydryl on Cys side chain.

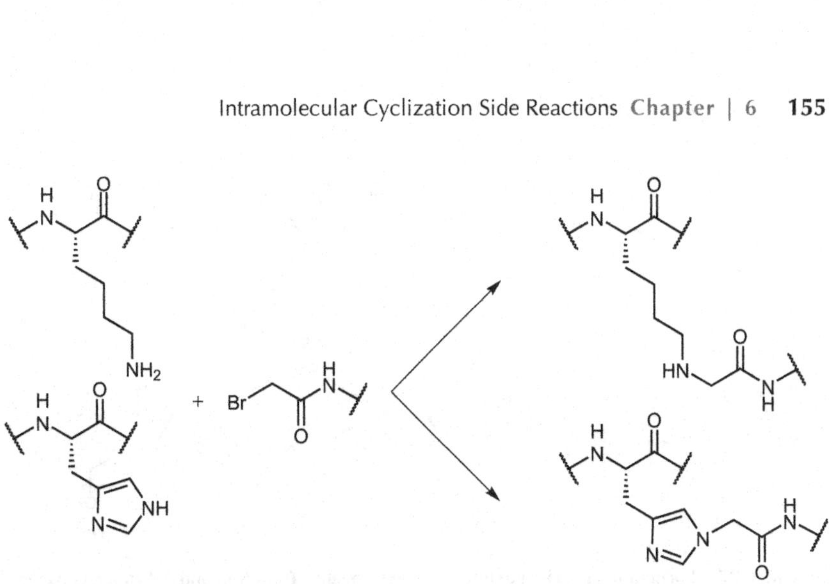

FIGURE 6.26 **Mono-substitution on Lys and His by bromoacetylated derivative.**

is also reactive towards bromoacetyl group to form either mono-substituted or di-substituted conjugate.[144] The process of monosubstitution on Lys and His is exhibited in Fig. 6.26.

The modification of Lys/His/Cys by bromoacetyl group occurs predominantly in basic conditions. In view of the fact that both side chain global deprotection of *N*-bromoacetylated peptide and subsequent reverse phase chromatographic purification is conducted in acidic milieu, the occurrence of the mismatched conjugation could be minimized by meticulous control of the pH values of the corresponding synthesis and purification processes. In spite of the tolerability of bromoacetyl group to Lys/His/Cys in the process of peptide synthesis side reactions could be provoked once Met residue is located in the parental *N*-bromoacetylated peptide due to the intrinsic features of the thioether substituent on Met. The nucleophilicity of thioether moiety on Met could be maintained within a broad pH spectrum. Met is able to rapidly react with many electrophiles at pH $>$ 1.7.[142,145] For instance, Met could be methylated by anisole in the presence strong acids like TFA or TFMSA.[146] As depicted in Fig. 6.27 the thioether moiety on Met could react with *N*-terminal bromoacetyl in the process of TFA- or HF-mediated peptide side chain global deprotection and give rise to carbamoylmethyl Met derivative **64** as a consequence. This cationic intermediate is unstable and might be subsequently subjected to decomposition or rearrangement in a plethora of manners. The side reactions affecting *N*-bromoacetylated peptide by Met is the most complicated ones compared with those induced by other amino acid residues. Of all the possible degradative substances of carbamoylmethyl Met derivative **64**, the demethylated product *S*-carbamoylmethyl homocysteine derivative **65** stands out as the most stable and dominant one. Carbamoylmethyl Met derivative **64** might also be decomposed to Met **66** or homoserine derivative **67** via hydrolysis.[142]

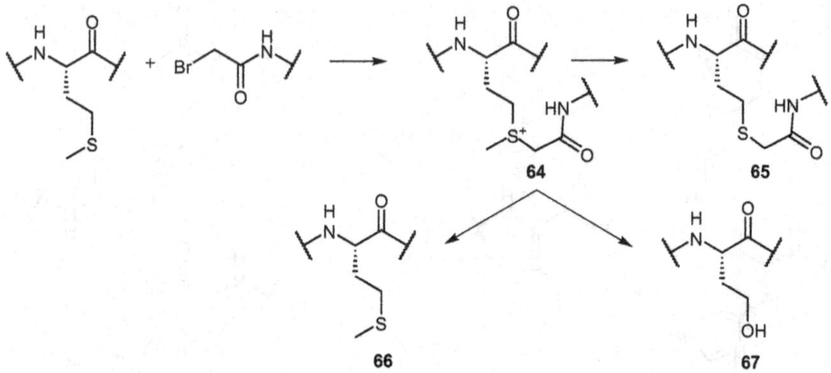

FIGURE 6.27 **Formation of carbamoylmethyl Met derivative from Met and *N*-bromoacetylated peptide and its degradation pathway.**

Aside from the synthetic and purification processes *N*-bromoacetylated peptides might also suffer from similar side reactions during lyophilization and storage, (Yang, Y., unpublished results) compromising the integrity of the target product as a consequence. In view of these challenges all the relevant processes with regard to the preparation of *N*-bromoacetylated peptides should be conducted with the utmost caution, pH control and adjustment in particular. The exclusion of mercaptan and thioether types of scavenger from the side chain global deprotection of *N*-bromoacetylated peptides is highly recommended. Moreover, due to the insufficiency of the systematic information in terms of the stability of *N*-bromoacetylated peptide materials, it is advised to lyophilize these products shielded from the light and the storage temperature is preferably kept below −20°C.

REFERENCES

1. (a) Carlescu I, Osborn HMI, Desbrieres J, Scutaru D, Popa M. *Carbohydr Res.* 2010;345:33–40. (b) Rivera-Fillat MP, Reig F, Martinez EM, Grau-Oliete MR. *J Pept Sci.* 2010;16:315–321. (c) Schon I, Kisfaludy L. *Int J Peptide Protein Res.* 1979;14:485–494. (d) Capasso S, Mazzarella L, Sica F, Zagari A, Cascarano G, Giacovazzo C. *Acta Crystallogr B.* 1992;48:285–290. (e) Capasso S, Mazzarella L, Zagari A. *Chirality.* 1995;7:605–609.
2. Bodanszky M, Tolle JC, Deshmane SS, Bodanszky A. *Int J Peptide Protein Res.* 1978;12: 57–68.
3. Yang Y, Sweeney WV, Schneider K, Thörnqvist S, Chait BT, Tam JP. *Tetrahedron Lett.* 1994;35:9689–9692.
4. Lauer JL, Fields CG, Fields GB. *Lett Pept Sci.* 1994;1:197–205.
5. Ryakhovsky VV, Khachiyan GA, Kosovova NF, Isamiddinova EF, Ivanov AS. *Beilstein J Org Chem.* 2008;4:39.
6. Zhu J, Marchant RE. *J Pept Sci.* 2008;14:690–696.
7. Radkiewicz JL, Zipse H, Clarke S, Houk KN. *J Am Chem Soc.* 1996;118:9148–9155.

8. Kaneshiro CM, Michael K. *Angew Chem Int Ed Engl*. 2006;45:1077–1081.
9. Sandmeier E, Hunziker P, Kunz B, Sack R, Christen P. *Biochem Biophys Res Commun*. 1999;261:578–583.
10. Nicolas E, Pedroso E, Giralt E. *Tetrahedron Lett*. 1989;30:497–500.
11. Bodanszky M, Sigler GF, Bodanszky A. *J Am Chem Soc*. 1973;95:2352–2357.
12. Mitsuyasu N, Waki M, Kato T, Izumiya N. *Mem Fac Sci Kyushu Univ Ser C*. 1970;73:88154.
13. Martinez J, Bodanszky M. *Int J Peptide Protein Res*. 1978;12:277–283.
14. (a) Zahariev S, Guarnaccia C, Pongor CI, Quaroni L, Cemazar M, Pongor S. *Tetrahedron Lett*. 2006;47:4121–4124. (b) Rabanal F, Pastor JJ, Nicolás E, Albericio F, Giralt E. *Tetrahedron Lett*. 2000;41:8093–8096.
15. Kostidis S, Stathopoulos P, Chondrogiannis N-I, Sakarellos C, Tsikaris V. *Tetrahedron Lett*. 2003;44:8673–8676.
16. Tam JP, Wong T-W, Riemen MW, Tjoeng FS, Merrifield PW. *Tetrahedron Lett*. 1979;42: 4033–4036.
17. Yamamoto N, Takayanagi A, Sakakibara T, Dawson PE, Kajihara Y. *Tetrahedron Lett*. 2006;47:1341–1346.
18. Dölling, R, Beyermann, M, Hänel, J, Kernchen, F, Krause, E, Franke, P, Brudel, M, Bienert, M. *Third International Innovation, Perspective in Solid-Phase Synthesis Symposium*. Oxford, U.K., 31st Aug–4th Sep, 1993:1, Poster 21.
19. Ruczyński J, Lewandowska B, Mucha P, Rekowski P. *J Pept Sci*. 2008;14:335–341.
20. Tam JP, Riemen MW, Merrifield RB. *Pept Res*. 1988;1:4033–4036.
21. Dixon MJ, Nathubhai A, Andersen OA, van Aalten DMF, Eggleston IM. *Org Biomol Chem*. 2009;7:259–268.
22. Stathopoulos P, Papas S, Kostidis S, Tsikaris V. *J Pept Sci*. 2005;11:658–664.
23. (a) Albericio F, Andreu D, Giralt E, et al. *Int J Peptide Protein Res*. 1989;34:124–128. (b) Dimarchi RD, Tam JP, Merrifield RB. *Int J Peptide Protein Res*. 1982;19:270–279.
24. Süli-Vargha H, Scholsser G, Ilaš J. *J Pept Sci*. 2007;13:742–748.
25. Flora D, Mo H, Mayer JP, Khan MA, Yan l Z. *Bioorg Med Chem Lett*. 2005;15:1065–1068.
26. Offer J, Quibell M, Johnson T. *J Chem Soc Perkin Trans*. 1996;1:175–182.
27. Cebrián J, Domingo V, Reig F. *J Peptide Res*. 2003;62:238–244.
28. Chen R, Tolbert TJ. *J Am Chem Soc*. 2010;132:3211–3216.
29. Michael, M, *Frontiers in Modern Carbohydrate Chemistry, ACS., Symposium Serie 960*. American Chemical Society: Washington, D.C., 2007:328–353.
30. Mergler M, Dick F, Sax B, Weiler P, Vorherr T. *J Pept Sci*. 2003;9:36–46.
31. Taichi T, Yamazaki T, Kimura T, Nishiuchi Y. *Tetrahedron Lett*. 2009;50:2377–2380.
32. Dölling, R, Beyermann, M, Hänel, J, Kernchen, F, Krause, E, Franke, P, Brudel, M, Bienert, M. *Twenty third European Peptide Symposium*, Braga, Portugal; 1994:Poster P061.
33. Schon I, Rill A. *Collect Czech Chem Commun*. 1989;54:3360–3373.
34. Dölling R, Beyermann M, Hänel J, et al. *J Chem Soc Chem Commun*. 1994:853–854.
35. Dölling, R, Beyermann, M, Hänel, J, Kernchen, F, Krause, E, Franke, P, Brudel, M, Bienert, M. *Peptides 1994, Proceedings of the 23rd European Peptide Symposium*. In: Maia, HLS, ed. ESCOM: Leiden, 1995:244–245.
36. Mergler M, Dick F. *J Pept Sci*. 2005;11:650–657.
37. Röder R, Henklein P, Weisshoff H, et al. *J Pept Sci*. 2010;16:65–70.
38. Goldschmidt S, Wick M. *Justus Liebigs Ann Chem*. 1952;575:217–231.
39. Goodman M, Stueben K. *J Am Chem Soc*. 1962;84:1279–1283.
40. Isidro-Llobet A, Guasch-Camell J, Álvarez M, Albericio F. *Eur J Org Chem*. 2005: 3031–3039.

41. Isidro-Llobet A, Just-Baringo X, Álvarez M, Albericio F. *Pept Sci*. 2008;90:444–449.
42. Cardona V, Eberle I, Barthelemy S, et al. *Int J Pept Res Ther*. 2008;14:285–292.
43. Megler M, Dick F, Sax B, Stähelin C, Vorherr T. *J Pept Sci*. 2003;9:518–526.
44. Baba T, Sugiyama H, Seto S. *Chem Pharm Bull*. 1973;21:207–209.
45. Yang, Y. In: *Doctoral Dissertation, Molecular recognition of integrin* $\alpha_3\beta_1$ *and inorganic compounds by tailor-made peptides*. Bielefeld University, 2008:216–217.
46. Palasek SA, Cox ZJ, Collins JM. *J Pept Sci*. 2007;13:143–148.
47. Löhr B, Orlich S, Kunz H. *Synlett*. 1999:1136–1138.
48. Blaskovich MA, Evindar G, Rose NGW, Wilkinson S, Luo Y, Lajoie GA. *J Org Chem*. 1998;63:3631–3646.
49. Fujii N, Nomizu M, Futaki S, et al. *Chem Pharm Bull*. 1986;34:864–868.
50. Vigil-Cruz SC, Aldrich JV. *Lett Pept Sci*. 1999;6:71–75.
51. Karlström A, Undén. *Tetrahedron Lett*. 1996;37:4243–4246.
52. Wade JD, Mathieu MN, Macris M, Tregear GW. *Lett Pept Sci*. 2000;7:107–112.
53. Li X, Kawakami T, Aimoto S. *Tetrahedron Lett*. 1998;39:8669–8672.
54. Weygand F, Steglich W, Bjarnason J, Akhtar R, Chytil N. *Chem Ber*. 1968;101:3623–3641.
55. Quibell M, Owen D, Peckman LC, Johnson T. *J Chem Soc Chem Commun*. 1994: 2343–2344.
56. (a) Wessely F, Komm E. *Z Physiol Chem*. 1928;174:306–318. (b) Wessely F, Kemm. E, Mayer J. *Z Physiol Chem*. 1929;180:64–74.
57. (a) Wieland T, Heinke B. *Justus Liebigs AnnChem*. 1956;599:70–80. (b) Kopple KD, Renick RJ. *J Org Chem*. 1958;23:1565–1567. (c) Schellenberg P, Ulrich J. *Chem Ber*. 1959;92: 1276–1287.
58. Milton RC de L, Milton SCF, Adams PA. *J Am Chem Soc*. 1990;112:6039–6046.
59. Toniolo C, Bonora GM, Mutter M, Pillai VNR. *Makromol Chem*. 1981;182:1997–2005.
60. (a) Mutter M, Nefzi A, Sato T, Sun X, Wahl F, Wöhr T. *Pept Res*. 1995;8:145–153. (b) Haack T, Mutter M. *Tetrahedron Lett*. 1992;33:1589–1592.
61. König W, Geiger R. *Chem Ber*. 1970;103:788–798.
62. Subirós-Funosas R, Prohens R, Barbas R, El-Faham A, Albericio F. *Chem Eur J*. 2009;15: 9394–9403.
63. Bodanszky M, Kwei JZ. *Int J Peptide Protein Res*. 1978;12:69–74.
64. Guibé F. *Tetrahedron*. 1998;54:2967–3042.
65. Isidro-Llobet A, Álvarez M, Albericio. *Tetrahedron Lett*. 2005;46:7733–7736.
66. Urge L, Otvos Jr L. *Lett Pept Sci*. 1995;1:207–212.
67. (a) Wong SY, Guile GR, Rademacher TW, Dwek RA. *Glycoconjug J*. 1993;10:227–234. (b) Vetter D, Tumelty D, Singh SK, Gallop MA. *Angew Chem Int Ed Engl*. 1995;34:60–63.
68. Anisfeld ST, Lansbury Jr PT. *J Org Chem*. 1990;55:5560–5562.
69. (a) Christian-Brams I, Jansson AM, Meldal M, Breddam K, Bock K. *Bioorg Med Chem*. 1994;2:1153–1167. (b) Urge L, Jackson DC, Gorbics L, Wroblewski K, Graczyk G, Otvos L. *Tetrahedron*. 1994;50:2373–2390. (c) Meinjohanns E, Meldal M, Paulsen H, Dwek RA, Bock K. *J Chem Soc Perkin Trans*. 1998;1:549–560.
70. Robinson AB, Scothler JW, McKerrow JH. *J Am Chem Soc*. 1973;95:8156–8159.
71. Meinwald YC, Stimson ER, Scheraga HA. *Int J Peptide Protein Res*. 1986;28:79–84.
72. Capasso S, Mazzarella L, Sica F, Zagari A, Salvadori S. *J Chem Soc Perkin Trans*. 1993;2: 679–682.
73. Capasso S, Salvadori S. *J Peptide Res*. 1999;54:377–382.
74. Capasso S, Balboni G, Di Cerbo P. *Biopolymers*. 2000;53:213–219.

75. Geiger T, Clarke S. *J Biol Chem*. 1987;262:785–794.
76. Robinson NE, Robinson AB. *Molecular clocks, deamidation of asparaginyl and glutaminyl residues in peptides and proteins*. Cave Junction, OR: Althouse Press; 2004:32.
77. Patel K, Borchardt RT. *Pharm Res*. 1990;7:787–793.
78. Flatmark T. *Acta Chem Scand*. 1966;20:1487–1496.
79. Robinson NE, Robinson AB. *J Peptide Res*. 2004;63:437–448.
80. Aswad DW. *J Biol Chem*. 1984;259:10714–10721.
81. Bornstein P, Balian G. *Methods Enzymol*. 1977;47:132–145.
82. Kwong MY, Harris RJ. *Protein Sci*. 1994;3:147–149.
83. Brennan TV, Clarke S. In: Asward DW, ed. *Deamidation and isoaspartate formation in peptides and proteins*. Boca Raton, FL: CRC Press; 1995:65–90.
84. Robinson NE, Robinson AB. *Proc Natl Acad Sci USA*. 2001;98:944–949.
85. Xie M, Schowen RL. *J Pharm Sci*. 1999;88:8–13.
86. Robinson NE, Robinson AB. *Molecular clocks, deamidation of asparaginyl and glutaminyl residues in peptides and proteins*. Cave Junction, OR: Althouse Press; 2004:42.
87. Li R, Topp EM, Hageman MJ. *J Peptide Res*. 2002;59:211–220.
88. McKerrow JH, Robinson AB. *Anal Biochem*. 1971;42:565–568.
89. Capasso S, Mazzarella L, Zagari A. *Pept Res*. 1991;4:234–238.
90. Brennan TV, Clarke S. *Protein Sci*. 1993;2:331–338.
91. (a) Quesnel A, Briand J-P. *J Peptide Res*. 1998;52:107–111. (b) Sieber P, Riniker B. *Tetrahedron Lett*. 1991;32:739–742.
92. Bedford J, Hyde C, Johnson T, et al. *Int J Peptide Protein Res*. 1992;40:300–307.
93. Levitt M. *Biochemistry*. 1978;17:4277–4285.
94. Boissonnas RA, Guttmann S, Jaquenoud P-A, Waller J-P, Cherbuliez E, Stoll A. *Helv Chim Acta*. 1955;38:1491–1501.
95. Panday SK, Prasad J, Dikshit DK. *Tetrahedron Asymm*. 2009;20:1581–1632.
96. Blombäck B. *Methods Enzymol*. 1967;11:398–411.
97. Schilling S, Wasternack C, Demuth HU. *Biol Chem*. 2008;389:983–991.
98. Kawasaki I, Itano HA. *Anal Chem*. 1972;48:546–556.
99. Bodanszky M, Martinez J. *Synthesis*. 1981:333–356.
100. Schnabel E, Klostermeyer H, Dahlmans J, Zahn H. *Justus Liebigs Ann Chem*. 1967;707: 227–241.
101. Moroder L, Göhring W, Lucietto P, et al. *Hoppe-Seylers Z Physiol Chem*. 1983;364: 1563–1584.
102. Kiso Y, Fujiwara Y, Kimura T, Nishitani A, Akaji K. *Int J Peptide Protein Res*. 1992;40: 308–314.
103. Takashima H, du Vigneaud V, Merrifield RB. *J Am Chem Soc*. 1968;90:1323–1325.
104. Beyerman HC, Lie TS, van Veldhuizen CH. In: Nesvadab H, ed. *Peptides 1971*. Amsterdam: North-Holland; 1973:162–164.
105. Dorman LC, Nelson DA, Chow RC. In: Landes S, ed. *Progress in peptide research*. New York: Gordon and Breach; 1972:65–68.
106. Maiorov VD, Kislina IS, Voloshenko GI, Librovich NB. *Russ Chem Bull*. 2000;49: 1526–1530.
107. Schnölzer M, Alewood P, Jones A, Alewood D, Kent SBH. *Int J Peptide Protein Res*. 1992;40:180–193.
108. Wendlberger G, Göhring W, Hübener G, et al. In: Marshall GR, ed. *Peptides: chemistry and biology*. Leiden: ESCOM; 1988:287.

109. Dimarchi RD, Tam JP, Kent SBH, Merrifield RB. *Int J Peptide Protein Res*. 1982;19:88–93.
110. Viau M, Létourneau M, Sirois-Deslongchamps A, Boulanger Y. *Pept Sci*. 2007;88:754–763.
111. García AF, Butz P, Trierweiler B, et al. *J Agric Food Chem*. 2003;51:8093–8097.
112. Fife TH, DeMark BR. *J Am Chem Soc*. 1977;99:3075–3080.
113. Wendlberger G, Moroder L, Thamm P, Wilschowitz L, Wünsch E. *Monatsh Chem*. 1979;110:1317–1330.
114. Orlowska A, Witkowska E, Izdebski J. *Int J Peptide Protein Res*. 1987;30:141–144.
115. Marshall GR. *J Org Chem*. 1971;36:3966–3970.
116. Mazurov AA, Andronati SA, Korotenko AT, Gorbatyuk VY, Shapiro YE. *Int J Peptide Protein Res*. 1993;42:14–19.
117. Bonora BM, Toniolo C, Fontana A, Di Bello C, Scoffone E. *Biopolymers*. 1974;13:157–167.
118. LeQuesne WJ, Young GT. *J Chem Soc*. 1950:1959–1963.
119. (a) Lopez-Rodriguez ML, Morcillo MJ, Fernandez E, Rosado ML, Pardo L, Schaper K-J. *J Med Chem*. 2001;44:198–207. (b) Stilz HU, Guba W, Jablonka B, et al. *J Med Chem*. 2001;44:1158–1176. (c) Nefzi A, Giulianotti M, Truong L, Rattan S, Ostresh JM, Houghten RA. *J Comb Chem*. 2002;4:175–178.
120. Zhang H-C, McComsey DF, White KB, et al. *Bioorg Med Chem Lett*. 2001;11:2105–2109.
121. Liley M, Johnson T. *Tetrahedron Lett*. 2000;41:3983–3985.
122. Limal D, Semetey V, Dalbon P, Jolivet M, Briand J-P. *Tetrahedron Lett*. 1999;40:2749–2752.
123. Quibell M, Turnell WG, Johnson T. *J Chem SocPerkin Trans*. 1993;1:2843–2849.
124. Xiao X-Y, Ngu K, Chao C, Patel DV. *J Org Chem*. 1997;62:6968–6973.
125. DeWitt SH, Kieley JS, Stankovic CJ, Schröder MC, Cody DMR, Pavia MR. *Proc Natl Acad Sci USA*. 1993;90:6909–6913.
126. (a) Vázquez J, Royo M, Albericio F. *Lett Org Chem*. 2004;1:224–226. (b) Colacino E, Lamaty F, Martinez J, Parrot I. *Tetrahedron Lett*. 2007;48:5317–5320. (c) Park K-H, Kurth MJ. *Tetrahedron Lett*. 2000;41:7409–7413.
127. Chong PY, Petillo PA. *Tetrahedron Lett*. 1999;40:4501–4504.
128. Peyman A, Volkmar W, Knolle J, et al. *Bioorg Med Chem Lett*. 2000;10:179–182.
129. Schöder E, Lübke K. *The Peptides*The Peptides, Vol 1 New York: Academic; 1965:55.
130. Schwenzer B, Weber E, Losse G. *J Prakt Chem*. 1985;327:479–486.
131. Völter W, Altenburg A. *Liebigs Ann Chem*. 1983:1641–1655.
132. Fischer E. *Ber Dtsch Chem Ges*. 1902;35:1095–1106.
133. (a) Schlögl H, Fabtschowitz H. *Monatsh Chem*. 1953;84:937–955. (b) MacLaren JA. *Aust J Chem*. 1958;11:360–365.
134. (a) Koedjikov AH, Blagoeva IB, Pojarlieff IG, Stankevic EJ. *J Chem Soc Perkin Trans*. 1984;2:1077–1081. (b) Kaneti J, Kirby AJ, Koedjikov AH, Pojarlieff IG. *Org Biomol Chem*. 2004;2:1098–1103.
135. Rajic Z, Zorc B, Raic-Malic S, et al. *Molecules*. 2006;11:837–848.
136. Geohegan KF, Hoth LR, Tan DH, Borzilleri KA, Withka JM, Boyd JG. *J Proteome Res*. 2002;1:181–187.
137. Schnolzer M, Kent SB. *Science*. 1992;256:221–225.
138. (a) Gibb BC, Mezo AR, Sherman JC. In: Kaumaya PTP, Hodges RS, eds. *Peptides: chemistry, structure and biology*. England: Mayflower Scientific Ltd; 1996:585–586. (b) Futaki S, Ishikawa T, Niwa M, Kitagawa K, Yagami T. *Tetrahedron Lett*. 1995;36:5203–5206.
139. (a) Linder W, Robey FA. *Int J Peptide Protein Res*. 1987;30:794–800. (b) Robey FA, Fields RL. *Anal Biochem*. 1989;177:373–377.

140. Robey FA. In: Pennington MW, Dunn BM, eds. *Methods in molecular biology*. Totowa, New Jersey: Humana Press; 1994:81–82.
141. Englebresten DR, Chomba CT, Robillard GT. *Tetrahedron Lett*. 1998;39:4929–4932.
142. Gurd FRN. *Methods Enzymol*. 1967;11:532–541.
143. Hermanson GT. *Bioconjugate techniques*. Academic Press; 2008:109–110.
144. Crestfield AM, Stein WH, Moores S. *J Biol Chem*. 1963;238:2413–2419.
145. Vithayathil PJ, Richards FM. *J Biol Chem*. 1960;235:2343–2351.
146. Kiso Y, Ito K, Nakamura S, Kitagawa K, Akita T, Moritoki H. *Chem Pharm Bull*. 1979;27: 1472–1475.

Chapter 7

Side Reactions on Amino Groups in Peptide Synthesis

Amino group represents the most versatile nucleophile in peptide synthesis. Its acylation reaction is the underlying principle of the amino acid coupling and peptide chain elongation. Due to the high reactivity the amino group could undertake a plethora of side reactions in the process of peptide synthesis. The most common side reactions affecting amino group include acetylation, trifluoroacetylation, formylation, and alkylation. These side reactions could take place at the steps of N^{α}-deprotection, amino acid coupling, side chain global deprotection, N^{α}-protection, and even chromatographic purification.

7.1 N^{α}-ACETYLATION SIDE REACTIONS

Peptide N^{α} acetylation is a frequently occurring side reaction in SPPS, leading to the termination of the peptide assembly and the formation of truncated peptide impurities. The origin of this side reaction could be ascribed to multiple incentives, one of which lies in the intentional peptide chain termination via acetic acid anhydride/pyridine (or DIEA) treatment. At certain amino acid coupling steps along SPPS this handling is purposely carried out so that the unacylated peptide chains on solid supports could be "quenched" by N^{α}-acetylation end-capping. The resultant N^{α}-acetyl truncated sequence might be eventually readily separated from the target peptide product by virtue of the distinct differences of chromatographic properties. Otherwise, the incomplete amino acid couplings would inevitably result in the formation of peptide impurities with deletion sequences that might possess analogous chromatographic properties to the target product. This situation will render the downstream purification tremendous challenges to remove the subjected deletion sequence impurities from the target product. In spite of the regular employment in SPPS the intentional N^{α}-acetylation treatment of peptidyl resin might give rise to the unexpected peptide chain termination. The residual acetic acid anhydride applied at the previous N^{α}-acetylation step might survive the subsequent resin rinse, and react with the liberated N^{α}-functional group to lead to the undesired acetylation side reaction, terminating by this way the growing peptide chains.

Side Reactions in Peptide Synthesis. http://dx.doi.org/10.1016/B978-0-12-801009-9.00007-0

Another pathway of unexpected peptide N^{α}-acetylation is facilitated by His residue located in the subject peptide chain.[1] Some protecting groups on the His imidazolyl side chain like Tos are chemically unstable in certain synthetic conditions, and could hence be cleaved prematurely. The intentional acetic acid anhydride treatment of unconverted peptide chains could simultaneously acetylate the prematurely liberated His imidazolyl side chains, and convert them to the corresponding amide moieties. Acyl imidazolides are intrinsically excellent acylating agents[2] and the formed Ac-imidazolide derivatives, in case not sufficiently hydrolyzed, would eventually dictate the acetyl $N^{im}{\rightarrow}N^{\alpha}$ migration process contingent on the primary and secondary structures of the concerned peptides. The rearrangement process would result in the shift of the acetyl group from the concerned His imidazolyl substituent to the N^{α}-group on peptide backbone, terminating in this manner the affected growing peptide chain. The similar process has been elaborated in Section 4.3. Besides acetyl $N^{im}{\rightarrow}N^{\alpha}$ migration, the Ac-Im derivative could also mediate the acetyl $N^{im}{\rightarrow}$Lys-N^{ε} shift, and give rise to the formation of a N^{ε}-acetyl Lys side product.[3]

The guanidino side chain on Arg could also be affected by the acetylation side reaction in the process of acetic acid anhydride treatment on peptidyl resin. Since the most frequently applied Arg building blocks for peptide synthesis are the derivatives with monoprotection at the N^{ω} of guanidino substituent, the unprotected $N^{\omega'}$- and N^{δ}- moieties are partially shielded from various electrophiles in the reaction system by virtue of the electron-withdrawing effect of the N^{ω}-protecting group as well as its steric hindrance. Monoprotected Arg derivatives such as Arg(Tos) or Arg(NO_2) are potentially susceptible to acetylation modification on their guanidino substituent upon acetic acid anhydride/triethylamine treatment (whereas principally resistant to Ac-Im-directed acetylation).[4] New types of guanidino-protecting groups like Pdf, bearing substantial steric hindrance could effectively shield the overall guanido substituent from the potential acetylation side reactions initiated by acetic acid anhydride/base treatment. It has been verified by the corresponding stress tests (Yang, Y., unpublished results) that Fmoc-Arg(Pbf)-OH immobilized on CTC resin does not suffer from detectable acetylation by 10 equiv. acetic acid anhydride/DIEA treatment within 5 h.

Peptide synthesis processes devoid of the intentional acetylation endcapping treatment could notwithstanding being subjected to acetylation side reactions, leading to unexpected peptide chain termination. These phenomena take place predominantly at the steps of difficult amino acid couplings[5] or upon the initiation of peptide chain aggregations on solid supports that is inherently interconnected with sluggish coupling kinetics.[6] The root causes of these acetylation side reactions do not lie in the corresponding synthetic routes or manufacturing processes but are attributed to the residual acetyl impurities in the reagents applied in the affected reactions. For example, some commercial amino acids might contain small quantities of acetic acids.[6–8] Utilization of such starting materials for the target peptide production could induce competitive reactions between the concerned amino acids and the residual acetic acid that results in the activation of acetic acids and the consequent undesirable acetylation of peptide N^{α}-groups by the activated acetic acid molecules. This aberrant process finally gives rise to the formation of backbone N^{α}-acetylated peptide side products that terminate the peptide chain elongation. Moreover, some amino acid materials could be contaminated by residual ethyl acetate (Yang, Y., unpublished results) that might be able to react with N^{α} group in the process of peptide synthesis by means of aminolysis and result in the formation of truncated N^{α}-acetyl peptide impurities. The elucidated mechanism of unexpected peptide N^{α}-acetylation side reaction is basically in compliance with the observation that acetylation side reactions frequently address the amino acid couplings with low reaction kinetics, since the competitive reactions of residual acetic acid/ethyl acetate against the amino acids to acylate the N^{α}-group on the peptide chain will be favored once the target amino acid coupling is sluggish. Meticulous quality control and the strict release of starting materials with regard to residual acetic acid/ethyl acetate contents are supposed to be conducted to these amino acid raw materials particularly if they are involved in sensitive couplings complicated by N^{α}-acetylation side reactions. Preventative strategies, e.g., lyophilization and recrystallization are effective to remove residual acetic acid/ethyl acetate from the corresponding amino acid materials in order to suppress the potential acetylation side reactions.

The intentional peptide N^{α}-acetylation end-capping processes could nevertheless be concomitantly accompanied with overacetylation side reactions that give rise to the formation of *N*-acetyl acetamide derivatives[9] (Fig. 7.1). Actually *N*-acyl amide type side products (imides) are frequently formed in the process of peptide synthesis. Aspartimide formation and hydantoin formation, which are regarded as one of the most significant side reactions in peptide synthesis, are triggered by the nucleophilic attack of an amide unit on an acylating agent. Double acetylation side reactions might take place under certain circumstances at the step of intentional peptide N^{α}-acetylation endcapping. Solutions such as lowering the stoichiometry of acetic acid anhydride, cooling-down the affected reaction, shortening the reaction duration, or application of weaker bases like pyridine (Yang, Y., unpublished results) in place of triethylamine – could effectively be employed to suppress the overacetylation side reactions.

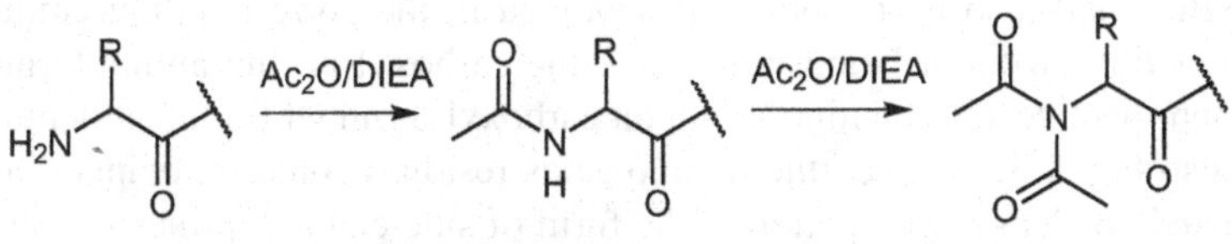

FIGURE 7.1 **Peptide N$^{\alpha}$-group overacetylation side reaction.**

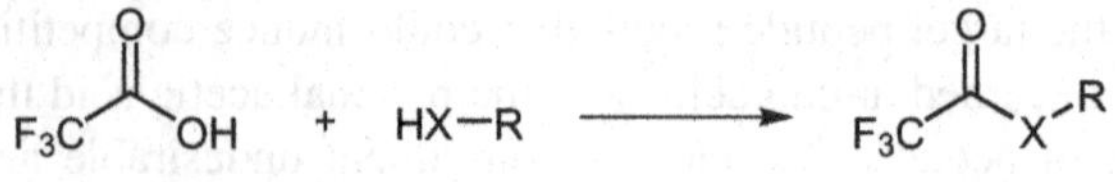

N = O or NH

FIGURE 7.2 Trifluoroacetylation on hydroxyl and amino groups.

7.2 TRIFLUOROACETYLATION SIDE REACTIONS

Trifluoroacetylation on amino groups are more frequently occurring than N^{α}-acetylation in peptide synthesis. This undesired process is mostly originated from the functions of TFA with peptide N^{α}-functional groups or hydroxyl substituent on Ser/Thr/Tyr that give rise to the formation of trifluoroacetamide or trifluoroacetate side products, respectively (Fig. 7.2).[10]

The N^{ε}-functional group on the Lys side chain could be affected by the trifluoroacetylation side reaction as well.[11] The outcome of this undesired process is the formation of trifluoroacetylated side products with a +96 amu molecular weight increase relative to the target products. The generated trifluoroacetylated impurities might be difficult to be separated from the products due to their potentially similar chromatographic properties that will inevitably bring forward substantial challenges to the affected peptide production.

Peptide trifluoroacetylation side reaction in the presence of TFA might be initiated at the following steps during peptide synthesis: (1) TFA-mediated cleavage of side chain-protected peptide from solid supports; (2) peptide side chain global deprotection; and (3) repetitive N^{α}-Boc deblocking in Boc-mode SPPS.

Convergent segment condensation is deemed as an effective synthetic strategy for peptide preparation,[12] which has found widespread applications especially for peptides/proteins with long sequences that are basically impractical for the routine stepwise synthetic method. Peptide segments involved in a convergent condensation reaction normally contain reciprocal reactive functional groups that could be spliced to form a new bond by appropriate coupling reagents, constructing by this means the target compact peptide product. Most peptide segment condensations take place between the amino and carboxyl groups since the generated amide bond is in principle compatible with peptide attributes. Peptide segment condensation could be conducted by "minimal protection"[13] or "maximal protection"[14] strategies. In the former process except Lys, Cys, Asp/Glu (under certain circumstances the carboxyl side chain on Asp/Glu could also be intentionally involved in the corresponding conjugation reactions without being protected), the carboxyl on the amino-terminal fragment peptide **1** and amino group on carboxyl-terminal fragment peptide **2** (see also Fig. 7.3), all the other amino acids residues could participate in the subjected condensation reaction in the form of side-chain unprotected derivatives. This synthetic strategy might circumvent the inherent problems, e.g., low solubility of the side chain-protected peptides that lead to excessively diluted

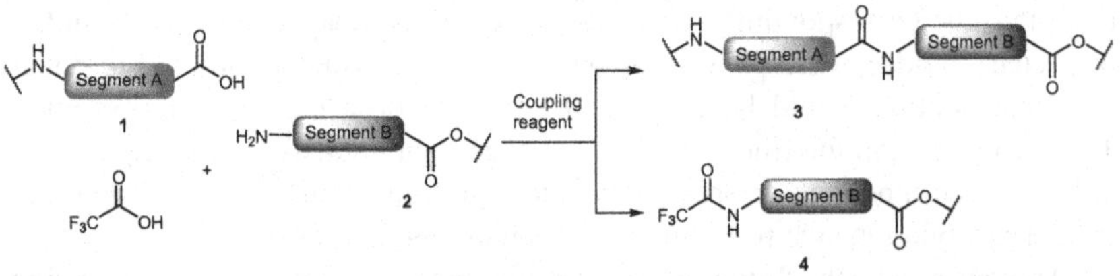

FIGURE 7.3 Peptide N$^{\alpha}$-trifluoroacetylation side reaction during segment condensation.

solution of the subjected fragment solution, and will consequently retard the target condensation reactions. In "maximal protection" strategy all the functional groups except the ones that are involved in the condensation reaction are basically protected in order to maximally prevent the occurrence of potential side reactions. The utilization of TFA is nevertheless unavoidable as to cleave the peptide segments from the corresponding solid supports no matter whether the side chain unprotected fragment (minimal protection) or side chain protected fragment (maximal protection) is desired as the precursor for the subsequent conjugation. The abundant TFA molecules participating in the concerned peptide fragment cleavage could be largely removed from the peptide system at the isolation step by tandem precipitation and filtration. In many cases, the addressed peptide segments are subjected to liquid chromatographic treatment in order to remove as much as possible the potential critical impurities that might interfere with the following segment condensation. However, if TFA are not sufficiently excluded from the target peptide fragments at the due isolation and purification steps, the survived TFA molecules are capable of initiating competitive reactions against the carboxyl derivative **1** and functioning with the amino compound **2** to generate trifluoroacetated side product **4** concomitantly with target peptide **3** (Fig. 7.3).

The formation of **4** might not cause overwhelming challenges for the subsequent chromatographic purification. However, the involvement of TFA in the concerned segment condensation would inevitably cause the loss of the carboxylate component **2** that in turn leads to the reduction of the reaction efficiency.

The competitive reaction of TFA against the amino-terminal fragment **1** in the process of segment condensation could predominate particularly in case of the sluggish kinetics of the target reaction. Utmost care should be exerted under such circumstances as to minimize the modifications of susceptible substituents by TFA in the condensation reaction. Since TFA could be readily carried over concomitantly into the peptide intermediate from the step of cleavage of the side chain-protected peptide fragment off the solid support, excessive survived TFA should therefore be maximally removed from peptide derivative, so as to prevent the interfering trifluoroacetylation side reaction. It is advisable to utilize tertiary bases such as pyridine or DIEA to neutralize the excessive TFA carried over from the side chain-protected peptide cleavage step in order to convert them into the corresponding TFA salts that could be readily removed from the protected peptides by simple workup process. N^{α}-group on the carboxyl-terminal fragment peptide would be associated with trifluoroacetate counterion upon TFA treatment. An intermediary salt exchange to nonreactive anion, e.g., Cl^{-}[15] is hence recommended prior to the condensation reaction so as to prevent the trifluoroacetylation side reaction under such circumstances.

Another practical solution to suppress the potential trifluoroacetylation side reaction during peptide segment condensation is to utilize another deblocking reagent in place of TFA to detach the immobilized protected peptide fragment from solid supports, and/or globally remove the side chain protecting groups. For example, Boc could be cleaved from the corresponding protected substituent by HCl/dioxane;[16] side chain-protected peptides could be detached from CTC resin by HFIP[17] or TFE solution; side chain protecting groups, e.g., Pbf, Trt, tBu, Boc as well as ordinary peptide resin linkers, e.g., Hydroxymethyl phenyl, HMPA Rink Amide, and PAL could be effectively cleaved by 0.1 N HCl in HFIP or TFE.[18] Employment of these technologies in the processes of concerned peptide cleavage and side chain global deprotection is devoid of TFA, and the potential trifluoroacetylation side reactions are hence thoroughly prevented.

Trifluoroacetylation side reactions might also be provoked in the absence of coupling reagents. For instances, the repetitive N^{α}-Boc cleavage and subsequent neutralization in Boc peptide chemistry is principally capable of inducing N^{α}-trifluroacetylation side reactions that result in peptide chain termination.[19] As is displayed in Fig. 7.4, the hydroxymethyl functional groups on solid supports for SPPS, e.g., Wang resin, could be insufficiently loaded at the due steps of Boc-type SPPS, and the survived hydroxyl substituents could accommodate an acid-catalyzed esterification reaction with TFA at the repetitive TFA-mediated N^{α}-Boc deblocking steps, giving rise to immobilized trifluoroacetate derivatives that undergo base-induced acyl *O*→*N* migration during the subsequent DIEA-mediated neutralization treatments. The concerned trifluoroacetyl is shifted from the affected hydroxyl group to the liberated and deprotonated N^{α}-group on a peptide chain, terminating in this

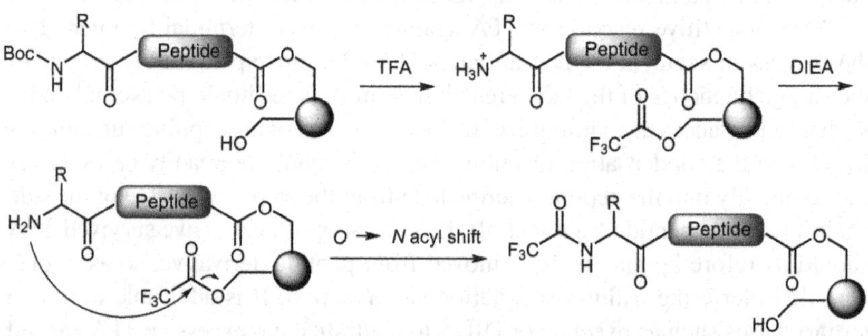

FIGURE 7.4 Trifluroacetylation induced by acyl *O*→*N* migration on hydroxymethyl resin.

manner the growing peptide chain as N^{α}-trifluoroacetylated end-capped truncated peptide impurity.

The undesired liberation of hydroxymethyl group in the process of Boc-mode SPPS could also be initiated by premature peptide acidolytic cleavage at steps of repetitive TFA-mediate N^{α}-Boc deblocking.[19,20] Peptide chains immobilized on hydroxymethyl-type resins could be prematurely detached from the stationary phase by TFA treatment in Boc SPPS. The inclination of this undesired process is partially dependent on the features of the SPPS resin applied in this process. In case of acid-sensitive hydroxymethyl linker and/or peptides with long sequences that have to undertake considerable rounds of N^{α}-Boc deblocking by repetitive TFA treatments, the corresponding ester bonds between peptide chains and resin linker will be more susceptible to premature acidolytic cleavage. The hydroxymethyl groups liberated in this process will be exposed to the subsequent SPPS processes, and their function with abundant surrounding TFA via acid-catalyzed esterification would give rise to trifluoroacetate derivatives that might undergo based-catalyzed $O \rightarrow N$ acyl migration at the steps of DIEA-mediated neutralization after N^{α}-Boc deblocking as described in Fig. 7.4. The released affected hydroxymethyl groups upon $O \rightarrow N$ acyl migration are capable of inducing further similar trifluoroacetylation side reactions.

It is to note that the aforementioned migration of trifluoroacetyl groups is especially prone to occur on peptides with *N*-terminal Pro residue,[21,22] which is principally ascribed to the increased nucleophilicity of the secondary N^{α}-group on Pro relative to those on other amino acids. Apart from Cys, Pro possesses the highest pK_a of N^{α}-group among the natural amino acids, which facilitates the underlying nucleophilic attack on the formed trifluoroacetate derivatives in the process of base-catalyzed acyl $O \rightarrow N$ migration.

Besides Boc-mode peptide synthesis trifluoroacetylation side reactions affecting peptides with *N*-terminal Pro could also readily take place in Fmoc-chemistry SPPS[23] predominantly at the steps of TFA-mediated peptide resin cleavage/side chain global deprotection. It has been concluded that trifluoroacetylation side reactions prevailing on peptides with N^{α}-Pro in the process of acid-mediated global deprotection could be significantly suppressed by replacing Fmoc-Pro-OH with Boc-Pro-OH as the *N*-terminal residue.[23] This preventive effect is attributed to the transient formation of carbamate intermediate from Boc-Pro during acidolysis, which is basically analogous to the degradative Trp carbamic acid intermediate Trp(CO_2) derived from Trp(Boc) acidolysis. Although the stability of the backbone Pro-carbamic acid is inferior to that of the Trp carbamic acid counterpart, its transient existence could nevertheless to some extent protect *N*-terminal Pro from trifluoroacetylation modification. It is, therefore, recommended in this connection to utilize Boc-Pro-OH in place of Fmoc-Pro-OH as the building block for the preparation of peptide bearing *N*-terminal Pro residue.

In the case of an extreme situation in which the trifluoroacetylation side reaction is extraordinarily prone to take place, it is necessary to enhance the acid stability of the ester bond between the peptide chain and the resin linker.

Hydroxymethyl-Pam resin,[20,24] for instance, is bestowed with reinforced resistance toward TFA-induced peptide premature cleavage. The trifluoroacetylation side reaction via the aforementioned mechanism is thus restrained in SPPS employing hydroxymethyl-Pam resin.

Since N^{α}-trifluoroacetylation could take place at the neutralization step after N^{α}-Boc deblocking in Boc-mode SPPS, the elongated base treatment of the subjected peptide resin would possibly facilitate the trifluoroacetyl migration that consequently aggravates the trifluoroacetylation-dictated peptide chain termination side reaction. It is therefore advisable to rationally shorten the duration of base neutralization of peptide resin in order to minimize the base-catalyzed trifluroacetylation.

The liability of trifluoroacetylation might be peptide sequence dependent, as it has been discovered that a single amino acid modification in a peptide bearing *N*-terminal Pro drastically alters its propensity toward trifluoroacetylation side reaction,[23] although the explicit correlation between peptide sequence and trifluoroacetylation inclination has not been unequivocally clarified as yet.

The hydroxyl substituents on Ser, Thr and Tyr could also suffer from trifluoroacetylation modification upon TFA treatment during peptide side chain global deprotection (Yang, Y., unpublished results), giving rise to the formation of impurities with a molecular weight increase of 96 amu. Under some circumstances when a strong acid, e.g., TFMSA, is added into the TFA solution to facilitate peptide side chain global deprotection the extent of trifluoroacetylation on Ser/Thr/Tyr side chain could be intensified. The mechanism for such phenomenon is the direct esterification between Ser/Thr/Tyr side chain hydroxyl groups and abundant TFA molecules, promoted by the charged strong acid. In case the affected Ser/Thr/Tyr residue is located in the *endo* sequence of the subjected peptide chain, the formed trifluoroacetate impurity could be reversed to the original unmodified hydroxyl derivative at the following steps through hydrolysis. Moreover, trifluoroacetylation could also affect the solid phase synthesis of peptide alcohol on CTC resin, converting the *C*-terminal backbone hydroxyl group to the corresponding trifluoroacetate side product at the step of peptidyl resin cleavage/side chain global deprotection.[25]

The more prevailing trifluoroacetylation addresses Ser/Thr residues in case they are located on peptide *N*-terminus.[10,26] The process of trifluoroacetylation by this means is schematized in Fig. 7.5. The *N*-terminal Ser/Thr upon

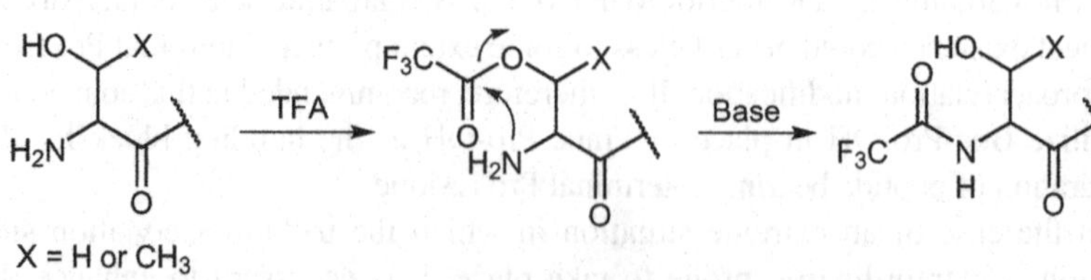

FIGURE 7.5 N-terminal Ser/Thr-induced N^{α}-trifluoroacetylation process.

TFA treatment would be modified at its hydroxyl side chain through TFA-catalyzed esterification to form the corresponding side chain-derivatized Ser/Thr(trifluoroacetyl). This kind of derivative might undergo acyl *O*→*N* migration in basic conditions at the following workup or purification steps. The trifluoroacetyl group covalently linked on the Ser/Thr hydroxyl side chain is thus shifted to the peptide backbone N^{α}-group, giving rise to the formation of the N^{α}-trifluoroacetated side product. Compared with *endo*-Ser/Thr residues the *N*-terminal Ser/Thr is more susceptible to the trifluoroacetylation side reaction facilitated by the hydroxyl substituents on the side chains of the *N*-terminal Ser/Thr. When the addressed *N*-terminal Ser/Thr is replaced by Ala the trifluoroacetylation side reaction will be suppressed accordingly.[26]

It has been verified by various relevant studies[25,26] that the extent of the peptide trifluoroacetylation side reaction occurred at the step of side chain global deprotection is evidently proportional to the duration of the reaction. Elongated acid treatment of the sensitive peptide in this respect would substantially aggravate the trifluroacetylation side reaction. It is advisable to rationally control the duration of the peptide global deprotection particularly for those susceptible to trifluoroacetylation. The concentration of TFA could also impact on trifluoroacetylation since increased TFA concentration would lead to the intensified side reaction in this connection.[25] Justified dilution of TFA solution would facilitate trifluoroacetylation suppression. Furthermore, lowering the reaction temperature could be conducted as well to minimize this side reaction.

Another source of trifluoroacetylation side reaction is the inferior quality TFE solvent. This topic will be separately discussed in Section 14.7.

7.3 FORMYLATION SIDE REACTIONS

Amino group formylation is regarded as a regular protein modification strategy that has been utilized in the research domains such as chemoattractants to phagocytic cells[27] and protein posttranslational modification.[28] However, aberrant formylation formations on amino groups originating from various processes pose undesirable side reactions to peptide synthesis. Reasonable synthetic routes and manufacturing processes are required to be designed and developed accordingly to suppress its occurrence.

The formyl functional groups that provoke peptide formylation are basically originated from the following sources: (1) protecting group on the indolyl side chain of Trp residue; (2) formic acid introduced in the processes of peptide synthesis, protein/peptide analysis and liquid chromatography; and (3) impurities in DMF solvent. The most vulnerable functional groups in peptide synthesis with regard to formylation modification include: (1) indolyl side chain on Trp; (2) Peptide backbone N^{α}-group, Lys side chain N^{ε}-group and His side chain N^{im} substituent; and (3) hydroxyl side chain on the Ser and Thr residues.

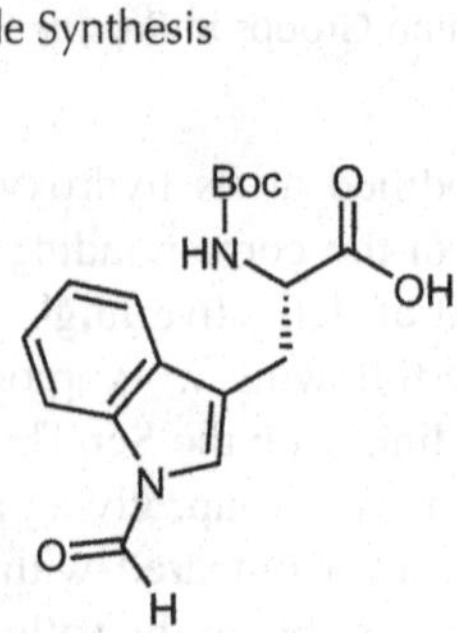

FIGURE 7.6 **Structure of Boc-Trp(For)-OH.**

7.3.1 Trp(For)-Induced Peptide Formylation

The indolyl side chain on Trp residue in Boc chemistry is normally protected by means of formylation. Boc-Trp(For)-OH is regularly utilized as an ordinary building block in the Boc chemistry-mode SPPS (see also Fig. 7.6). N^{im}-formyl Trp derivative possesses extraordinary acid stability that could even resist the treatment by HF. It could, however, be cleaved in strong acid-weak nucleophile mixture, e.g., HF/EDT[29] or HF/2-thiol ethanol[30] through the so-called "push–pull" mechanism.[31] Moreover, Trp could also be regenerated from Trp(For) protected precursor in basic conditions, e.g., hydrazine, NaOH, $NaHCO_3$ or piperidine/DMF.[29,32,33] If the formyl side chain protecting group on Trp is not quantitatively cleaved in the corresponding acidic deblocking solution, which is not an uncommon phenomenon in peptide synthesis,[34] impurities containing Trp(For) residue with a molecular weight increase by 28 amu will be generated.

The emergence of peptide formylation, apart from an insufficient removal of formyl protecting group from Trp side chain upon HF treatment, could also be induced in the process of base-mediated deblocking of formyl protecting group. The cleaved formyl derivatives, in case not timely and sufficiently quenched, might be able to irreversibly modify Lys-N^{ε}[34] or peptide backbone N^{α}-group.[34] Addressing this formylation mechanism basic deblocking reagents for formyl cleavage preferably possess appreciable formyl-quenching capacity[35] in order to protect the liberated amino group from being formylated. *N,N'*-dimethylethylenediamine, for instance, is an appropriate candidate in this connection as it could both efficiently regenerate Trp from its side chain-protected precursor Trp(For) and effectively transform the cleaved formyl group *in situ* to the corresponding *N*-formyl-*N,N'*-dimethylethylenediamine **5** (Fig. 7.7). Although this methodology could not assure of the absolute insusceptibility of Lys-N^{ε} toward formylation in the affected process, it could effectively enhance the resistance of the addressed peptide to the undesired *N*-formylation modification compared with other deblocking bases, e.g., hydroxylamine or $NaHCO_3$.[35]

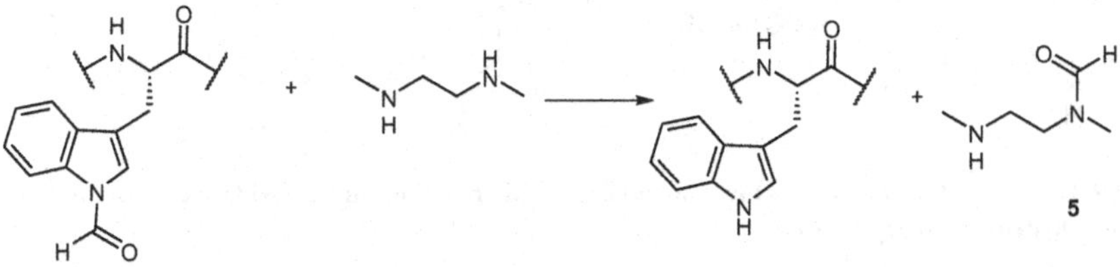

FIGURE 7.7 ***N,N'*-Dimethylethylenediamine-directed Trp(For) deblocking.**

7.3.2 Formic Acid-Induced Peptide Formylation

The second major source of formylation side reaction in peptide synthesis is formic acid that might be utilized both in upstream and downstream processes. Formic acid[36] or HCl/formic acid mixture[33] is frequently used in Boc chemistry as reagents for N^{α}-Boc deblocking. HCl/formic acid is verified to be more efficient relative to HCl/acetic acid mixture in terms of N^{α}-Boc protecting group removal.[33] However, the involvement of formic acid in the peptide synthesis process could be detrimental to the integrity of the side chain-unprotected Trp residue since formic acid could potentially lead to formylation modification on Trp indolyl side chain. As indicated in Fig. 7.8, HCl/formic acid solution that is supposed to direct the peptide N^{α}-Boc deblocking process, could bring about formylation on the Trp side chain and the resultant formation of Trp(For) side product.[33] This formylation process is fortunately reversible, and Trp could be regenerated upon appropriate workup. Trp normally participates in Boc-mode peptide synthesis in the manner of side-chain protected derivative Trp(For) which could be efficiently converted to Trp by appropriate bases, e.g., piperidine, hydrazine or hydroxylamine.

Under some extreme circumstances formic acid itself could mediate formylation on amino groups in the absence of catalyst. An investigation indicates that aniline and its derivatives could function with pure formic acid at 60°C, and be transformed into the corresponding formamide compounds.[37]

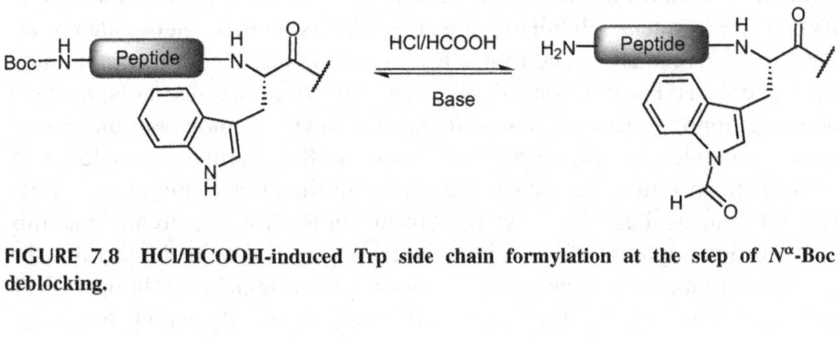

FIGURE 7.8 HCl/HCOOH-induced Trp side chain formylation at the step of N^{α}-Boc deblocking.

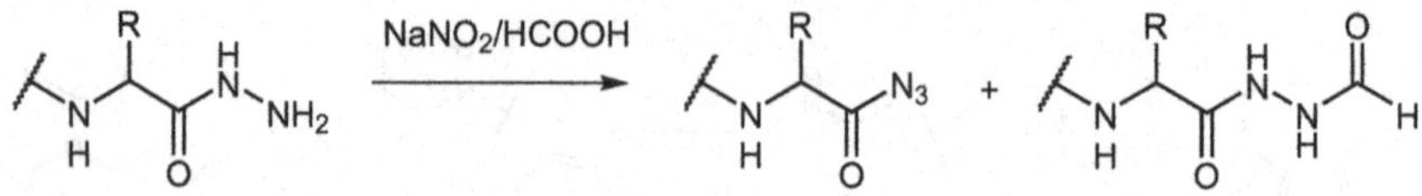

FIGURE 7.9 Formic acid-induced formylation side reaction during oxidative conversion of acyl hydrazine to acyl azide.

Apart from N^{α}-Boc deblocking reagent formic acid could also serve as catalyst in the process of peptide synthesis and trigger peptide formylation side reaction as a consequence. For example, a chemical preparation of peptide acyl azide could be realized through acyl hydrazine intermediate that is converted to the target product by $NaNO_2$ treatment in acidic condition.[38] HCl is generally utilized as the catalyst for this oxidation process. Organic acids like formic acid or acetic acid could also be employed for the same purpose. However, if they are adopted as catalysts the formed acyl hydrazine moiety on peptide *C*-terminus might suffer from formylation and acetylation modification, and be transformed into the corresponding formylated and acetylated peptide acyl hydrazine, respectively, as indicated in Fig. 7.9.[39] In view of this undesired process employment of formic/acetic acid as catalyst should be avoided for the acid-catalyzed oxidation of peptide acyl hydrazine to the corresponding acyl azide product. Formylated acyl hydrazine impurity could undergo deformylation upon hydrazine treatment.[39] However, this process might potentially arouse side reactions, e.g., aspartimide formation, DKP formation, as well as racemization, and should be hence conducted with utmost caution.

Among numerous protein analysis technologies the interpretation of peptide/protein sequence aided by qualitative analysis of peptide/protein segments derived from specific fragmentation of the target analyte has emerged as a promising method, among which CNBr-mediated specific cleavage of -Met-Xaa-[40] peptide bond as well as formic acid assisted -Asp-Pro- acidolysis[41] is routinely performed. Both of the aforementioned processes would utilize formic acid. A 70% of formic acid serves as buffer solution for CNBr directed -Met-Xaa- degradation reaction (0.1 N HCl could also be used for the same purpose, nevertheless formic acid is preferred in this connection due to its enhanced protein-solubilizing effect as well as reduced Met-oxidation induction relative to HCl), and formic acid is used as cleavage reagent for -Asp-Pro- acidolysis. Formylation side reactions have been detected in both of the aforementioned processes. It was found that bovine hemoglobin underwent evident modification after 5 days incubation in 88% formic acid solution at ambient temperature,[42] which was ascribed to the esterification process between formic acid and hydroxyl substituents on Ser/Thr. It is to note that this side reaction might interfere with the protein analysis despite the fact that the generated formate could be readily reduced to the original Ser/Thr upon EDT treatment.[42] Besides Ser/Thr formic acid could also modify peptide N^{α}-group,

Lys-N^{ε} and Trp-N^{in} giving rise to the formation of corresponding formylated side products.[43]

On top of the roles as reagent/catalyst in the upstream process formic acid could also be utilized as cosolvent or buffer for the mobile phase of liquid chromatography in the downstream process. Formylation modification could be induced in the process of peptide purification as well.[42] Due to its significant protein/peptide solubilizing effect formic acid is routinely employed as solvent for otherwise sparsely soluble analytes.[44] Peptides solubilized in formic acid might suffer from a formylation side reaction which is per se analogous to peptide trifluoroacetylation in TFA solution. In reverse phase liquid chromatography 0.1% formic acid is commonly used as buffer for the mobile phase and peptide elution with higher formic acid concentration could even employed as appropriate.[42,44] Nevertheless, all these performances could initiate undesired formylation side reactions. Formic acid at 0.1% concentration is principally incapable of inducing severe formylation side reaction; however, the collected RP-HPLC fraction pools might be subsequently addressed to condensation process in which the concentration of formic acid could be substantially increased, and the susceptibility of peptide to formylation modification will be enhanced accordingly.[45] Target product could be regenerated through 1M HCl treatment which could hydrolyze N^{α}-formylated side products. However, it is to note that this process might cause peptide acidolysis if not meticulously conducted.

7.3.3 DMF-Induced Peptide Formylation

Another major source of formylation comes from DMF that is utilized as one of the most common organic solvents in peptide synthesis. DMF is normally stable in most conditions but might undergo various degradation when mixed with strong acids or bases, especially under high temperature, and be hydrolyzed to formic acid/dimethylamine or dimethylamine formate.[46] DMF with inferior quality is therefore potentially capable of inducing peptide formylation side reactions, particularly at high temperature, and/or in the presence of strong acid or strong base. Besides, coupling reactions between amino and carboxyl group in DMF could also be affected by undesired formylation as the free formic acid in DMF, if any, could be activated by the coupling reagent and acylate the corresponding amino group in competition with the carboxyl reactant, giving rise to the corresponding formyl amide impurity.[47]

DMF itself instead of the degraded formic acid could become the direct source of formylation side reaction in peptide synthesis through transamidation process in which amino group is converted to the corresponding formyl amide by DMF. An investigation in an effort to study the stability of coupling additive Oxyma indicates that when a certain peptide immobilized on solid support is stressed overnight in DMF at ambient temperature in the presence of Oxyma, approximately 2% N^{α}-formylated side product will be formed as a consequence

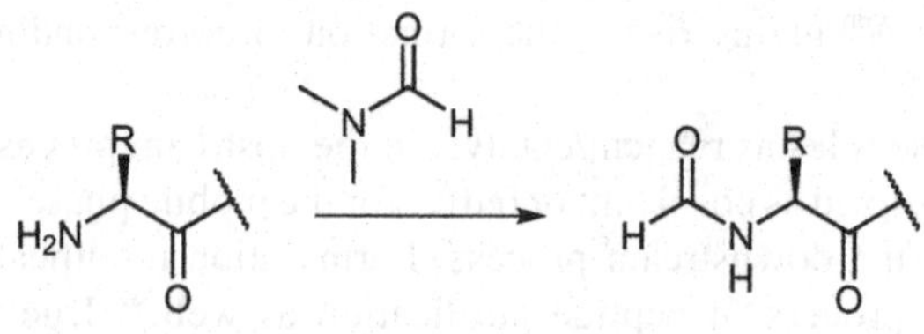

FIGURE 7.10 Peptide N^{α}-formylation directly induced by DMF.

(see also Fig. 7.10). Moreover, if the same procedure is performed at 80°C under microwave irradiation, the content of the formylated impurity would be increased to 48% after 10 min. However, the contrast experiments conducted in NMP do not generate detectable formylated side products even under microwave irradiation.[48]

DMF-induced formylation could also take place at the steps of peptidyl resin cleavage and peptide side chain global deprotection. Catalytic hydrogenation for Merrifield resin cleavage, and Z protecting deblocking could be mediated by $Pd(OAc)_2$ in DMF solution.[49] This process could be nevertheless interfere with formylation side reactions addressing Lys-N^{ε} or His-N^{im}. It has been verified by systematic contrast experiments that the subjected formylation process is not induced by residual formic acid in DMF, but by DMF itself.[4] If NMP in place of DMF is opted as solvent for the addressed catalytic hydrogenation reaction, the formation of formylated side product will be suppressed. Lys-N^{ε} functional group could also be susceptible to DMF-induced formylation side reaction in the process of peptidyl-Pam resin aminolysis.[50]

Amino acid or its ester could function directly with DMF at 60°C in the presence of imidazole and give rise to the corresponding formamide derivatives.[51] This reaction is illustrated in Fig. 7.11. DMF serves as a formyl donor in this process while imidazole is known as an excellent catalyst for acyl transfer reactions.[52] Since imidazole could potentially catalyze DMF-induced formylation, it is to note that peptide synthesis in DMF directed by CDI coupling reagent could be potentially interfered with the undesired formylation side reaction on amino functional groups, facilitated by the occurrence of imidazole molecules released upon the target activation of carboxyl group by CDI. Although relevant side reactions have not yet been published, the potential formylation process is

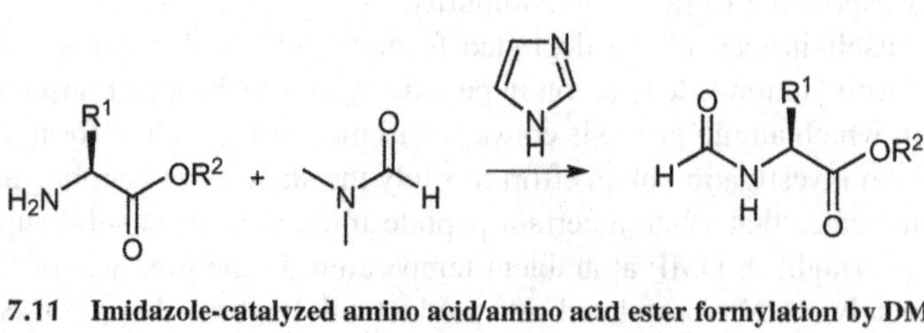

FIGURE 7.11 Imidazole-catalyzed amino acid/amino acid ester formylation by DMF.

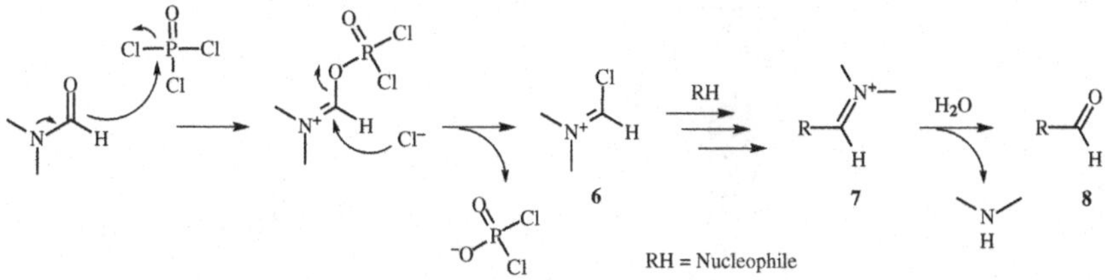

FIGURE 7.12 Mechanism of the Vilsmeier–Haack reaction.

noteworthy particularly in case of CDI-mediated sluggish and difficult coupling reactions. Since imidazole serves as an outstanding catalyst for formyl transfer reactions, the imidazolyl substituent on His side chain will be potentially susceptible to formylation modification[4] in spite of the inherent instability of the resultant His(For) derivative.

Another unnegligible pathway of DMF-mediated peptide formylation side reaction complies with the mechanism analogous to the Vilsmeier–Haack reaction.[53] Vilsmeier–Haack reaction normally utilizes DMF and phosphorus oxychloride to generate the active chloroiminium intermediate as a synthon of formyl in an effort to formylate a variety of nucleophilic substrates.[54] The mechanism of the Vilsmeier–Haack reaction is illustrated in Fig. 7.12. Firstly, DMF is activated by $POCl_3$ or other equivalent reagents, e.g., phosgene, oxalyl chloride and thionyl chloride to form chloroiminium intermediate **6** which reacts subsequently with a nucleophilic substrate to yield iminium salt **7**. Subsequently, this iminium salt undergoes hydrolysis to give rise to formyl derivative **8**.

Vilsmeier–Haack reaction undoubtedly offers a versatile synthetic strategy for the formylation of reactive aromatic and heteroaromatic substrates, as well as the construction of heterocyclic compounds.[55] However, DMF-induced formylation side reaction might be aroused in this process in a manner analogous to the Vilsmeier–Haack reaction as illustrated in Fig. 7.13. In an attempt to prepare the active peptide amide *N*-acyl-benzimidazolinone (Nbz) derivative **10** which serves as a precursor to the corresponding peptide thioester **11** addressed to a subsequent native chemical ligation, peptide intermediate **9** is supposed to undertake consecutive treatments by *p*-nitrophenyl chloroformate/DMF and DIEA/DMF solution.[56,57] However, it is detected that formylated peptide-Nbz derivative **12** is generated as an impurity in the aforementioned process, presumably attributed to the presence of DMF and the activating reagent in favor of Nbz formylation by means of the Vilsmeier–Haack equivalent reaction.[57]

Indeed, variations of the Vilsmeier–Haack reaction have been accomplished with DMF and the activating reagent other than $POCl_3$, e.g., phenyl chloroformate,[58] alkyl chloroformates[59] and benzoyl halides.[58,60] Corresponding iminium derivatives are formed in these reactions, and they function readily with the corresponding nucleophiles to facilitate the formylation processes.

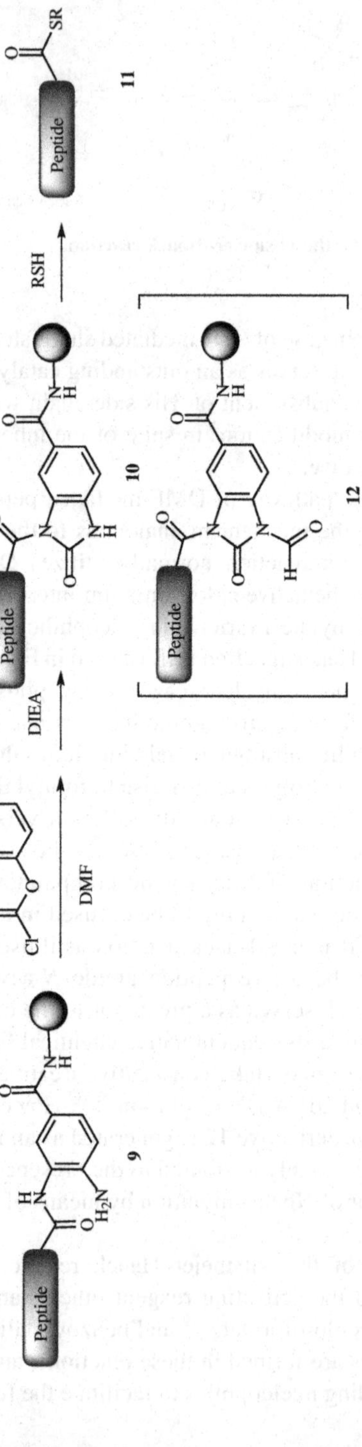

FIGURE 7.13 Formylation side reaction on peptide Nbz derivative.

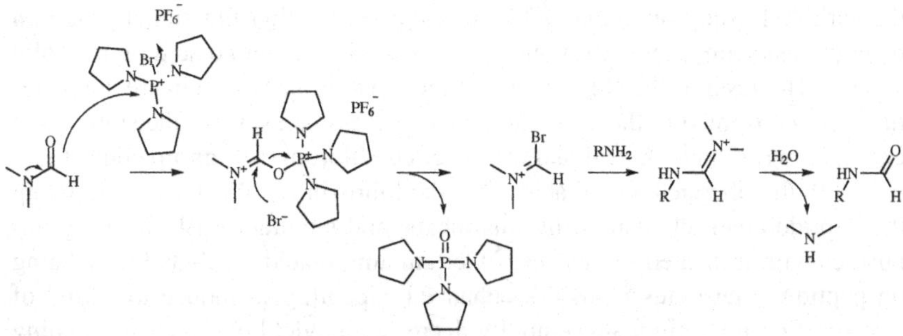

FIGURE 7.14 Proposed mechanism of peptide formylation by PyBroP/DMF.

Therefore, peptide preparation that is susceptible to DMF-induced formylation side reactions are advised to be carried-out in solvent systems devoid of tertiary amide compounds.

Halophosphonium coupling reagents like BroP, PyBroP and PyCloP exhibit excellent performances with respect to difficult couplings, e.g., *N*-methyl amino acids[61] and α,α-disubstituted amino acids.[62] In spite of these advantages halophosphonium coupling reagents might cause some side reactions. It has been figured out that PyBroP/DMF solution could modify a free amino group by converting it to the corresponding dimethyliminium salt that is hydrolyzed to the resultant *N*-formyl compound.[63] This process might also follow the mechanism analogous to the Vilsmeier–Haack reaction proposed in Fig. 7.14. If this reaction takes place on the *N*-terminus of the affected peptide chain in the process of a PyBroP/DMF-mediated amino acid coupling, the addressed peptide chain elongation will be unexpectedly terminated by means of N^{α}-formylation as a consequence.

7.4 PEPTIDE *N*-ALKYLATION SIDE REACTIONS

N-alkylation is regarded as a common side reaction in peptide synthesis which could address peptide backbone N^{α} or corresponding side chain functional groups on Lys, Arg, His and Trp. The *N*-alkylation process could occur at steps of amino acid coupling, deblocking of N^{α}-protecting groups or peptide side chain global deprotection. Alkylating agents that provoke these side reactions might originate from a variety of sources, e.g., functional groups on SPPS resin, degradative urethane protecting groups for the N^{α}-group, decomposed side chain protecting groups or even organic solvents employed in the corresponding peptide synthesis.

7.4.1 Chloromethyl Resin Induced Peptide *N*-Alkylation Side Reactions

At the early era, SPPS Merrifield resin (chloromethyl resin) was frequently adopted as the stationary phase for peptide synthesis.[64] The reaction between the carboxyl group on amino acid and the chloromethyl functional group on resin in basic condition will immobilize the subjected amino acid on the solid support. However, if the chloromethyl functional groups on resin are not quantitatively consumed in the aforementioned process and survive the subsequent end-capping, if any, the residual unconverted chloromethyl group could function with the liberated amino acid N^{α} at the following SPPS steps, terminating the peptide chain elongation by this means and simultaneously forming impurities with truncated sequences.[65] Certain amino acids such as Pro residing on peptide *N*-terminus[66] possess enhanced capability to induce this kind of side reaction that might substantially decrease the yield of the corresponding peptide synthesis.

7.4.2 Peptide *N*-Alkylation During Deblocking of N^{α}-Urethane Protecting Group

Cleavage of peptide N^{α}-urethane protecting groups, e.g., Fmoc, Boc, and Z could potentially induce *N*-alkylation side reactions. Removal of N^{α}-Fmoc is normally realized by secondary amine like piperidine. The degradative byproduct DBF **13** will be instantaneously entrapped by piperidine to give rise to DBF-piperidine adduct **14**[67] that is subsequently removed from the reaction system through appropriate workup. However, if the Michael addition reaction between DBF and piperidine does not go to completion, the unquenched DBF is able to irreversibly modify the liberated peptide N^{α}-group to give N^{α}-Fm peptide impurity **15** (Fig. 7.15).

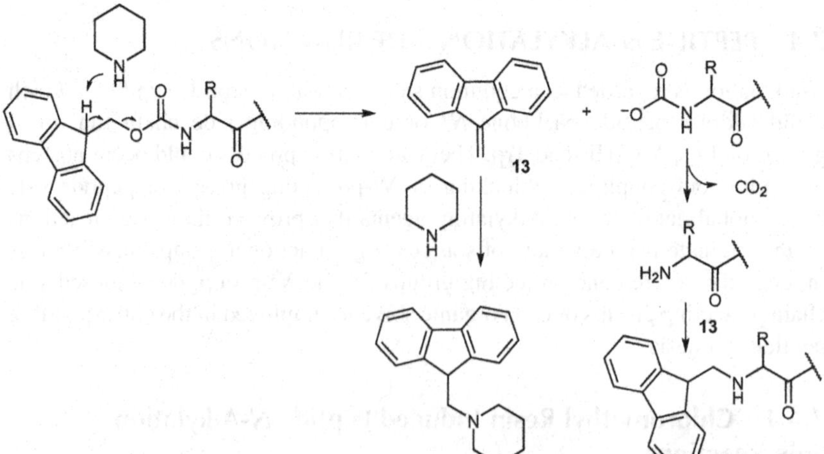

FIGURE 7.15 **Peptide N^{α}-Fm alkylation side reaction in the process of Fmoc deblocking.**

This side reaction will be exaggerated in case the quenching of DBF by the corresponding Fmoc deblocking reagent is proceeding sluggishly.[67] The kinetics of this process is significantly correlated to the features of the subjected base and solvent.[68] Nonnucleophilic bases like DBU could only be utilized as the Fmoc deblocking reagent in continuous flow systems instead of batch-wise reactions[69] in order to minimize peptide N^{α}-alkylation by DBF due to the inherent incompetence of DBU with respect to quench DBF. Sufficient removal of DBF in liquid phase peptide synthesis poses a more challenging task compared with that in SPPS, since it is imperative to evaluate the overall performances of amine derivatives both as a Fmoc deblocking reagent and as a nucleophilic scavenger for the released DBF. In this connection the efficiency of excessive amine removal, and their capabilities to sufficiently scavenge DBF should be comprehensively taken into account (diethylamine and dimethylamine possess advantages over other amines in this connection due to their high volatility). Octanethiol could serve as an excellent scavenger for DBF.[70] Its immobilization on appropriate solid supports would facilitate DBF scavenging in liquid phase peptide synthesis.[71] In case the resultant DBF-amine adduct could be readily separated from the product, the subjected amine could be taken advantage of as the appropriate reagent for N^{α}-Fmoc deblocking in liquid phase peptide synthesis. 4-(Aminomethyl)pyridine (4-AMP)[72] and tris(2-aminoethyl)amine (TAEA)[73] meet the aforementioned criteria, and could be hence utilized in place of piperidine to mediate N^{α}-Fmoc deblocking reaction in liquid phase peptide synthesis. The formed corresponding DBF adducts will be readily removed from the reaction system by aqueous extraction.

Acid labile urethane protecting groups could also initiate amino alkylation in the process of acidolytic deprotection. This kind of side reaction was initially detected during the acidolysis of *N*-Bpoc protected hydroxylamine **16**[74] as indicated in Fig. 7.16. *N*-2-(4-Biphenyl)isopropylamine side product **18** is generated concomitantly with the target product hydroxylamine **17**. This alkylation side reaction is attributed to the release of 2-(4-biphenyl)isopropyl cation upon acidolytic removal of Bpoc and its subsequent addition to the liberated amino group.

Similar to Bpoc, Boc is another acid-labile urethane protecting group. Boc is routinely deblocked from the corresponding shielded N^{α}-functional group by 50% TFA/DCM solution. If acidolysis of Boc could induce *tert*-butylation side

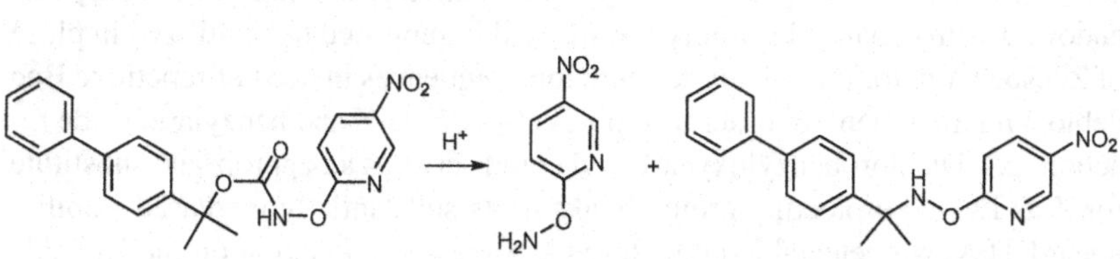

FIGURE 7.16 ***N*-alkylation side reaction during acidolysis of N-Bpoc protected hydroxylamine.**

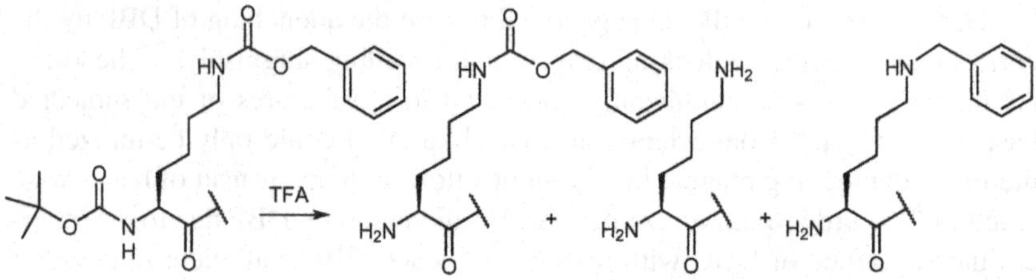

19 20 21

FIGURE 7.17 Nᵋ-de-Z and benzylation side reaction during TFA treatment of Boc-Lys(Z) derivative.

reaction on the subjected peptide, a plethora of peptide impurities with N^{α}-*tert*-butylated truncated sequences are doomed to be generated in peptide synthesis adopting Boc chemistry. Fortunately, according to a dedicated investigation[75] no significant *N-tert*-butylation side reaction will be aroused in the process of TFA-mediated peptide N^{α}-Boc deprotection.

Although Boc protecting group is basically incapable of inducing significant peptide N^{α}-alkylation side reactions, acidolysis of Boc-Lys(Z)-resin by TFA could lead to the formation of impurity N^{ε}-modified-Lys.[75] As illustrated in Fig. 7.17, when treated with 50% TFA solution compound bearing Boc-Lys(Z) moiety is converted to the target product H-Lys(Z) derivative **19** as well as N^{ε}-de-Z impurity **20**[76] and N^{ε}-benzylated compound **21**. The root cause of the formation of these two impurities lies in the intrinsic instability of the Z protecting group on Lys-N^{ε} during N^{α}-Boc acidolytic cleavage, giving rise to the premature deblocking of N^{ε}-Z and the subsequent benzylation of the liberated Lys-N^{ε} by Z-degradative benzyl cation. The formation of N^{ε}-de-Z derivative **20** could probably provoke undesired acylation side reaction on the liberated Lys-N^{ε} at the following amino acid coupling steps, and consequently give rise to side products with peptide graft on the affected Lys side chain.[76] Addition of a variety of scavengers into 50% TFA/DCM deblocking solution could not efficiently suppress this side reaction.[76] Taking into consideration that N^{α}-Z-Lys could be transformed to N^{α}-Bzl-Lys in TFA whereas Lys(Z) is converted to Lys(Bzl),[76] it implies that the process depicted in Fig. 7.17 might proceed via an intramolecular mechanism. In view of the inherent instability of Ly(Z) during Boc acidolysis alternative protecting group on Lys-N^{ε} endowed with enhanced stability toward acid is supposed to be utilized in place of Z, especially for those peptides with long sequences in need of repetitive Boc deblocking treatment, in order to suppress Lys-N^{ε} de-Z and benzylation side reaction. 2,4-Dichlorobenzyloxycarbonyl could serve as an appropriate substitute for Z as Lys-N^{ε} protecting group thanks to its substantially increased stability toward TFA. No detectable premature cleavage and alkylation on the Lys side chain is induced with the employment of 2,4-dichlorobenzyloxycarbonyl protecting group.[75,76]

7.4.3 Peptide *N*-Alkylation During Global Deprotection

Peptide *N*-alkylation at the step of side chain global deprotection is generally triggered by various cation species released by the acidolytic cleavage of amino acid side chain protecting groups or resin linkers. In case they are not efficiently scavenged these cationic derivatives could attack a variety of nucleophilic functional groups on amino acids to give rise to diverse side products. This kind of side reaction has been systematically introduced in Chapter 3 and will not be reiterated herein. Some protecting groups on amino acids could be degraded into other reactive species upon cleavage, and initiate *N*-alkylation side reactions. Alloc on Lys, All on Asp/Glu and Bum/Bom on His are potentially capable of inducing peptide *N*-alkylation during their respective deblocking processes.

7.4.3.1 Formaldehyde-Induced Peptide N-Alkylation During Side Chain Global Deprotection

Formaldehyde, either released in the process of product preparation or introduced as residual impurity from organic solvent, is one of the major root causes of *N*-alkylation side reactions in peptide chemistry. Formaldehyde serves as a common crosslinking agent that mediates diverse protein modification.[77] It could function with peptide/protein on N^{α}-group or the corresponding functional groups on Arg, His, Cys, Lys or Trp.[78] Formaldehyde can not only modify a specific amino acid but also crosslink two amino acids or two functional groups within one amino acid by means of methylene bridge.[79–81]

The reaction between formaldehyde and peptide/protein is a complicated process. Modifications of formaldehyde on peptides/proteins are principally classified into three categories: (1) formation of *N*-hydroxymethylamine derivative from formaldehyde and amino acids; (2) the generated hydroxymethylamine intermediate could undergo further dehydration to produce imine derivative; and (3) crosslinking of two amino acids or two functional groups within one amino acid by means of methylene bridge. As is depicted in Fig. 7.18, formaldehyde reacts with amino group, e.g., N^{α} or Trp/Lys/Arg/His side chain in acidic milieu to form *N*-hydroxymethylamine derivative with a molecular weight increase by 30 amu.[82,83] Some amide group is also capable of functioning with formaldehyde to give rise to *N*-hydroxymethylamide derivative.[84] The generated *N*-hydroxymethylamine intermediate could be further subjected to the dehydration process in acid condition, and converted to imine counterpart with a molecular weight increase by 12 amu. This process might address peptide N^{α} or side chain substituents on Trp or Lys (see also Fig. 7.18).

Imine derivatives formed in the aforementioned process could further react with amino acids, e.g., Gln, Asn, Trp, His, Arg, Cys, or Tyr, establishing by this means methylene-bridged crosslinking.[80] Moreover, peptide N^{α}-group could be derivatized by formaldehyde, and transformed into substituted imidazolidin-4-one via an imine intermediate.[85,86] This process is described in Fig. 7.19.

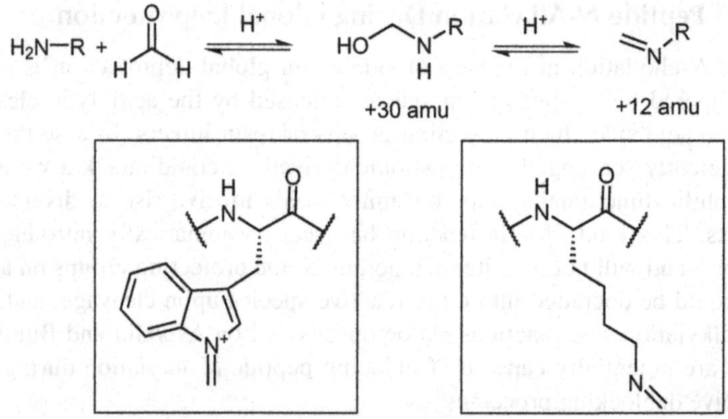

FIGURE 7.18 *N*-hydroxymethylamine and imine formation from amine and formaldehyde.

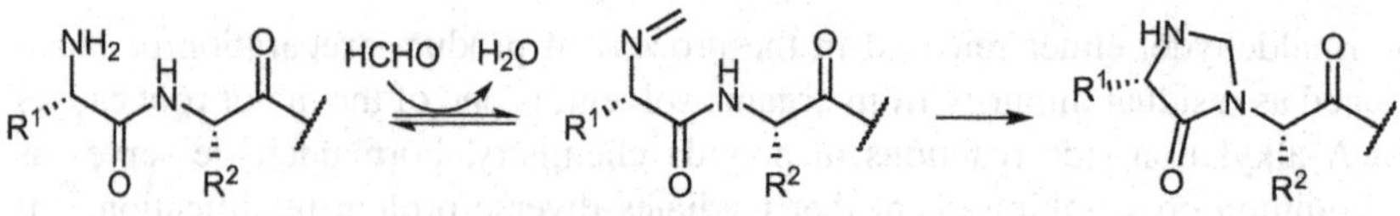

FIGURE 7.19 Formaldehyde-induced imidazolidinone formation on peptide *N*-terminus.

Formaldehyde-mediated methylene bridge crosslinking process could also take place between Lys-N^{ε} and His-N^{im}, or even between Lys-N^{ε} and Arg-$N^{\delta,\omega,\omega'}$, simultaneously establishing two methylene bridges in the latter case[87] (Fig. 7.20).

If the formaldehyde-crosslinking affected Arg and Lys originate from same peptide molecule, the aforementioned process would generate a cross-linked derivative with molecular weight increase by 24 amu compared with the target

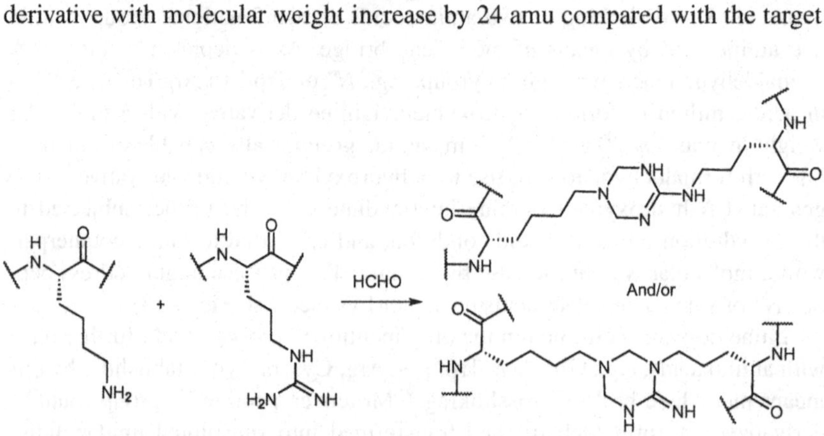

FIGURE 7.20 Formaldehyde-induced crosslinking between Lys and Arg.

peptide product. Similar reaction could also address Lys and deoxyguanosine.[88] Formaldehyde-induced crosslinking process is highly probably initiated by the reaction between formaldehyde and Lys, while the generated reactive imine intermediate further reacts with appropriate functional groups on other amino acids, and gives rise to the formation of cross-linked adducts.

Only primary amines participate in the aforementioned formaldehyde-initiated crosslinking process. In another word, only in case of the formation of imine derived from primary amine and formaldehyde could the further function with other amino acids by means of crosslinking be realized. Both N^{α} and Lys-N^{ε} could dictate this cascade process. A pertinent investigation has been conducted in this connection.[87] H-Gly-OH is added together with formaldehyde to N^{α}-acetylated peptide bearing Asn, Gln, His, Trp, Tyr, Arg, respectively, (the purpose of N^{α}-acetylation is to exclude the possible occurrence of imidazolidinone formation depicted in Fig. 7.19) as well as peptides with unacetylated N^{α}-terminus in an effort to illuminate the crosslinking pattern between Gly-N^{α} and the corresponding reciprocal substrate. The products derived from the aforementioned crosslinking processes are summarized in Fig. 7.21. It is to note that the subjected reactions have been stressed for 48 h, and no quantitative conversions have been achieved. The crosslinking between H-Gly-OH and N^{α}/Arg/Tyr is relative faster while the kinetics of crosslinking with Asn/Gln/His/Trp is low.

Besides Lys-N^{ε} and peptide-N^{α} the sulfhydryl side chain on Cys is also capable of triggering formaldehyde-mediated methylene-bridge crosslinking process. H-Cys can be subjected to the formation of five-member-ring thiazolidine through formaldehyde treatment.[89,90] Moreover, Cys can be cross-linked to guanidino substituent on Arg.[87] Products of this process are exhibited in Fig. 7.22. Treatment of Cys/Arg containing peptide by formaldehyde might splice Cys sulfhydryl group with Arg-$N^{\delta,\omega,\omega'}$ and give rise to different methylene-bridged crosslinking products. It is implied by these phenomena that formaldehyde-induced crosslinking normally occur between primary amine or Cys-sulfhydryl and substituents on His, Trp, Tyr, Asn, Gln side chains.[87]

Formaldehyde-mediated amino acid crosslinking by means of methylene-bridge is apparently pH-dependent. Although the formation of N-hydroxymethylamine and imine intermediates could be realized in acidic condition, crosslinking between diverse amino acid is in need of high pH[91] and this process will be ceased once the pH value is decreased below a certain threshold.

The formaldehyde-mediated crosslinking process between primary amino group/sulfhydryl and other reactive amino acids is not limited to an intermolecular pattern. Under certain favorable circumstances analogous reactions can proceed within one peptide by means of intramolecular crosslinking. Formaldehyde-initiated crosslinking of peptide N^{α} with reciprocal reactive amino acid residue within the same peptide molecule is a common phenomenon in peptide synthesis. In case of Cys or Trp that is located on the N-terminal of a certain peptide molecule formaldehyde in the system will be

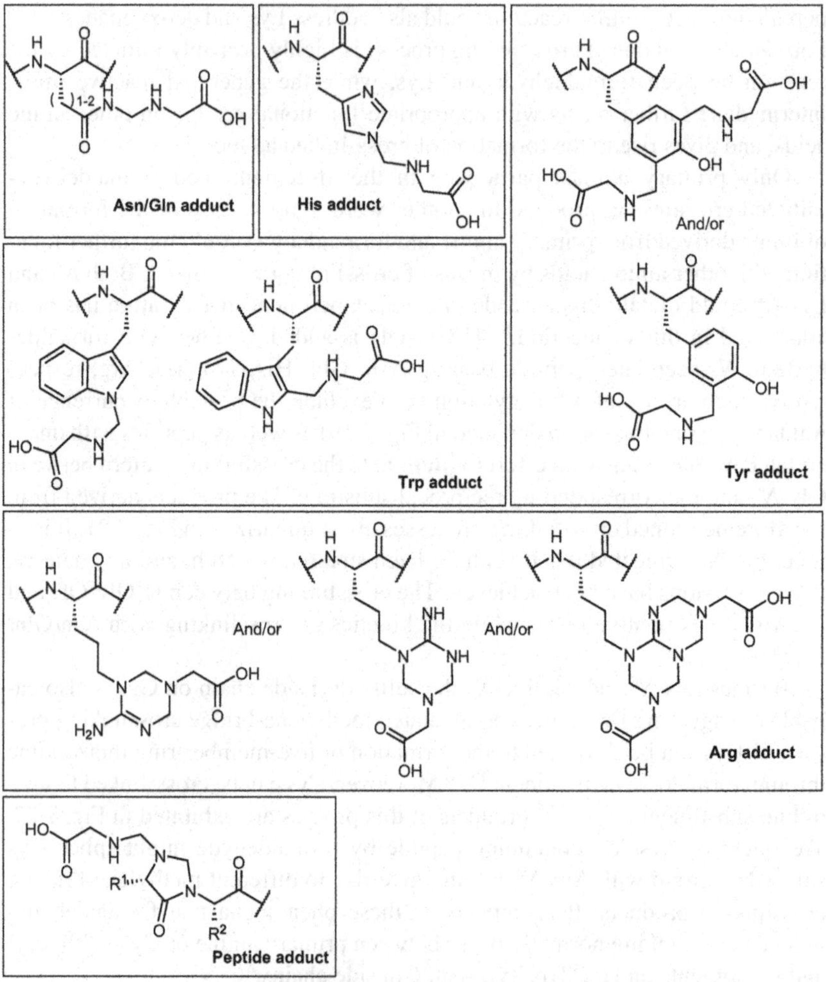

FIGURE 7.21 **Crosslinked derivatives between formaldehyde/H-Gly-OH and various amino acids.**

able to trigger crosslinking between the N^{α}-group and the sulfhydryl/indolyl substituent on the Cys/Trp side chain, and give rise to the formation of the corresponding cyclic products, i.e., Cys is converted to Thz derivative **22**; Trp to tetrahydro-β-carboline carboxylic acid derivative **23**; Lys(Nma) to dihydroquinazolinone derivative **24**;[92] and His to tetrahydropyrido-3,4-imidazole-6-carboxylic acid **25**[93] (Fig. 7.23).

Since formaldehyde-induced methylene-bridge crosslinking predominantly takes place at high pH, these side reactions are basically suppressed in acid-mediated peptide side chain global deprotection but are contingent in the work-up processes,[94] which might be ascribed to *N*-hydroxymethyl protecting groups

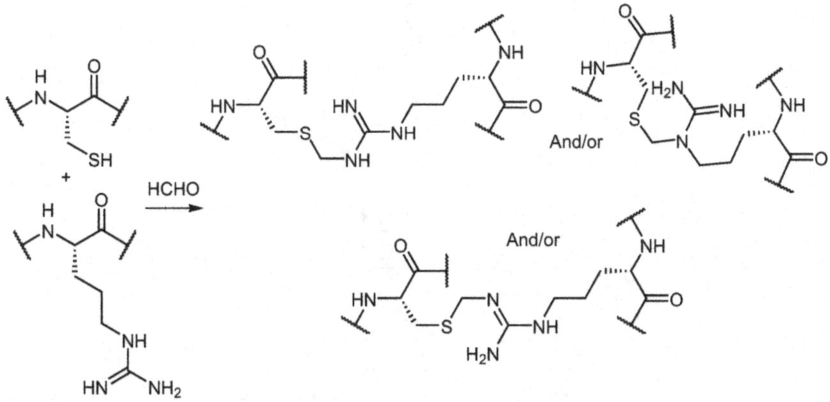

FIGURE 7.22 Crosslinking between Cys-sulfhydryl and Arg-guanidino through formaldehyde treatment.

that originally shield His-N^{im}. The subjected cleaved *N*-hydroxymethyl compound could undergo degradation during the peptide purification process, and release formaldehyde as a byproduct (hydroxymethylation of amino group in the presence of formaldehyde is a reversible process) that is capable of initiating the crosslinking process on peptide *N*-terminal Cys or Trp.

Formaldehyde and formic acid could synergistically induce methylation on amino groups following the mechanism of the Eschweiler–Clark reaction.[95] In this process primary amine or secondary amine is subjected to methylation in the presence of formaldehyde and formic acid by means of reductive amination. The mechanism of this reaction is depicted in Fig. 7.24. Firstly, the addressed primary amine reacts with formaldehyde to form *N*-hydroxymethylamine derivative that is readily subjected to the dehydration process and converted to the corresponding imine compound. The latter undertakes subsequently proton transfer from the coexisting formic acid and is transformed into iminium salt. The concomitantly generated formate anion functions as a reducing agent at the following reaction step to reduce the iminium salt to the corresponding secondary amine. The release of CO_2 in this process endows the equilibrium the driving force to the direction of secondary amine formation. The intermediate secondary amine could further react with formaldehyde/formic acid in the identical manner, and this cascade process will be terminated at the formation of the corresponding tertiary amine. Similar side reaction can affect Fenfluramine (3-trifluoromethyl-*N*-ethylamphetamine) on its secondary amino substituent in the presence of formaldehyde/formic acid. As a consequence *N*-methyl fenfluramine is formed as an impurity.[96]

When formaldehyde is present in the reaction system it could be oxidized to formic acid by air.[84] The Eschweiler–Clark reaction synergistically induced by formaldehyde and formic acid can affect peptide N^{α}-group and Lys-N^{ε} by means of amino monomethylation and dimethylation.

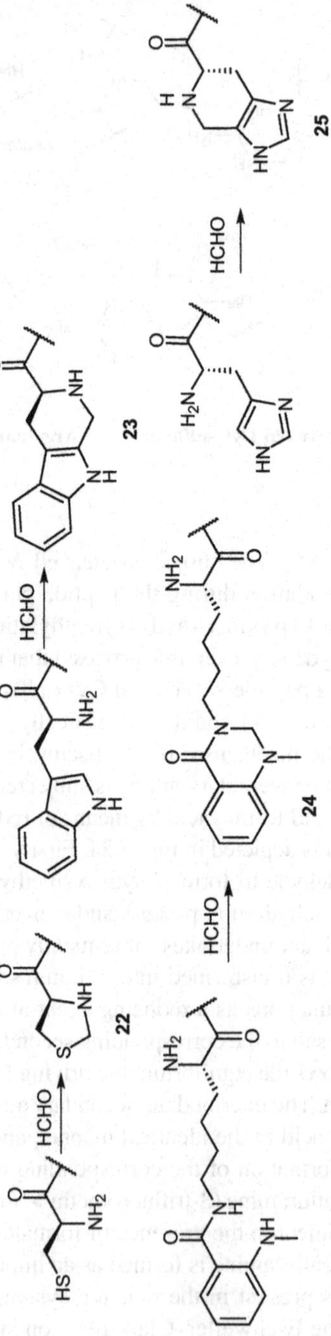

FIGURE 7.23 **Formaldehyde-induced crosslinking on peptide N-terminal Cys, Trp, Lys(Nma), and His.**

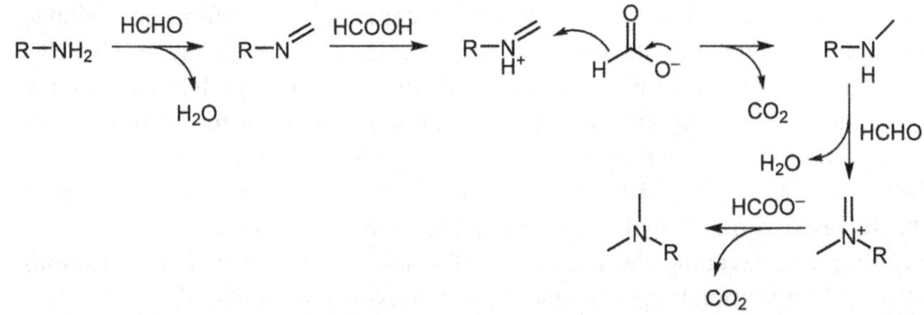

FIGURE 7.24 **Mechanism of the Eschweiler–Clarke reaction.**

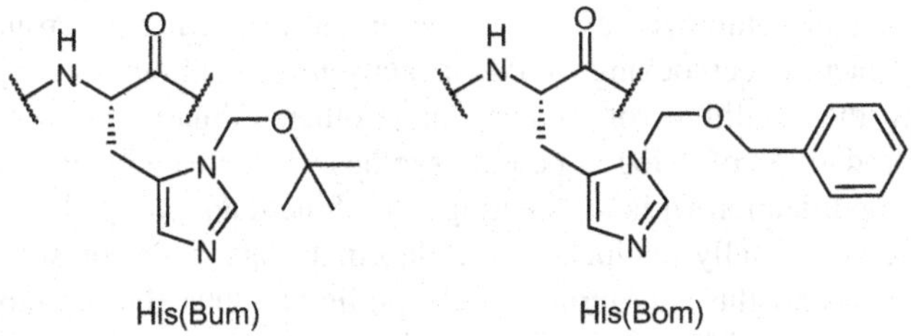

FIGURE 7.25 **Structures of His(Bum) and His(Bom).**

Aldehyde that affects peptide synthesis might be originated from residual impurities contaminating certain organic solvents, e.g., methanol[97] and PEG.[98] Moreover, utilization of certain protected amino acid residues such as His(Bum/Bom) (Fig. 7.25) can lead to formaldehyde formation at the side chain deprotection step. Employment of His(Bum/Bom) as building blocks for peptide synthesis is favored in view of the high tendency of His residue to undergo racemization at the coupling step, predominantly due to the direct H^{α} abstraction by the unprotected N^{π} moiety on His imidazolyl side chain in the process of His carboxyl-activation. This process gives rise to the formation of enol intermediate that regulates the configuration conversion of the chiral C^{α}.[99] The strategy of π-N^{im} protection on His imidazolyl side chain is developed in an effort to suppress the distinctive racemization of His residue during carboxyl-activation. His(Bum) and His(Bom) are invented in light of this idea, and have been utilized for peptide synthesis both in Fmoc and Boc chemistry. The advantageous effects of these two His-N^{im} protecting groups against racemization have been verified compared with N^{τ}-protection strategy.[100,101]

One of the major limitations of His(Bum) and His(Boc) utilization in peptide synthesis lies in the release of formaldehyde upon deblocking these protecting groups [89,101] and the resultant alkylation side reactions. Formaldehyde produced from degraded Bum and Bom protecting groups, in case of insufficient quenching in due reactions, is potentially capable of causing a variety of alkylation side reactions. Efficient formaldehyde scavengers are, therefore, necessitated to avoid potential modifications of amino acids by formaldehyde. Cys·HCl is frequently employed as a scavenger for the side chain global deprotection of peptides containing His(Bum/Bom) residue by virtue of the high tendency of Cys-sulfhydryl to entrap formaldehyde. Hydroxylamine compounds, e.g., $HONH_2 \cdot HCl$, $MeONH_2 \cdot HCl$ and $BuONH_2 \cdot HCl$ could also effectively scavenge the released formaldehyde during peptide side chain global deprotection,[92] avoiding or alleviating the undesirable formaldehyde-mediated modifications of peptide *N*-terminal Cys, Trp and other reactive amino acids.

7.4.3.2 Peptide N-alkylation during Pd(0)-catalyzed N-Alloc deblocking

Alloc is routinely employed as a versatile amino-protecting group in peptide synthesis. Since its deblocking by Pd(0) catalyst/nucleophilic scavenger is orthogonal to the conditions of cleaving most other ordinary amino-protecting groups, introduction of Alloc in peptide synthesis has created new dimensions for amino-modification without affecting other functional groups.[102]

N-Alloc is normally deblocked in mild conditions in the presence of the Pd(0) catalyst and the appropriate nucleophilic reagents that entrap the degraded allyl species. The mechanism of this process is described in Fig. 7.26. Deblocking of the *N*-Alloc protecting group proceeds via Pd(0)-regulated η^3-allyl palladium complex intermediate.[103] The Pd(0) catalyst, e.g., $Pd^0(PPh_3)_4$ (the actual catalyst species might be $Pd^0(PPh_3)_2$ from $Pd^0(PPh_3)_4$ degradation), firstly attacks the target allyl group to release the corresponding carbamate anion, giving rise to the formation of cationic π-allyl-palladium complex **26** and anionic carbamate **27**. From the aspect of organometallic chemistry this reaction belongs to the category of oxidative addition,[104] since the oxidation state of the palladium is increased from 0 to +2 and its coordination number is accordingly increased. Carbamate derivative **27** undergoes spontaneous degradation subsequently to release the corresponding amine compound, whereas π-allyl-palladium complex **26** undertakes the nucleophilic attack from a nucleophile (regarded as allyl scavenger) and is transformed to allyl-Nu derivative **28** upon releasing Pd(0) from the complex. The latter step is an example of reductive elimination in organometallic chemistry[104] in that the oxidation state

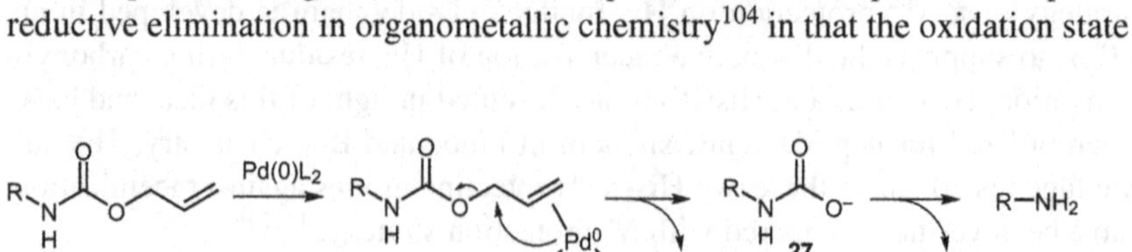

FIGURE 7.26 Mechanism of Pd(0)-catalyzed N-Alloc deblocking.

of palladium decreases from +2 to 0, and the $Pd^0(PPh_3)_2$ catalyst is regenerated by this means.

The prerequisite for the Pd(0)-catalyzed *N*-Alloc deprotection reaction is the sufficient leaving tendency of allyl-bonded group in the parental molecule in order to readily form the π-allyl-palladium complex cation **26**. Carboxylic acid, phenol, and phosphonic acid are all endowed with this attribute. On the contrary, allyl ether, allyl thioether and allyl amine are resistant to the Pd(0)-catalyzed deallylation due to the inferior leaving tendencies of the corresponding hydroxyl, sulfhydryl and amino groups. In light of this consideratoin the appropriate protecting group for the parental alcohol, mercaptan and amine compounds is supposed to be Alloc in place of allyl, thanks to the enhanced leaving tendency of carbamate in case of the *N*-Alloc derivative.

In spite of the important roles of allyl and Alloc as orthogonal protecting groups in peptide synthesis, they are potentially capable of inducing allylation side reactions on susceptible amino groups in the process of Pd(0)-catalyzed deblocking.[105] The subjected *N*-allylation side reactions might follow a variety of mechanisms of under such circumstances:

1. The liberated amino substituent regenerated from *N*-Alloc shielded derivative can direct a competitive reaction against the nucleophilic allyl scavenger to react with the formed cationic π-allyl-palladium complex, and be transformed to the corresponding *N*-allyl derivative **29** (Fig. 7.27). This undesired allylation process can be effectively suppressed by the protonation of the regenerated amino group, restraining by this means its nucleophilic attack on the cationic π-allyl-palladium complex that leads to the concerned *N*-allylation side reaction. Alternatively, utilization of the large excessive allyl scavenger can also help minimize the competitive amino-allylation.
2. *N*-allylation side reaction could also be provoked in the hydride-involved Alloc-deblocking process as depicted in Fig. 7.28. Differing from the aforementioned *N*-allylation mechanism, the subjected hydride-mediated *N*-allylation takes place upon the post-formation of allyl derivative **30**

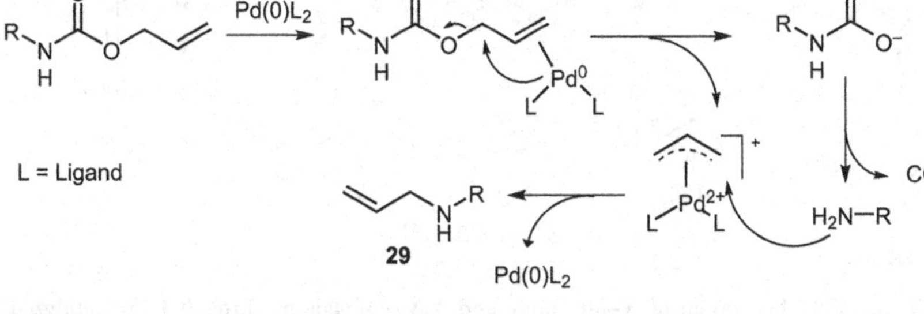

FIGURE 7.27 *N*-allylation side reaction through the nucleophilic attack of the liberated amino group on cationic π-allyl-palladium complex.

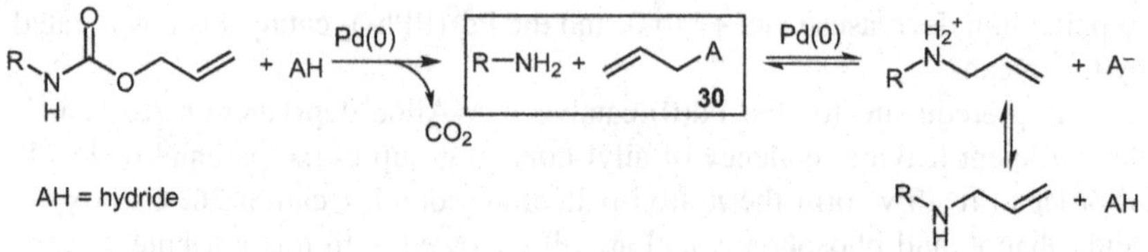

FIGURE 7.28 Reversible hydride allyl scavenger-induced *N*-allylation during Alloc deblocking.

obtained from cationic π-allyl-palladium complex entrapping by hydride allyl-scavenger AH. Due to the relatively high leaving tendency of substituent A in allyl derivative **30** the regenerated amine can initiate a nucleophilic attack on **30**, and the allyl group is consequently shifted to the affected amine.

3. *N*-Alloc protected amine compounds can undergo decarboxylation-rearrangement triggered by Pd(0)-catalyst in the absence of an allyl scavenger, and be converted to the corresponding *N*-allyl derivative. The susceptibility to this side reaction is largely dependent on the inherent properties of the addressed amine. Generally speaking, secondary amines undergo faster decarboxylation-rearrangement than primary amines.[104,106] The scheme of this side reaction is described in Fig. 7.29. The carbamate anion intermediate formed in the process of Pd(0)-catalyzed degradation of *N*-Alloc amine could function with cationic π-allyl-palladium complex in an intramolecular manner (intra-ion pair), giving rise to the corresponding allylamine by releasing CO_2 and $Pd(0)L_2$. The newly formed allylamine in this process could further react with π-allyl-palladium complex cation to give *N,N*-diallylamine derivative. Similar to *N*-Alloc protected amine (allyl carbamate) *O*-Alloc protected alcohol (allyl carbonate) could undergo analogous decarboxylation-rearrangement, and be converted to the corresponding allylether. However, alcohol derivatives are normally not profoundly affected by this side reaction thanks to the weaker nucleophilicity of hydroxyl group.

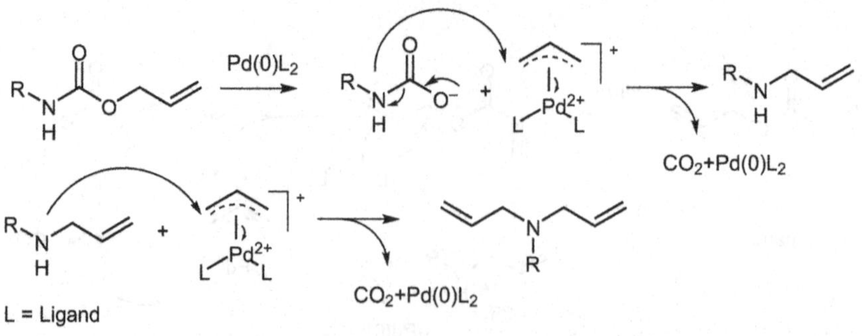

FIGURE 7.29 Formation of *N*-allylamine and *N,N*-diallylamine through Pd(0)-catalyzed decarboxylation-rearrangement of allyl carbamate.

In light of the aforementioned mechanisms of amine allylation it is advisable to opt the appropriate nucleophilic allyl scavenger to suppress this undesirable side reaction. The most commonly utilized ally scavengers in peptide synthesis are generally categorized into the following classes:

1. *Oxygen-type nucleophile*: Carboxylic acid or carboxylate derivatives could not effectively restrain *N*-allyllation side reactions.[107] Acetic acid has also been disqualified as an efficient allyl scavenger.[108] Other *O*-nucleophiles for allyl scavenging include HOSu[109] and HOBt[110] that have been routinely employed in peptide synthesis.
2. *Nitrogen-type nucleophile*: morpholine has ever been applied as an allyl scavenger for the deallylation of amino acids and glycopeptides.[111] Pyrroline and piperidine could also be utilized in an effort to entrap the released allyl groups.[112,113] Even tertiary amine, e.g., NMM, is applicable to serve as an allyl scavenger in the Pd(0)-mediated deallylation reaction.[114]
3. *Carbon-type nucleophile*: weakly acidic β-bis(carbonyl) compounds, e.g., 5,5-dimethylcyclohexane-1,3-dione (dimedone)[115] and 1,3-dimethylbarbituric acid[116] can be used as an allyl scavenger in peptide synthesis. Their function with cationic π-allyl-palladium complex is regarded as irreversible.[116] During the due scavenging process β-bis(carbonyl) compound firstly transfers a proton to the carbamate-π-allyl-palladium complex intermediate, and it is converted simultaneously to an enol derivative which regulates an irreversible function with the π-allyl-palladium complex, scavenging by this means the allyl species and regenerating the Pd(0) catalyst. Pd(0)/β-bis(carbonyl) compounds have already been widely used in a variety of deallylation reactions. Due to the distinctive irreversibility of the reaction, as well as advantages that the liberated amine derivative will be protonated by the excessive acidic β-bis(carbonyl) compound, the potential *N*-allylation side reaction in the process of the N-Alloc deblocking could be largely suppressed.
4. *Sulfur-type nucleophile*: 2-mercaptobenzoic acid can be utilized as allyl scavenger.[117]
5. *Silyl nucleophile*: *N*-trimethylsilylamine derivatives have found applications as proper scavengers in the deallylation process.[106,118] The formed intermediate trimethylsilyl carboxylate or trimethylsilyl carbamate is hydrolyzed in neutral condition to give the corresponding carboxylic acid and amine, respectively. Thanks to the fact that the *N*-Alloc protected amine is firstly transformed to silyl carbamate instead of free amine, the shielded amino group will hence be unable to participate the competitive nucleophilic attack on the π-allyl-palladium complex cation, and the potential allylation side reaction could be thus avoided. In spite of the alleged advantages of silyl-scavengers allylation modification on amino group could still be detected in some reactions. Nonetheless, the addition of mild silanyl nucleophiles, e.g., trimethylsilyl acetate, trimethylsilyl trifluoroacetate and trimethylsilyl methansulfonate, is verified to be able to thoroughly suppress

N-allylation side reactions. This merit might be attributed to the mechanism that the added silanyl reagents will promptly function with the liberated carbamate anion to generate trimethylsilyl carbamate via silyl exchange, while the formed π-allyl-palladium complex cation could still be trapped by the concomitant *N*-silylamine scavenger. The potential *N*-allylation side reaction is thus suppressed by this synergistic effect. One of the intrinsic drawbacks *N*-trimethylsilylamine lies in the premature deblocking of Fmoc protecting groups that limits its application for *N*-Alloc cleavage in the presence of Fmoc groups.

6. *Hydride nucleophile*: Conventional hydrides employed as allyl scavengers in deallylation reaction include silane, tributyltin hydride, and borohydride. Pd(II)-type catalysts, e.g., $PdCl_2(PPh_3)_2$ and $Pd(OAc)_2$ could be utilized for deallylation in the presence of these hydrides, as they could be reduced *in situ* to the corresponding Pd(0) counterparts that catalyze the deallylation process.[119]

Tributyltin hydride reduces the degraded allyl group to propene, and the liberated amino group is simultaneously transformed to the corresponding *N*-tributyltin derivative, that regenerates the target amine product by releasing the tributyltin species in an acidic milieu. Tributyltin hydride-facilitated deallylation reaction is, therefore, advised to proceed in the presence of a weak acid, e.g., acetic acid, in order to obtain the amine product directly.[120] Differing from *N*-trimethylsilylamine tributyltin hydride is compatible with Fmoc, and will not cause premature Fmoc deblocking in the subjected reaction conditions.

Pd(0)/$NaBH_4$ or Pd(0)/$NaBH_3CN$ mixture could be applied for deallylation process.[121] It has been disclosed in a study[122] that the primary/secondary amine–borane system can effectively suppress *N*-allylation side reactions occurred in Pd(0)-mediated deallylation processes. $NH_3{\cdot}BH_3$ and $Me_2NH{\cdot}BH_3$ have also exhibited excellent performances in this connection.

The most frequently employed deallylation reagents in peptide synthesis are Pd(0)/$PhSiH_3$ that is capable of quantitatively removing Alloc protecting groups from the shielded amine within a couple of minutes at ambient temperature.[123] If Pd(0)/$PhSiH_3$-regulated *N*-Alloc removal proceeds in the presence of acylating agents like amino acid active ester, the addressed *N*-Alloc precursor can be converted *in situ* to the corresponding *N*-acyl derivative.[123] Thanks to the inherent properties of the Pd(0)/$PhSiH_3$ system with regard to the superior reaction kinetics, adaptability to one spot deblocking/acylation synthetic strategy, compatibility with the other functional moieties/protecting groups, as well as the evident suppression of *N*-allylation side reaction, Pd(0)/$PhSiH_3$ has gained profound applications in peptide synthesis for *N*-Alloc removal.

It could be summarized from the aforementioned descriptions that Pd(0)-mediated *N*-Alloc removal could induce undesirable *N*-allylation side reactions which mostly address secondary amine derivatives. This side reaction could be, nevertheless, suppressed or alleviated through the utilization of excessive, and/or highly effective allyl scavengers in the affected reaction systems.

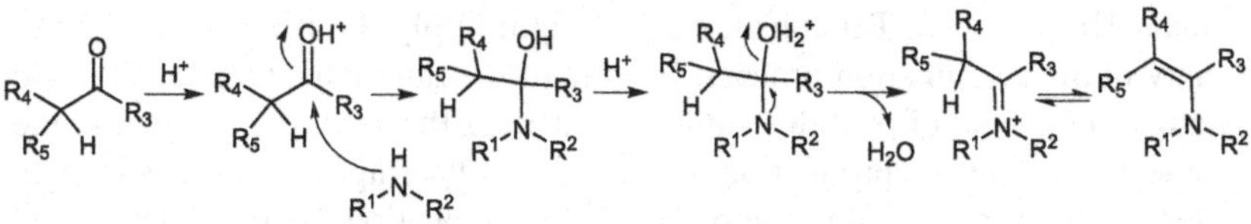

FIGURE 7.30 Mechanism of enamine formation.

7.4.4 *N*-Alkylation Side Reaction on *N*-Terminal His via Acetone-Mediated Enamination

Enamine formation is a common reaction in organic synthesis that normally takes places between secondary amine and H^{α}-bearing carbonyl compounds, e.g., ketone or aldehyde. The function between the two reactants under acidic condition gives rise to enamine formation through nucleophilic addition and dehydration intermediary steps (Fig. 7.30). No imine could be generated under such circumstances due to the fact that the participating secondary amine possesses merely one H atom bonded to the amino group.

The peptide could also be subjected to an enamination modification on the His imidazolyl side chain during preparative or purification processes (Yang, Y., unpublished results). These side reactions are generally addressing His residue located on peptide *N*-terminus (see also Fig. 7.31). Acetone frequently induces enamination on His side chain in the process of peptide production. The contamination of peptide material by acetone is normally attributed to the inherently deficient cleaning process of reaction vessels. Acetone is routinely utilized as detergent in industrial chemical production. Nevertheless, if residual acetone is not sufficiently removed from the subjected vessel and is carried over to the next manufacturing step, they might be capable of initiating modifications of the *N*-terminal His residue on its imidazolyl side chain. The outcome of this process is the generation of derivative **31** bearing His-N^{im}-(prop-1-en-2-yl) moiety whose molecular weight is increased by 40 amu compared with the target His product. No dedicated research has been conducted to investigate the rationale of the selectivity of this modification on *N*-terminal His instead

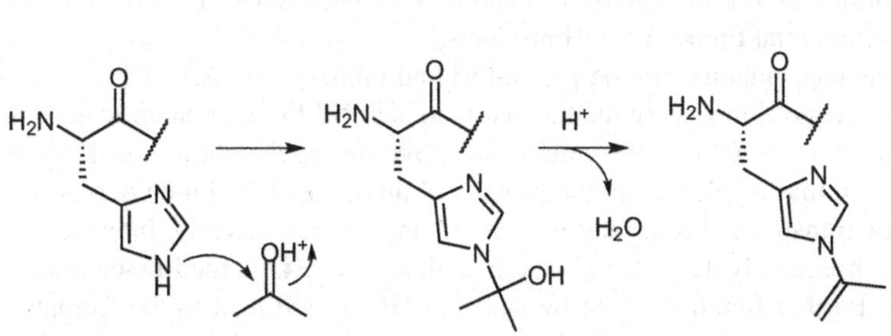

31

FIGURE 7.31 Acetone-induced N-terminal His enamination.

of *endo*-His residues. This phenomenon might imply the participation of N^{α} moiety in the enamination process. In spite of the rare indications of this side reaction in the field of peptide synthesis it justifies the necessity to exercise the utmost caution for the production of peptide API bearing *N*-terminal His in an industrial environment from the preparation, purification, vessel cleaning and storage aspects.

7.5 SIDE REACTIONS DURING AMINO ACID N^{α}-PROTECTION (Fmoc-OSu INDUCED Fmoc-β-Ala-OH AND Fmoc-β-Ala-AA-OH DIPEPTIDE FORMATION)

A variety of side reactions could affect the process of amino acid N^{α}-Fmoc derivatization. The dipeptide Fmoc-Xaa-Xaa-OH formation triggered by Fmoc-Cl mediated N^{α}-Fmoc functionalization will be elaborated dedicatedly in Section 10.1. In this section Fmoc-OSu-induced Fmoc-β-Ala-OH and Fmoc-β-Ala-AA-OH dipeptide formation are described.

Since Fmoc-chemistry is predominantly followed for peptide manufacturing the strategy of amino acid N^{α}-Fmoc derivatization has attracted substantial attentions. A plethora of Fmoc derivatization reagents, e.g., Fmoc-Cl, Fmoc-N_3 and Fmoc-OSu have been developed and utilized. Applicability of Fmoc-Cl and Fmoc-N_3 in large-scale manufacturing is anyhow restricted due to their respective intrinsic limitations in that Fmoc-Cl will inevitably result in dipeptide formation whereas Fmoc-N_3 leads to sluggish reactions and low yields.[124] Thanks to the excellent performances with regard to yield, reaction kinetics and product purity Fmoc-OSu has obtained intensive applications in the field of Fmoc-amino acid manufacturing. Amino acid N^{α}-Fmoc derivatization by Fmoc-OSu is generally conducted with standard procedures in dioxane/H_2O,[125] ACN/H_2O[126] or DMF/H_2O[127] mixed solvents facilitated by $NaHCO_3$,[128] Na_2CO_3,[125] or triethylamine[126] as auxiliary base.

In spite of the overall superior attributes of Fmoc-OSu over Fmoc-Cl and Fmoc-N_3 the succinimide moiety in Fmoc-OSu molecule can induce a series of distinctive side reactions. One of the most predominant challenges associated with the employment of Fmoc-OSu for amino acid N^{α}-Fmoc derivatization is the formation of Fmoc-β-Ala-OH and Fmoc-β-Ala-AA-OH as contaminants in the commercial Fmoc-AA-OH products.[129]

The mechanism of Fmoc-β-Ala-OH and Fmoc-β-Ala-AA-OH formation in the process of Fmoc-OSu mediated amino acid N^{α}-Fmoc protection is depicted in Fig. 7.32.[129] Fmoc-OSu might undertake the nucleophilic attack by water or HOSu that is released in the process of amino acid N^{α}-Fmoc derivatization, and be transformed into intermediate **32** and **33**, respectively. Intermediate **33** will subsequently degrade to isocyanate derivative **34** via the Lossen rearrangement. Further function of **34** by water or HOSu will lead to the formation of intermediary carbamic acid **36** and carbamate **37**, respectively, both of which will be decomposed via different pathway to H-β-Ala-OSu **38**. In the presence

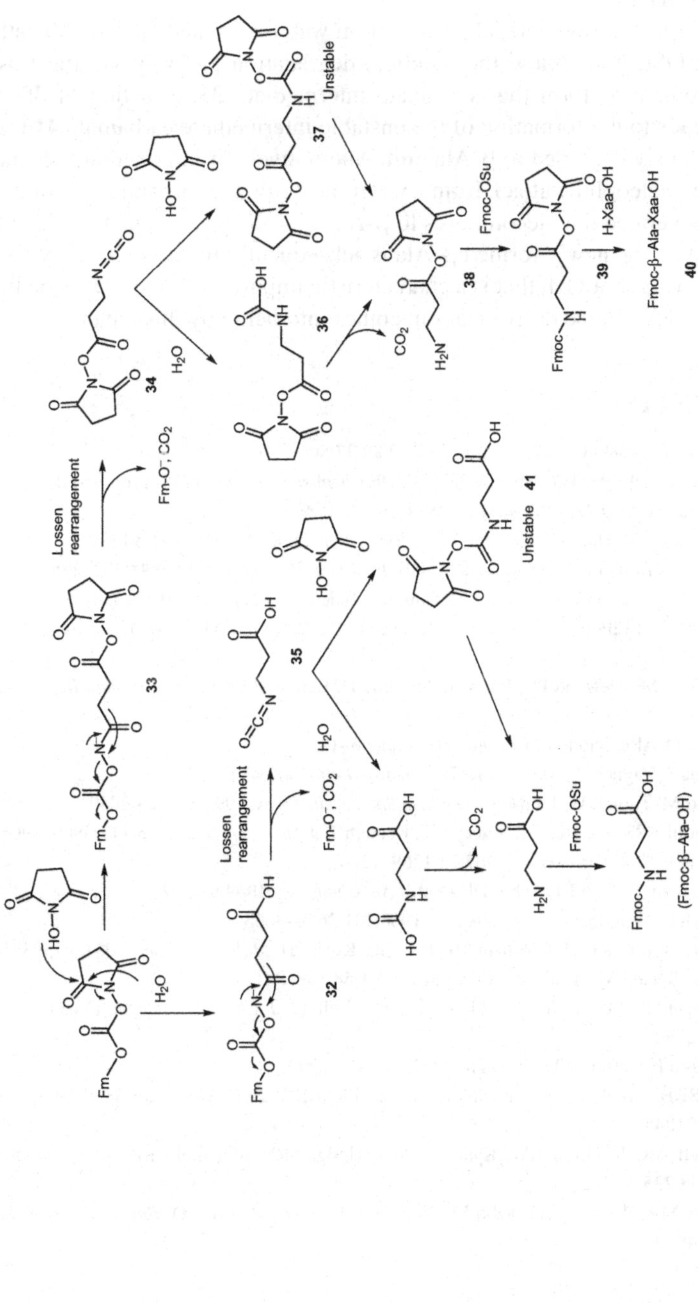

FIGURE 7.32 Mechanism of Fmoc-β-Ala-OH/Fmoc-β-Ala-Xaa-OH dipeptide formation during Fmoc-OSu mediated N$^{\alpha}$-Fmoc derivatization on Xaa.

of excessive Fmoc-OSu derivative **38** will be inevitably subjected to N^{α}-Fmoc derivatization, and be consequently converted to the corresponding active ester Fmoc-β-Ala-OSu **39**. The underivatized amino acid H-Xaa-OH in the reaction system will react with Fmoc-β-Ala-OSu **39** to give rise to dipeptide Fmoc-β-Ala-Xaa-OH **40**.

Similarly, intermediate **32** derived from water-mediated nucleophilic attack at Fmoc-OSu will follow the identical degradation pathway via the Lossen rearrangement to form the isocyanate intermediate **35**. Reaction of **35** with HOSu leads to the formation of the unstable intermediate carbamate **41** that is spontaneously degraded to β-Ala unit. Alternatively, if intermediate **35** undertakes the nucleophilic attack from a water molecule the carbamic acid of β-Ala will be generated that decomposes to β-Ala as well by releasing CO_2. The free N^{α} group on the newly formed β-Ala is subsequently functionalized by Fmoc-OSu. Fmoc-β-Ala-OH, that is a characteristic impurity of Fmoc-OSu mediated amino acid N^{α}-Fmoc derivatization, comes into being by this means.

REFERENCES

1. Ishiguro T, Eguchi C. *Chem Pharm Bull.* 1989;37:506–508.
2. (a) Riordan JF, Wacker WEC, Vallee BL. *Biochemistry.* 1965;4:1758–1765. (b) de Satz VB, Santome JA. *Int J Pept Protein Res.* 1981;18:492–499.
3. Macdonald JM, Haas AL, London RE. *J Biol Chem.* 2000;275:31908–31913.
4. Hsieh KH, Demaine MM, Gurusiadaiah S. *Int J Pept Protein Res.* 1996;48:292–298.
5. Coïc Y-M, Le Lan C, Neumann J-M, Jamin N, Baleux F. *J Pept Sci.* 2010;16:98–104.
6. Schieck A, Müller T, Schulze A, Haberkorn U, Urban S, Mier W. *Molecules.* 2010;15:4773–4783.
7. Schnölzer M, Alewood PE, Jones A, Alewood D, Kent SBH. *Int J Pept Protein Res.* 1992;40:180–193.
8. Marder O, Albericio F. *Chim Oggi.* 2003;June:6–11.
9. Yamada S, Yaguchi S, Matsuda K. *Tetrahedron Lett.* 2002;43:647–651.
10. Quibell M, Turnell W, Johnson T. *J Chem Soc Perkin Trans.* 1993;1:2843–2849.
11. Youngquist RS, Fuentes GR, Lacey MP, Keough T. *J Am Chem Soc.* 1995;117:3900–3906.
12. Wünsch E. *Z Naturforsch B.* 1967;22:1269–1276.
13. Hirschmann R, Nutt RF, Veber DF, et al. *J Am Chem Soc.* 1969;91:507–508.
14. Wünsch E, Wendlberger G. *Chem Ber.* 1968;101:3659–3663.
15. Burkhart DJ, Kalet BT, Coleman MP, Post GC, Koch TH. *Mol Caner Ther.* 2004;3:1593–1604.
16. Han G, Tamaki M, Hruby VJ. *J Pept Res.* 2001;58:338-341.
17. Bollhagen R, Schmiedberger M, Barlos K, Grell E. *J Chem Soc Chem Commun.* 1994:2559–2560.
18. Palladino P, Stetsenko D. *Org Lett.* 2012;14:6346–6349.
19. Kent SBH, Mitchell AR, Engelhard M, Merrifield RB. *Proc Natl Acad Sci USA.* 1979;76:2180–2184.
20. Mitchell AR, Erickson BW, Ryabtsev MN, Hodges RS, Merrifield RB. *J Am Chem Soc.* 1976;98:7357–7362.
21. Ondetti MA, Williams NJ, Sabo EF, Pluscec J, Weaver ER, Kocy O. *Biochemistry.* 1971;10:4033–4039.

22. Bush ME, Alkan SS, Nitecki DE, Goodman JW. *J Exp Med.* 1972;136:1478–1483.
23. Haase C, Burton MF, Agten SM, Brunsveld L. *Tetrahedron Lett.* 2012;53:4763–4765.
24. Tam J, Kent SBH, Wong TW, Merrifeld RB. *Synthesis.* 1979:955–957.
25. Kocsis L, Ruff F, Orosz G. *J Pept Sci.* 2006;12:428–436.
26. Hübener G, Göhring W, Musiol H-J, Moroder L. *Pept Res.* 1992;5:287–292.
27. Rabiet M-J, Huet E, Boulay F. *Eur J Immunol.* 2005;35:2486–2495.
28. Wisniewski JR, Zougman A, Mann M. *Nucleic Acids Res.* 2008;36:570–577.
29. Matsueda GR. *Int J Pept Protein Res.* 1982;20:26–34.
30. Kisfaludy L, Schön I, Náfrádi J, Varga L, Varró V. *Hoppe-Seyler's Z Physiol Chem.* 1979;359: 887–895.
31. Kiso Y, Ukawa K, Namamura S, Ito K, Akita T. *Chem Pharm Bull.* 1980;28:673–676.
32. Yamashiro D, Li CH. *J Org Chem.* 1973;38:2594–2597.
33. Ohno M, Tsukamoto S, Makisumi S, Izumiya N. *Bull Chem Soc Jpn.* 1972;45:2852–2855.
34. Chowdhury SK, Chait BT. *Anal Biochem.* 1989;180:387–395.
35. Odagami T, Tsuda Y, Kogami Y, Kouji H, Okada Y. *Chem Pharm Bull.* 2009;57:211–213.
36. Halpern B, Nitecki DE. *Tetrahedron Lett.* 1967;31:3031–3033.
37. Bandgar BP, Kinkar SN, Chobe SS, Mandawad GG, Yemul OS, Dawane BS. *Arch Appl Sci Res.* 2011;3:246–251.
38. Curtius T. *Ber Dtsch Chem Ges.* 1902;35:3226–3228.
39. Hojo K, Maeda M, Smith TJ, Kawasaki K. *Chem Pharm Bull.* 2002;50:140–142.
40. Gross E, Witkop B. *J Biol Chem.* 1962;237:1856–1860.
41. Piszkiewicz D, Landon M, Smith EL. *Biochem Biophys Res Commun.* 1970;40:1173–1178.
42. Tarr GE, Crabb JW. *Anal Biochem.* 1983;131:99–107.
43. Shively JE, Pande H, Yuan PM, Hawke D. In: Elzinga M, ed. *Methods in Protein Sequence Analysis IV.* Clifton, New Jersey: Hamana Press; 1982:447–454.
44. Heukeshoven J, Dernick R. *J Chromatogr.* 1985;326:91–101.
45. Viville R, Scarso A, Durieux JP, Loffet A. *J Chromatogr.* 1983;262:411–414.
46. Meglitskii VA, Kvasha NM. *Fibre Chem.* 1971;3:327–329.
47. Effenberger F, Mück AO, Bessey E. *Chem Ber.* 1980;113:2086–2099.
48. Subirós-Funosas R, Prohens R, Barbas R, El-Faham A, Albericio F. *Chem Eur J.* 2009;15: 9394–9403.
49. Schlatter JM, Mazur RH. *Tetrahedron Lett.* 1977;18:2851–2852.
50. Lelièvre D, Turpin O, El Kazzouli S, Delmas A. *Tetrahedron.* 2002;58:5525–5533.
51. Suchý M, Elmehriki AAH, Hudson RHE. *Org Lett.* 2011;13:3952–3955.
52. Wieland T, Vogeler K. *Angew Chem Int Ed Engl.* 1963;2:42.
53. Vilsmeier A, Haack A. *Ber Dtsch Chem Ges A/B.* 1927;60:119–122.
54. (a) Jones G, Stanforth SP. *Org React.* 1997;49:1–330. (b) Jones G, Stanforth SP. *Org React.* 2000;56:355–686.
55. Mahata PK, Venkatesh C, Syam Kumar UK, Ila H, Junjappa H. *J Org Chem.* 2003;68:3966–3975.
56. Mahto SK, Howard CJ, Shimko JC, Ottesen JJ. *ChemBioChem.* 2011;12:2488–2494.
57. Siman P, Blatt O, Moyal T, et al. *ChemBioChem.* 2011;12:1097–1104.
58. Koganty RR, Shambhu MB, Digenis GA. *Tetrahedron Lett.* 1973;45:4511–4514.
59. Richter R, Tucker B. *J Org Chem.* 1983;48:2625–2627.
60. Barluenga J, Campos PJ, Gonzalez-Nuñez E, Asensio G. *Synthesis.* 1985:426–428.
61. Coste J, Frérot E, Jouin P. *J Org Chem.* 1994;59:2437–2446.
62. Frérot E, Coste J, Pantaloni A, Dufour M-N, Jouin P. *Tetrahedron.* 1991;47:259–270.
63. Stierandová A, Šafář P. Peptide 1994. In: Maria HLS, ed. *Proceedings of the Twenty-Third Peptide Symposium.* Leiden: ESCOM; 1995:183.

64. Merrifield RB. *Angew Chem Int Ed Engl.* 1985;24:799–810.
65. Christensen M, Schou O. *Int J Pept Protein Res.* 1978;12:121–129.
66. Schou O, Bucher D, Nebelin E. *Hoppe-Seyler's Z Physiol Chem.* 1976;357:103–106.
67. Carpino LA. *Acc Chem Res.* 1987;20:401–407.
68. Carpino LA, Mansour EME, Knapczyk J. *J Org Chem.* 1983;48:666–669.
69. Wade JD, Bedford J, Sheppard RC, Tregear GW. *Pept Res.* 1991;4:194–199.
70. Ueki M, Nishigaki N, Aoki H, Tsurusaki T, Katoh T. *Chem Lett.* 1993;22:721–724.
71. Sheppeck JE. *Tetrahedron Lett.* 2000;41:5329–5333.
72. Beyermann M, Bienert M, Niedrich H, Carpino LA, Sadat-Aalaee D. *J Org Chem.* 1990;55: 721–728.
73. Carpino LA, Sadat-Aalaee D, Beyermann M. *J Org Chem.* 1990;55:1673–1675.
74. Sheradsky T, Salemnick G, Frankel M. *Isr J Chem.* 1971;9:263.
75. Mitchell AR, Merrifield RB. *J Org Chem.* 1976;41:2015–2019.
76. Erickson BW, Merrifield RB. *J Am Chem Soc.* 1973;95:3757–3763.
77. Metz B, Jiskoot W, Hennink WE, Crommelin DJA, Kersten GFA. *Vaccine.* 2003;22:156–167.
78. (a) Fraenkel-Conrat H, Olcott HS. *J Biol Chem.* 1948;174:827–843. (b) Fraenkel-Conrat H, Olcott HS. *J Am Chem Soc.* 1948;70:2673–2684.
79. Means GE, Feeney RE. *Anal Biochem.* 1995;224:1–16.
80. Kelly DP, Dewar MK, Johns RB, Wei-Let S, Yates JF. *Adv Exp Med Biol.* 1977;86:641–647.
81. Kallen RG. *J Am Chem Soc.* 1971;93:6236–6248.
82. Kallen RG, Jencks WP. *J Biol Chem.* 1966;241:5864–5878.
83. da Silva RA, Estevam IHS, Bieber LW. *Tetrahedron Lett.* 2007;48:7680–7682.
84. Wu Y, Levons J, Narang A, Raghavan K, Rao VM. *AAPS PharmSciTech.* 2011;12:1248–1263.
85. Fowles LF, Beck E, Worrall S, Shanley BC, de Jersey J. *Biochem Pharmacol.* 1996;51: 1259–1267.
86. Braun KP, Cody RB, Jones DR, Peterson CM. *J Biol Chem.* 1995;270:11263–11266.
87. Metz B, Kersten GFA, Hoogerhout P, et al. *J Biol Chem.* 2004;279:6235–6243.
88. Lu K, Ye W, Zhou L, et al. *J Am Chem Soc.* 2010;132:3388–3399.
89. Mitchell MA, Runge TA, Mathews WR, et al. *Int J Pept Protein Res.* 1990;36:350–355.
90. Gesquièe J-C, Diesis E, Tartar A. *J Chem Soc Chem Commun.* 1990:1402–1403.
91. Gold TB, Smith SL, Digenis GA. *Pharm Dev Technol.* 1996;1:21–26.
92. Taichi M, Kimura T, Nishiuchi Y. *Int J Pept Res Ther.* 2009;15:247–253.
93. Kitamoto Y, Maeda H. *J Biochem.* 1980;87:1519–1530.
94. Kumagaye KY, Inui T, Nakajima K, Kimura T, Sakakibara S. *Pept Res.* 1991;4:84–87.
95. Eschweiler W. *Chem Ber.* 1905;38:880–882.
96. Gannett P, Hailu S. *J Anal Toxicol.* 2001;25:88–92.
97. Morton AA, Mark JG. *Ind Eng Chem Anal.* 1934;6:151–152.
98. Hamburger R, Azaz E, Donbro M. *Pharm Acta Helv.* 1975;50:10–17.
99. Veber DF. In: Walter R, Meienhofer J, eds. *Peptides: Chemistry, Structure and Biology.* Ann Arbor, MI: Ann Arbor Science; 1975:307.
100. Brown T, Jones JH, Richards JD. *J Chem Soc Perkin Trans.* 1982;1:1553–1561.
101. Mergler M, Dick F, Sax B, Schwindling J, Vorherr T. *J Pept Sci.* 2001;7:502–510.
102. Guibé F. *Tetrahedron.* 1998;54:2967–3042.
103. Jolly PW. *Angew Chem Int Ed Engl.* 1985;24:283–295.
104. Collman JP, Hegedus LS, Norton JR, Finke RG. *Principles and Applications of Organotransition Metal Chemistry.* Mill Valley, USA: University Science Book; 1987.
105. Minami I, Ohashi Y, Shimizu I, Tsuji J. *Tetrahedron Lett.* 1985;26:2449–2452.
106. Merzouk A, Guibé F, Loffet A. *Tetrahedron Lett.* 1992;33:477–480.

107. Jeffrey PD, McCombie SW. *J Org Chem*. 1982;47:587–590.
108. Kolasa T, Miller MJ. *J Org Chem*. 1990;55:1711–1721.
109. Kinoshita H, Inomata K, Kameda T, Kotake H. *Chem Lett*. 1985;14:515–518.
110. (a) Blankemeyer-Menge B, Frank R. *Tetrahedron Lett*. 1988;29:5871–5874. (b) Handa BK, Keech E. *Int J Pept Protein Res*. 1992;40:66–71.
111. Kunz H, Waldmann H. *Angew Chem Int Ed Engl*. 1984;23:71–72.
112. Baldwin JE, Moloney MG, North M. *Tetrahedron*. 1989;45:6319–6330.
113. Stanley MS. *J Org Chem*. 1992;57:6421–6430.
114. Bloomberg GB, Askin D, Gargaro AR, Tanner MJA. *Tetrahedron Lett*. 1993;34:4709–4712.
115. Kunz H, Unverzagt C. *Angew Chem Int Ed Engl*. 1984;23:436–437.
116. Nilsson YIM, Andersson PG, Baeckvall JE. *J Am Chem Soc*. 1993;115:6609–6613.
117. Genêt J-P, Blart E, Savignac M, Lemeune S, Lemaire-Audoire S, Bernard J-M. *Synlett*. 1993:680–682.
118. Garro-Hélion F, Merzouk A, Guibé F. *J Org Chem*. 1993;58:6109–6113.
119. Tsuji J, Mandaï T. *Synthesis*. 1996:1–24.
120. Dangles O, Guibé F, Balavoine G, Lavielle S, Marquet A. *J Org Chem*. 1987;52:4984–4993.
121. Hutchins RO, Learn K, Fulton RP. *Tetrahedron Lett*. 1980;21:27–30.
122. Gomez-Martinez P, Dessolin M, Guibé F, Albericio F. *J Chem Soc Perkin Trans*. 1999;1: 2871–2874.
123. Dessolin M, Guillerez M-G, Thieriet N, Guibé F, Loffet A. *Tetrahedron Lett*. 1995;32: 5741–5744.
124. Tessier M, Albericio F, Pedroso E, et al. *Int J Pept Protein Res*. 1983;22:125–128.
125. Sigler GF, Fuller WD, Chaturvedi NC, Goodman M, Verlander M. *Biopolymers*. 1983;22: 2157–2162.
126. deL Milton RC, Becker E, Milton SCF, Baxter JEJ, Elsworth JF. *Int J Pept Protein Res*. 1987;30:431–432.
127. Lapatsanis L, Milias G, Froussios K, Kolovos M. *Synthesis*. 1983:671–673.
128. Paquet A. *Can J Chem*. 1982;60:976–980.
129. Isidro-Llobet A, Just-Baringo X, Ewenson A, Álvarez M, Albericio F. *Pept Sci*. 2007;88: 733–737.

Chapter 8

Side Reactions on Hydroxyl and Carboxyl Groups in Peptide Synthesis

Hydroxyl and carboxyl plays critical roles in peptide synthesis. The carboxyl group is the underlying basis for peptide bond construction. It could, nonetheless, be entangled in a variety of side reactions in the process of peptide synthesis that are characteristic of a carboxyl group, e.g., imide formation, transesterification, other undesired esterification, and so forth. Hydroxyl group could also induce a plethora of side reactions in peptide synthesis due to its evident nucleophilicity that mediates acyl migration and consequent peptide fragmentation. Some of the carboxyl and hydroxyl-related side reactions, such as aspartimide formation, acidolysis of the peptide C-terminal *N*-Me-AA, *O*-acyl isopeptide-induced des-Ser/Thr formation, sulfonation and acyl migration have been dedicatedly introduced in other sections and will not be reiterated in this chapter.

8.1 SIDE REACTIONS ON Asp/Glu SIDE CHAIN AND PEPTIDE BACKBONE CARBOXYLATE

8.1.1 Base-Catalyzed Asp/Glu(OBzl) Transesterification Side Reaction During the Loading of Chloromethyl Resin

The most detrimental side reaction addressing Asp/Glu residues in peptide synthesis is doubtless the formation of Aspartimide/Glutarimide that has been described in depth in Section 6.1. In addition, the carboxylate side chains on Asp/Glu in the form of protected esters are also susceptible to transesterification side reactions.

In the process of chloromethyl resin loading such as Boc-Xaa-OH mediated Merrifield resin derivatization the incoming amino acids are immobilized on the solid support by means of an ester bond through a base-catalyzed substitution reaction. This process could be promoted by catalysts like Cs salt,[1] triethylamine (TEA) or tetramethylammonium hydroxide (TMAH).[2] It has been concluded through contrast experiments that TMAH outperforms routinely employed Cs salt

Side Reactions in Peptide Synthesis. http://dx.doi.org/10.1016/B978-0-12-801009-9.00008-2

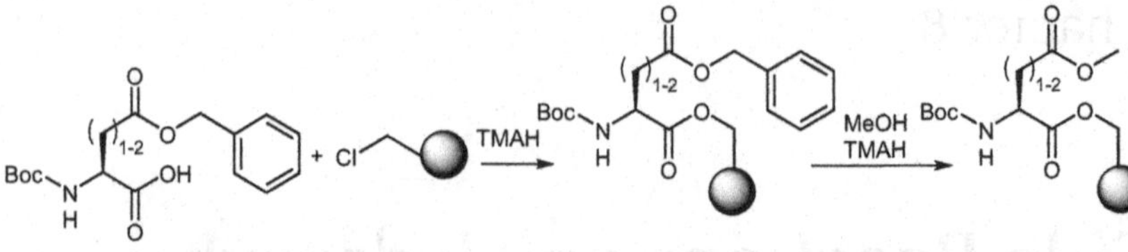

FIGURE 8.1 **Transesterification side reaction during Merrifield loading by Boc-Asp(OBzl)/Glu(OBzl)-OH.**

with regard to catalysis effects.[3] However, an inherent deficiency of this process is detected in that in case of the base-catalyzed immobilization of Asp(OBzl) or Glu(OBzl) the target substitution reactions between Boc-Asp(OBzl)/Glu(OBzl)-OH and chloromethyl substituents on the solid support will be interfered with concomitant transesterification side reactions on the Asp/Glu side chain ester.[3] As is depicted in Fig. 8.1, the benzyl ester moiety on the protected-Asp/Glu side chain undergoes TMAH-catalyzed transesterification in methanol at the step of Boc-Asp(OBzl)/Glu(OBzl)-OH directed chloromethyl resin loading. The subject benzyl ester could be transformed to the corresponding methyl ester in this process which would not regenerate the unshielded Asp/Glu derivative whereas Asp(OBzl)/Glu(OBzl) is capable of being sufficiently deblocked at the step of the HF-mediated side chain global deprotection. Asp(OMe) impurity with 14 amu molecular weight increase is generated by this means. TMAH is in principle an excellent catalyst for the esterification process so that Asp(OBzl)/Glu(OBzl) can be converted to Asp(OMe)/Glu(OMe) and Asp(OtBu)/Glu(OtBu) in methanol and *tert*-butanol, respectively, driven by TMAH. Routinely utilized Cs salt catalysts for chloromethyl resin loading will induce less extent of the transesterification side reaction compared with that caused by TMAH.[3]

The above transesterification process could nevertheless be taken advantage of when peptidyl Merrifield resin is treated by TMAH/*tert*-butanol at high temperature (70°C). Not only will Asp(OBzl)/Glu(OBzl) residues be transformed to Asp(OtBu)/Glu(OtBu) but also the peptide *C*-terminus is simultaneously converted to the corresponding *tert*-butyl ester with concomitant peptide chain immobilization from the solid supports (see also Fig. 8.2).[3] This process is substantially meaningful in peptide synthesis, particular for in-solution segment condensation since the protected carboxyl-component peptide *tert*-butyl ester could be derived from this one-pot transesterification/cleavage synthetic strategy.

8.1.2 Esterification Side Reactions on Asp/Glu During Peptidyl Resin Cleavage and Product Purification

Methanol and acetonitrile are frequently utilized as organic eluents in reverse-phase liquid chromatographic purification of peptides. In spite of various advantages of acetonitrile over methanol in terms of RP-HPLC performances such as

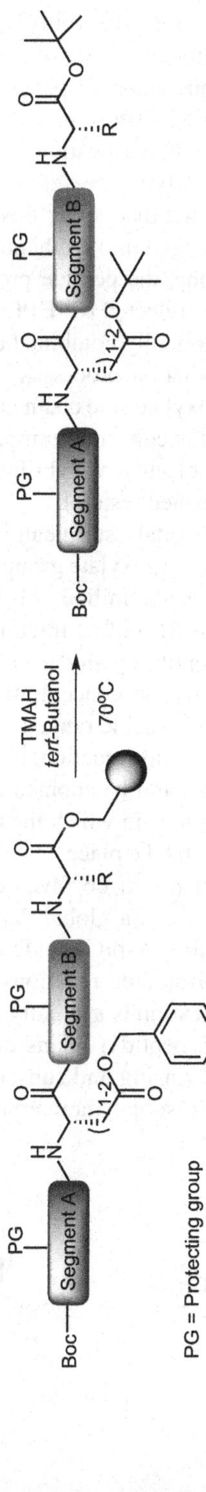

FIGURE 8.2 **TMAH/*tert*-butanol mediated peptide-Merrifield resin cleavage through transesterification and simultaneous transformation of Asp/Glu(OBzl) to Asp/Glu(OtBu).**

elution strength, peak symmetry and reproducibility[4] methanol is favored under certain circumstances predominantly by virtue of its evident cost-advantage. Utilization of methanol as an organic eluent, especially in the industrial fields, has been performed by numerous RP-HPLC processes. An obvious inherent limitation of methanol-directed HPLC purification is reflected by the potential methyl esterification on Asp/Glu carboxylate side chains and/or peptide carboxylate backbone. Owing to the fact that acidic modifiers, e.g., TFA, formic acid or acetic acid are routinely charged into methanol/H_2O eluent systems for reverse-phase liquid chromatography, the peptide product fractions pooled in this process is normally acidic. The collected RP-HPLC fractions in compliance with the target specifications are normally combined and concentrated prior to lyophilization or other relevant isolation processes. The subjected esterification modification on Asp/Glu carboxylate side chain and/or peptide carboxylate backbone could be intensified in this concentration process (Fig. 8.3). Methyl ester impurities with a +14 amu molecular weight increase are thus generated and accumulated (Yang, Y., unpublished results).

In view of the potential detrimental esterification modifications on Asp/Glu side chain or peptide backbone carboxylate group in the process of peptide methanol solution condensation in acidic milieu, it is necessary to strictly control the process of concentration of RP-HPLC fractions eluted from methanol systems, particularly for the susceptible peptide individuals. Parameters like concentration temperature, vacuum, and concentration ratio are supposed to be rationally defined with proven acceptable range (PAR) in order to minimize the extent of the methyl esterification side reaction, if any, to acceptable ranges. Acetonitrile in place of methanol should be adopted as an organic component for RP-HPLC in the case of extremities in which the methyl esterification side reactions are extraordinarily prone to take place.

Analogously if methanol happens to be involved in the acid-mediated peptidyl resin cleavage and/or side chain global deprotection processes, the peptide backbone carboxylate and/or Asp/Glu side chain carboxylate might be entangled in methyl esterification side reactions. The residual methanol that triggers this undesired modification is normally originated from peptidyl resin rinsing. For certain reasons peptidyl resins derived from solid phase assembly are treated by methanol rinsing and subsequent drying prior to the next handling or storage. If the residual methanol (not necessarily trivial

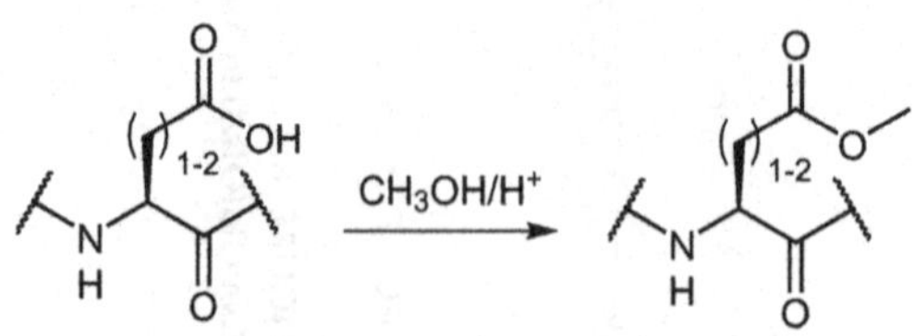

FIGURE 8.3 **Methyl esterification on carboxylate side chain of Asp/Glu.**

when molar ratio to peptide is accounted) is not sufficiently removed at the drying step the survived methanol which is carried over to the peptidyl resin cleavage/side chain global deprotection is capable of modifying peptide carboxylate substituents in the acidic milieu, and giving rise to the formation of methyl ester impurities (Yang, Y., unpublished results). Peptidyl resin rinsing with methanol should, therefore, be avoided in the case of esterification-susceptible peptides. Or alternatively, stringent and meticulous control of the drying process is necessary to be conducted.

8.2 SIDE REACTIONS ON Ser/Thr SIDE CHAIN HYDROXYL GROUPS

8.2.1 Alkylation Side Reactions on Ser/Thr Side Chain Hydroxyl Groups

Ser/Thr residues are susceptible to a variety of side reactions in peptide synthesis due to the nucleophilicity of their β-hydroxyl groups. It has been introduced in Section 3.9 about the formation of Ser(Acm) impurity in the process of the disulfide bridge boding between two Cys(Acm) residues. Moreover, Ser/Thr might also be subjected to other alkylation side reactions in peptide synthesis. The intrinsically distinctive property of secondary-β-hydroxyl substituent on Thr shields its nucleophilicity, and the severity of the side reactions addressing Thr is hence less than that of Ser under comparable conditions.

The acyl-type protecting groups bonded to hydroxyl side chains on Ser/Thr residues might be subjected to a variety of side reactions such as *O*→*N* acyl migration[5] and β-elimination[6] in the process of peptide synthesis. Although some Ser/Thr protecting groups such as *O*-tosyl, *O*-acyl, *O*-benzyloxycarbonyl, have ever been employed as derivatives to shield the β-hydroxyl substituents on the Ser/Thr side chains, their applications in the fields of peptide synthesis nowadays are evidently limited due to the inherently unavoidable characteristic side reactions. One of the few exceptions comes from the Alloc protecting group on Ser/Thr thanks to its unique orthogonality to other amino acid protecting groups that facilitates the selective removal. Alloc-protected Ser/Thr derivatives are widely employed as building blocks in glycopeptide synthesis[7] and the preparation of super-acid/base-sensitive peptides.[8]

The allyl group is inappropriate for hydroxyl-protection since allylether is incapable of regenerating the parental hydroxyl derivative by Pd(0) catalyst.[9] Utilization of Alloc as a feasible protecting group for Ser/Thr is therefore rationalized. Similar to *N*-Alloc derivatives, Alloc-protected hydroxyl compounds potentially suffer from an *O*-allylation side reaction in the process of Pd(0)-catalyzed Alloc removal reaction.[9] This process is depicted in Fig. 8.4. Owing to the chemical stability of side product Ser(All) toward Pd(0) catalyst, the target Ser product with free β-hydroxyl group could not be regenerated by elongated Pd(0) treatment.

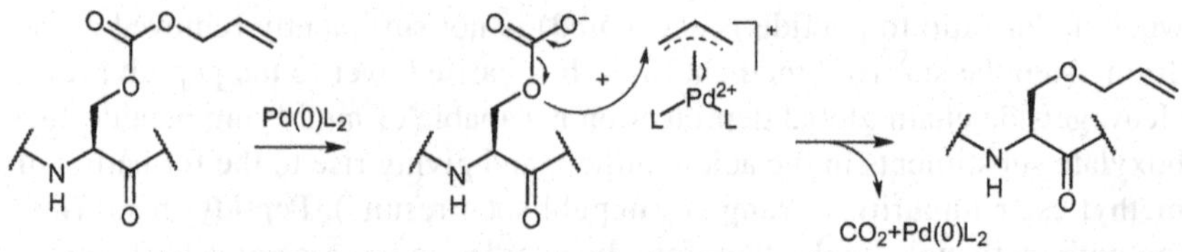

FIGURE 8.4 ***O*-allylation side reaction on Ser in Pd(0)-catalyzed Alloc deblocking reaction.**

It is to note that the extent of the above decarboxylation-rearrangement process addressing allyl carbonate is insignificant relative to that of the analogous side reactions on allyl carbamate. Principally the Pd(0)-mediated de-Alloc process would not result in severe *O*-allylation side reactions in the presence of effective allyl scavengers.[9]

8.2.2 Acylation Side Reactions on Ser/Thr Side Chain Hydroxyl Groups

8.2.2.1 Acylation Side Reactions on Ser/Thr Side Chain Hydroxyl Groups During Amino Acid Coupling

Owing to the relatively insignificant nucleophilicity of the β-hydroxyl substituents on the Ser/Thr residues side chain-unprotected Ser/Thr species could be participated in the peptide assembly as building blocks under certain circumstances, reflected as exemplars of "minimal protection" strategy in peptide chemistry.[10] The peptide intermediates derived in this synthetic route are generally bestowed with enhanced solubility and opportunities to undergo side chain global deprotection in relatively milder conditions.[11] This synthetic strategy is especially compatible with amino acid derivatives bearing secondary hydroxyl side chains,[12] e.g., Thr, Hyp, 3-phenylserine, 3-hydroxynorvaline, 3-hydroxynorleucine, and so forth, thanks to the shielding effects of the corresponding β-substituents. In spite of various advantages of the utilization of unprotected hydroxyl derivatives in peptide synthesis the occurrence of undesired modifications on the concerned unshielded hydroxyl group in the process of peptide acylation reaction cannot be thoroughly excluded, particularly in case the acylating agent is used in large excess, and/or the target amino acylation proceeds sluggishly which necessitates an elongated reaction time or harsher reaction conditions. Amino acylating reactions between amino acids/peptides with considerable steric hindrance might reduce the nucleophilic differences between hydroxyl and the amino group so that the undesired hydroxyl acylation side reaction could be substantially promoted under such circumstances.[13] In view of the undesired acylation on hydroxyl group the employment of the hydroxyl side chain unprotected amino acid/peptide as building blocks in peptide synthesis is supposed to be evaluated meticulously, and the reaction parameters should be qualified carefully.

Rational choice of appropriate coupling reagents could exert profound impacts on the extent of hydroxyl acylation side reactions.[14]

The detrimental effects of the undesired acylation on Ser hydroxyl group might not be explicitly reflected by the apparent hydroxyl modification per se. The side chain acylated Ser/Thr impurities are capable of inducing further peptide modifications at the subsequent reaction steps, i.e., N^{α}-protecting group removal and the base-catalyzed acyl *O*→*N* migration (see also Section 4.2) which might result in the unexpected peptide chain termination (see also Fig. 4.5), or alternatively, repetitive couplings of the affected amino acid (see also Fig. 4.8). Another possible consequence of the Ser/Thr side chain-hydroxyl acylation is the enhanced likelihood of β-elimination upon base treatment of the affected side chain-acylated Ser/Thr, giving rise to the formation of an ΔAla impurity which might accommodate a tandem nucleophilic addition by piperidine at the step of the N^{α}-Fmoc deblocking, and be converted to side product bearing 3-(1-piperidinyl)alanine moiety. This type of side reaction will be elaborated in the following section.

8.2.2.2 Acylation on Ser/Thr β-Hydroxyl Groups in Acidic Condition

Carboxylic acid and alcohol derivatives can react in a strong acidic environment to give the corresponding ester. In view of this mechanism side chain-unprotected Ser/Thr is susceptible to acylation reactions by carboxylic acid derivatives in an acidic milieu. For example, Ser/Thr could be transformed into the corresponding formate, acetate and trifluoroacetate in formic acid, acetic acid and TFA, respectively.

8.2.2.3 Acylation Side Reactions on Ser/Thr Side Chain Hydroxyl Groups Induced by Acid-Catalyzed Acyl N→O Migration

Due to the intrinsic property of acyl moiety to undergo migration between *O* and *N* atoms at different pH values, the hydroxyl side chains on Ser/Thr might serve as a nucleophilic acyl receptor for *N*-acyl moiety at appropriate pH value and under favorable atomic steric alignment to mediate acyl *N*→*O* shift via hydroxyoxazolidine intermediate.[15] The above process is generally triggered in acidic condition. Peptides bearing the *N*-terminal acyl-Ser/Thr moieties such as Ac-Ser/Thr, For-Ser/Thr and Palm-Ser/Thr could be subjected to *N*→*O* acyl migration at the step of the side chain global acidolytic deprotection, giving rise to the formation of side products with *N*-terminal H-Ser/Thr(Ac/For/Palm) structures. These impurities could suffer from β-elimination upon base treatment and be converted to ΔAla derivatives that might be further transformed to 3-(1-piperidinyl)alanine side products by piperidine.

Due to the susceptibilities of N^{α}-acyl-Ser/Thr-peptides to the aforementioned side reactions it is necessary to scrutinize the obtained products to detect the potential existence of side chain-acylated Ser/Thr isomers. It is noted that under certain circumstances Acyl-Ser/Thr and H-Ser(Acyl)/Thr(Acyl) isomers might possess very similar chromatographic properties that requests dedicatedly developed analytical methods to obtain an acceptable RP-HPLC resolution. Since the last upstream step of peptide synthesis, i.e., global deprotection, and the downstream process (chromatographic purification) is frequently conducted in acidic conditions, it is recommended to evaluate of the possibility of the potential acyl migration at these steps. Product treatment by weak bases, if possible, could be implemented in the production process to reverse the acyl migration to the desirable direction of N^{α}-acyl peptide formation.

8.2.3 β-Elimination Side Reactions on Ser/Thr

Amino acids bearing β-hydroxyl or β-sulfhydryl group are susceptible to base-catalyzed elimination side reactions that have been described in Chapter 2 with Ser/Thr(PO_3H_2) as substrate. Ser/Thr derivatives that are potentially subjected to β-elimination side reactions are nonetheless not limited to the side chain phosphorylated species. A variety of Ser/Thr compounds could be affected by β-elimination in basic conditions, especially in case the side chain hydroxyl group is bonded to an electron-withdrawing group, the susceptibility to β-elimination would be substantially enhanced. It is implied from the mechanism of the β-elimination process depicted in Fig. 8.5 that the electron-withdrawing moiety on the Ser/Thr side chain β-hydroxyl functional group would promote H^{α} abstraction and in turn the lability of C^{β}—O bond in basic environments, resulting in intensified β-elimination by this means.

A number of Ser/Thr derivatives like Ser/Thr(Tos),[16] Ser/Thr(Mesyl),[17] Ser/Thr(Glc),[18] Ser(SO_2Bzl)[19] and Ser(Boc)[20] could be affected by base-catalyzed β-elimination side reactions in the presence of NaOH, diethylamine[16] or TEA[17] during peptide synthesis, generating by this means side products with ΔAla moiety or ΔAla adducts via derivatization by various nucleophiles. It is noted that side chain underivatized Ser/Thr residues are not absolutely exempted from β-elimination side reactions.[21] Factors like concentration of bases, temperature, reaction time[22] and peptide sequence[23] could evidently regulate the tendencies of

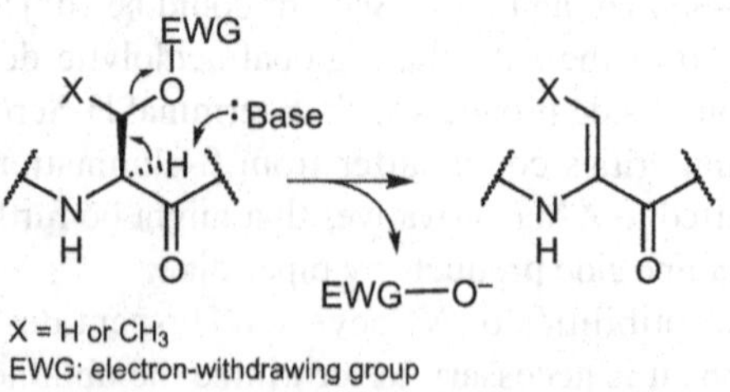

FIGURE 8.5 β-Elimination on Ser/Thr derivatives.

the concerned Ser/Thr-containing peptides to undergo the β-elimination process. The susceptibility of Thr to β-elimination is much less significant than that of Ser under identical conditions.

Analogous β-elimination side reactions could also interfere with the process of the activation of *O*-acylisodipeptide building blocks (Boc-Ser/Thr(Fmoc-Xaa)-OH) that have already obtained profoundly wide applications in the fields of peptide synthesis, giving rise to the formation of corresponding des-Ser/Thr impurities. This process has been described dedicatedly in Section 1.2. An alternative pathway of β-elimination on free Ser/Thr is regulated by activation of the addressed β-hydroxyl group with coupling reagents. As has been described previously the unprotected hydroxyl side chain on Ser/Thr could be activated by condensation agents, e.g., DSC[24] and CDI[25] in the process of peptide synthesis adopting "minimal protection" strategy. The generated activated hydroxyl intermediates by this means are prone to undergo β-elimination at the following synthetic steps, and be converted to side products with ΔAla or ΔAbu moiety. This process is illuminated in Fig. 8.6.

As one of the most frequently utilized coupling reagent types in peptide synthesis carbodiimide derivatives can also provoke β-elimination on Ser/Thr, whereas CuCl is capable of catalyzing this process.[26] The mechanism is indicated in Fig. 8.7.

In fact, the aforementioned condensation agents-dictated β-elimination processes on Ser/Thr are deliberately resorted to in an effort to produce ΔAla

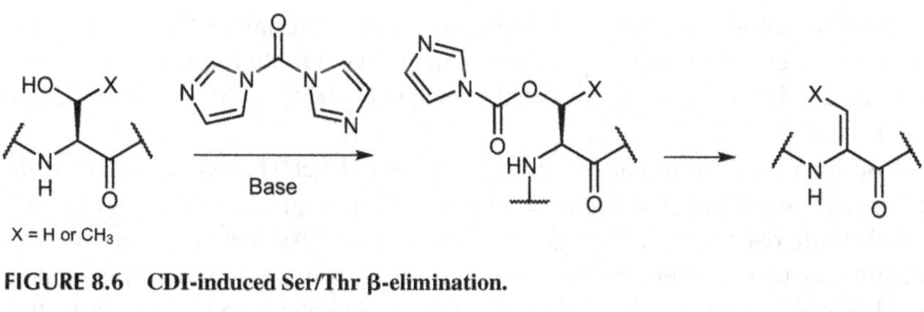

FIGURE 8.6 CDI-induced Ser/Thr β-elimination.

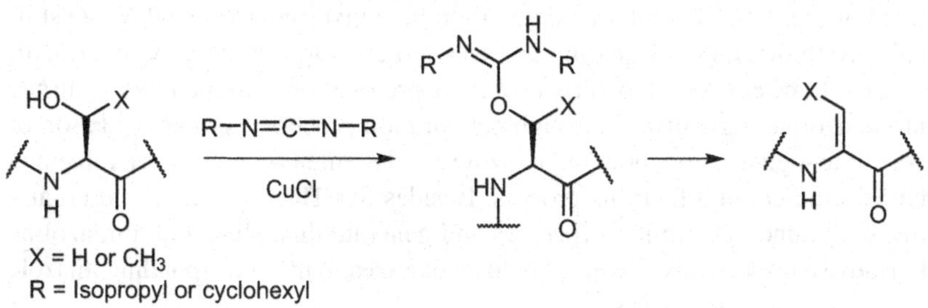

FIGURE 8.7 Carbodiimide-induced Ser/Thr β-elimination.

or ΔAbu derivatives. Treatment of Ser/Thr by an acetic acid anhydride/base could induce β-elimination and give rise to the target ΔAla/ΔAbu products.[27] These condensation agents activate the addressed hydroxyl group by bonding an electron-withdrawing moiety that could in turn substantially facilitate the resultant β-elimination process. Decent yields of the production of ΔAla/ΔAbu-derivatives could be acquired by this means. On the other hand, the undesired functions between unprotected β-hydroxyl on Ser/Thr with condensation agents might potentially induce β-elimination side reactions, and give commensurate impurities in the peptide synthesis.

8.2.4 *N*-Terminal Ser/Thr-Induced Oxazolidone Formation Side Reactions

N-terminal side chain unprotected Ser/Thr, in the event of peptide N^{α}-urethane protection, could initiate in the basic condition the process of β-hydroxyl nucleophilic attack on the backbone carbamate moiety to give rise to the formation of oxazolidone side product.[28] Peptide ester bearing N^{α}-Z, Moz, or Boc-protected, O^{β}-unprotected-Ser/Thr moiety that is originally supposed to undergo base-catalyzed *C*-terminal hydrolysis might suffer from the oxazolidone formation in this process.[28] The proposed mechanism of oxazolidone formation on N^{α}-Moz-Ser/Thr peptides in basic condition is displayed in Fig. 8.8.

Due to the distinctive tendency of the N^{α}-urethane-protected-Ser/Thr peptide to undergo oxazolidone formation, it is understandably advised to protect the concerned β-hydroxyl group on the *N*-terminal Ser/Thr in case base treatment of the target peptide is imperative. Or alternatively, relatively milder condition could be resorted to when hydrolysis of peptide ester is concerned.[29,30]

Besides the *N*-terminal N^{α}-urethane protected Ser/Thr, *endo* Ser/Thr residues with unprotected side chains in the peptide sequence could also trigger similar side reactions via their β-hydroxyl groups.[31] As is depicted in Fig. 8.9, peptide containing -Ser/Thr- moiety could suffer from an β-hydroxyl mediated nucleophilic attack on the *N*-terminal amide backbone, and give rise to the formation of five-member ring intermediate **1** similar to the counterpart illustrated in Fig. 8.8. This intermediate might be transformed via acyl *N*→*O* shift pathway to derivative **2** bearing an ester moiety. Or alternatively, intermediate **1** could be converted to derivative **3** via protonation, and the latter is turned into oxazoline derivative **4** through dehydration process. Further oxidation of oxazoline **4** gives corresponding oxazole **5**.[32] Parameters such as pH value and temperature could affect this process. Besides Ser/Thr, Cys and β-amino alanine could undergo similar processes and generate thiazoline and imidazoline derivatives, respectively,[31] which could be oxidized to the corresponding thiazole and imidazole counterparts.[32]

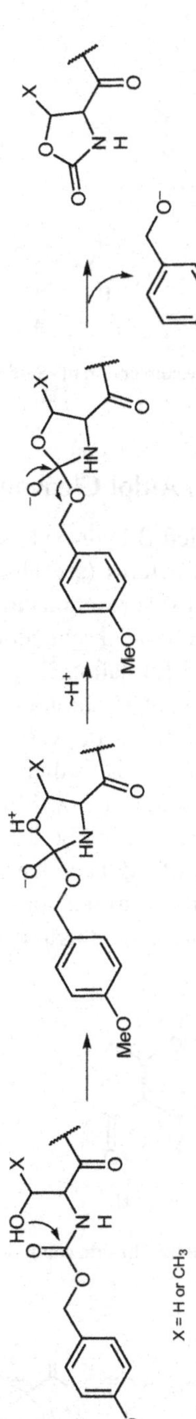

FIGURE 8.8 Proposed mechanism of oxazolidone formation on N^{α}-Moz-Ser/Thr peptide.

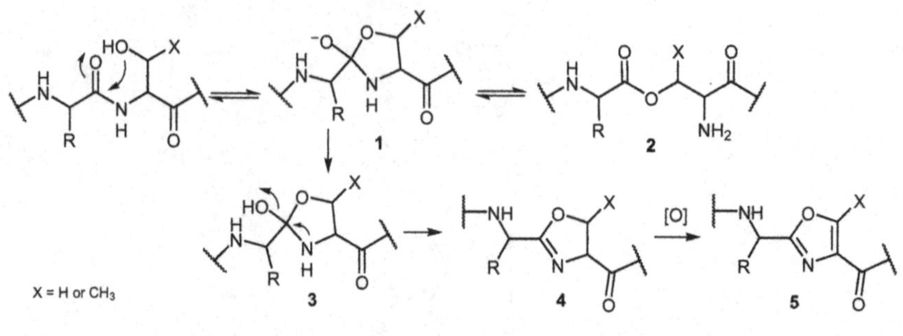

FIGURE 8.9 Proposed mechanism of the conversion of -Ser/Thr- peptide to isomeric peptide ester or oxazoline/oxazole.

8.2.5 Ser/Thr-Induced Retro Aldol Cleavage Side Reaction

Ser/Thr residues bearing unprotected β-hydroxyl side chain substituents contain per se β-hydroxy carbonyl structures (see also Fig. 8.10) that might be potentially entangled in a variety of side reactions in peptide synthesis.

Adol condensation represents a crucial synthetic strategy in organic synthesis in the context of C—C bond formation.[33] The reaction originates from the condensation between enol or enolate anion derived from α-H containing carbonyl compound, and another carbonyl derivative, giving rise to the formation of β-hydroxyl aldehyde/ketone that is further converted to enal/enone via dehydration. The above process can be catalyzed by both acid and base. Base catalyzed adol condensation is illustrated in Fig. 8.11.

Due to the inherent reversibility of aldol condensation, enone or β-hydroxyl carbonyl derivatives could be degraded to the corresponding aldehyde/ketone precursors under certain circumstances, and this process is termed as retro aldol

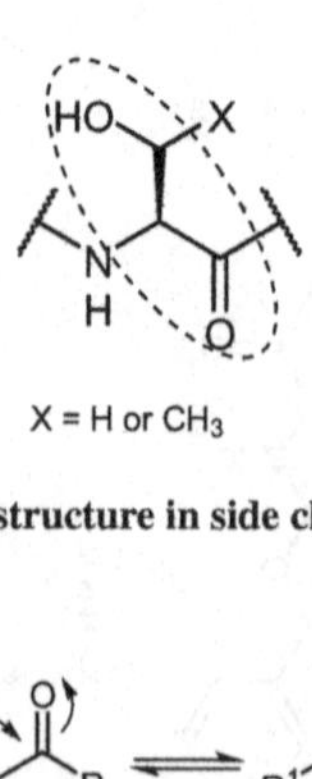

FIGURE 8.10 β-Hydroxy carbonyl structure in side chain unprotected Ser/Thr.

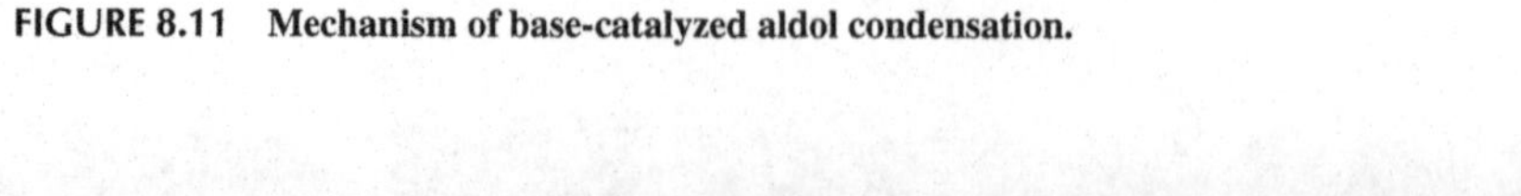

FIGURE 8.11 Mechanism of base-catalyzed aldol condensation.

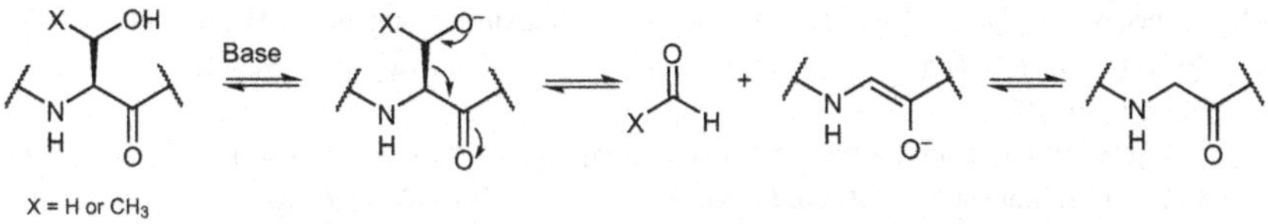

FIGURE 8.12 **Base-catalyzed retro aldol cleavage of Ser/Thr peptide.**

cleavage. Glucose is regarded as β-hydroxyl carbonyl compound, and its biological metabolism process undergoes retro aldol cleavage step at which glucose is degraded into two triose units by aldolase catalysis.[34] It is regarded as an important step in the circle of glycolysis. Similar to glucose and fructose, peptides bearing side chain unprotected Ser unit also contain β-hydroxyl carbonyl structure (see also Fig. 8.10) that implies the tendency to undergo retro aldol cleavage during peptide synthesis, or other relevant processing. Identical to aldol condensation, retro aldol cleavage could take place in both acidic and basic environments. Peptide ester bearing side chain unprotected Ser/Thr that is supposed to undergo base-catalyzed ester hydrolysis could accommodate a retro aldol cleavage side reaction. As is indicated in Fig. 8.12, the side chain unprotected Ser/Thr residue could undergo partial deprotonation by base treatment, and the resultant anionic β-hydroxyl carbonyl intermediate will be consequently susceptible to C^{α}—C^{β} bond cleavage, and be degraded into the corresponding formaldehyde/acetaldehyde and peptide enolate fragments. The latter is rearranged in a basic condition to the corresponding carbonyl compound, namely Gly derivative. Although this retro aldol cleavage on Ser/Thr peptides is not prevalent,[35,36] serine hydroxymethyltransferase-catalyzed conversion of Ser to Gly by means of retro aldol cleavage is regarded nonetheless as an important reaction type in physiological process.[37] The potential retro aldol cleavage side reaction on free Ser/Thr peptides is advised to be taken into account, particularly in case the concerned peptide esters have to undergo base-catalyzed hydrolysis.

REFERENCES

1. Gisin BF. *Helv Chim Acta*. 1973;56:1476–1482.
2. Loffet A. *Int J Pept Protein Res*. 1971;3:297–299.
3. Hsieh K-H, Demaine MM, Gurusidaiah S. *Int J Pept Protein Res*. 1996;48:292–298.
4. Kromidas S. *Practical Problem Solving in HPLC*. Weinheim: Wiley-VCH Verlag; 2000:26–27.
5. Sheehan JC, Goodman M, Hess GP. *J Am Chem Soc*. 1956;78:1367–1369.
6. Bodanszky M. *The Principles of Peptide Synthesis*. Berlin: Springer; 1993:135.
7. Kunz H, Waldmann H. *Angew Chem Int Ed Engl*. 1984;23:71–72.
8. Gothe R, Seyfarth L, Schumann C, et al. *J Prakt Chem*. 1999;341:369–377.
9. Guibé F. *Tetrahedron*. 1998;54:2967–3042.
10. Stewart JM. In: Gross E, Meienhofer J, eds. *The Peptides. Vol. 3. Protection of Functional Groups in Peptide Synthesis*. New York: Academic Press; 1981:169.

11. (a) Fischer R. In: Scoffone E, ed. *Peptides 1969*. Amsterdam: North-Holland; 1971:138. (b) Finn FM, Hofmann K. Neurath H, Hill RL, eds. *The Proteins. Vol. 2*. New York: Academic Press; 1976:105.
12. (a) Fischer PM, Retson KV, Tyler MI, Howden MEH. *Int J Pept Protein Res*. 1991;38:491–493. (b) Coy DH, Branyas N. *Int J Pept Protein Res*. 1979;14:339–343. (c) Reissmann S, Schwuchow C, Seyfarth L, et al. *J Med Chem*. 1996;39:929–936. (d) Ottl J, Musiol HJ, Moroder L. *J Pept Sci*. 1999;5:103–110.
13. Bodanszky M, Fink ML, Klausner YS, et al. *Int J Pept Protein Res*. 1977;42:149–152.
14. (a) Arold H, Reissmann S. *J Prakt Chem*. 1970;312:1130–1144. (b) Adamson JG, Blaskovich MA, Groenevell H, Lajoie GA. *J Org Chem*. 1991;56:3447–3449.
15. Bergmann M, Brand E, Weimann F. *Z Physiol Chem*. 1923;131:1–17.
16. Photaki I. *J Am Chem Soc*. 1963;85:1123–1126.
17. Chung YJ, Kim Y-C, Park H-J. *Bull Korean Chem Soc*. 2002;23:1481–1482.
18. Downs F, Peterson C, Murty VLN, Pigman W. *Int J Pept Protein Res*. 1977;10:315–322.
19. Ako H, Foster RJ, Ryan CA. *Biochem Biophys Res Commun*. 1972;47:1402–1407.
20. Schnabel E, Stoltefuss J, Offe H, Klauke E. *Justus Liebigs Ann Chem*. 1971;743:57–68.
21. Li W, Backlund PS, Boykins RA, Wang G, Chen H-C. *Anal Biochem*. 2003;323:94–102.
22. Aitken A, Howell S, Jones D, Madarzo J, Patel Y. *J Biol Chem*. 1995;270:5706–5709.
23. Mega T, Nakamura N, Ikenaka T. *J Biochem*. 1990;107:68–72.
24. Ogura H, Sato O, Takeda K. *Tetrahedron Lett*. 1981;22:4817–4818.
25. Andruszkiewicz R, Czerwi ski A. *Synthesis*. 1982:968–969.
26. Miller MJ. *J Org Chem*. 1980;45:3131–3132.
27. Kato T, Higuchi C, Mita R, Yamaguchi T. JP. 1985: 60190749; *Chem Abstr*. 1986;104:109267.
28. Chen S-T, Lo L-C, Wu S-H, Wang K-T. *Int J Pept Protein Res*. 1990;35:52–54.
29. Chen S-T, Wang K-T, Wong CH. *J Chem Soc Chem Commun*. 1986:1514–1516.
30. Furuta T, Matuso J, Kurata T, Yamamoto Y. *High Pressure Res*. 1993;11:93–106.
31. Paulus T, Henle T, Haeβner R, Klostermeyer H. *Z Lebensm Unters Forsch A1*. 1997;204: 247–251.
32. Paulus T, Riemer C, Beck-Sickinger AG, Henle T, Klostermeyer H. *Eur Food Res Technol*. 2006;222:242–249.
33. Brückner R. *Reaktionsmechanismen, 3. Auflage*. München: Elsevier GmbH, Spektrum Akademischer Verlag; 2004:561–566.
34. Romanto AH, Conway T. *Res Microbiol*. 1996;147:448–455.
35. Goodman M, Kenner GW. *Adv Protein Chem*. 1951;12:465–638.
36. Bodanszky M, Martinez J. *Synthesis*. 1981:333–356.
37. Nelson DL, Cox MM. In: Ahr K, ed. *Lehninger Principles of Biochemistry*. New York: W. H. Freeman and Company; 2008:693–694.

Chapter 9

Peptide Oxidation/Reduction Side Reactions

Some amino acid residues in peptide molecules such as Cys, Met, Trp, His and Tyr could be subjected to oxidation in the process of peptide synthesis, and be transformed into their oxidized counterparts. The oxidation process could be mediated by air or another oxidant species. On the contrary, certain amino acids, e.g., Nle(N_3), Trp and Cystine could suffer from undesired reduction side reactions induced by reductive scavengers utilized in peptide side chain global deprotection reactions. This chapter will be focused on the possible oxidation/reduction side reactions occurred in peptide synthesis.

9.1 OXIDATION SIDE REACTIONS ON Cys

Cell redox states impose crucial regulatory impacts on many biological phenomena, e.g., cell growth and apoptosis, while their redox states are largely governed by the oxidation–reduction equilibrium between Cysteine and Cystine.[1] The reductive sulfhydryl side chain on Cys could be oxidized to the corresponding disulfide derivative by appropriate oxidants. Interconversion between sulfhydryl/disulfide is entrusted with at least two biological significances: (1) disulfide bond formation is crucial for the establishment and stabilization of the structures of proteins; (2) bioactivities of many proteins are regulated by the redox status of the key Cys residues. Some *in vivo* interconversion of sulfhydryl/disulfide mediated by thiol-disulfide oxidoreductase[2] or by reactive oxygen species (ROS) in a nonenzymatic oxidation manner.[3] Selective construction of disulfide bridges between various Cys residues with a chemical synthetic strategy has already been established as one of the investigation focuses in peptide chemistry.

A variety of peptidyl pharmaceutical products bear nevertheless free Cys residues, which might suffer from undesired oxidation by air or another oxidant. Diverse Cys derivatives with different oxidation states are generated from the Cys-oxidation processes among which disulfide derivatives represent one of the most predominant species. The interconversion between Cysteine and Cystine will hence be introduced with emphasis in this chapter.

The most common species derived from Cys oxidation in chemical synthesis is as displayed in Fig. 9.1. Cys(−2) might be oxidized to Cys sulfenic acid (0),

Side Reactions in Peptide Synthesis. http://dx.doi.org/10.1016/B978-0-12-801009-9.00009-4

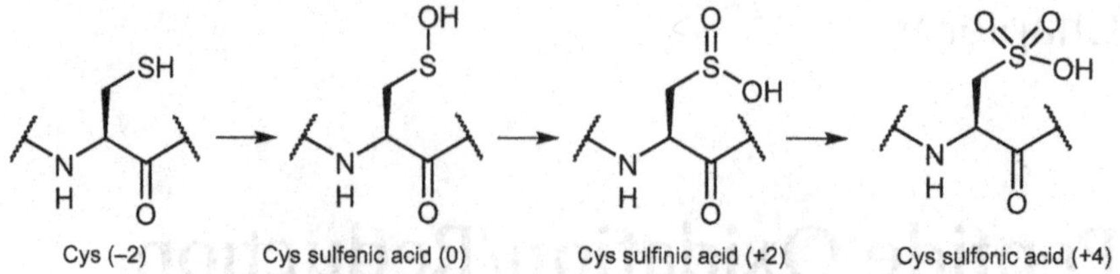

FIGURE 9.1 Cys derivatives with different oxidation states in peptide synthesis. Sulfur chemical valence in the bracket.

Cys sulfinic acid (+2) or Cys sulfonic acid (+4). These processes are realized in a cellular environment by means of posttranslational oxidation,[4] bestowing prominent significances on the regulatory functions of diverse proteins.[5]

Cys residue could be readily oxidized to Cys sulfenic acid,[6] while the latter is normally unstable, and is further oxidized to Cys sulfinic acid which is relatively more stable. Nevertheless, Cys sulfinic acid could also be oxidized to Cys sulfonic acid under certain circumstances.[7] Cys sulfenic acid might function with the sulfhydryl substituent on Cys to generate the corresponding disulfide derivative.[8] This process is generally taken as the basis of the disulfide bridge formation via the Cys oxidation by H_2O_2, DMSO and other oxidizing agents. That is to say, Cys is firstly oxidized to the Cys sulfenic acid counterpart, and the latter reacts subsequently with another molecule of sulfhydryl compound to give rise to the formation of a disulfide bond (see also Fig. 9.2). The oxidation of Cys to Cys sulfenic acid increases the electrophilicity of the addressed sulfur atom which is more prone to accommodate a nucleophilic attack from sulfhydryl or thiolate to construct the disulfide moiety.[9]

Besides the function with sulfhydryl derivative to generate a disulfide bond, Cys sulfenic acid could also react with another molecule of Cys sulfenic acid, and give rise to the formation of corresponding thiosulfinate[5,10] by virtue of the simultaneous nucleophilicity and electrophilicity attributes of Cys sulfenic acid.[4] Noncovalent dimer of Cys sulfenic acid[4,11] probably firstly comes into being via hydrogen bonds, and is subsequently transformed to the commensurate thiosulfinate. The proposed mechanism of this process is illustrated in Fig. 9.3.

Moreover, Cys sulfenic acid is capable of reacting with amino groups[12] on peptides to give sulfenamide, and even with the backbone amide located

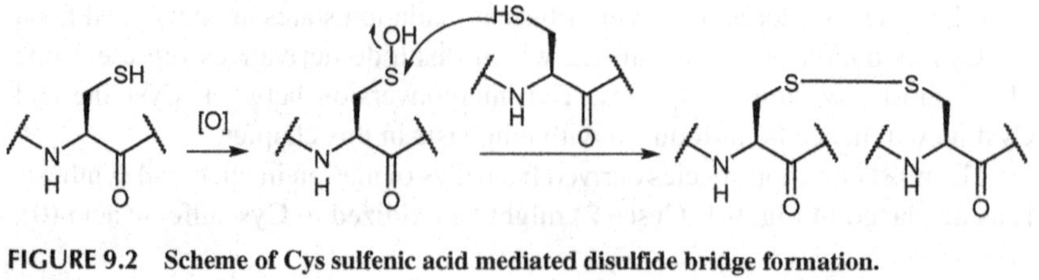

FIGURE 9.2 Scheme of Cys sulfenic acid mediated disulfide bridge formation.

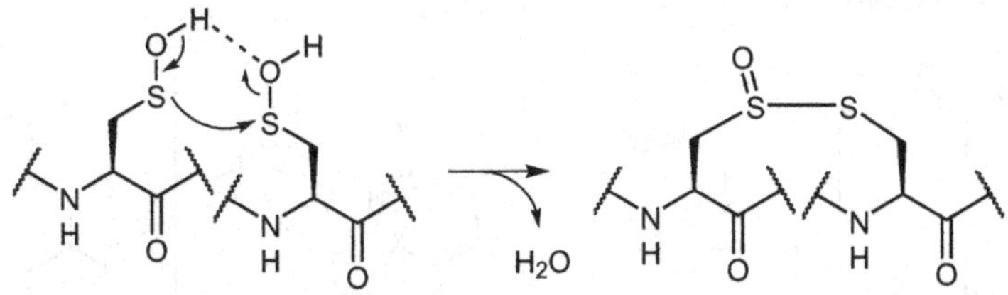

FIGURE 9.3 Thiosulfinate formation from two molecules of Cys sulfenic acid.

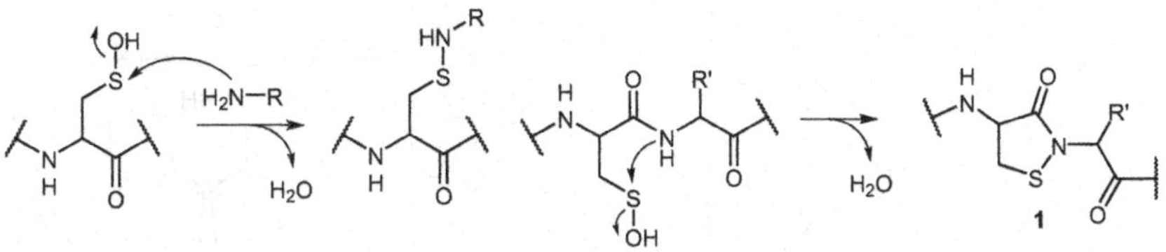

FIGURE 9.4 Sulfenamide/sulfenyl amide formation from Cys sulfenic acid and amine/amide.

neighboring to the subject Cys(O) on its *C*-terminal to form five-member ring sulfenyl amide derivative **1**.[13,14] These processes are depicted in Fig. 9.4. The sulfur atom on Cys sulfenic acid possesses sufficient electrophilicity[15] to accommodate the nucleophilic attacks from amine derivatives. Owing to the advantageous steric effect the amide backbone could also initiate the nucleophilic attack on the subjected Cys sulfenic acid to form sulfenyl amide moiety. The kinetics of the sulfenamide formation from the Cys sulfenic acid and amine is much slower than that of the disulfide bond formation from Cys sulfenic acid and sulfhydryl compounds.

Cys sulfenic acid could serve as nucleophile in an organic reaction to function with another Cys sulfenic acid to generate thiosulfinate (as indicated in Fig. 9.3). Moreover, Cys sulfenic acid might also lead to a S_N2 substitution reaction with NBD-Cl **2**; additive reaction with olefin **3** and alkyne **4**, as well as being reduced by 2 equiv. PPh_3 **5** (see also Fig. 9.5). Cys sulfenic acids act in the manner of the nucleophile in all these processes.[16]

DMSO-induced Cys sulfenic acid formation and consequent disulfide formation constitute one of the major sources of Cys oxidation in peptide synthesis. It is noted that this process might take place in a wide pH spectrum of 1–8.[17] In spite of the fact that this distinctive attribute has been utilized in DMSO-regulated peptide disulfide bridge formation,[18] the undesired Cys oxidation by DMSO in the process of peptide synthesis, particularly in acidic condition, could lead to various side reactions. DMSO is able to convert a sulfhydryl compound to the corresponding disulfide derivative in the presence of halogen acid.[19] Under a harsher reaction conditions, e.g., high temperature DMSO could even oxidize

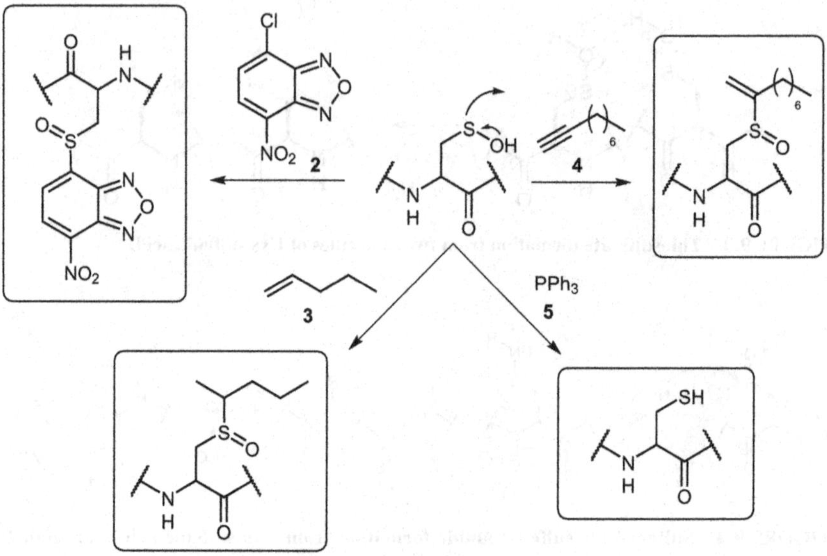

FIGURE 9.5 **Nucleophilic Cys sulfenic acid mediated reactions.**

sulfhydryl or disulfide derivative to their corresponding sulfonic acid counterparts (Eqs (9.1) and (9.2)) catalyzed by HCl, HBr, or HI.[20]

$$RSH + 3CH_2S(O)CH_3 \rightarrow RSO_3H + 3CH_3SCH_3 \quad (9.1)$$

$$RS\text{-}SR + 5CH_2S(O)CH_3 + H_2O \rightarrow 2RSO_3H + 5CH_3SCH_3 \quad (9.2)$$

Oxygen-regulated Cys oxidation is generally regarded as a radical mechanism.[21] There is nonetheless an assertion that oxygen-directed Cys disulfide formation in the presence of residual metal is interconnected with the Cys sulfenic acid intermediate formation.[8]

The disulfide bond formed in peptide synthesis could be subjected to a further oxidation process, and give rise to thiosulfinate **6**. This highly reactive species is capable of reacting with sulfhydryl compound to form a new disulfide derivative and the reciprocal Cys sulfenic acid.[22] These processes are schematized in Fig. 9.6. Hydrolysis of thiosulfinate leads to sulfenic acid,[23] and this process is deemed as the reverse reaction of dimerization of Cys sulfenic acid into thiosulfinate (as depicted in Fig. 9.3). Besides the oxidative formation of thiosulfinate from disulfide, the oxidation stress of the disulfide derivative under certain circumstances could also end up with fragmentation, e.g., treatment of Cys disulfide by NaOCl[24] or performic acid[25] will generate Cys sulfonic acid.

Further oxidation of thiosulfinate **6** gives rise to the corresponding thiosulfonate **7**. This process generally necessitates an high concentration of oxidizing

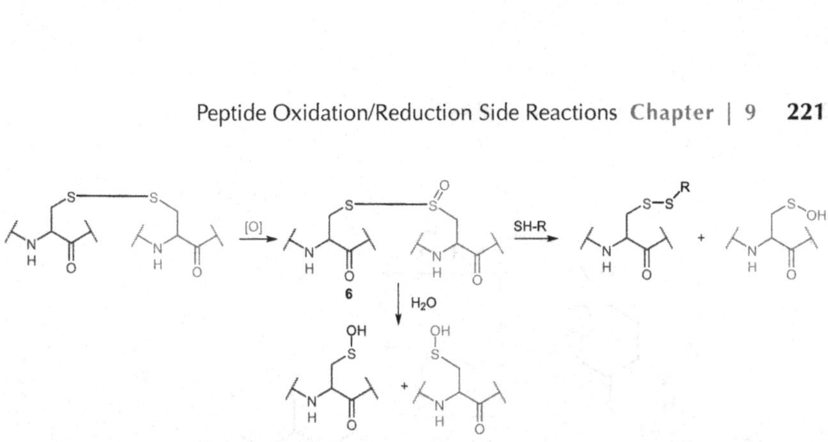

FIGURE 9.6 Process of thiosulfinate formation via oxidation of disulfide and hydrolysis/mercaptolysis of thiosulfinate.

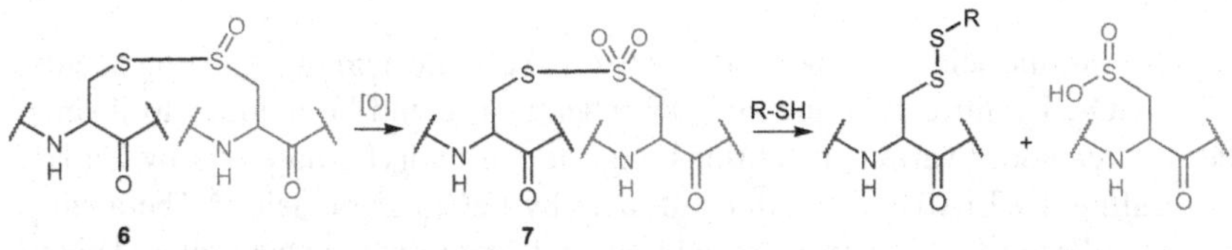

FIGURE 9.7 Process of thiosulfinate oxidation to thiosulfonate and thiosulfonate mercaptolysis.

agents. Thiosulfonate reacts with sulfhydryl compound to produce disulfide and sulfinic acid.[26] These processes are illustrated in Fig. 9.7.

Oxidation of Cys sulfenic acid could lead to the formation of Cys sulfinic acid, the further oxidation of the latter will result in Cys sulfonic acid. Cys sulfenic acid and Cys sulfinic acid could be subjected to a disproportionation process due to their distinctive oxidation states. The former will give rise to Cys sulfinic acid and Cys upon disproportionation (Eq. (9.3))[27] whereas the latter is converted to the corresponding Cys sulfonic acid (Eq. (9.4)).[28]

$$2\text{R-S-OH} \rightarrow \text{R-SH} + \text{RSO}_2\text{H} \tag{9.3}$$

$$4\text{R-SO}_2\text{H} + \text{H}_2\text{O} \rightarrow 2\text{RSO}_3\text{H} + \text{R-S(O)}_2\text{-S-R} \tag{9.4}$$

The chemical stability of Cys sulfinic acid is superior to that of Cys sulfenic acid. The pK_a of the former is around 2, and the deprotonated Cys sulfinic acid can lead the nucleophilic attack on halohydrocarbons or the Michael addition reaction with α,β-unsaturated compounds,[29] which results in the formation of sulfone derivatives **8** and **9**, respectively (Fig. 9.8). In most cases it is the sulfur atom on Cys sulfinic acid that functions as the nucleophile but under certain circumstances the oxygen atom could also regulate the aforementioned nucleophilic reactions.[30]

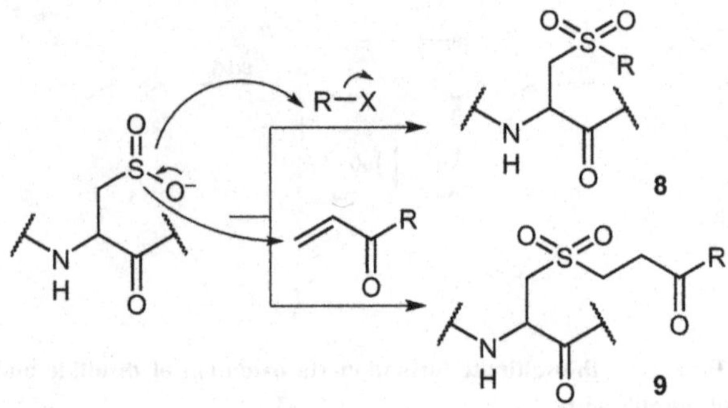

FIGURE 9.8 **Cys sulfinic acid-mediated nucleophilic reactions.**

Cys sulfinic acid could be oxidized to Cys sulfonic acid by oxidizing agents, e.g., H_2O_2, I_2, nitric acid and HClO[28] whereas it could be reduced to diverse derivatives under various conditions, e.g., to sulfhydryl compound by Zn/H^+, to disulfide by $LiAlH_4$, or to thiosulfonate by $FeCl_2$/acetic acid.[28] The acidity of Cys sulfonic acid is significant while its nucleophilicity is negligible. Due to the tremendous leaving tendency of the sulfonate, it can serve as a good leaving group in S_N1, S_N2 addition reactions as well as E1 and E2 elimination reactions.[4] Cys sulfonic acid is susceptible to β-elimination in basic conditions to form dehydroalanine and sulfinic acid.[31] Thiosulfonate derivatives **7** (Fig. 9.7) might suffer from diverse side reactions, e.g., hydrolysis and β-elimination due to the high-leaving tendency of the sulfonate group.

Other oxidation side reactions addressing Cys include degradation of disulfide and subsequent oxidation processes. Cys disulfide could undergo β-elimination in basic milieu.[30] Peptides or proteins bearing disulfide moieties are prone to suffer from this side reaction in basic condition[32] or under high temperature.[33] Dehydroalanine **10** and thiocysteine derivative **11**[5,30,32] are formed in this process, and the latter could be further oxidized to *S*-sulfocysteine compound **12** (Fig. 9.9).[34,35]

Besides the aforementioned ordinary oxidation side reactions on Cys such as the oxidative conversion of Cys to Cys sulfenic acid, sulfinic acid and sulfonic acid, sulfhydryl side chain protected Cys derivatives are also subjected to various oxidation side reactions. Cys[Bzl(4-Me)],[36] Cys(Bzl), Cys(Mob),[37]

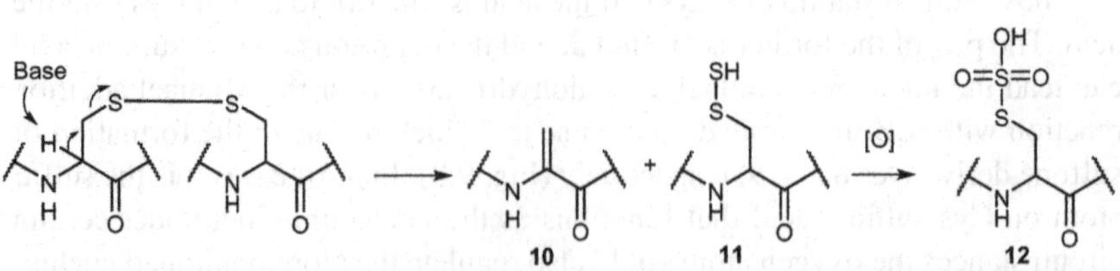

FIGURE 9.9 **β-Elimination of Cys disulfide and oxidation of thiocysteine.**

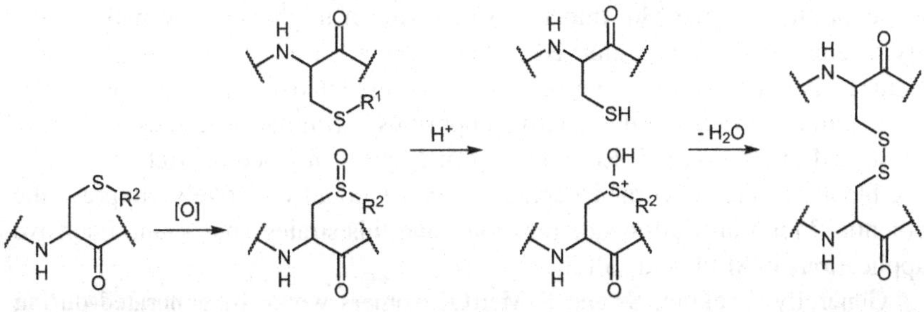

FIGURE 9.10 Disulfide formation from side chain protected Cys via Cys sulfoxide intermediate.

Cys(Acm),[38] and unsymmetrical disulfide[5] can be oxidized to the commensurate sulfoxide derivatives. These sulfoxides are capable of reacting with the free sulfhydryl group on Cys, and even with side chain protected species, e.g., Cys(tBu), Cys(Mob), and Cys(1-Ada) in acidic milieu to produce the corresponding disulfide compounds.[39,40] This process is illustrated in Fig. 9.10.[40] This type of reaction has been designedly exploited for the purpose of regioselective constructions of disulfide bridges in peptide synthesis.[40] Nevertheless, from the perspective of a side reaction, the undesired conversion of the side chain protected Cys species to Cys sulfoxide derivatives might interfere with the target regioselective constructions of multiple-disulfide network.

9.2 OXIDATION SIDE REACTIONS ON Met

Met oxidation is a notorious side reaction in peptide synthesis. The thioether moiety on the Met side chain could be oxidized to corresponding sulfoxide **13** or even sulfone **14** (see also Fig. 9.11) in the process of peptide synthesis, purification, storage and usage. These oxidative conversions, formation of sulfoxide in particular, exert significant impacts on protein physiological functions.[41]

Although Met oxidation poses an undesired side reaction to peptide synthesis, deliberate derivatization of Met to Met(O), and the utilization of the latter as a building block is sometimes designedly adopted in certain peptide synthesis. This seeming paradox is rationalized in view of the concern that due to the

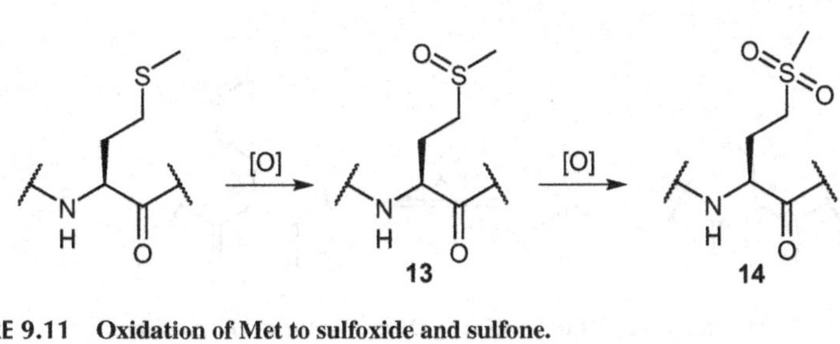

FIGURE 9.11 Oxidation of Met to sulfoxide and sulfone.

nucleophilicity of the Met thioether side chain, it might function with a variety of electrophiles, e.g., cleaved carbocations released from the peptide side chain global deprotection, to give rise to Met-alkylated side products as well as the potential subsequent degradative impurities.[42] Similar processes have been illustrated in Section 6.5. Employment of Met(O) in place of Met as a building block for the concerned peptide synthesis could effectively suppress the potential Met *S*-alkylation side reactions, and this strategy has found intensive applications in SPPS and LPPS.[42,43]

Generally speaking, S- and R-Met(O) isomers would be generated during the oxidation of L-Met, and their ratio is dependent on the specific conditions of the subjected oxidation process. R-Met(O) predominates in H_2O_2 mediated oxidation.[44] Among various agents $NaBO_3$ is verified to be one of the most convenient and effective oxidants in terms of Met oxidization, and it is noted that the Met(O) species is dominant in the product mixture derived from this process, and only an insignificant amount of Met-sulfone is generated.[45]

From the aspect of side reactions Met is vulnerable to various oxidants during peptide synthesis, and could be transformed to side product Met(O). For example, Met containing peptide in a solution exposed to air can be oxidized,[46] and this side reaction might prevail no matter whether in a preparative[47] or in a purification/isolation process.[48] Besides oxygen compounds capable of oxidizing Met to Met(O) include H_2O_2, NBS (*N*-bromosuccinimide)[49] and *N*-chlorosuccinimide,[50] etc. DMSO-mediated Met oxidation is a noteworthy side reaction in peptide synthesis, especially in the presence of HX catalysts whose presence exaggerates the extent of the oxidation side reaction.[51] It is a reversible process, and the equilibrium is driven to the direction of Met(O)/dimethyl thioether formation when DMSO is in significant excess (see also Fig. 9.12).[52] Peptides upon DMSO/HCl treatment could suffer from oxidation side reactions not only addressing Met, but also Trp that is converted to oxindolyl alanine derivative,[51] and Cys that is transformed to the corresponding Cystine.[19] Other oxidizable amino acids such as His and Tyr could resist DMSO-regulated oxidization.

In view of the evident susceptibility of Met to oxidation side reactions in peptide synthesis, especially in acidic conditions[53] as well as the potential alkylation side reactions addressing Met residues, intentional utilization of Met(O) as the protected synthon of Met, and the reductive regeneration of Met

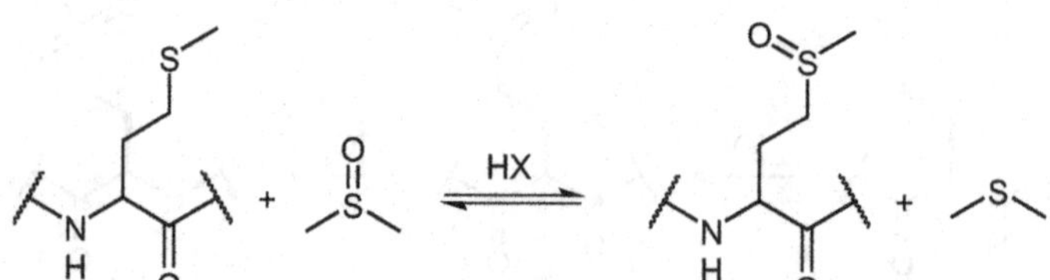

FIGURE 9.12 Oxidation of Met to Met(O) by DMSO in the acidic condition.

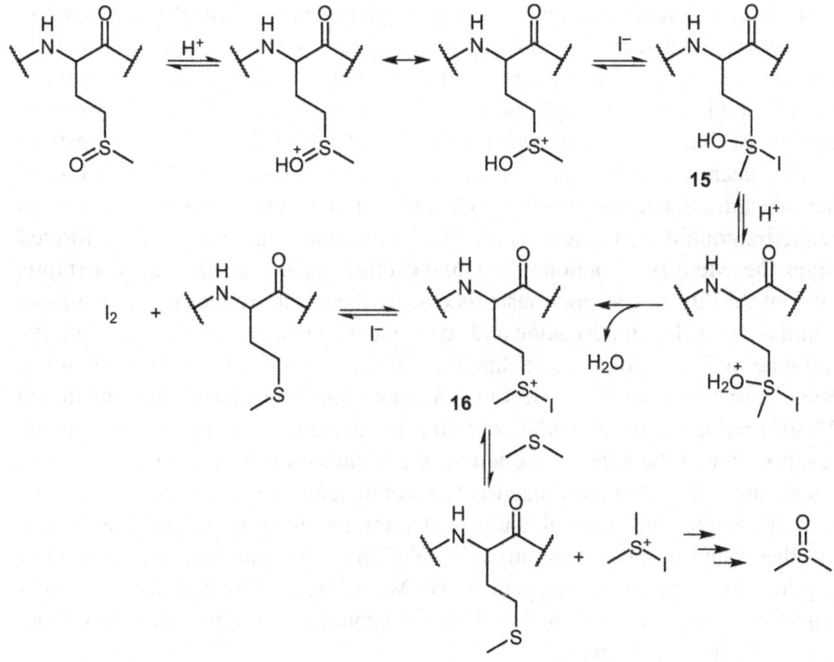

FIGURE 9.13 NH_4I/Me_2S mediated Met(O) reduction.

at an appropriate step has been established as a standard strategy in peptide synthesis.

Regeneration of Met from Met(O) in peptide synthesis is generally achieved with NH_4I/Me_2S treatment in acidic condition.[54] This process is described in Fig. 9.13.[55] The sulfoxide moiety in Met(O) is firstly protonated by acid, enhancing by this means the electrophilicity of the sulfur atom that undertakes subsequently the nucleophilic attack from I^-. Intermediate **15** is generated in this process, which undergoes protonation and subsequent dehydration to give rise to the formation of iodosulfonium intermediate **16**. The latter will accommodate the nucleophilic attack from another I^- anion to be transformed to Met upon releasing I_2. This Met(O) reduction process can be accelerated by Me_2S which reduces iodosulfonium intermediate **16** to Met. Me_2S is oxidized to DMSO in this process.[56] It has been reported that a Met(O) reduction mediated solely by NH_4I might induce concomitant hydrolysis of the thioester moiety while the addition of Me_2S to the subjected reaction could reinforce the stability of the thioester in the reductive environment.[57]

NH_4I/Me_2S-mediated Met(O) reduction is a relatively mild reaction, and is deemed as compatible with disulfide derivative.[58] Compounds bearing a free sulfhydryl functional group should be avoided in this process in order to prevent the occurrence of potential disulfide-thiol exchange processes.[59] The Cys(Acm) derivative also remains stable under the conditions of Met(O) reduction[56] in spite of the formation of I_2 in this process. Side chain-unprotected Cys will be converted to the corresponding disulfide counterpart in the referred process of Met(O) reduction. This might be due to the generation of an iodosulfonium intermediate that is able to activate the sulfhydryl side chain of Cys and construct by this means a disulfide bond between two free Cys residues.[58] The imidazolyl and phenolic functional groups on His and Tyr are stable under the NH_4I/Me_2S reductive condition, and remain inert to the iodosulfonium intermediate formed upon the Met(O) reduction.[56,60] On the other hand, Trp derivative exhibits evident lability in the previous process, as the acidic reductive environment could result in Trp dimerization and consequent formation of either 2,2'-indole–indoline or 2,2'-indole–indole dimeric Trp derivative.[58] This phenomenon has been separately described in Section 3.4. Shortening of the NH_4I/Me_2S-mediated Met(O) reduction in an acidic condition is advisable as to minimize this side reaction. Since the Met(O) reduction at a dedicated step in peptide synthesis could cause yield drop, this reaction is generally conducted concomitantly at the step of peptide side chain global deprotection by charging NH_4I/Me_2S as well as other relevant scavengers into TFA solution for the simultaneous removal of peptide side chain protecting groups and Met(O) reduction. This process might sometimes even be combined with the concomitant disulfide formation from free sulfhydryl derivatives.[54]

TMSBr could also regenerate Met from Met(O). The possible mechanism is proposed in Fig. 9.14.[61] Thanks to the Lewis acid property of TMSBr its silylation on the Met(O) sulfoxide side chain simultaneously activates the affected sulfoxide substituent, facilitating by this means the nucleophilic attack of Br^- on the concerned sulfur atom to give the intermediate **17**. This intermediate undergoes subsequently further silylation by another molecule of TMSBr to give rise to derivative **18** whose hexamethyldisiloxy moiety functions as a good leaving group upon the nucleophilic attack of the Br^-. Br_2 is released in this process, and the reduced Met are regenerated as a consequence. Reducing agents, e.g., thioanisole or EDT are added to this reductive reaction in order to scavenge the *in situ* formed Br_2 so as to minimize its potential modifications on sensitive amino acid residues such as Tyr.[61,62] The TMSBr-mediated

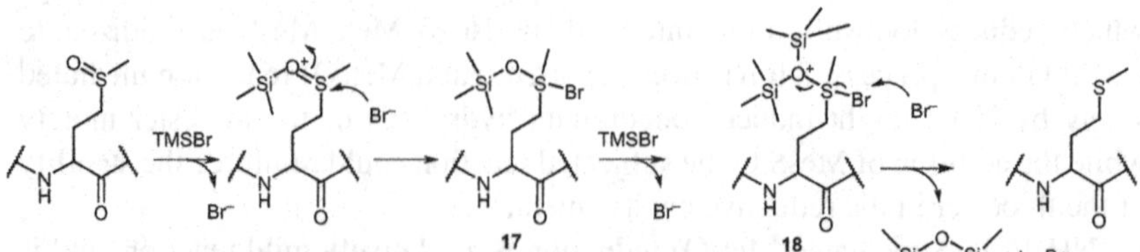

FIGURE 9.14 Proposed mechanism of TMBr-mediated Met(O) reduction.

Met(O) reduction can also be combined with TFA-directed peptide side chain global deprotection reaction. The analogous Met(O) reduction reactions could also regulated by $MeSiCl_3/NaI$[63] and TMSCl/thiophenol.[64]

Bu_4NBr is another appropriate reducing agent for Met regeneration from its oxidized counterpart, and it could be added into a TFA solution in an effort to conduct a concomitant peptide side chain global deprotection. Relative to NH_4I this compound possesses enhanced solubility in TFA, and could direct a more sufficient and accelerated Met(O) reduction reaction. In some cases it is capable of realizing a quantitative regeneration of Met from Met(O) within 5 min.[65] The Bu_4NBr/TFA/Anisole system for Met(O) reduction, in the event of the occurrence of the Cys residue in the subjected peptide, might result in the formation of an undesired Cys disulfide side product. The addition of EDT and thioanisole to the previous TFA solution can evidently suppress this side reaction.[65] As for the process development of the Bu_4NBr/TFA-mediated Met(O) reduction, it is advisable to charge the Bu_4NBr to peptide/TFA solution just prior to the end of the addressed peptide side chain global deprotection reaction, in order to shorten the period of the reduction reaction, and to minimize the potential side reactions in this process.

Sulfhydryl or thioether reagents such as DTT,[66] thioacetic acid, thioethanol and *N*-(methyl)mercaptoacetamide[67] are endowed with advantages to suppress Met oxidation in the process of peptide side chain global deprotection. The drawbacks of these compounds in this connection lie in the low reaction kinetics and an unsatisfactory conversion. In Boc chemistry Me_2S is regularly added to HF for peptide global deprotection to regenerate Met from Met(O).[68] In addition, 2-PySH can also participate in the HF- or TfOH-directed peptide side chain global deprotection that suppresses the formation of Met(O) in this process. HF/2-PySH system exhibits superior efficiency compared with TfOH/2-PySH in this connection. Moreover, 2-PySH could efficiently suppress Met alkylation side reactions initiated by Bzl and MeBzl in acidic conditions.[69]

9.3 OXIDATION SIDE REACTIONS ON Trp

Similar to Met and Cys, oxidation of Trp is a common oxidation side reaction in peptide synthesis. Oxidants such as H_2O_2,[70] O_2,[71] O_3[72] and DMSO/HCl[51] could initiate this undesired process. Oxidation of Trp might take place both under physiological conditions and at certain chemical synthesis steps. This process is relatively complicated, and various oxidized species could be derived. Under different conditions Trp could be oxidized to Kyn (Kynurenine), NFK (*N*-formylkynurenine),[73] hydroxy-*N*-formylkynurenine, 3-hydroxykynurenine, 5-hydroxy-Trp, 7-hydroxy-Trp,[74] Oia (oxindolylalanine) or DiOia (dioxindolylalanine).[75] Moreover, Trp derivatives might also be subjected to photooxidation and be converted to the corresponding oxidized species.[76]

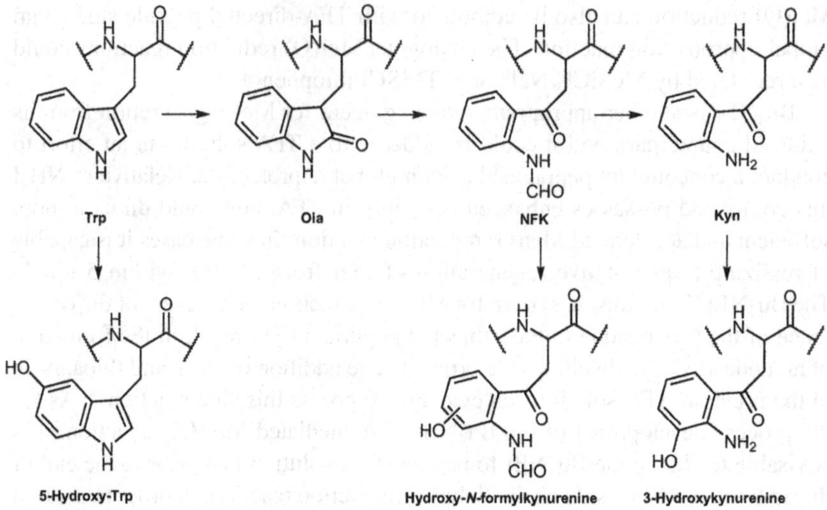

FIGURE 9.15 Scheme of Trp oxidation.

Trp is extraordinarily prone to undergo oxidation in acidic conditions, especially at the step of an acid-mediated peptide side chain global deprotection. Common Trp oxidation processes are described in Fig. 9.15. Oxidation predominantly addresses the pyrrole ring of the Trp indolyl side chain,[75] transforming the affected Trp to the corresponding Oia derivatives upon H_2O_2 treatment. Oia could be further oxidized to NFK[3] which is either in tandem oxidized to hydroxy-*N*-formylkynurenine, or converted to Kyn via deformylation process.[77] Oxidation of Kyn will give rise to 3-hydorxykynurenine. Oxidative modifications on Trp are nonetheless not limited to the pyrrole ring moiety whereas the phenyl group is also susceptible to oxidative degradation to give 5-hydroxy-Trp.

In consideration of the inherent susceptibility of the Trp residue to suffer from oxidative modifications in the process of an acid-mediated peptide side chain global deprotection, additives, like DTT or methylindole could be charged to the concerned reaction system in an effort to minimize the extents of Trp oxidation side reactions.[78]

9.4 OXIDATION SIDE REACTIONS ON OTHER AMINO ACIDS AND AT NONSYNTHETIC STEPS

His could also be subjected to an oxidation modification in the process of peptide synthesis. His-containing peptide or protein could be transformed to the corresponding 2-oxo-His counterparts *in vivo* and *in vitro* by means of photocatalytic oxidation or metal-catalyzed oxidation.[79,80] The proposed mechanism

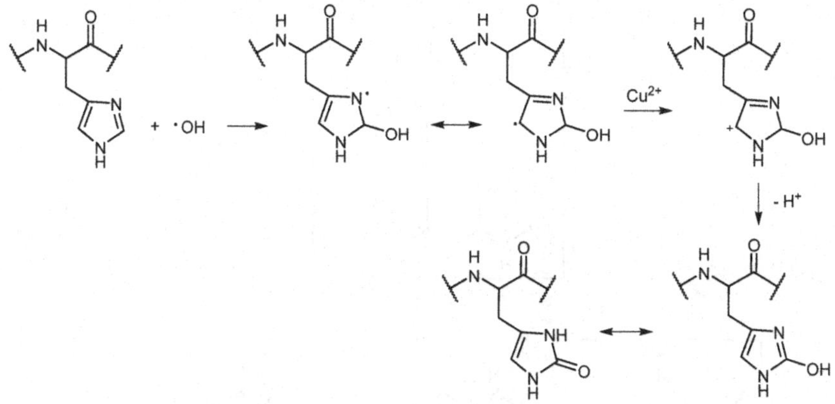

FIGURE 9.16 Metal-catalyzed His oxidation to 2-oxo-His.

of this conversion is displayed in Fig. 9.16. This process could be regulated by Cu^{2+}-catalyzed and radical-involved oxidation.[81]

The phenolic side chain on Tyr could be modified by a variety of oxidants, and be converted to the corresponding oxidized Tyr derivatives through either radical or nonradical routes. Oxidized Tyr species include DOPA (3,4-dihydroxyphenylalanine), 3-nitro Tyrosine, 3-Cl Tyrosine, 3,5-dichloro Tyrosine, 3-Br Tyrosine, 3,5-dibromo Tyrosine and dimeric Tyrosine[82] (see also Fig. 9.17). Phe residues in peptide and protein could also be addressed to oxidation by air or relevant radicals, and give rise to the formation of *ortho*/*meta*-Tyrosine[82] derivatives or DOPA.[83]

On top of the aforementioned oxidizable derivatives, all amino acids in peptide/protein are actually theoretically susceptible to oxidative stress in the air, and could be transformed to the corresponding peroxide radicals via their backbone C^{α} radicals, finally leading to peptide/protein backbone fragmentation.[84] The specifics of this process will not be elaborated herein.

Excipients such as polysorbate or PEG for the purpose of formulation of peptide/protein drug substances might cause undesired amino acid oxidation owing to the potential peroxide release. Peroxide derivatives in Tween-80 could trigger intensive peptide oxidation.[85] Polymer materials for the peptide packaging including a prefilled syringe, might also release peroxide substances, which result in peptide oxidation as a consequence. Antisolvents such as ether derivatives for the purpose of product precipitation/isolation in the process of peptide manufacture are also capable of oxidizing the sensitive peptides due to the peroxide contaminants. Ether species that are less prone to form peroxides, e.g., MTBE (methyl *tert*-butyl ether) or CPME (cyclopentyl methyl ether) are recommended as appropriate antisolvents for the process of peptide precipitation that could minimize the oxidation side reactions addressing sensitive peptides at this step.

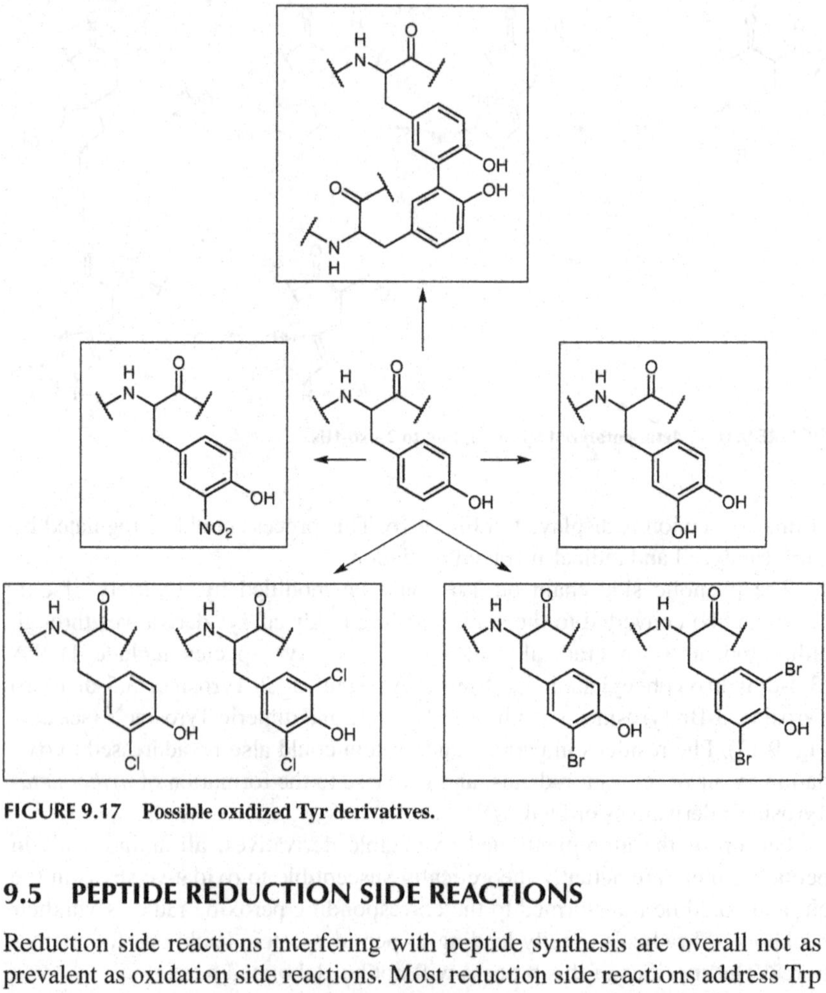

FIGURE 9.17 **Possible oxidized Tyr derivatives.**

9.5 PEPTIDE REDUCTION SIDE REACTIONS

Reduction side reactions interfering with peptide synthesis are overall not as prevalent as oxidation side reactions. Most reduction side reactions address Trp whose indolyl side chain could be reduced to the corresponding indoline derivative by reducing agents, like TES employed as a scavenger for peptide side chain global deprotection. (This phenomenon has been dedicatedly introduced in Section 3.5 and so will not be reiterated here.) Peptides containing disulfide bridges are also evidently subjected to reduction side reactions in the presence of mercaptan compounds or other reducing agents. This category of side reaction will be elaborated in a separate section.

Thanks to the unique properties of the azide group to the domains, like click chemistry[86] and Staudinger Ligation,[87] peptides bearing azide functional group are playing more and more important roles in bioconjugation.[88] Some azide amino acids [89] or saccharide derivatives[90] are utilized as building blocks in an effort to prepare the corresponding azide peptide precursor products. In spite of their widespread applications in bioconjugation technologies azide peptides are

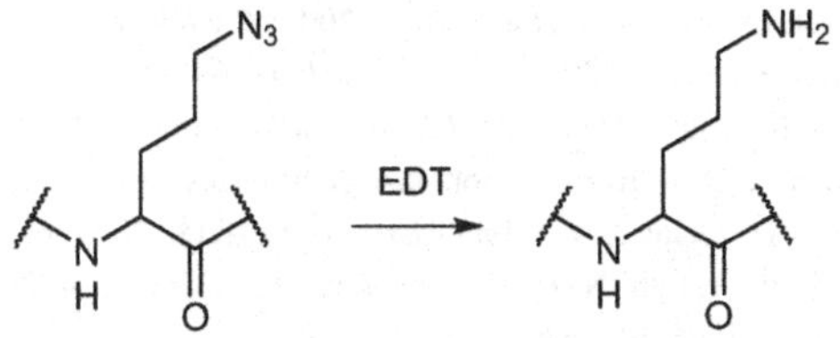

FIGURE 9.18 Reduction of Nva(N_3) to Orn by EDT.

susceptible to a reduction side reaction in the synthetic process due to the presence of certain reductive scavengers involved in peptide side chain global deprotection. It has been discovered that the addition of EDT or DTT as scavenger to the side chain global deprotection of the Nva(N_3)-containing peptide results in the undesired reduction of azide to amine (Fig. 9.18).[91] This phenomenon is ascribed to the evident reducing capacity of EDT/DTT, while EDT could cause much more intensive azide reduction relative to DTT under the same conditions. It has been confirmed that sulfhydryl compounds could quantitatively reduce the azide substituent on AZT (azidothymildine) to the corresponding amino group in basic conditions,[92] and convert the N^{α}-azido peptide to an amino counterpart in the basic milieu.[93] It is implied simultaneously from these investigations that azide groups are also subjected to reduction side reactions by sulfhydryl compounds, and are transformed to the corresponding amines in the TFA-mediated peptide side chain global deprotection.

Nle(N_3)-containing peptide could be intensively reduced to the Lys counterpart in TFA/EDT (Yang, Y., unpubished results). This is, nevertheless, contrary to the conclusion from Schneggenburger et al.[91] that neither Fmoc-Nle(N_3)-OH nor Nle(N_3)-containing peptide suffered from reduction under identical conditions, except under the condition that EDT is substituted by thioanisole.

REFERENCES

1. (a) Nagahara N. *Amino Acids*. 2011;41:59–72. (b) Paget MS, Buttner MJ. *Annu Rev Genet*. 2003;37:91–121. (c) Wouters MA, Fan SW, Haworth NL. *Antioxid Redox Signal*. 2010;12:53–91.
2. (a) Dorenbos R, Stein T, Kabel J, et al. *J Biol Chem*. 2002;277:16682–16688. (b) Möller M, Hederstedt L. *Antioxid Redox Signal*. 2006;8:823–833.
3. Berlett BS, Stadtman ER. *J Biol Chem*. 1997;272:20313–20316.
4. Reddie KG, Carroll KS. *Curr Opin Chem Biol*. 2008;12:746–754.
5. Jeong J, Jung Y, Na S, et al. *Mol Cell Proteom*. 2011;10(3):M110.000513.
6. Seo YH, Carroll KS. *Proc Natl Acad Sci USA*. 2009;106:16163–16168.
7. Aversa MC, Barattucci A, Bonaccorsi P, Giannetto P. *Curr Org Chem*. 2007;11:1034–1052.
8. Rehder DS, Borges CR. *Biochemistry*. 2010;49:7748–7755.
9. Allison WS. *Acc Chem Res*. 1976;9:293–299.
10. Claiborne A, Miller H, Parsonage D, Ross RP. *FASEB J*. 1993;7:1483–1490.
11. Davis FA, Jenkins LA, Billmers RL. *J Org Chem*. 1986;51:1033–1040.
12. Allison WS, Benitez LV, Johnson CL. *Biochem Biophys Res Commun*. 1973;52:1403–1409.

13. Salmeen A, Andersen JN, Myers M, et al. *Nature*. 2003;423:769–773.
14. Yang J, Groen A, Lemeer S, et al. *Biochemistry*. 2007;46:709–719.
15. Sivaramakrishnan S, Keerthi K, Gates KS. *J Am Chem Soc*. 2005;127:10830–10831.
16. Patai S. *Sulfenic Acids and Derivatives*. Hoboken, New Jersey: John Wiley & Sons; 1990.
17. Tam JP, Wu C-R, Liu W, Zhang J-W. *J Am Chem Soc*. 1991;113:6657–6662.
18. Otaka A, Koide T, Shide A, Fujii N. *Tetrahedron Lett*. 1991;32:1223–1226.
19. Lowe OG. *J Org Chem*. 1975;40:2096–2098.
20. Lowe OG. *J Org Chem*. 1976;41:2061–2064.
21. Tarbell DS. Kaarasch N, ed. *Organic Sulfur Compounds*, Vol. 1. New York: Pergamon; 1961:97.
22. Batista-Viera F, Manta C, Carlsson J. *Appl Biochem Biotechnol*. 1994;44:1–14.
23. Nagy P, Ashby MT. *Chem Res Toxicol*. 2007;20:1364–1372.
24. Jong LI, Abbott NL. *Langmuir*. 2000;16:5553–5561.
25. Cremlyn RJ. *An Introduction to Organosulfur Chemistry*. Chichester, England: John Wiley & Sons Ltd; 1996:51.
26. Paulsen CE, Carroll KS. *Chem Rev*. 2013;113:4633–4679:Article ASAP.
27. Abraham RT, Benson LM, Jardine I. *J Med Chem*. 1983;26:1523–1526.
28. Cremlyn RJ. *An Introduction to Organosulfur Chemistry*. Chichester, England: John Wiley & Sons Ltd; 1996:93–94.
29. Ooata Y, Sawaki Y, Isono M. *Tetrahedron*. 1970;26:731–736.
30. (a) Jönsson TJ, Johnson LC, Lowther WT. *J Biol Chem*. 2009;284:33305–33310. (b) Wang Z, Rejtar T, Zhou ZS, Karger BL. *Rapid Commun Mass Spectrom*. 2010;24:267–275.
31. Dai J, Zhang Y, Wang J, et al. *Rapid Commun Mass Spectrom*. 2005;19:1130–1138.
32. De Marco C, Coletta M, Cavallini D. *Arch Biochem Biophys*. 1963;100:51–55.
33. Volkin DB, Klibanov AM. *J Biol Chem*. 1987;262:2945–2950.
34. Shimizu A. *Rinsho Byori*. 2006;54:924–934.
35. Mader P, Volke J, Kuta J. *Collect Czech Chem Commun*. 1970;35:552–564.
36. Heath WF, Tam JP, Merrifield RB. *Int J Pept Protein Res*. 1986;28:498–507.
37. Funakoshi S, Fujii N, Akaji K, Irie H, Yajima H. *Chem Pharm Bull*. 1979;27:2151–2156.
38. Yajima H, Akaji K, Funakoshi S, Fujii N, Irie H. *Chem Pharm Bull*. 1980;28:1942–1945.
39. Haruaki Y, Fujii N, Funakoshi S, Watanabe T, Murayama E, Otaka A. *Tetrahedron*. 1988;44: 805–819.
40. Fujii N, Otaka A, Watanabe T, et al. *J Chem Soc Chem Commun*. 1987:1676–1678.
41. Heinrikson RL, Kramer KJ. In: Kaiser ET, Kezdy FJ, eds. *Progress in Bioorganic Chemistry*. New York: Wiley; 1974:141.
42. Bodanszky M, Martinez J. Gross E, Meienhofer J, eds. *The Peptides, Analysis, Structure, Biology*, Vol. 5. New York: Academic Press; 1983:111.
43. Barany G, Merrifield RB. Gross E, Meienhofer J, eds. *The Peptides, Analysis, Structure, Biology*, Vol. 2. New York: Academic Press; 1980:217.
44. Iselin B. *Helv Chim Acta*. 1961;44:61–78.
45. Fujii N, Sasaki T, Funakoshi S, Irie H, Yajima H. *Chem Pharm Bull*. 1978;26:650–653.
46. Glazer AN. Neurath H, Hill RL, eds. *The Proteins*, Vol. 2. New York: Academic Press; 1976:2.
47. Hofmann K, Finn FM, Linetti M, Montibeller J, Zanetti G. *J Am Chem Soc*. 1966;88:3633–3639.
48. Harris JI, Roos P. *Biochem J*. 1959;71:434–445.
49. Witkop B. *Adv Protein Chem*. 1962;16:221–321.
50. Shechter Y, Burstein Y, Patchornik A. *Biochemistry*. 1975;14:4497–4503.
51. Savige WE, Fontana A. *J Chem Soc Chem Commun*. 1976:599–600.
52. Lipton S, Bodwell CE. *J Agric Food Chem*. 1976;24:26–31.
53. Hofmann K, Haas W, Smithers MJ, et al. *J Am Chem Soc*. 1965;87:620–631.

54. Huang H, Rabenstein DL. *J Pept Res*. 1999;53:548–553.
55. Landini D, Montanari F. *Tetrahedron Lett*. 1964;5:2691–2696.
56. Vilaseca M, Nicolás E, Capdevila F, Giralt E. *Tetrahedron*. 1998;54:15273–15286.
57. Hackenberger CPR. *Org Biomol Chem*. 2006;4:2291–2295.
58. Nicolás E, Vilaseca M, Giralt E. *Tetrahedron*. 1995;51:5701–5710.
59. Andreu D, Albericio F, Solé NA, Munson MC, Ferrer M, Barany G. In: Pennington MW, Dunn BM, eds. *Methods in Molecular Biology. Vol. 35. Peptide Synthesis Protocols*. Towata, New Jersey: Humana Press; 1994:91–169.
60. Büllesbach EE, Schwabe C. *J Biol Chem*. 1991;266:10754–10761.
61. Beck W, Jung G. *Lett Pept Sci*. 1994;1:31–37.
62. Fujii N, Otaka A, Sugiyama N, Hanato M, Yajima H. *Chem Pharm Bull*. 1987;35:3880–3883.
63. Olah GA, Husain A, Singh BP, Mehrotra AK. *J Org Chem*. 1983;48:3667–3672.
64. Numata T, Togo G, Oae S. *Chem Lett*. 1979;8:329–332.
65. Taboada L, Nicolás E, Giralt E. *Tetrahedron Lett*. 2001;42:1891–1893.
66. Polzhofer KP, Ney KH. *Tetrahedron*. 1971;27:1997–2001.
67. Houghten RA, Li VH. *Anal Biochem*. 1979;98:36–46.
68. Tam JP, Heath WF, Merrifield RB. *J Am Chem Soc*. 1983;105:6442–6455.
69. Taichi M, Kimura T, Nishiuchi Y. *Int J Pept Ther*. 2009;15:247–253.
70. (a) Nielsen HK, de Weck D, Finot PA, Liadon R, Hurrell RF. *Br J Nutr*. 1985;53:281–292. (b) Kell G, Steinhart H. *J Food Sci*. 1990;55:1120–1123.
71. Holt LA, Milligan B, Rivett DE, Stewart FHC. *Biochim Biophys Acta*. 1977;499:131–138.
72. Okajima T, Kawata Y, Hamaguchi K. *Biochemistry*. 1990;29:9168–9175.
73. Krogull MK, Fennema O. *J Agric Food Chem*. 1987;35:66–70.
74. Suto D, Ikeda Y, Fujii J, Ohba Y. *J Clin Biochem Nutr*. 2006;38:1–5.
75. Simat TJ, Steinhart H. *J Agric Food Chem*. 1998;46:490–498.
76. Pattison DI, Rahmanto AS, Davies MJ. *Photochem Photobiol Sci*. 2012;11:38–53.
77. Takikawa O, Yoshida R, Kido R, Hayaishi O. *J Biol Chem*. 1986;261:3648–3653.
78. Pennington MW. In: Pennington MW, Dunn BM, eds. *Methods in Molecular Biology. Vol. 35. Peptide Synthesis Protocols*. Towata, New Jersey: Humana Press; 1994:44–45.
79. Li S, Schöneich C, Borchardt RT. *Biotechnol Bioeng*. 1995;48:490–500.
80. Uchida K, Kawakishi S. *J Biol Chem*. 1994;269:2405–2410.
81. Schöneich C. *J Pharm Biomed Anal*. 2000;21:1093–1097.
82. Davies MJ, Fu S, Wang H, Dean RT. *Free Radic Biol Med*. 1999;27:11–12.
83. Blanchard M, Bouchoule C, Djaneye-Boundjou G, Canesson P. *Tetrahedron Lett*. 1988;29: 2177–2178.
84. Headlam HA, Mortimer A, Easton CJ, Davies MJ. *Chem Res Toxicol*. 2000;13:1087–1095.
85. Herman AC, Boone TC, Lu HS. In: Pearlman R, Wang YJ, eds. *Formulation, Characterization, and Stability of Protein Drugs: Case Histories*. New York, NY: Plenum Press; 1996:324–325.
86. Kolb HC, Finn MG, Sharpless KB. *Angew Chem Int Ed Engl*. 2001;40:2004–2021.
87. Köhn M, Breinbauer R. *Angew Chem Int Ed Engl*. 2004;43:3106–3116.
88. Moses JE, Moorhouse AD. *Chem Soc Rev*. 2007;36:1249–1262.
89. Link AJ, Vink MKS, Tirrell DA. *J Am Chem Soc*. 2004;126:10598–10602.
90. Grandjean C, Boutonnier A, Guerreiro C, Fournier J-M, Mulard LA. *J Org Chem*. 2005;70: 7123–7132.
91. Schneggenburger PE, Worbs B, Diederichsen U. *J Pept Sci*. 2010;16:10–14.
92. Handlon AL, Oppenheimer NJ. *Pharm Res*. 1988;5:297–299.
93. Tornoe CW, Davis P, Porreca F, Meldal M. *J Pept Sci*. 2000;6:594–602.

Chapter 10

Redundant Amino Acid Coupling Side Reactions

Incomplete amino acid couplings in the process of the peptide stepwise assembly normally lead to the formation of side products with deletion sequences. This phenomenon is particularly prone to affect couplings of amino acids with evident steric hindrances, and/or when the concerned peptide chains have already adopted certain secondary conformations that impede the efficient incorporation of the forthcoming amino acid residues.

In spite of the abundant occurrences of impurities with deletion sequences in peptide synthesis, redundant amino acid couplings are also regarded as a type of common side reactions in peptide chemistry that give rise to the formation of peptide impurities with undesired repetitive amino acid sequences. Some redundant amino acid couplings such as that of Gly are potentially capable of introducing insurmountable challenges at the subsequent product purification steps. In-depth investigations on the root causes of the ordinary redundant amino acid couplings are deemed as a rational tactic to tackle these side reactions in peptide synthesis.

10.1 DIPEPTIDE FORMATION DURING AMINO ACID N^{α}-Fmoc DERIVATIZATION

Redundant amino acid couplings in peptide synthesis might be partially ascribed to the unacceptable quality of the amino acid raw materials due to the inherently deficient process design and/or insufficient purification. Manufacturing of Fmoc-protected amino acids might resort to the derivatization reagent Fmoc-Cl[1] which functions with N^{α}-unprotected amino acids under Schotten–Baumann conditions to give the corresponding N^{α}-Fmoc protected amino acid derivatives (Fig. 10.1). The pure products are normally derived from crystallization and applied in peptide synthesis as raw materials.

One of the most severe problems implicated in the Fmoc-Cl regulated amino acid N^{α}-Fmoc derivatization is the formation of dipeptide or even tripeptide impurities. The liability of such a side reaction is partially dependent on the intrinsic properties of the concerned amino acid.[2] Apart from the Fmoc-Cl mediated amino acid derivatization, Alloc-Cl/Z-Cl directed preparation of the Alloc-Xaa-OH/Z-Xaa-OH might induce commensurate dipeptide impurities as

Side Reactions in Peptide Synthesis. http://dx.doi.org/10.1016/B978-0-12-801009-9.00010-0

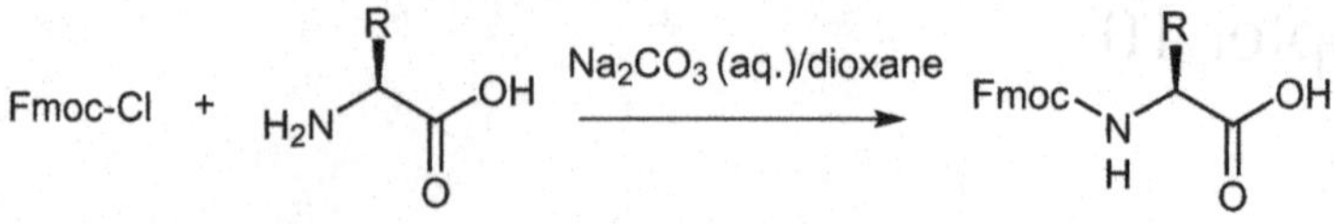

FIGURE 10.1 Fmoc-Cl mediated amino acid N^{α}-Fmoc derivatization.

well.[3] The mechanism of this kind of side reaction is displayed in Fig. 10.2.[4] Fmoc-Xaa-OH derived from Fmoc-Cl and H-Xaa-OH could further function with the excess Fmoc-Cl to give mixed anhydride **1**, while the latter could undertake nucleophilic attack from underivatized H-Xaa-OH in the reaction system to form dipeptide Fmoc-Xaa-Xaa-OH **2**. Or alternatively Fmoc-Cl firstly reacts with H-Xaa-OH to give a mixed anhydride **3** that is subsequently derivatized by Fmoc-Cl and converted subsequently to dipeptide Fmoc-Xaa-Xaa-OH **2**. Mixed anhydride **3** might also be evolved to a N^{α}-underivatized dipeptide H-Xaa-Xaa-OH **4** through the reaction with H-Xaa-OH. Dipeptide **4** is subsequently capable of undertaking derivatization by Fmoc-Cl to give dipeptide Fmoc-Xaa-Xaa-OH **2**.

If the formed dipeptide impurities are not sufficiently separated from the target product Fmoc-Xaa-OH via crystallization or an equivalent process, they will be carried over to the corresponding peptide synthesis as contaminants of the affected raw materials, potentially resulting in the formation of an *endo*-Xaa side product as a consequence. Amino acids with less steric hindrances, e.g., Gly and Ala are more susceptible to dipeptide formation in this manner.[5]

In light of the intrinsic dipeptide formation during amino acid N^{α}-Fmoc derivatization more applicable synthetic strategies have been developed in this connection. Less reactive amino acid N^{α}-Fmoc derivatization reagents are utilized in an effort to reduce the extent of mixed anhydride formation, and in turn to minimize dipeptide impurities. Given the increased nucleophilicity of

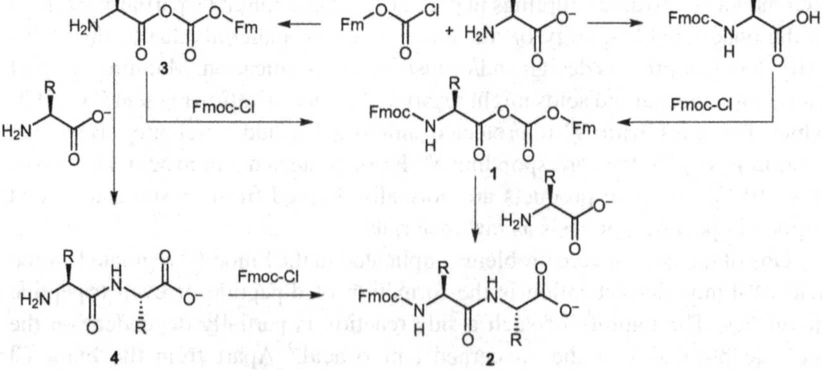

FIGURE 10.2 Dipeptide formation during Fmoc-Cl mediated amino acid N^{α}-Fmoc derivatization.

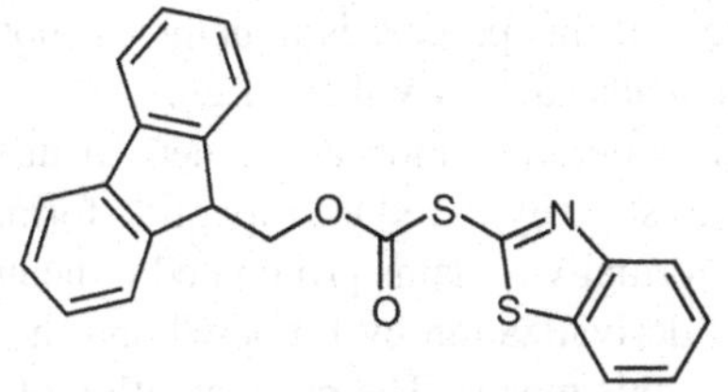

FIGURE 10.3 **Fmoc-2-MBT.**

the amino group relative to the carboxylate, the selectivity of the amino group over the carboxylate with regard to accommodating acylation by N^{α}-Fmoc derivatization reagents with restrained electrophilicity will be increased. Commensurately, the undesired dipeptide formation in the process of amino acid N^{α}-Fmoc derivatization is diminished. Fmoc-N_3[2] and Fmoc-OSu[4] have already been employed in place of Fmoc-Cl for N^{α}-Fmoc derivatization of amino acids. Both of these reagents have exhibited improved properties as to suppress dipeptide formation. However, the utilization of Fmoc-N_3 is limited by its evidently low reactivity toward amino acids,[6] whereas Fmoc-OSu could result in the formation of Fmoc-β-Ala-OH and Fmoc-β-Ala-Xaa-OH impurities during amino acid N^{α}-Fmoc derivatization via the Lossen rearrangement.[7] Relevant side reactions have been introduced in Section 5.6 and Section 7.5.

In consideration of the inherent deficiencies of Fmoc-Cl, Fmoc-N_3 and Fmoc-OSu as to derivatize amino acids, new types of N^{α}-Fmoc derivatization reagents have been developed in an effort to replace the conventional ones. Fmoc-2-MBT (Fig. 10.3) exhibits evident advantages in terms of amino acid N^{α}-Fmoc derivatization.[8] In spite of its substantially decreased activity relative to Fmoc-OSu (pK_a of 2-MBT is higher than that of HOSu)[8] Fmoc-2-MBT can effectively suppress or even avoid the formation of dipeptide thanks to its reduced electrophilicity. The released 2-MBT byproduct could be readily separated from the product via workup process.

In addition, immobilized HOBt could be functioned with Fmoc-Cl, and converted to corresponding solid phase Fmoc-OBt derivative. It is capable of reacting with the addressed amino acids to produce and release the target Fmoc-Xaa-OH product. Dipeptide formation could be effectively avoided by this means.[9]

Synthetic strategies of amino acid N^{α}-Fmoc derivatization can be rationally modified with the purpose of addressing the root cause of dipeptide formation. Due to the fact that this side reaction comes into being via a mixed anhydride intermediate, a modest increase of the ratio amino acid/Fmoc derivatization reagent could effectively restrain dipeptide formation. For example, employment of 20–25% excessive amino acid cf. Fmoc-Cl and 4 equiv. Na_2CO_3 cf. amino acid as well as minimal volume of dioxane as solvent mixture (dioxane/H_2O, 1:10, v/v) under vigorous agitation at room temperature could substantially suppress dipeptide formation during the amino acid Fmoc derivatization process.[10]

It is noted that the effect of this process is nonetheless not sufficiently evident for hydrophobic amino acids such as Val and Leu.

Amino acid N^{α}-Fmoc derivatization under neutral instead of basic conditions is another practical strategy to reduce dipeptide formation. The carboxylate on the amino acid normally remains protonated in neutral environment that prevents the undesired derivatization by Fmoc-Cl and the formation of mixed anhydride is blocked by this means. The prerequisition of Fmoc derivatization of an amino acid under such conditions is the timely removal of released HCl generated in the process. This can be realized through the addition of Zn powder into the reaction system to restrain the rising of the pH in an effort to facilitate smooth N^{α}-Fmoc derivatization of the deprotonated amino groups, and prevent the undesired reaction between the protonated carboxylate and Fmoc-Cl.[11]

Addition of TMSCl to the reaction system of amino acid and Fmoc-Cl could also suppress dipeptide formation in the process of amino acid N^{α}-Fmoc derivatization. TMSCl reacts with the free amino acid to form the corresponding *N,O*-bis(trimethylsilyl)amino acid intermediate. The transient protection on carboxylate could effectively suppress the undesired derivatization by Fmoc-Cl and consequently the formation of a mixed anhydride intermediate. The dipeptide impurity could be hence evidently diminished.[5] The regeneration of Fmoc-Xaa-OH from its silyl ester precursor is readily realized in the workup process through hydrolysis treatment.

In view of the fact that dipeptide Fmoc-Xaa-Xaa-OH could remain in the corresponding Fmoc-Xaa-OH raw material as contaminant it is imperative to dedicatedly develop/validate the corresponding analytical methods to detect and quantify dipeptide impurity, and implement them to the process of raw material control and release for peptide API manufacturing. Rational definition of the specification of dipeptide impurity in the corresponding amino acid raw material, strict control, and release of these materials could effectively address the root cause of the redundant amino acid couplings induced by this means in peptide synthesis.

10.2 REDUNDANT AMINO ACID COUPLING VIA PREMATURE Fmoc DEPROTECTION

Another major source of redundant amino acid coupling in peptide synthesis is the premature deblocking of the N^{α}-Fmoc protecting group either from the incoming amino acid in solution phase, or the immobilized peptide on the solid support (see also Fig. 10.4). If the undesired premature Fmoc deblocking addresses an amino acid that is supposed to be assembled into the peptide chain the consequence will be the formation of two types of amino acid species in solution, i.e., N^{α}-Fmoc protected derivative **5** and N^{α}-unprotected counterpart **6**. Their couplings to the immobilized peptide chain on the solid support will give the target N^{α}-Fmoc protected peptide **7** and peptide derivative **8** with free N^{α} group, respectively. The latter is in general capable of further reacting with

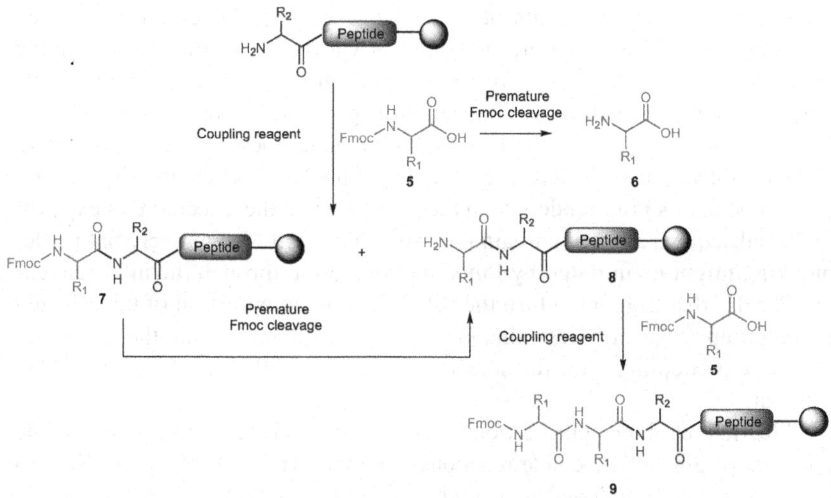

FIGURE 10.4 Redundant amino acid coupling initiated by premature deblocking of N^{α}-Fmoc.

amino acid **5** or **6** in the presence of a coupling reagent, thanks to its unprotected N^{α}-functional group, to generate side product **9** bearing a redundant amino acid in the sequence. A premature Fmoc deblocking process could also address the immobilized peptide chain. When N^{α}-Fmoc protected peptidyl resin **7** suffers from premature Fmoc cleavage it will be converted to the N^{α}-unprotected counterpart **8**. If this process takes place in the presence of activated amino acid species, the exposed free N^{α} on the solid support might be acylated by the activated amino acid residue to give side product **9** with a repetitive amino acid in the sequence.

There are a variety of factors that might contribute to the occurrence of N^{α}-Fmoc premature cleavage from the bonded amino acid/peptide, and their origins will be elaborated in the following sections.

10.2.1 Lys-N^{ε}-Induced Fmoc Premature Cleavage

Lysine plays crucially important roles in peptide derivatization processes, e.g., peptide cyclization,[12] peptide dendrimers,[13] peptide labelling,[14] and so on. Thanks to its two amino groups Lys derivatives are frequently serving as scaffolds for the lead screenings of small molecule drugs.[15] Orthogonal protections on these amino groups afford the feasibility of selective modifications of the side chain N^{ε} and backbone N^{α} on concerned Lys residue.[3] Alloc[16] and Mtt[17] are orthogonal amino-protecting groups that are frequently utilized in Fmoc-mode peptide chemistry. *N*-Alloc could be removed by Pd(0) catalysis via a transallylation process, and scavenged by an allyl receptor like $PhSiH_3$.[18] Mtt is generally cleaved off the amino group by dilute TFA solution.

Site-selective modification of Lys on a certain peptide could be realized by taking advantage of the orthogonality of various protecting groups on the addressed Lys residue. However, redundant amino acid incorporation side reaction might be potentially induced in this process,[19] indicating that the N^{α}-Fmoc protecting group on the Lys backbone undergoes undesired premature cleavage during the selective deprotection of the Lys side chain. This premature Fmoc deblocking renders both the N^{α} and N^{ε} on the affected Lys exposed to the subsequent acylation treatment. The concerned N^{α}-Fmoc premature deblocking might be initiated by Fmoc deprotonation through the allyl scavenger $PhSiH_3$ that triggers in turn the subsequent decomposition of the affected Fmoc group.[20] Nevertheless, this mechanism could not explain the redundant amino acid coupling after the selective removal of Mtt from Fmoc-Lys(Mtt) substrate.

Albericio et al. designed rational systematic experiments to investigate the potential origins of these side reactions.[19] Peptide H-Lys(H-Phe)-Ala-NH_2 **13** was designed and employed as a template in this connection, and Fmoc-AM-resin was utilized as the solid support for the corresponding peptide assembly. The synthetic scheme is displayed in Fig. 10.5. Peptides Fmoc-Lys(X)-Ala- were firstly assembled on AM resin (**10a**: X = Mtt; **10b**: X = Alloc) and the Alloc/Mtt protecting groups on Lys-N^{ε} side chain were subsequently removed by commensurate deblocking methods. Mtt was cleaved off by TFA/TES/DCM (1:1:98) and the released amino group was then neutralized by DIEA/DCM(1:19); while Alloc was deblocked by $Pd(PPh_3)_4/PhSiH_3$. The liberated Lys-N^{ε} from each peptide was subsequently acylated by Ddz-Phe-OH[21] in the presence of DIC/HOBt to give the target intermediate Fmoc-Lys(Ddz-Phe)-Ala-AM resin **11**. N^{α}-Fmoc protecting group on **11** was firstly deblocked by 20% piperidine/DMF, and the Ddz protecting group was subsequently cleaved by treatment of 95% TFA/DCM solution that simultaneously detached the target peptide H-Lys(H-Phe)-Ala-NH_2 **13** from the solid support. Side product H-Phe-Lys(H-Phe)-Ala-NH_2 **14** was detected in the crude product implying the undesired incorporation of Phe to the peptide backbone. The percentage ratio of side product **14**/product **13** derived from the synthetic strategy employing Fmoc-Lys(Mtt)-OH as a starting material was 20:80 which was comprehensively determined by the combination of analytical methods AAA, RP-HPLC, and MALDI-TOF-MS; while this value was 35:65 for Fmoc-Lys(Alloc)-OH-mediated peptide synthesis. It was proposed that the occurrence of the referred side products was due to the premature N^{α}-Fmoc cleavage by the liberated Lys-N^{ε} on the respective Lys(Mtt/Alloc) precursor. The concerned N^{α} liberated by this means was subsequently acylated by the incoming amino acid Ddz-Phe-OH. Simultaneous acylation of Lys-N^{ε} and N^{α} will unavoidably result in the formation of a side product **14** bearing redundantly incorporated amino acid.

Peptide H-Ala-Lys(H-Phe)-Ala-NH_2 derived from intermediate Fmoc-Ala-Lys(Alloc)-Ala-AM resin has been prepared in a similar manner with an effort to test the impact of Lys location on the referred side reaction. 14% of

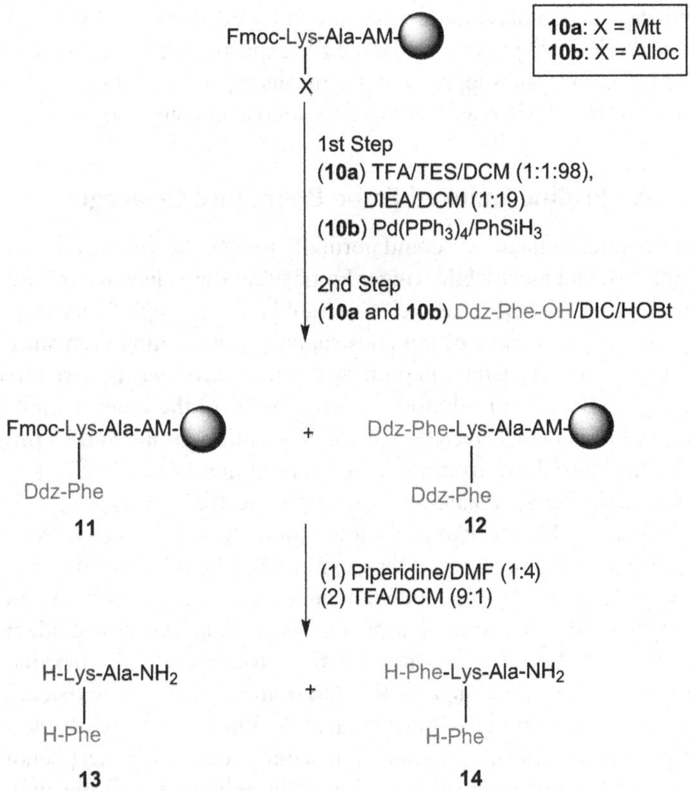

FIGURE 10.5 Synthetic scheme of Lys side chain derivatized peptide and the concurrent formation of side product.

side product H-Phe-Ala-Lys(H-Phe)-Ala-NH_2 was finally detected in the crude product that is to some extent comparable with 20% from the previous trial. This result partially accounts for the tentative conclusion that the extent of N^{α}-Fmoc premature cleavage by Lys-N^{ε} is independent on the spatial proximity between Fmoc and Lys-N^{ε}. In another word, this process might follow an intermolecular instead of intramolecular mechanism.

Fmoc premature deblocking by Lys-N^{ε} could be alleviated or even prevented by rational optimization of the corresponding synthetic route. As for the peptide precursor containing Lys(Mtt) residue, the DIEA/DCM neutralization step after Mtt removal is supposed to be skipped so that the liberated Lys-N^{ε} remains protonated which deprives it of the ability to trigger Fmoc deblocking. In the subsequent amino acid coupling step the addressed Lys-N^{ε} can be neutralized *in situ* by an appropriate base in the presence of the coupling reagent like PyAOP in order to restrain the undesired Fmoc cleavage by the exposed amino groups. Similar tactics have also been applied to alleviate DKP side reaction.[22] As for the treatment of Lys(Alloc) containing peptide, concomitant Ddz-Phe-F mediated *in situ* condensation could be conducted during Alloc removal by $Pd(PPh_3)_4/PhSiH_3$[18,19] in order to reduce the exposure time of Fmoc to the liberated amino group, and suppress by this means the undesired Fmoc premature cleavage as well as the tandem redundant amino acid coupling.

10.2.2 N^{α}-Proline-Induced Fmoc Premature Cleavage

"Difficult peptide sequences" could normally reduce the kinetics of the amino acid couplings, and meanwhile trigger Fmoc premature cleavage off the to-be-assembled Fmoc-Xaa-OH[23] as well as redundant coupling of this amino acid residue. There are a variety of moieties capable of mediating premature Fmoc deblocking[19,24] among which peptide *N*-terminal Pro and its derivatives are deemed as significant contributors.[25] Differing from the other natural amino acids N^{α} on Pro is categorized as a secondary amino group, and its pK_a is ca. 10.6,[26] higher than those from the other natural amino acids (pK_a 8.9–9.8)[26] except Cys (pK_a 10.8). Analogous to piperidine (pK_a 11.2)[27] Pro-N^{α} is capable of inducing Fmoc cleavage via β-elimination mechanism. Moreover, its intrinsic property as a secondary amine also facilitates Fmoc cleavage. These factors account for the redundant amino acid couplings in the presence of N^{α}-unprotected Pro. From the aspect of reaction kinetics the evident steric hindrance afforded by Pro five-member ring structure could retard its acylation process that in turn elongates the duration of the Pro-N^{α} existence. This effect reinforces accordingly Pro-regulated N^{α}-Fmoc cleavage off the incoming Fmoc-Xaa-OH that is principally in compliance with the phenomenon that the redundant coupling side reaction preferably addresses the amino acid located neighboring Pro on its *N*-terminal side in the peptide sequence, particularly when the concerned Pro is suffering from a sluggish acylation reaction during the peptide assembly.

As indicated in Fig. 10.6 Moroder et al. carried out a dedicated investigation[28] in which Fmoc-Gly-Pro-Hyp-OH **17** was prepared from the condensation reaction between Fmoc-Gly-Pro-OSu **15** and H-Hyp-OH **16** in solution. It turned out, nonetheless, that ca. 5% side product Fmoc-Gly-Pro-Gly-Pro-OH **20** was formed in the crude product. The referred condensation reaction proceeded sluggishly owing to the relatively weak reactivity of succinimidyl ester on Fmoc-Gly-Pro-OSu **15** fragment in combination with the low nucleophilicity and evident steric hindrance of H-Hyp-OH **16** (analogous to Pro, Hyp is also regarded as secondary amine but its basicity is weaker than that of Pro, $pK_a = 9.73$).[26] The Hyp-N^{α} could simultaneously initiate the nucleophilic attack on succinimidyl ester (path A), and the process of Fmoc cleavage off Fmoc-Gly-Pro-OSu **15** (path B). The latter performance results in premature Fmoc deblocking, and gives rise to the formation of H-Gly-Pro-OSu **18**. If degraded peptide fragment **18** is further condensed with another molecule of Fmoc-Gly-Pro-OSu **15**, the dimeric derivative Fmoc-Gly-Pro-Gly-Pro-OSu **19** would be formed as a consequence. Hydrolysis of **19** leads to the formation

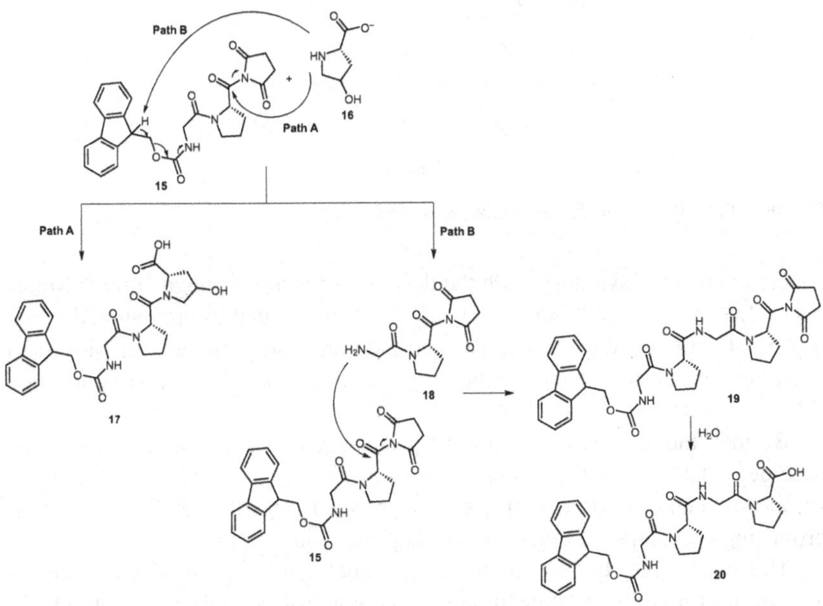

FIGURE 10.6 Hyp-induced premature cleavage of Nα-Fmoc and the redundant amino acid coupling.

of side product Fmoc-Gly-Pro-Gly-Pro-OH **20**. This observation sustains the claim that Pro and Pro derivatives are capable of inducing premature Fmoc deprotection.

A variety of synthetic strategies have been developed to address the redundant coupling side reactions originated from the Fmoc premature cleavage. *N*α-protection with groups orthogonal to the labile Fmoc, e.g., Bpoc, Trt, or Ddz[29] (Fig. 10.7) could be adopted to fundamentally avoid the Pro-mediated Fmoc premature deblocking. These protecting derivatives on amino groups remain stable in basic conditions, and could hence adapt to *N*α-Pro treatment. Bpoc, Trt and Ddz are generally cleaved off the bonded-*N*α by dilute acid solution (Bpoc: 0.2–0.5% TFA; Trt: 1%TFA/DCM; Ddz: 1–5% TFA/DCM)[29] in which tBu-type protecting groups are not affected, endowing by this means the

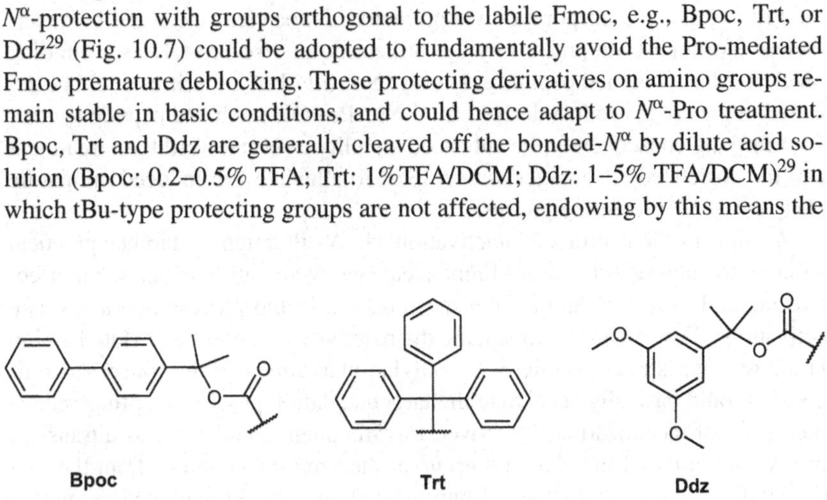

FIGURE 10.7 Acid labile Nα-protecting groups Bpoc, Trt, and Ddz.

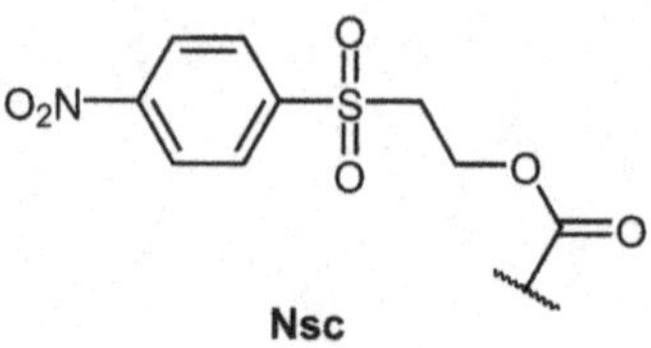

Nsc

FIGURE 10.8 **Base labile N^{α}-protecting group Nsc.**

concerned peptide synthesis with the desired protection orthogonality.[21] Moreover, Ddz can also undergo photolysis at 280 nm,[30] that affords an additional degree of orthogonality. Bpoc, Trt, and Ddz are also preferably employed as N^{α}-protecting group for the synthesis of peptides that are susceptible to DKP side reactions.[22]

Besides the above acid labile N^{α}-protecting groups, new generations of base-degradable protecting groups such as Nsc[31] (Fig. 10.8) have also been developed and employed in peptide synthesis. Nsc is regarded as the one of most promising alternative to Fmoc for amino protection.[32–34]

The mechanism of Nsc deblocking is analogous to that of Fmoc cleavage. Both of these base labile amino-protecting groups could be quantitatively cleaved off the corresponding shielded amino groups by 20% piperidine/DMF or 1% DBU/DMF.[31] The stability of Nsc toward base is nonetheless 3–10 folds higher than of that of Fmoc,[33] rendering it less susceptible to premature cleavage side reactions. Nsc exhibits extraordinarily improved resistance against Pro-induced premature cleavage. In a dedicatedly designed contrast experiment[25] redundant Pro couplings were detected in the preparation of target peptide H-Pro-Pro-Pro-Pro-Pro-Pro-Ala-NH_2, indicated by the formation of the *endo*-Pro side product with an increased molecular weight by 97 amu. The root cause of the Pro redundant incorporation via premature Fmoc cleavage could be nevertheless thoroughly suppressed by the utilization of Nsc-Pro-OH as a building block for the concerned peptide assembly. No *endo*-Pro impurities could be detected in the crude product derived from Nsc-Pro-OH mediated peptide synthesis, implying consequently the enhanced stability of Nsc toward base treatment and the correlated advantages with respect to alleviate Pro-induced premature cleavage of N^{α}-protecting groups.

Amino transient protection/activation via *N*-silylation is another practical strategy to suppress premature Fmoc cleavage. *N*-silylation is per se an effective method to reduce amino acid racemization in the process of carboxylate activation[23,35] by means of enhancing the reactivity of the to-be-acylated amino group with weak nucleophilicity.[36] *N*-silylation treatment of the concerned amino acid could basically accelerate the target acylation process, and thus reduce the extent of racemization. Moreover, this treatment could serve as a transient protection on the affected amino group at the same time, and restrain the detrimental effect with regard to N^{α}-unprotected amino acid-mediated premature Fmoc cleavage.[23]

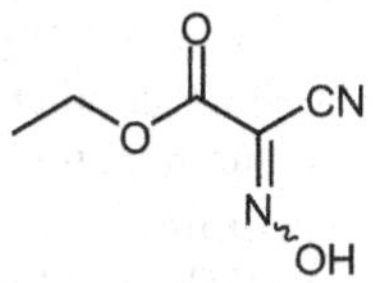

FIGURE 10.9 **Structure of Oxyma.**

Fine-tuning of the reaction acidity has also been justified to suppress premature cleavage of N^{α}-protecting groups. Some weakly acidic *N*-hydroxylamine coupling additives, e.g., Oxyma[37] (Fig. 10.9) turn out to be effective in restraining the redundant coupling of amino acid Xaa in the condensation reaction between Fmoc-Xaa-OH and the peptide bearing *N*-terminal Pro.[38] The performance of Oxyma in this regard is superior to that of other *N*-hydroxylamine analogs like HOBt and HOAt.

10.2.3 DMF/NMP-Induced Fmoc Premature Cleavage

DMF and NMP are widely used as organic solvents for peptide synthesis. In spite of the fact that these solvents remain stable in common conditions DMF could be subjected to various degradations under certain circumstances, and release diverse byproducts that could interfere with peptide synthesis.

DMF could be degraded to CO and dimethylamine under reflux,[39] and this process will be substantially accelerated by bases, e.g., NaOH and CaH_2. In some chemical productions in which CO gas is necessitated as inputting starting material DMF could even be employed as CO source.[40] In consideration of the formation of dimethylamine upon the DMF degradation this organic solvent has also been deliberately utilized in many chemical productions as the precursor of an amination reagent.[41] Upon heating, DMF can react in the presence of inorganic acids[42] with acid chloride,[42] ester, or acid anhydride to give the corresponding amide derivatives. It is noted that the formation of dimethylamine from the function of DMF with acid chloride could be realized even in the absence of catalyst. Aryl halides could also react with DMF upon heating, and release dimethylamine derivatives.[43,44] One possible mechanism of this process is that DMF firstly degrades into CO and dimethylamine, the latter reacts *in situ* with the aryl halide to give the commensurate dimethylamine.[45] Degradation of DMF could also take place sluggishly at ambient temperature. Nevertheless, in the presence of acid or base DMF itself it can be degraded to CO and dimethylamine without the involvement of aryl halide[46] and the released dimethylamine will be accumulated in the DMF. DMF is also degraded under the UV irradiation into formaldehyde and dimethylamine. Due to the inherent hygroscopicity of DMF the increased water content could promote its decomposition and the release of dimethylamine. Moreover, dimethylamine is employed as a starting material for the industrial manufacturing of DMF and

DMAc; residual unconverted dimethylamine might remain as a contaminant in these organic solvents.[47] Similar to DMF another routinely used organic solvent for peptide synthesis NMP might contain methylamine since the latter is one of the raw materials for NMP manufacturing.[48]

The residual dimethylamine in DMF could trigger undesired N^{α}-Fmoc deblocking during peptide production. Actually diethylamine has been utilized as replacement of piperidine for Fmoc cleavage in some industrial peptide manufacturing thanks to the cost advantage of the former. It has been confirmed by dedicated experiments that N^{α}-Fmoc protected amino acids could suffer from Fmoc cleavage in DMF at an ambient temperature (Yang, Y. unpublished results),[49] and the extent thereof is evidently interrelated with the quality of the DMF solvent. The presence of dimethylamine in DMF directly results in the premature cleavage of N^{α}-Fmoc off the corresponding amino acid or peptide that in turn triggers the redundant amino acid/peptide coupling as indicated in Fig. 10.4. It is not uncommon that the redundant amino acid coupling is induced by the superfluous recoupling reflected by the positive Kaiser Test results. Some operators have simply ignored the possibility of the positive Kaiser Test owing to DMF-provoked premature Fmoc cleavage, especially when the concerned Fmoc-protected peptide resin is incubated in DMF overnight, and the Kaiser Test is only conducted after prolonged (mostly unnecessary) contact of the Fmoc-protected peptide resin with DMF. The positive result induced by the DMF-triggered Fmoc premature cleavage is frequently misinterpreted as an incomplete amino acid coupling and recoupling is implemented which finally leads to redundant amino acid incorporation.

In view of the premature Fmoc cleavage induced by dimethylamine/methylamine contaminating DMF/NMP and the consequent amino acid redundant coupling, it is imperative, especially in industrial peptide manufacturing, to strictly control the quality of the DMF/NMP, and to setup a rational specification to this end. The bromophenol blue indicator or IC could be utilized to qualitatively or quantitatively analyze amine derivatives, respectively. Moreover, DMF/NMP could be subjected to vacuum degassing, nitrogen gas flow bubbling or aluminum oxide treatment to remove the contaminating dimethylamine/methylamine prior to usage.

10.2.4 Residual Piperidine-Induced Fmoc Premature Cleavage

Immobilized peptide chains on solid support in SPPS normally undergo piperidine-directed N^{α}-Fmoc cleavage after the addressed amino acid is assembled. The excess piperidine is removed from the peptidyl resin by washing. DMF or NMP is generally utilized in this process as a rinsing solvent. The referred washing could either be performed in a continuous or batch-wise manner. A new circle of amino acid couplings is conducted afterwards. However, if the concerned peptidyl resin does not undertake sufficiently washing treatment residual piperidine might survive and remain in the reaction system. Moreover, if no analysis was conducted with respect to the existence of piperidine in the washing solution or no qualified washing process parameters are developed and piperidine contaminated peptidyl resin was subjected to the next amino acid coupling step, the affected amino acid coupling will be carried out in the presence of a residual piperidine that might result in an undesired N^{α}-Fmoc cleavage in parallel with the amino acid coupling. Under such circumstances a redundant amino acid incorporation will be triggered in light of the mechanism elucidated in Fig. 10.4.

In consideration of the redundant amino acid coupling brought about by deficient peptidyl resin washing treatment after-N^{α}-Fmoc deprotection, effective washing process should be developed as to ensure the sufficient removal of excess piperidine prior to the following amino acid coupling. Furthermore, a reliable analytical method to control the existence of piperidine in the washing solution is supposed to be developed and implemented in the peptide manufacturing. The addressed peptidyl resin should not be released to the subsequent amino acid coupling in case the piperidine content detected in the washing solution exceeds the predetermined specification. Owing to its inherent property as a secondary amine piperidine it could be analyzed by colorimetric methods, e.g., a chloranil test, which could also be adopted as an alternative to the ninhydrin test to detect the unacylated Pro in SPPS.[50] Only in case of a negative result of the chloranil test of the washing solution could the material be released to the next coupling step, otherwise the washing step must be repeated until no piperidine can be detected in order to avoid the potential premature N^{α}-Fmoc cleavage and the consequent amino acid redundant coupling.

10.2.5 DMAP/DIEA-Induced Fmoc Premature Cleavage

DMAP is routinely used as a catalyst for an acylation reaction in peptide synthesis,[51] especially applicable to difficult amino/hydroxyl acylation reactions, difficult amino acid coupling,[52] preparation of depsipeptide,[53] as well as the loading of the hydroxyl-type resin.[54] In spite of its high potency to catalyze acylation reactions, DMAP could meanwhile trigger some side reactions such as enhanced racemization of an amino acid in the process of carboxylate activation.[55] Moreover, DMAP could also promote premature N^{α}-Fmoc cleavage off the corresponding shielded amino acid or peptide owing to the inherent lability of Fmoc toward DMAP. Even a catalytical amount of DMAP is capable of triggering the Fmoc deblocking process.[56] Although it was claimed that DMAP did not cause any detectable premature Fmoc cleavage in certain peptide synthesis,[57] this kind of side reaction is prone to interfere with less sterically hindered amino acids like Gly.[58] DMAP-induced premature Fmoc cleavage is substantially magnified when the target acylation proceeds slowly. In an investigation DMAP was added to the reaction system as to accelerate the process of the convergent condensation of two peptide fragments, it was nonetheless concluded that premature Fmoc deprotection was initiated upon the addition of DMAP, and a redundant condensation of the affected fragment was consequently induced (Yang, Y. unpublished results). It is hence advisable to evaluate the potential detrimental effects of the DMAP-regulated Fmoc premature cleavage and the resultant redundant couplings of the affected amino acid/peptide, particularly under certain special circumstances.

Premature N^{α}-Fmoc deblocking might also be provoked in the presence of a tertiary amine other than DMAP. In spite of the drastically enhanced stability of Fmoc toward tertiary amines relative to primary and secondary amines, premature Fmoc cleavage could still be initiated by some tertiary amines like DIEA, particularly in the process of the condensation of amino acids or peptides with overwhelming steric hindrances.[23] It was reasoned in an investigation that kinetics Fmoc cleavage was evidently enhanced upon the addition of DIEA to the Fmoc-amino acid DMF solution (Yang, Y. unpublished results). For the tertiary amine-participating amino acid/peptide condensation, if the target reaction proceeds sluggishly the premature N^{α}-Fmoc deblocking will be intensified, and the resultant redundant amino acid/peptide coupling could be facilitated as a consequence.

10.2.6 Hydrogenation-Induced Fmoc Premature Cleavage

Although N^{α}-Fmoc was once deemed as sufficiently stable under hydrogenation conditions in some literature;[59] this shaky conclusion was scrutinized later on, and it has since been realized that catalytic hydrogenation in the presence of acids could result in undesired Fmoc cleavage.[24,49] The observed lability of Fmoc toward hydrogenation is believed to be attributable to its distinctive β-phenylethyloxy moiety, although much less sensitive than arylmethoxy derivatives, e.g., benzyloxycarbonyl. Moreover, it has been discovered that the Fmoc cleavage under neutral Pd/C-catalyzed hydrogenation conditions could be significantly accelerated in the presence of 2–5 equiv. CH_3CN, or merely by increasing the H_2 pressure without the addition of CH_3CN.[60] Liberation of the amino group originally shielded by a protecting derivative, e.g., Z via hydrogenation treatment could induce simultaneous premature Fmoc deblocking. The subsequent amino acid coupling upon the completion of the hydrogenation treatment will inevitably lead to the redundant incorporation of the concerned amino acid onto the affected liberated amino group that is supposed to be shielded by Fmoc.

10.2.7 Fmoc Deblocking in the Starting Material

The spontaneous degradation of Fmoc-amino acid during storage is another source of premature Fmoc deblocking. The N^{α}-Fmoc protecting group is susceptible to thermal cleavage. It has been reasoned that the N^{α}-Fmoc protecting group on amino acid could be quantitatively cleaved in neat DMSO, DMF or NMP after 15 min at 120°C even in the absence of base.[61] The inappropriate conditions for a Fmoc amino acid storage or lengthy preservation could trigger Fmoc cleavage. In case the batch of the Fmoc amino acid that is suffered from the Fmoc deblocking is not subject to retest, and is utilized as a raw material in certain peptide manufacturing, the occurrence of redundant amino acid couplings will be hence induced.

Predefined regular retests of the Fmoc amino acid starting materials should be conducted prior to their release to the corresponding production. The content of N^{α}-unprotected free amino acid impurity is supposed to be analyzed, and the release of the concerned material should be regulated by the corresponding specification. Starting material-dictated redundant amino acid couplings, that give rise to the formation of critical impurities, should be analyzed with utmost attention. Under special circumstances tightened specifications should be set up for these amino acid raw materials in order to minimize contents of the critical impurities owing to the redundant amino acid coupling.

10.3 REDUNDANT AMINO ACID COUPLING INDUCED BY NCA FORMATION

It has been introduced in Section 5.4 that the NCA derivative could be produced from the intensively activated derivatives of the Boc/Z-amino acid, e.g., Boc/Z-amino acid chloride. The occurrence of this undesired process is attributed to: (1) the innate lability of the Boc/Z-amino acid chlorides; (2) the low activity of the reciprocal nucleophile toward the concerned Boc/Z-amino acid chlorides. The additive effects of these factors render the addressed Boc/Z-amino acid chloride derivative sufficient time to undergo an intramolecular cyclization process to give rise to the corresponding NCA compounds. The process of this side reaction is referred to Fig. 5.7.

It is notable that the sequential side reactions would basically not detain at the step of NCA formation, since NCA is per se an active species of the corresponding amino acid that is capable of mediating further acylation reactions. The intentional utilization of NCA derivatives as building blocks for the construction of polypeptides in a controlled polymerization manner has been established as a common practice in this field (for reviews see the relevant research[62]). However, if the formation of NCA is realized unexpectedly, its occurrence in the process of peptide synthesis might induce redundant amino acid incorporation. The mechanism of this undesired process is illustrated in Fig. 10.10. The Boc protected amino acid **21** is firstly converted to the active counterpart **22** upon activation, and the latter is supposed to react subsequently with peptide resin **23** to give the target product **24**. However, if the formation of NCA **25** is derived from **22** via intramolecular cyclization it might function as well with peptide resin **23** to form the carbamate peptide intermediate that undergoes spontaneous decarboxylation and gives rise to the N^{α}-unprotected peptide species **26**. The unshielded N^{α} functional group on **26** is susceptible to further electrophilic attack from either NCA **25** or an activated amino acid **22** so that a redundant amino acid residue is squeezed into the target peptide chain. This kind of side reaction is not limited to the Boc-amino acid chloride. An active ester-dictated Boc amino acid coupling can also result in NCA formation and consequent

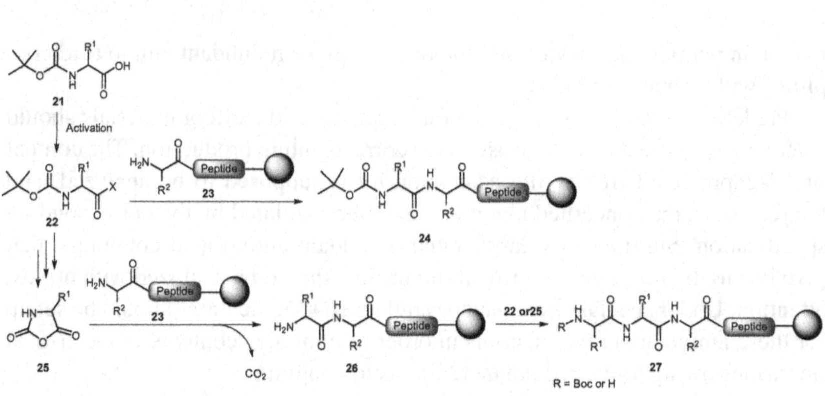

FIGURE 10.10 Redundant amino acid coupling initiated by NCA formation.

redundant amino acid incorporation, particularly when the coupling reagent is EDC and the solvent is DCM.[63]

In light of the inherent factors that are in favor of NCA formation from Boc-Xaa-Cl, Fmoc-amino acid chloride should be utilized in place of the Boc/Z counterpart since the former is significantly less susceptible to NCA formation under identical conditions. This synthetic dogma should be accentuated especially when difficult amino acid couplings are concerned since the retarded amino acylation will inevitably enhance the competitive NCA formation and the consequent redundant amino acid coupling. Utilization of the Fmoc-amino acid chloride in preference to Boc/Z-amino acid chloride could effectively alleviate this side reaction.

10.4 His-N^{im} PROMOTED Gly REDUNDANT INCORPORATION

The imidazolyl side chain substituent on His, once unprotected or unintentional deblocked at its N^{π} functional group, could serve as an acyl transfer catalyst in the process of peptide assembly, functioning with amino acids and rearranging them from His side chain to the peptide backbone. Amino acid derivatives with significantly less steric hindrance such as Gly could undergo aminolysis at its carboxylate by His-N^{im} and subsequent $N^{im}\rightarrow N^{\alpha}$ acyl migration at the tandem TFA-mediated N^{α}-Boc deblocking and base neutralization steps. Redundant Gly incorporation into the peptide backbone is thus coming into being. This process has been illustrated in detail in Section 4.3 in the context of "peptide rearrangement side reaction."

10.5 REDUNDANT COUPLING INDUCED BY THE UNDESIRED AMINO ACID-CTC RESIN CLEAVAGE

One frequently underestimated issue in the context of peptide assembly on CTC resin is the distinctive acid lability of the first amino acid immobilized on the solid support. The concerned 2-chlorotrityl ester bond between the first amino acid and the spacer on the CTC resin is especially susceptible to acid treatment. Its stability is somehow reinforced by the subsequent amino acid incorporation via the inductive effects of the peptide backbone on the addressed ester bond. In view of this phenomenon, the coupling of the second amino acid onto the amino acid derivatized CTC resin should be carried out with the utmost caution with respect to choose the appropriate coupling reagent. Since DIC is utilized in the absence of the base (coupling additive HOBt, if any, is even acidic), the coupling reaction condition (acidic to neutral) could be detrimental to the amino acid-CTC resin and lead to the premature acidolytic cleavage of the loaded amino acid from the solid support. The affected amino acid released into the coupling solution in this process could also participate in the ongoing peptide assembly and be re-immobilized to the resin facilitated by the coupling reagent DIC, leading to redundant amino acid incorporation. It was figured out that the Gly-CTC resin is particularly liable to suffer from this undesired premature acidolytic cleavage side reaction (Yang, Y. unpublished results). The handling of the Gly-CTC resin in the peptide production requires, therefore, the utmost attention in terms of the coupling of an incoming amino acid to the Gly-CTC resin.

The process of this side reaction is illustrated in Fig. 10.11. Fmoc-amino acid **29** is supposed to be assembled to H-Gly-CTC resin **28** by DIC/HOBt to form the target intermediate **30**. However, the immobilized Gly could be cleaved from the resin in this process due to the innate susceptibility of the addressed 2-chlorotrityl ester bond to acidolysis. The cleaved Gly might be reattached to underivatized H-Gly-CTC resin **28** in the presence of DIC/HOBt, giving rise to H-Gly-Gly-CTC resin **31**. Owing to the excessive Fmoc-amino

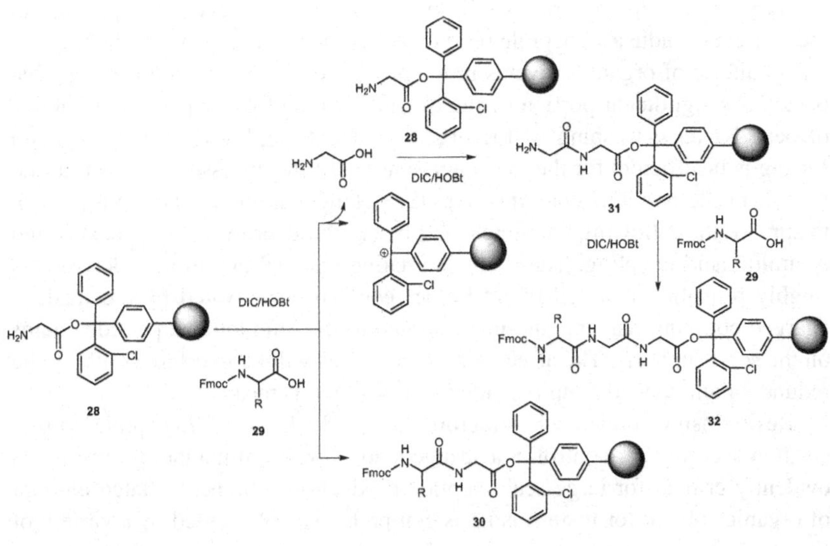

FIGURE 10.11 Redundant Gly incorporation by premature acidolytic cleavage of H-Gly-CTC resin.

acid **29** and DIC/HOBt added to the reaction system intermediate **31** could be readily acylated by **29** and converted to the corresponding side product **32**. The redundant Gly residue is thus squeezed into the backbone of the peptide chain.

In consideration of the redundant amino acid coupling in this manner the rational choice of coupling reagent for the concerned amino acid will be of importance to alleviate the formation of peptide impurity with redundantly incorporated amino acid residue. Under such circumstances TOTU stands out as the appropriate alternative to DIC due to the intrinsic attributes of the former, and the fact that the tertiary base will have to be utilized on this occasion, minimizing the risk of the potential acidolytic degradation of the concerned amino acid-CTC resin. The stability of the 2-chlorotriytyl ester bond toward acidolysis is evidently enhanced by the incorporation of the second amino acid. That is to say, the anchored dipeptide is not as acidic labile as the single amino acid, and the introduction of DIC/HOBt as the coupling reagent could be well tolerated for the following amino acid assemblies.

10.6 REDUNDANT AMINO ACID COUPLING INDUCED BY INSUFFICIENT RESIN RINSING

One of the distinctive advantages of SPPS over LPPS lies in the fact that the excessive starting materials or reagents could be readily removed from the reaction systems by sufficient rinsing of the peptidyl resin once the target reaction is completed. This straightforward process tremendously facilitates the convenient and effective peptide assembly, and circumvents the cumbersome intermediary isolation/purification steps. One of the focuses in SPPS, especially in the context of industrial peptide production, is the balance between the rational mass balance of organic solvents in the process of peptide manufacturing, that occupies a significant portion of the overall cost, and the requested efficiency of peptidyl resin washing. If the organic solvent employed in peptidyl resin rinsing is insufficient for the thorough removal of the excessive leftover amino acid from the previous coupling step, the detained amino acids might be able to survive the following handlings, and be carried over even to the next round of amino acid coupling. These contaminating residual amino acid derivatives (highly probably deprived of N^{α}-Fmoc) could be reactivated by the freshly-charged coupling reagent, and incorporated to the immobilized peptide chains on the solid supports. The affected peptide chains will be interfered with by the redundant amino acid coupling side reactions by this means.

Resin rinsing efficiency is sometimes an overlooked topic in peptide synthesis, however, a good control in a stringent and consistent manner to this end is evidently critical for large-scale peptide production. The necessitated amount of organic solvent for resin rinsing is comprehensively decided by a variety of factors such as solvent type, resin volume, resin backbone material, peptides immobilized on the resin, properties of the starting materials/reagents to be removed from the system, temperature, drainage rate, rinsing manner (continuous or portion-wise), geometry of the reaction vessel, type of impellor, and so on. A sufficient resin washing taking these parameters into account could effectively minimize the redundant amino acid incorporation induced in this manner. Generally speaking, in industrial production a well-tested and validated organic solvent/resin ratio is implemented in an effort to realize efficient but economical resin washings, rendering the content of residual amino acid postrinsing within an acceptable range.

10.7 REDUNDANT AMINO ACID COUPLING INDUCED BY OVERACYLATION SIDE REACTION

Thanks to the shielding effects of various protecting groups on the nucleophilic N^{α} on amino acids, either by means of the electron-withdrawing effect or steric hindrance, most of the protected amino acids are predominately exempted from overacylation side reactions on their N^{α} at the activation steps. Nevertheless, some certain amino acids or their analogs, owing to their distinctive properties, might be entangled in overacylation process on their N^{α} at the step of carboxylate activation in spite of the protection effects. Urethane-protected aminooxy acetic acids, for instance, are known to suffer from this kind of side reaction that in consequence results in the redundant residue incorporation into the target peptide chains.[64,65]

Aminooxy peptides have been endowed with profound applications in oxime ligation[66] in which an aminooxy and a reciprocal ketone/aldehyde carbonyl is ligated into an oxime moiety in a chemoselective manner. The introduction of aminooxy group into the target ligation fragment is frequently realized by the utilization of commercially available Boc-aminooxyacetic acid (Boc-Aoa-OH) as building block.[67] As indicated previously, Boc-Aoa-OH as well as other protected derivatives of Aoa, e.g., Aloc-Aoa-OH, might suffer from overacylation that affects the nitrogen from -NH-O- moiety upon carboxylate activation, leading to the formation of Aoa(Aoa)-peptide as a side product. The enhanced nucleophilicity of the nitrogen on Boc-Aoa-OH and the increased susceptibility to overacylation as a consequence might be explained by the α-effect[68] of the neighboring oxygen atom that affects both the nucleophilicity and basicity attributes of the referred nitrogen. In a dedicated investigation[69] it is discovered that the addressed overacylation process occurs predominately at the step of Boc-Aoa-OH carboxylate activation, whereas the incorporated Boc-Aoa residue is much less unaffected (Fig. 10.12). Apparently the nucleophilicity of the nitrogen on Boc-Aoa would be altered once it is assembled into the peptide chain, thus decreasing its susceptibility to overacylation.

It is discovered that the extent of Boc-Aoa overacylation is directly correlated to the type and amount of base utilized in the corresponding Boc-Aoa-OH coupling. Weaker bases, e.g., collidine, in place of DIEA could significantly suppress the overacylation while decrease of base equivalent might also reduce

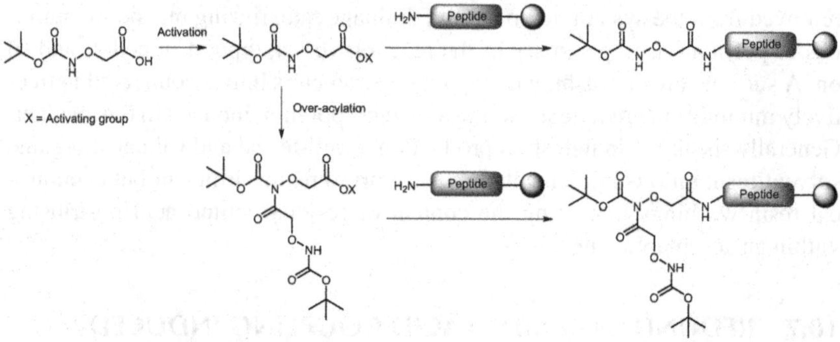

FIGURE 10.12 Overacylation on Boc-Aoa-OH during carboxylate activation and consequent redundant Boc-Aoa incorporation.

the extent of this side reaction. The option of carbodiimide coupling reagents that are independent on base-catalysis is beneficial for overacylation suppression. Moreover, thorough protection of Boc-Aoa-OH nitrogen in the form of *N,N*-di-Boc protected aminooxyacetic acid Boc_2-Aoa-OH[65] addresses the root causes of the overacylation process, and could eventually thoroughly prevent this side reaction.

REFERENCES

1. Chang C-D, Waki M, Ahmad M, Meienhofer J, Lundell EO, Haug JD. *Int J Pept Protein Res.* 1980;15:59–66.
2. Tessier M, Albericio F, Pedroso E, et al. *Int J Pept Protein Res.* 1983;22:125–128.
3. Albericio F. *Biopolymers.* 2000;55:123–139.
4. Ten Kortenaar PBW, van Dijk BG, Peeters JM, Raaben BJ, Adams PJHM, Tesser GI. *Int J Pept Protein Res.* 1986;27:398–400.
5. Bolin DR, Sytwu I-I, Humiec F, Meienhofer J. *Int J Pept Protein Res.* 1989;33:353–359.
6. Carpino LA, Han GY. *J Org Chem.* 1972;37:3404–3409.
7. Hlebowicz E, Andersen AJ, Andersson L, Moss BA. *J Pept Res.* 2005;65:90–97.
8. Isidro-Llobet A, Just-Baringo X, Ewenson A, Álvarez M, Albericio F. *Pept Sci.* 2007;88:733–737.
9. Dendrinos K, Kalivretenos AG. *J Chem Soc Perkin Trans.* 1998;1:1463–1464.
10. Sigler GF, Fuller WD, Chaturvedi NC, Goodman M, Verlander M. *Biopolymers.* 1983;22: 2157–2162.
11. Gopi HN, Babu VVS. *J Pept Res.* 2000;55:295–299.
12. Rovero P. In: Kate SA, Albericio F, eds. *Practical Solid-Phase Synthesis: A Book Companion.* New York: Marcel Dekker; 2000:331–364.
13. Kates SA, Daniels SB, Albericio F. *Anal Biochem.* 1993;212:303–310.
14. Braunitzer G, Schrank B, Petersen S, Petersen U. *Hoppe-Seyler's Z Physiol Chem.* 1973;354:1563–1566.
15. (a) Obrecht D, Villalgordo JM, eds. *Solid-Supported Combinatorial and Parallel Synthesis of Small-Molecular-Weight Compound Libraries.* Oxford: Pergamon; 1998. (b) Bannwarth W, Felder E, eds. *Combinatorial Chemistry: A Practical Approach.* Weinheim: Wiley-VCH; 2000. (c) Kates SA, Albericio F, eds. *Practical Solid-Phase Synthesis: A Book Companion,.* New

York: Marcel Dekker; 2000. (d) Seneci P. *Solid-Phase Synthesis and Combinatorial Technologies*. New York: John Wiley & Sons; 2001.
16. Guibé F. *Tetrahedron*. 1997;53:13509–13556; 1998;54:2967–3042.
17. Aletras A, Barlos K, Gatos D, Koutsogianni S, Mamos P. *Int J Pept Protein Res*. 1995;45: 488–496.
18. Thieriet N, Alsina J, Giralt E, Guibé F, Albericio F. *Tetrahedron Lett*. 1997;38:7275–7278.
19. Farrera-Sinfreu J, Royo M, Albericio F. *Tetrahedron Lett*. 2002;43:7813–7815.
20. Kumar V, Aldrich JV. *Abstr Pap Am Chem Soc*. 2000;220:ORGN-226.
21. Birr C. In: Epton R, ed. *Innovation, Perspective in Solid-Phase Synthesis*. Birmingham, UK: SPCC (UK) Ltd; 1990:155–181.
22. Alsina J, Giralt E, Albericio F. *Tetrahedron Lett*. 1996;37:4195–4198.
23. Wenschuh H, Beyermann M, Winter R, Bienert M, Ionescu D, Carpino L. *Tetrahedron Lett*. 1996;37:5483–5486.
24. Martinez J, Tolle JC, Bodanszky M. *J Org Chem*. 1979;44:3596–3598.
25. Carreño C, Méndez ME, Kim Y-D, et al. *J Pept Res*. 2000;56:63–69.
26. Dawson RMC, Elliott DC, Elliot WH, Jones KM. *Data for Biochemical Research*. Oxford: Clarendon Press; 1959.
27. Hall Jr HK. *J Am Chem Soc*. 1957;79:5441.
28. Ottl J, Jürgen H, Moroder L. *J Pept Sci*. 1999;5:103–110.
29. Isidro-Llobet A, Álvarez M, Albericio F. *Chem Rev*. 2009;109:2455–2504.
30. Birr C, Lochinger W, Stahnkc G, Lang P. *Liebigs Ann Chem*. 1972;763:162–172.
31. Samukov VV, Sabirov A, Pozdnyakov PI. *Tetrahedron Lett*. 1994;35:7821–7824.
32. Sabirov AN, Kim Y-D, Kim H-J, Samukov VV. *Protein Pept Lett*. 1997;4:307–312.
33. Ramage R, Jiang L, Kim Y-D, Shaw K, Park J-L, Kim H-J. *J Pept Sci*. 1999;5:195–200.
34. Maier TC, Podlech J. *Adv Synth Cat*. 2004;346:727–730.
35. Romani S, Moroder L, Göhring W, et al. *Int J Pept Protein Res*. 1987;29:107–117.
36. Behrendt R, Schenk M, Musiol H-J, Moroder L. *J Pept Sci*. 1999;5:519–529.
37. Subirós-Funosas R, Prohens R, Barbas R, El-Faham A, Albericio F. *Chem Eur J*. 2009;15: 9394–9403.
38. Subirós-Funosas R, El-Faham A, Albericio F. *Biopolymers*. 2012;98:89–97.
39. Muzart J. *Tetrahedron*. 2009;65:8313–8323.
40. Wan Y, Alterman M, Larhed M, Hallberg A. *J Org Chem*. 2002;67:6232–6235.
41. (a) Tsai J-Y, Chang C-S, Huang Y-F, et al. *Tetrahedron*. 2008;64:11751–11755. (b) Goswami S, Das NK. *J Heterocycl Chem*. 2009;46:324–326. (c) Watanabe T, Tanaka Y, Sekiya K, Akita Y, Ohta A. *Synthesis*. 1980:39–41.
42. Coppinger GM. *J Am Chem Soc*. 1954;76:1372–1373.
43. Agarwal A, Chauhan PMS. *Syn Commun*. 2004;34:2922–2930.
44. Kadish KM, Han BC, Franzen MM, Araullo-McAdams C. *J Am Chem Soc*. 1990;112: 8364–8368.
45. Sharma A, Mehta VP, Van der Eycken E. *Tetrahedron*. 2008;64:2605–2610.
46. Perrin DD, Armarego WLF, Perrin DR. *Purification of Laboratory Chemicals*. Oxford: Pergamon; 1966:143.
47. Weissermel K, Arpe H-J. *Industrial Organic Chemistry: Important Raw Materials and Intermediates*. Weinheim: WILEY-VCH Verlag; 2003:45–46.
48. Harreus AL, Backes R, Eichler J-O, et al. *2-Pyrrolidone. Ullmann's Encyclopedia of Industrial Chemistry*; Wiley-VCH Verlag GmbH & Co. KGaA, Weinheim; 2011.
49. Atherton E, Bury C, Shepppard RC, Williams BJ. *Tetrahedron Lett*. 1979;20:3041–3042.
50. Christensen T. *Acta Chem Scand B*. 1979;33:763–766.

51. Berry DJ, DiGiovanna CV, Metrick SS, Murugan R. *Arkivoc*. 2001;i:201–226.
52. Wang SS, Tam JP, Wang BSH, Merrifield RB. *Int J Pept Protein Res*. 1981;18:459–467.
53. Jou G, González I, Albericio F, Lloyd-Williams P, Giralt E. *J Org Chem*. 1997;62:354–366.
54. Wang SS, Kulesha ID. *J Org Chem*. 1975;40:1227–1234.
55. Atherton E, Benoiton NL, Brown E, Sheppard RC, Williams BJ. *J Chem Soc Chem Commun*. 1981:336–337.
56. Atherton E, Logan CJ, Sheppard RC. *J Chem Soc Perkin Trans*. 1981;1:538–546.
57. Jensen KJ, Barany G. *J Pept Res*. 2000;56:3–11.
58. Gradas A, Jorba X, Giralt E, Pedroso E. *Int J Pept Protein Res*. 1989;33:386–390.
59. (a) Carpino LA, Han GY. *J Am Chem Soc*. 1970;92:5748–5749. (b) Kelly RC, Gebhard I, Wicnienski N. *J Org Chem*. 1986;51:4590–4594.
60. Maegawa T, Fujiwara Y, Ikawa T, Hisashi H, Monguchi Y, Sajiki H. *Amino Acids*. 2009;36:493–499.
61. Höck S, Marti R, Riedl R, Simeunovic M. *CHIMIA*. 2010;64:200–202.
62. (a) Deming TJ. *Adv Polym Sci*. 2006;202:1–18. (b) Kricheldorf HR. *Angew Chem Int Ed Engl*. 2006;45:5752–5784. (c) Hadjichristidis N, Iatrou H, Pitsikalis M, Sakellariou G. *Chem Rev*. 2009;109:5528–5578. (d) Habraken GJM, Peeters M, Dietz CHJT, Koning CE, Heise A. *Polym Chem*. 2010;1:514–524.
63. Benoiton NL, Lee YC, Chen FMF. *Int J Pept Protein Res*. 1991;41:587–594.
64. Wahl F, Mutter M. *Tetrahedron Lett*. 1996;37:6861–6864.
65. Brask J, Jensen KJ. *J Pept Sci*. 2000;6:290–299.
66. Ulrich S, Boturyn D, Marra A, Renaudet O, Dumy P. *Chem Eur J*. 2014;20:34–41.
67. Offord RE, Gaertner HF, Wells NC, Proudfoot EI. *Methods Enzymol*. 1997;287:348–369.
68. Fina NJ, Edwards JO. *Int J Chem Kinet*. 1973;Vol. V:1–26.
69. Decostaire IP, Lelièvre D, Zhang H, Delmas AF. *Tetrahedron Lett*. 2006;47:7057–7060.

Chapter 11

Peptide Racemization

Amino acid racemization is one of the most detrimental side reactions in chemical peptide synthesis. It is known that the biological activities of the peptide/protein molecules are directly correlated with their dimensional atomic alignments while the inversion of the configuration at a certain peptide C^{α} chiral center might cause the local spatial rearrangement of critical functional groups. Besides, racemization of a single amino acid in a certain peptide chain could exert a significant impact on the overall conformation of the affected molecule.[1] Both of these factors are capable of substantially influencing the biological activities of concerned peptide molecules.[2]

Taking the consequences of amino acid racemization into account it is understandably necessary to keep the chiralities of the assembled amino acids intact during the peptide chemical synthesis. Nevertheless, this norm is sometimes difficult to be complied with due to the intensive inclination of certain amino acids to undergo racemization predominantly at the step of amino acid/peptide carboxylate activation. The mechanism of amino acid racemization, racemization-prone peptide synthesis, as well as solutions in an effort to alleviate the racemization side reaction will be dedicatedly discussed in this chapter.

11.1 PEPTIDE RACEMIZATION MECHANISM

11.1.1 Peptide Racemization via Oxazol-5(4*H*)-one Formation

Strong activation of the backbone carboxylate group on an amino acid or a peptide might induce the formation of oxazol-5(4*H*)-one intermediate provided that the N^{α} group on the same amino acid molecule is acylated. The concerned oxazol-5(4*H*)-one intermediate formed in this process is regarded as the most predominant source of amino acid racemization in peptide synthesis.[3] The mechanism of oxazol-5(4*H*)-one regulated amino acid racemization is illustrated in Fig. 11.1.

In this mechanism the peptide or *N*-acylated amino acid **1** is activated on its carboxylate group by the corresponding coupling reagent, and converted to its active counterpart **2**. If this active compound does not undergo a rapid aminolysis process with the incoming N^{α}-unprotected amino acid derivative **4** due to the slow aminolysis kinetics correlated with the inherent weak nucleophilicity of the compound **4**, steric hindrance of either reactant, low reaction

Side Reactions in Peptide Synthesis. http://dx.doi.org/10.1016/B978-0-12-801009-9.00011-2

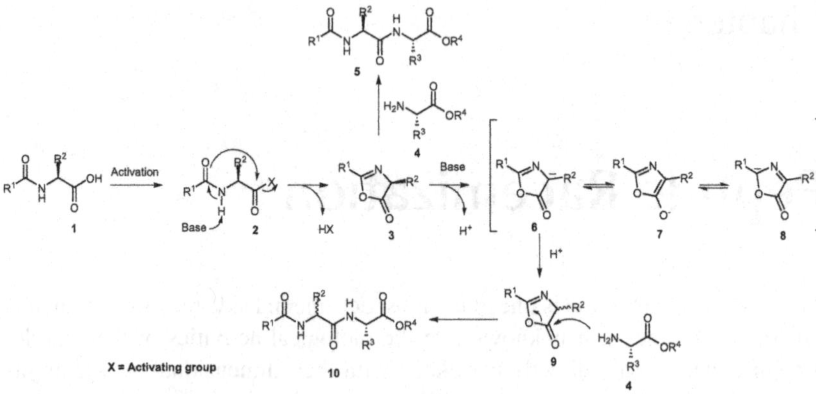

FIGURE 11.1 **Mechanism of amino acid racemization via oxazol-5(4*H*)-one intermediate.**

temperature or low reactant concentration, it will be partially converted to the corresponding oxazol-5(4*H*)-one derivative **3** with retained chirality on its C^{α}. Although oxazolone intermediate **3** possesses reactivity as well toward amino acid **4** to construct the target peptide **5**, racemization of this derivative occurs at a much faster rate relative to the aminolysis.[4] H^{α} on oxazolone **3** is abstracted in this process by base and oxazolone **3** is converted to the corresponding oxazolone anion **6** which could be transformed to **7** and **8** via resonance. Racemization takes place when the oxazolone anion **6** undergoes protonation to give rise to the racemic mixture **9** that could be subjected to further aminolysis by an amino acid derivative **4** to generate peptide diastereomers **10**. It is implied in Fig. 11.1 that one of the prerequisites for the occurrence of racemization via an oxazolone intermediate is the presence of acyl substituent (acetyl, benzoyl, peptidyl, etc.) on N^{α} of the activated amino acid. It is indicated that the urethane-protected oxazol-5(4*H*)-one is less likely to undergo deprotonation, and hence racemization.[5] Practically the thorough suppression of racemization by N^{α}-urethane protection, especially for those residues with a high tendency for racemization, is nonetheless unrealistic.

11.1.2 Peptide Racemization via Enolate Formation

An alternative mechanism of amino acid racemization to oxazolone formation, which is normally less highlighted, does affect carbodiimide-mediated amino acid/peptide activation and its liability to racemization. Amino acid racemization following this mechanism is illustrated in Fig. 11.2. Amino acid/peptide **11** is firstly activated on its backbone carboxylate by carbodiimide and transformed to the corresponding *O*-acylisourea derivative **12**. The latter could be subjected to enolate formation and subsequent intramolecular proton abstraction to generate the zwitterionic enolate derivative **13** that is tautomerized to its enol counterpart **14** stabilized by an intramolecular hydrogen bond.[6] This process finally

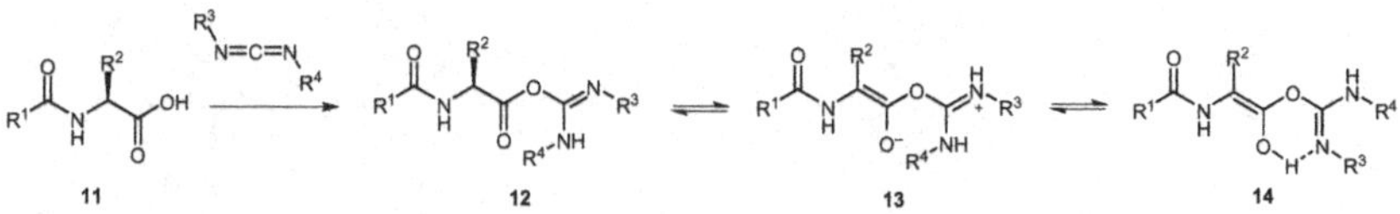

FIGURE 11.2 **Racemization of *O*-acylisourea via enolate formation.**

leads to the racemization of the affected amino acid during the carbodiimide-mediated coupling reaction.

11.1.3 Peptide Racemization via Direct H^{α} Abstraction

Amino acid racemization via direct H^{α} abstraction by base in the process of peptide synthesis is possible, especially to those amino acid residues with relatively more acidic H^{α}.[7] This mechanism could be regarded as a spin-off of the enolate-induced amino acid racemization pathway. The re-protonation of the generated anionic amino acid intermediate by solvent or other proton resources leads to amino acid racemization (Fig. 11.3). Actually amino acid racemization following this pathway seldom occurs when the backbone carboxylate of the concerned amino acid/peptide is in the form of free carboxylic acid (X = OH) thanks to the shielding effect of the carboxylate anion against the proton abstraction on the C^{α} stereogenic center. However, when the addressed carboxylate is transformed to the corresponding ester at the amino acid/peptide activation step, racemization through direct H^{α} abstraction would be evidently activated under some circumstances.

Compared with the oxazol-5(4*H*)-one formation mechanism racemization-induced by direct proton abstraction in peptide synthesis does not prevail. Normally it only addresses the most susceptible amino acid residues (e.g., aryl glycine like Phg) since the engendered anion upon H^{α} abstraction could be preferably stabilized by the aryl side chain group.[8] Amino acid residues with electron-withdrawing groups on side chain C^{β} position such as Cys[9] also tend to undergo racemization via this pathway thanks to the stabilization effect of these substituents on the anionic amino acid derivatives formed upon H^{α} abstraction. Other residues like His could facilitate intramolecular H^{α} abstraction by its imidazolyl side chain and readily suffer from racemization.[10] Besides, the saponification of amino acid or peptide esters by bases could also trigger H^{α} abstraction and subsequent racemization on the affected C^{α} stereogenic center.

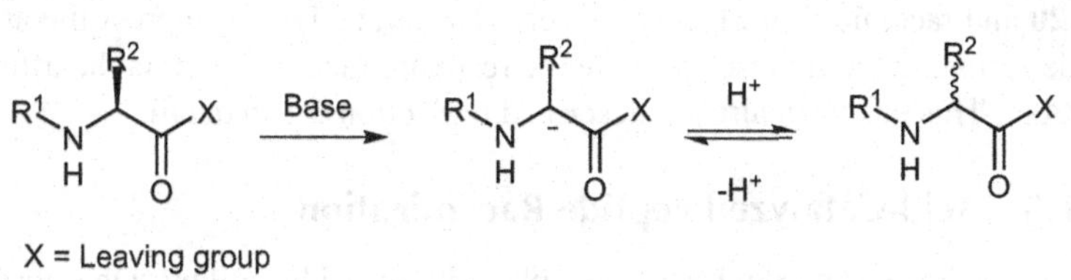

FIGURE 11.3 **Amino acid racemization via direct H^{α} abstraction.**

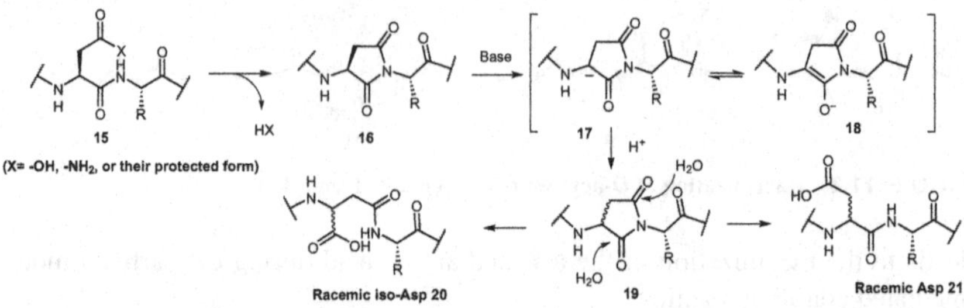

FIGURE 11.4 **Aspartimide formation-induced racemization.**

11.1.4 Peptide Racemization via Aspartimide Formation

Aspartimide formation has been separately elaborated in Section 6.1 as a common side reaction in peptide synthesis. One of the major consequences of aspartimide formation is the racemization on the affected Asp/Asn residue in this process. It has been figured out that the rate of aspartimide residue to undergo racemization in a model hexapeptide H-Val-Tyr-Pro-Asu-Gly-Ala-OH (Asu: aspartimide residue) at pH 7.4 is significantly much faster than that of the Asp-containing counterpart under similar conditions.[11] This phenomenon implies that besides the chemical synthesis process aspartimide formation and concomitant racemization, it could also address the relevant peptides/proteins during storage and aging. Indeed although the formation of D-isomers from other amino acid residues have been detected, the kinetics of the formation of D-Asp *in vivo* via racemization is generally more exacerbated than in other species[12] and this phenomenon is predominantly attributed to the aspartimide formation from which these residues are prone to suffer.[13]

The mechanism of aspartimide formation-induced Asp/Asn racemization is illustrated in Fig. 11.4. Asp/Asn-containing peptide **15** with its side chain in the form of either free carboxylate/amide, or its corresponding protected species could accommodate the nucleophilic attack of the backbone amide nitrogen from its *C*-terminal neighboring amino acid and give rise to the formation of chiral aspartimide derivative **16**. The acidity of H^{α} on aspartimide **16** is enhanced relative to their Asp/Asn precursor due to the stabilization effect of the deprotonated anionic aspartimide derivative **17** and **18** via resonance.[13] The reprotonation of aspartimide anion engenders the racemic aspartimide derivative **19** that could undertake subsequent hydrolysis at its two imide carbonyl sites, leading to the ring opening of aspartimide and the formation of racemic iso-Asp **20** and racemic Asp **21**, respectively. The strategies to suppress the aspartimide formation could also alleviate the resultant racemization on the affected Asp/Asn. This has been already described in Section 6.1 in detail.

11.1.5 Acid-Catalyzed Peptide Racemization

Peptide racemization could also take place in an acidic milieu. One possible mechanism is the racemization via the enolate formation. This phenomenon is more likely to occur in strongly acidic conditions.[14] Although the extents of peptide racemization in strong acids regularly utilized in peptide synthesis, e.g., HF or HBr are verified to be mostly negligible, potential tendency of racemization at a higher temperature should not be ignored.

Another acid-catalyzed peptide racemization prevails in 6N HCl-directed peptide hydrolysis for amino acid analysis.[15] In order to differentiate the *D*-amino acid generated in the process of acidolysis from those formed in other processes special methods such as the employment of deuterated hydrochloric acid for acidolysis is developed and applied.[16]

11.2 RACEMIZATION IN PEPTIDE SYNTHESIS

11.2.1 Amino Acids with a High Tendency of Racemization in Peptide Synthesis

Some amino acids have a relatively high tendency to undergo racemization through various mechanisms. The incorporation of these amino acid derivatives in the process of peptide synthesis will understandably mandate meticulous screening of the appropriate reaction conditions in order to minimize the extent of racemization.

11.2.1.1 Histidine

His racemization in peptide chemical synthesis is characterized as one of the most serious side reactions occurring to this residue, both in Boc-[17] and Fmoc-mode[18] chemistry. Besides the oxazol-5(4*H*)-one formation-induced racemization pathway which prevails in the most chiral amino acids during peptide chemical synthesis, His could also suffer from facile racemization via a distinctive mechanism. Indeed, Histidine is regarded one of the most racemization-prone amino acids owing to its intrinsic chemical structure.

The imidazolyl substituent on the His side chain bears two nitrogen atoms (N^{τ} and N^{π}) that could function as intramolecular base catalysts. The Histidine building blocks utilized in the peptide chemistry nowadays are predominantly the ones whose N^{τ} is protected by an appropriate shielding group thanks to its relatively less steric hindrance compared with its N^{π} counterpart. There are indeed some His derivatives protected at their N^{π} functional groups,[19] nevertheless their applications in peptide manufacturing are not regarded as routine strategies, and are hence merely resorted to under some extreme circumstances. Moreover, their introductions into peptide synthesis might induce some other side reactions as consequences.[20] It is believed that the reactivity of the unprotected N^{τ} on the concerned His residue is largely restrained thanks to either the shielding or electron-withdrawing effect introduced by the protecting group on the N^{π}.[21] However, the protection of His-N^{τ} could not thoroughly deactivate its N^{π} which is able to trigger an autocatalysis process during peptide synthesis. The unshielded N^{π} atom could initiate H^{α} abstraction via an intramolecular

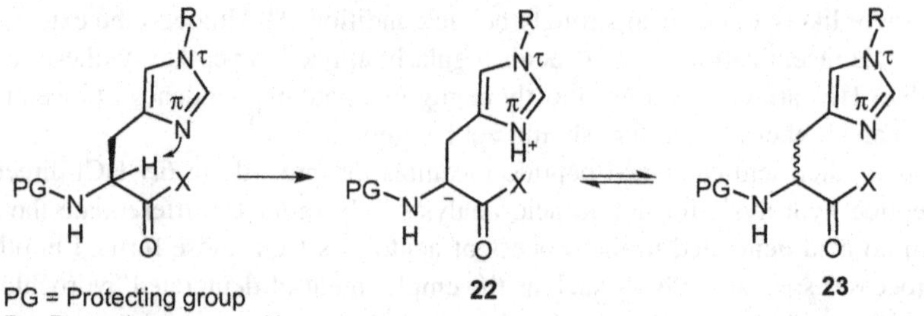

FIGURE 11.5 His racemization via H^{α} abstraction by side chain N^{π}.

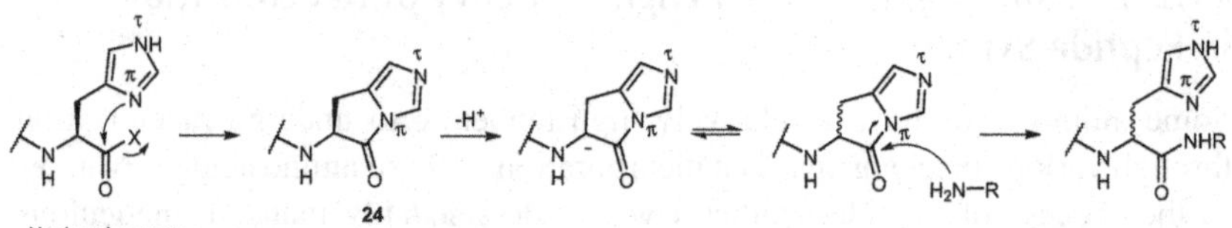

FIGURE 11.6 Proposed mechanism of His racemization via imidazolide formation.

autocatalysis process, as depicted in Fig. 11.5.[10] The spatial alignment advantages of N^{π} in terms of H^{α} abstraction at the step of His activation will lead to the formation of the cationic His derivative **22**, that could either tautomerize to its corresponding enolate species, or undergo reprotonation to give rise to His racemate **23**.

Another Histidine racemization mechanism addresses the side chain unprotected His derivative in the process of activation on its backbone carboxylate. The unshielded His-N^{π} might function as a nucleophile to induce an intramolecular attack on the activated carboxylate moiety to give rise to the optically labile cyclic carbonyl imidazolide derivative **24** that undergoes racemization at the subsequent synthetic steps (Fig. 11.6).

The inclination of His to undergo racemization through the aforementioned distinctive mechanism on top of the regular oxazolone formation at the step of carboxylate activation renders His one of the most susceptible amino acid residues to racemization in peptide synthesis. The corresponding solutions addressing these root causes will be elaborated in the following sections.

11.2.1.2 Cysteine

Regardless of the types of protecting groups on the Cys sulfhydryl side chain, cysteine derivatives are extraordinarily prone to undergo racemization during peptide synthesis, and this inherent attribute has already been recognized in the early days of peptide chemistry.[22] The consensus has been reached that the extraordinary susceptibility of *N*,*S*-protected cysteine active ester to racemization is predominantly attributed to the direct H^{α} abstraction in the presence of base, as depicted in Fig. 11.3.[23] It has been verified by a plethora of investigations that the Cys activation in the absence of base[24] or with a significantly reduced amount of base[25] could evidently decrease the extent of the racemization side reaction. It is to note that the severity of the cysteine racemization in peptide synthesis is markedly intensified during immobilization of the cysteine derivatives onto the hydroxyl-type resin[26] particularly in the presence of the DMAP catalyst.[27] In accordance with the high susceptibility of the cysteine ester to racemization, the peptide chain immobilized on the solid support via the cysteine ester bond could even suffer from cysteine racemization during peptide assembly through a repetitive piperidine treatment for N^{α}-Fmoc removal.[27] On the contrary, there is no apparent cysteine racemization in this process when the concerned cysteine residue is anchored to the solid support via an amide bond[28] or once the cysteine building block has been incorporated into the peptide chain at the locations other than *C*-terminus.[9]

Apparently, factors such as the type of cysteine sulfhydryl protecting group, coupling reagent, solvent, base, temperature and amino acid preactivation exert substantial impacts on the susceptibility of the cysteine residue to racemization in the process of peptide synthesis. They will be dedicatedly elaborated in thc following sections.

11.2.1.3 Glycosylated Amino Acid

The chemical conjugation between protein and sugar units that is characterized as protein glycosylation represents an important process of protein co- or posttranslational modification that possesses crucial biological significances.[29] Generally speaking protein/peptide glycosylation could be classified into *N*-glycosylation where the sugar unit is conjugated to the Asn side chain via a β-*N*-glycosidic bond[30] or *O*-glycosylation in which the sugar unit is bonded to the Ser/Thr/Tyr/Hyp/Hyl hydroxyl side chain by means of an *O*-glycosidic bond.[31] Besides, *C*-[32] and *S*-glycopeptide derivatives[33] also exist.

In view of the significant importance of glycosylated peptides in a variety of relevant domains chemical synthesis of glycosylated peptide derivative has been established as a crucial strategy to facilitate the acquisition of a deeper understanding of the biological functions of peptide/protein glycosylation in these fields.[34] Similar to peptide phosphorylation, peptide glycosylation could also be realized via the incorporation of preformed glycosylated amino acid building blocks or by posttranslational global glycosylation strategy.[35] Compared with the global glycosylation tactics the employment of glycoamino acid building blocks in an effort to regio-selectively incorporate the glycosylation modification has been widely practiced as a routine synthetic strategy in peptide chemistry,[36] especially for the formation of *O*-glycosidic bond that is relatively less compatible with the postsynthetic glycosylation strategy. Advantages of this synthetic manner could also be inherently manifested by the enhanced regioselectivity and anomeric selectivity. The versatile glycoamino acid building strategy could effectively evade the inherent synthetic limitations with respect to the posttranslational incorporation of the oligosaccharide unit. The assembly of the preformed glycosylated residues into the target glycopeptide chain will be in most cases compatible with the regular peptide chemistry. This synthetic strategy has found its widespread application both for *N*- and *O*-glycopeptide synthesis.

Nonetheless, *O*-glycosylated Ser/Thr building blocks are inherently correlated with some side reactions among which racemization stands out as a serious problem. Actually Ser is identified as a racemization-prone amino acid residue,[37] predominantly attributed to its electron-withdrawing hydroxyl substituent on the C^{β} of the side chain[38] as well as the solvation effect of hydroxyl group on water or the hydroxide ion that are capable of initiating an attack on the concerned Ser-H^{α}.[39] Glycoserine derivatives are even more susceptible to racemization compared with their unglycosylated precursor. Under certain extreme conditions like drastically prolonged preactivation the racemized derivative could even predominate in the product mixture.[40] The extraordinarily high inclination of glycoserine to suffer from racemization in the process of peptide assembly seems to be regulated by a variety of factors. Firstly, due to the significant steric hindrance the kinetics of activated glycoserine derivatives to undergo aminolysis would be markedly decreased. The consequent prolonged existence of an unconverted Fmoc-glycoserine active species will lead it astray from the target aminolysis pathway; whereas this will intensify the oxazolone formation that triggers tandem H^{α} abstraction and racemization. Secondly, it is observed via the comparison of the conformation of the oxazolone derivative of Ser(Trt) and Ser(Ac3GalNAc$_{\alpha}$) that the H^{α}—C^{α} σ-bond and H^{β}—C^{β} σ*-antibonding orbital of the latter are much more overlapping than the former, facilitating by this means the readily H^{α} abstraction by bases.[40] Thirdly, the reactivity discrimination of certain peptides N^{α} toward D- and L-glycoserine acylation could be more evident than toward the common amino acid derivatives,[40] presumably due to the enhanced reacting preferences of glycoserine residues correlated with their significant steric hindrances. High tendencies of glycoserine amino acids to undergo racemization during its activation and assembly into the peptide chains are therefore mutually influenced by the aforementioned factors as well as other potentially effective ones.

11.2.1.4 N-Alkyl Amino Acid and $C^{\alpha,\alpha}$-Disubstituted Amino Acid

N-Alkyl amino acids have already been identified at the early era of peptide chemical synthesis as a class of amino acids that is prone to undergo racemization during the carboxylate activation process.[41] The extraordinary tendency of *N*-alkyl amino acids to suffer from racemization during peptide assembly has become one of the most challenging obstacles for the chemical synthesis of *N*-alkyl amino acid-containing peptides. Moreover, *N*-alkyl amino acid derivatives are also liable to substantial racemization during saponification and acidolysis under mild conditions while ordinary peptides are mostly unaffected.[42]

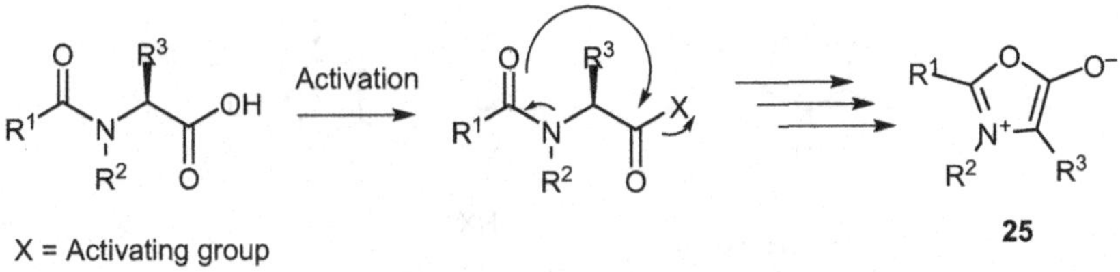

25

FIGURE 11.7 Formation of oxazolonium ion upon *N*-alkyl amino acid activation.

The intensive susceptibility of *N*-alkyl amino acids to racemization at the step of carboxylate activation is attributed to the strong tendencies of these derivatives to transform into the corresponding oxazolonium ion **25** in this process (Fig. 11.7).[41,43] It is analogous to the oxazol-5(4*H*)-one intermediate formation occurring in the process of the carboxylate activation of normal amino acids. The existence of oxazolonium ion **25** has been unambiguously confirmed by NMR analysis.[44] The electron-donating effect of the *N*-alkyl substitution on the addressed *N*-alkyl amino acid facilitates the ring closure and the formation of the corresponding oxazolonium ion. Differing from ordinary amino acids the activated N^{α}-acyl-*N*-alkyl amino acid derivatives could be subjected to the oxazolonium ion formation and racemization even in the absence of the base.[45]

It has been confirmed that the inclination of the transformation of *N*-alkyl amino acid to its corresponding oxazolonium ion and racemization is temperature-dependent. The extent of racemization could be reduced by performing the activation of *N*-alkyl amino acid at low temperature, e.g., −5°C.[43] It is indicated that the enhanced C^{α} chiral stability of *N*-alkyl amino acid upon carboxylate activation in the presence of *N*-hydroxylamine additives under low temperature is attributed to the suppression of the oxazolonium intermediate formation.[44] Utilization of appropriate coupling reagents could also be adopted with the aim to suppress the racemization of *N*-alkyl amino acids in the process of peptide assembly. For instance, triphosgene has been successfully utilized as the coupling reagent for the chemical preparation of cyclosporin and omphalotin derivatives that contain multiple or even consecutive *N*-Me amino acid residues.[46] Active Fmoc-*N*-Me amino acid chloride is formed *in situ* in this process, and racemization is significantly suppressed. The addition of *N*-hydroxylamine additives to the amino acid activation system has also been verified to affect the susceptibility of *N*-alkyl amino acids to racemization at the step of carboxylate activation.[44] Moreover, the addition of $CuCl_2$ to the activation reaction of *N*-alkyl amino acid has been justified to effectively suppress the occurrence of the racemization in this process.[43]

$C^{\alpha,\alpha}$-disubstituted amino acids are frequently incorporated into the peptide chains to investigate their inducing impacts on peptide conformation[47] as well as the correlated biological significance.[48] $C^{\alpha,\alpha}$-disubstituted amino acids are considerably liable to oxazol-5(4*H*)-one formation owing to the *gem*-diaklyl effect[49] (Fig. 11.8).

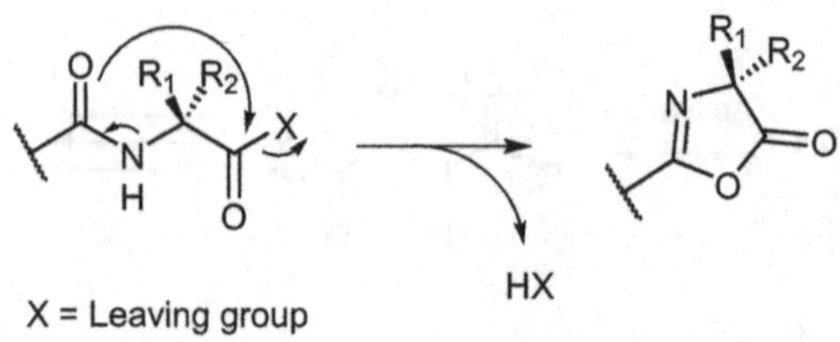

FIGURE 11.8 Oxazol-4(5*H*)-one formation from $C^{\alpha,\alpha}$-disubstituted amino acid/peptide.

The racemization of $C^{\alpha,\alpha}$-disubstituted amino acid at the step of carboxylate activation is nonetheless unlikely since racemization of the formed $C^{\alpha,\alpha}$-disubstituted oxazol-5(4*H*)-one derivative would require C—C or C—N bond scission. Actually the shelf-stability of oxazol-5(4*H*)-one species from the $C^{\alpha,\alpha}$-disubstituted amino acid precursor is intentionally taking advantages for certain peptide synthesis. Oxazol-5(4*H*)-one of the $C^{\alpha,\alpha}$-disubstituted amino acids could be prepared and isolated in an effort to incorporate the concerned $C^{\alpha,\alpha}$-disubstituted amino acid into the target peptide chain that might be otherwise difficult by means of ordinary coupling strategies due to its significant steric hindrance.[50] The low reactivity of the oxazol-5(4*H*)-one could be counterbalanced by increasing the reaction temperature and/or elongating the reaction time. The exploit of oxazol-5(4*H*)-one of $C^{\alpha,\alpha}$-disubstituted amino acid as an activated amino acid species verifies the chiral stability of such a derivative. Nevertheless, it has been reported that the peptide with $C^{\alpha,\alpha}$-disubstituted amino acid as the *C*-terminal residue could suffer from racemization on the penultimate residue in its backbone carboxylate activation process (Fig. 11.9).[50] Basically racemization via this mechanism could also affect peptide with ordinary amino acid located on the *C*-terminus.[51] However, it does not pose a pronounced problem under such circumstance for a normal amino acid. A peptide with $C^{\alpha,\alpha}$-disubstituted amino acid on the *C*-terminus is more susceptible to racemization on the penultimate reside at the step of backbone carboxylate activation, owing to the relatively slow coupling kinetics that facilitates the racemization via the indicated mechanism.

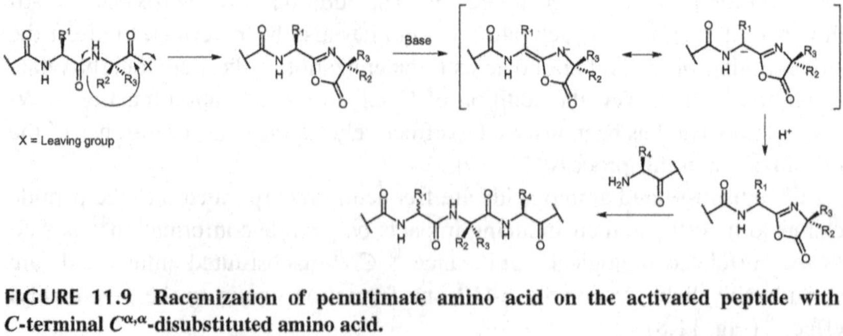

FIGURE 11.9 Racemization of penultimate amino acid on the activated peptide with *C*-terminal $C^{\alpha,\alpha}$-disubstituted amino acid.

11.2.1.5 Aryl Glycine Derivatives

Aryl glycine derivatives are regarded as important nonproteinogenic amino acids that have been detected in some natural small molecules[52] as well as antibiotics like vancomycin.[53] The chemical synthesis of aryl glycine-containing peptides via a conventional strategy poses formidable challenges predominantly attributed to the extraordinarily high tendency of these amino acid building blocks to undergo racemization upon the carboxylate activation in the presence of bases.[54] Different from canonical amino acids that predominantly racemize via oxazol-5(4*H*)-one intermediate mechanism aryl glycine derivatives are frequently involved in an intensive racemization through direct H^{α} abstraction by base at the step of the backbone carboxylate activation and subsequent aminolysis. The distinctive attribute of aryl glycine compounds in the context of high liability to racemization is owing to the high acidity of their H^{α} relative to ordinary amino acids. The generated anionic intermediate of aryl glycine upon H^{α} abstraction could be substantially stabilized by the aryl side chain.

In view of the unacceptably high degree of aryl glycine racemization during peptide assembly[55] special synthetic strategies like Umpolung amide formation[56] could be adopted to circumvent the formidable racemization side reaction addressing these amino acids.

11.2.2 DMAP-Induced Racemization

4-Dialkylaminopyridine derivatives such as DMAP have found widespread applications in organic synthesis, e.g., acylation, alkylation, esterification, lactamization, silylation, sulfonation, etc.[57] In peptide chemistry DMAP is routinely utilized for difficult couplings such as acylation of the hydroxyl group in the process of depsipeptide synthesis,[58] immobilization of the first amino acid to hydroxymethyl spacers on solid supports,[59] coupling of $C^{\alpha,\alpha}$-disubstituted amino acid,[60] acylation of the *N*-alkyl amino acid,[61] peptide cyclization, and so on.

In spite of the versatility of DMAP to promote acylation reactions, it is known that this catalyst enhances the racemization side reaction of N^{α}-urethane protected amino acid, e.g., Boc-, Fmoc-, and Z-amino acids in their backbone carboxylate activation process.[27] In view of this side reaction stimulated by DMAP, the amount of DMAP catalyst should be meticulously controlled under sensitive circumstances[27] since decrease of DMAP stoichiometry in corresponding reactions could effectively alleviate the severity of amino acid racemization.[62] Temperature is also decisively crucial for the extent of DMAP-induced racemization, and rational decrease of reaction temperature could drastically diminish the degree of the amino acid racemization in this process.[63] For extreme cases in which racemization is extraordinarily liable to occur at the step of the DMAP-catalyzed esterification reactions an alternative amino acid coupling strategy could be adopted instead such as a Mitsunobu reaction[64] or esterification with amino acid fluorides.[65] Employment of a highly effective coupling reagent additive such as HOAt in place of DMAP could also contribute to lower the extent of racemization in the process of amino acid coupling.[66]

11.2.3 Microwave Irradiation-Induced Racemization

Since its introduction into the realm of peptide synthesis microwave-assisted SPPS has quickly found widespread application particularly for difficult peptide assemblies, e.g., *N*-methyl amino acid-rich peptides,[67] glycopeptides,[68] glycopeptoids,[69] phosphopeptides,[70] as well as other categories of difficult peptides.[71]

In spite of the valuable contributions that microwave irradiation has made in the context of difficult peptide synthesis its introduction has inevitably induced various side reactions such as aspartimide formation and racemization.[72] Sensitive amino acid residues like Cys and His are especially susceptible to racemization upon the introduction of microwave irradiation. It is believed that at least the high temperature employed in microwave-assisted SPPS at the steps of amino acid coupling and Fmoc deprotection partially accounts for the aggravated racemization.[73] In light of this concern, it is advisable to decrease the temperature of amino acid activation/coupling in order to alleviate microwave irradiation-induced amino acid racemization.[72] Although this strategy could not thoroughly circumvent the inherent racemization problem associated with the introduction of microwave irradiation, its employment in combination with other racemization-suppressing tactics[55] could to some extent reduce the racemization extent when microwave-assisted SPPS technology is concerned.

11.2.4 Racemization During Peptide Segment Condensation

Convergent peptide segment condensation represents an important synthetic strategy for the construction of large peptide/protein that is hardly achievable by a classic stepwise synthetic tactic. This technology utilizes complementary peptide fragment precursors synthesized separately as building units that are spliced in a convergent manner either in solution[74] or on solid support,[75] giving rise to the target product with high purity that is much more amenable for chromatographic purification. A variety of technologies such as the incorporation of backbone-protected amino acid, pseudoproline, and isodipeptide have been developed to address the inherent solubility limitation of peptide segments, rendering this synthetic strategy a versatile and practical alternative to the classic stepwise peptide assembly.

One of the most formidable challenges associated with the employment of peptide segment condensation is the facile racemization on the *C*-terminal amino acid during its backbone carboxylate activation process. It is known that activated N^{α}-acyl amino acids are susceptible to undergo racemization that is attributed to the facile formation of an optically labile oxazolone intermediate. Analogous to N^{α}-acyl amino acid the *C*-terminal amino acid on a peptide molecule bears N^{α}-acyl structure. This structural attribute offers the concerned peptide high tendency to suffer from racemization at its C^{α} on the *C*-terminal amino acid at the step of carboxylate activation. The *C*-terminal activated amino acid could be readily converted to the corresponding oxazol-5(4*H*)-one intermediate that undergoes facile racemization as a consequence. The severity of the racemization in the process of peptide segment condensation could even be exaggerated when the disadvantageous effects are taken into account such as low solubility of the peptide fragments and the resultant slow reaction kinetics.

In light of this limitation meticulous design is requested when peptide segment condensation strategy is employed to construct the target peptide molecule. Parameters such as coupling reagent, additive, solvent and temperature should be carefully and correlatively evaluated in the context of their influences on racemization. More importantly, racemization-prone amino acid residues such as His, Cys, Ser, aryl glycine and *N*-alkyl amino acid should not be located on the *C*-terminus position of the amino-component peptide segment. Amino acids that are devoid of, or resistant to racemization such as Gly or Pro are preferably opted as the *C*-terminal residue on the amino-component peptide fragment when convergent segment condensation strategy is adopted.

11.3 STRATEGIES TO SUPPRESS RACEMIZATION IN PEPTIDE SYNTHESIS

11.3.1 Amino Acid N^{α}-Protecting Group

Since the oxzazol-5(4*H*)-one formation from the activated N^{α}-derivatized amino acid precursor constitutes a significant pathway for the amino acid racemization during peptide synthesis, the type of N^{α}-protecting group on the concerned amino acid will understandably exert tremendously important impacts on the susceptibility of the amino acid derivative to undergo racemization in the carboxylate activation process. Intensive investigation of exploiting appropriate amino acid N^{α}-protecting groups that are helpful to resist amino acid racemization has been systematically conducted, and the repertoire of the N^{α}-protecting groups has been markedly enriched ever since.

It is known that N^{α}-acyl protected amino acids are extremely prone to cyclize to the corresponding oxzazol-5(4*H*)-one intermediate during backbone carboxylate activation. The employment of N^{α}-acyl-type protecting groups such as formyl, acetyl and benzoyl for amino acids has already been regarded as obsolete, and abandoned in the peptide chemistry.

11.3.1.1 N^{α}-Urethane Protecting Group

The intensive utilization of N^{α}-urethane protecting groups, e.g., Z, Boc, and Fmoc in place of N^{α}-acyl group for amino acids represents a milestone in the history of peptide chemistry. This momentum not only drastically facilitated

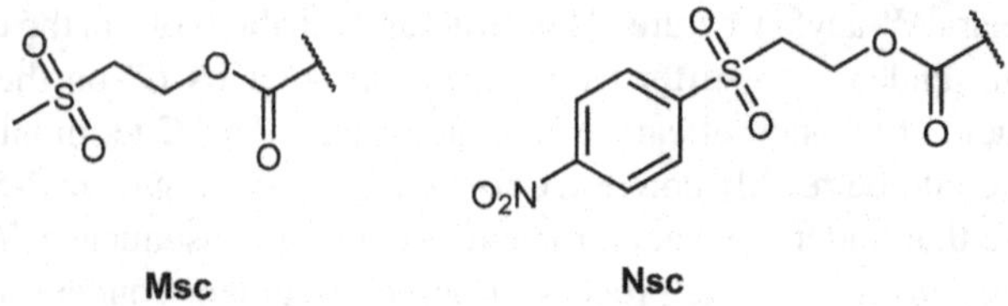

FIGURE 11.10 **Msc and Nsc protecting groups.**

the readily removal of the N^{α}-protecting group but also significantly suppressed the racemization of the concerned amino acid during its activation and coupling process. Nevertheless N^{α}-urethane protecting groups could not thoroughly avoid the occurrence of amino acid racemization, especially when highly potent coupling reagents are employed for the activation of N^{α}-urethane amino acid. Due to the increased acidity of the H^{α} on the activated amino acid species the racemization could also be induced at moderate to severe level on the N^{α}-urethane protected amino acid via oxazolone intermediate. Intensive racemization of Fmoc-Cys(Trt)-OH,[76] Fmoc-Cys(StBu)-OH,[24] Fmoc-Cys(Xan)-OH, Fmoc-Cys(Tmob)-OH, Fmoc-Cys(Acm)-OH,[9] Fmoc-Ser(tBu)-OH[37] and many other N^{α}-urethane protected amino acids has been detected in peptide assembly.

In addition to the canonical Fmoc protecting group other types of base-labile N^{α}-urethane protecting groups have also been designed, among which N^{α}-sulfonylethoxycarbonyl derivatives have been widely utilized in peptide synthesis. Currently 2-(methylsulfonyl)ethoxycarbonyl (Msc)[77] and 2-(4-nitrophenylsulfonyl)ethoxycarbonyl (Nsc)[78] are regarded as appealing candidates in the category of sulfonylethoxycarbonyl protecting group (Fig. 11.10). They are similar to Fmoc group in that they could be cleaved under alkaline condition via β-elimination mechanism[79] but the deprotection kinetics is slower than that of Fmoc. The enhanced stability of N^{α}-Msc and N^{α}-Nsc in DMF bestows them notable advantages over Fmoc under circumstances where premature Fmoc cleavage and consequent side reactions such as redundant amino acid coupling described in Section 10.2 could take place. Moreover, differing from Fmoc protection tactics N^{α}-Msc is stable under catalytic hydrogenation conditions and does not inactivate the hydrogenation catalysts.[80] These attributes introduce enhanced orthogonality for the selective removal of protecting groups. The optimal removal of N^{α}-Msc could be completed with 10–20 s with DMF/MeOH/4M NaOH (14:5:1)[81] while N^{α}-Nsc cleavage strategy is identical to that of Fmoc but with slower kinetics.[82]

One of the appreciable properties of the N^{α}-Nsc protecting group is reflected by the fact that the extent of the racemization of N^{α}-Nsc-protected His,[83,84] Cys and Ser[84] derivatives during TBTU/DIEA or HBTU/DIEA mediated activation is significantly reduced relative to those of the corresponding N^{α}-Fmoc protected counterparts.

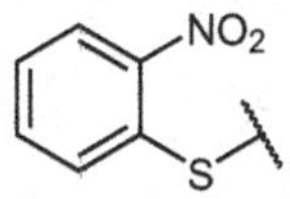

FIGURE 11.11 **Nps protecting group.**

11.3.1.2 *N*α-Sulfanyl Protecting Group

Now that the carbonyl functionality on the protected amino acid derivative plays a pivotal role in the oxazolone intermediate formation and the subsequent racemization, a plethora of *N*α-protecting groups that are void of carbonyl moiety have been designed and applied to amino acid *N*α-protection to address racemization side reactions.

*N*α-sulfanyl protecting groups such as Nps (Fig. 11.11) has been designed and utilized as an alternative *N*α-protecting group. *N*α-sulfanyl protected amino acid derivatives are not subject to oxazolone intermediate-induced racemization during amino acid activation owing to the absence of carbonyl moiety imperative for the oxazolone formation. Moreover, the *N*α-sulfanyl protecting group does not afford the subjected amino acid disadvantageous steric hindrance effect compared with bulkier protecting groups, e.g., *N*α-trityl or *N*α-tritylsulfanyl.[85]

*N*α-Nps could be selectively removed by 1–2 equiv. of HCl or HBr in protic[86] and aprotic solvents[85] while not affecting tBu, Boc and Bzl side chain protecting groups in the latter conditions.[87] Alternatively, *N*α-Nps protecting group could be cleaved by nucleophiles such as thioacetamide, thiourea or thiol.[88] Triphenylphospine treatment could also cleave the sulfenamide bond and liberate the shielded *N*α group from Nps protecting group.[89] Coupling of *N*α-Nps amino acids is generally conducted by the DCC/HOSu method,[90] since more acidic additives such as HOBt, HOAt, and HOOBt could lead to an undesired premature *N*α-Nps cleavage.[91]

One of the most outstanding merits of *N*α-Nps protection strategy is that the derived *N*α-Nps amino acids are free of oxazolone-induced racemization at the carboxylate activation step. Although *N*α-Nps protected amino acids have not found wide applications in SPPS, these derivatives offer a promising strategy to immobilize the first amino acid on the hydroxymethyl type resin since the derivatization of hydroxymethyl resin with *N*α-Nps amino acids by DIC/HOBt could be realized without racemization[92] while under identical reaction conditions *N*α-urethane protected amino acid might suffer from a pronounced extent of racemization side reaction depending on the type of addressed amino acid and the coupling conditions.

11.3.1.3 *N*α-Sulfonyl Protecting Group

*N*α-Tos is regarded as the prototype of the *N*α-sulfonyl protecting groups for amino acid. However, its actual application in peptide chemistry as *N*α-protecting group has been evidently limited due to the intrinsically cumbersome

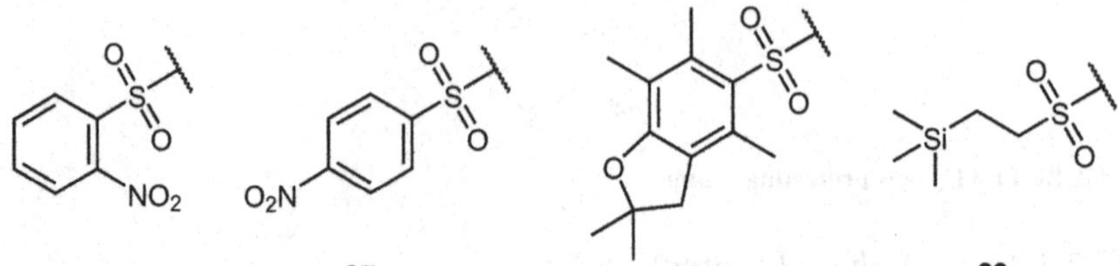

26 27 28 29

FIGURE 11.12 **N^{α}-sulfonyl protecting groups.**

removal condition (Na in liquid NH_3[93]) and the highly reactive sulfonamide nitrogen that could induce a variety of side reactions like *N*-alkylation.[94] In view of these limitations a plethora of new N^{α}-sulfonyl protecting groups in place of N^{α}-Tos has been developed and utilized in peptide chemistry. Representative candidates include 2-nitrobenzenesulfonyl (*o*Nbs) **26**, 4-nitrobenzenesulfonyl (*p*Nbs) **27**, Pbf **28** and 2-(trimethylsilyl)ethylsulfonyl (SES) **29** (Fig. 11.12).

Cleavage of N^{α}-Pbf protecting group is realized with 10% DMS/TFA.[95] *o*Nbs and *p*Nbs could be removed via nucleophilic aromatic substitution mechanism[96] by either thiophenol/K_2CO_3 in DMF or 2-mercaptoethanol/DBU in DMF, depending on whether the shielded N^{α} group is a primary or *N*-alkyl amine.[97] N^{α}-SES is readily cleaved by CsF to liberate the amino group, and give rise to TMSF, ethane and SO_2.[98] The introduction of the sulfonyl protecting group on N^{α} significantly increased the acidity of the hydrogen linked to the affected N^{α} group, rendering it possible to carry out selective *N*-alkylation on N^{α}-sulfonyl protected amino acids[99] or peptides.[100]

Owing to their intrinsic chemical structures N^{α}-sulfonyl protected amino acids will not be subjected to oxazolone-induced racemization.[100] Taking advantage of this distinctive property it is feasible to readily convert N^{α}-sulfonyl amino acids to the corresponding highly-activated derivatives like acid chlorides and employed in formidably difficult coupling reactions such as assembly of *N*-allylated peptides[100] or coupling of MeAib to MeAib[95] that are otherwise exceedingly susceptible to racemization with canonical N^{α}-urethane protected amino acids. In addition, thanks to the increased inductive effect of the sulfonyl group N^{α}-sulfonyl protected amino acid halides are more reactive than their *N*-urethane counterparts,[94] entrusting additional advantages to N^{α}-sulfonyl protected amino acids in case of sterically hindered amino acid couplings.

11.3.1.4 N^{α}-Alkyl Protecting Group

Alkyl type N^{α}-protecting groups have not been extensively utilized in peptide synthesis, especially in SPPS, partially due to the difficulty to remove them from the protected amino acid precursors. In spite of this limitation, N^{α}-benzyl, N^{α}-benzhydryl, and N^{α}-trityl derivatives have attracted considerable attentions as the former could be cleaved by hydrogenation, and the latter two are removed by acidolysis.

Among these protecting groups trityl has found wide application although its predominant utilization in peptide synthesis is limited to the side-chain protection on Cys, His, Asn and Gln instead of N^{α}-group. N^{α}-trityl protecting group could be readily removed by 1% TFA/DCM or 0.1 M HOBt/TFE.[101] It can be cleaved even under milder conditions like 0.2% TFA, 1% H_2O in DCM,[102,103] rending it compatible with acid labile resin.

One of the most distinctive advantages of N^{α}-trityl protection strategy is attributed to the preventative effect on base-induced DKP formation at the step of N^{α}-protecting group removal during peptide synthesis,[103] thanks to the protonation of the liberated N^{α} upon the acidolytic removal of the trityl protecting group. It restrains the occurrence of DKP formation that is induced by the nucleophilic attack of the unshielded N^{α} on the backbone amide or ester bond.

Another notable advantage of N^{α}-Trt protection strategy is the reduced racemization of the concerned amino acid at the step of backbone carboxylate activation. This is partially attributed to the pronounced shielding effect of the bulky N^{α}-Trt against the abstraction of H^{α} by base. More importantly, the intrinsic structure of Trt precludes the oxazolone intermediate formation from the activated N^{α}-Trt amino acid. This attribute ensures the activate species N^{α}-Trt amino acid-(1-benzotriazolylester) exemption from racemization via oxazolone intermediate even at elevated temperature (30–80°C) and in the presence of excessive triethylamine.[104] N^{α}-Trt NCAs have also been utilized for peptide assembly in THF at 40°C or under reflux for 6–20 h without racemization.[105]

11.3.1.5 N^{α},N^{α}-bis Protection Strategy

In view of side reactions induced by the acidic NH proton from N^{α}-urethane-protected amino acids including racemization via oxazolone intermediate formation, dual protection on the concerned N^{α} group is occasionally adopted to suppress correlated side reactions. N^{α},N^{α}-bis-Boc protected amino acid (Fig. 11.13) for instance, is derived from N^{α}-Boc-amino acid benzyl ester through the treatment by Boc_2O in the presence of DMAP catalyst in ACN, followed by catalytic hydrogenation.[106] N^{α},N^{α}-bis-Boc amino acid can be converted to the corresponding N^{α}-Boc amino acid NCA in $SOCl_2$/DMF,[107] offering an alternative synthetic strategy for the preparation of amino acid UNCA.

Due to the absence of NH proton N^{α},N^{α}-bis-Boc amino acid does not racemize under standard coupling conditions via the formation of oxazol-5(4*H*)-one

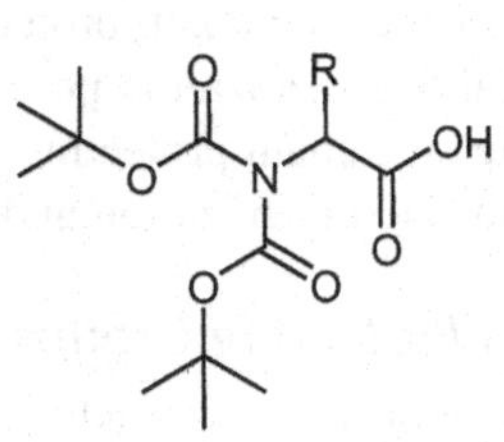

FIGURE 11.13 N^{α},N^{α}-bis-Boc amino acid.

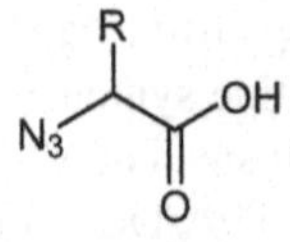

FIGURE 11.14 Azido acid derivative.

intermediate. However, racemization of these derivatives in the process of carboxylate activation could not be excluded since the introduction of the additional electron-withdrawing Boc protecting group on N^α would increase the acidity of the subjected H^α that consequently facilitates the base-directed H^α abstraction and consequent racemization. This side reaction could, nevertheless, be diminished through rapid coupling by N^α,N^α-bis-Boc amino acid fluoride as the acylating agent.[107,108]

11.3.1.6 α-Azido Acid as Synthon of Amino Acid

α-azido acids (Fig. 11.14) can be utilized in peptide chemistry as the synthon of the corresponding amino acid. The transient azido group is reduced to amino group by methods like catalytic hydrogenation,[109] phosphine treatment followed by hydrolysis (Staudinger reaction)[110] or DTT reduction.[111]

The chemical structure of azido acid precludes the racemization of its activated species via the oxazolone formation pathway. This attribute has been taken advantage for some special couplings in which the extent of racemization is unacceptably high with ordinary amino acid building block. The azido acids could be assembled into the peptide chain by DCC/HOBt with little or no racemization.[112] For difficult couplings in which steric hindrance effects become predominant azido acids could be converted to the highly activated acid chlorides[109] or even acid bromides[113] (on the contrary, it is impractical to isolate the stable Fmoc/Boc/Z-protected amino acid bromide) and incorporated into the peptide chains without detectable racemization.[114]

11.3.2 Amino Acid Side Chain Protecting Group

The types of amino acid side chain protecting groups, if any, might notably impact on the inclination of the subjected amino acids to suffer from racemization at the step of carboxylate activation and coupling. Regardless of the racemization mechanism (oxazolone formation, enolation, or direct H^α abstraction) the inductive and steric effects of the side chain protecting group could interfere with the steps that are pivotal to racemization process. Consequently, meticulous screening of appropriate side chain protecting groups could be of help to minimize the extent of amino acid racemization under certain circumstances.

11.3.2.1 Cys Side Chain Protecting Groups

As cysteine is regarded as one of the most readily racemized amino acids the rational selection of the protecting group on its sulfhydryl side chain will be of significance when the susceptibility of cysteine derivatives to racemization are taken into account. A consensus has been established that *N,S*-protected Cys derivatives undergo racemization predominantly via direct H^{α} abstraction pathway in the presence of excess bases.[115] Cys is extremely prone to undergo racemization in the process of immobilization to hydroxymethyl-type resin through the esterification reaction. Moreover, the extent of *C*-terminal Cys ester racemization could even be further increased during the repetitive N^{α}-Fmoc removal upon piperidine treatment (this phenomenon is exclusively addressing *C*-terminal Cys ester, while *C*-terminal Cys amide and internal Cys residue are unaffected with respect to racemization by piperidine treatment).[27] Detailed studies clearly revealed that the type of the Cys side chain protecting group would pronouncedly affect Cys racemization in this process, with an decreased racemization order S-StBu > S-Trt > >S-Tacm > S-Acm > S-MeBzl > S-tBu.[116,117] Normally very little or no racemization is observed in the synthesis of peptide with *C*-terminal Cys(StBu) amide or Cys(StBu) located in the internal position of the peptide sequence.[117]

11.3.2.2 His Side Chain Protecting Groups

The selection of protecting group on the imidazolyl side chain has magnificent impacts on the tendency of His to undergo racemization at the carboxylate activation and coupling step. As elucidated in Section 11.2.1.1 activated His residue could suffer from facile racemization partially due to the catalysis effect of the unshielded N^{π} from its imidazolyl side chain (Fig. 11.5). This mechanism verifies the evident importance of the choice of His side chain protecting group, both from electron-withdrawing and steric shielding aspects on the readiness of the concerned His to be affected by racemization side reaction.

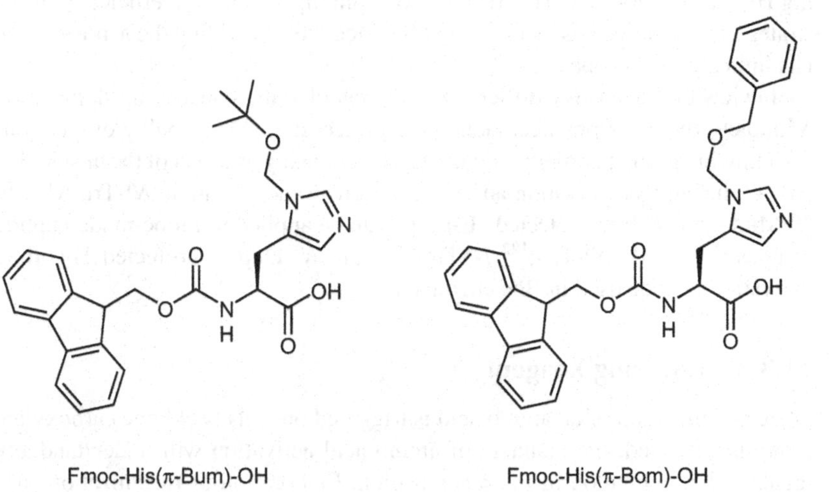

FIGURE 11.15 **Fmoc-His(π-Bum)-OH and Fmoc-His(π-Bom)-OH.**

Site selective protection on His-N^{im} poses a challenge since N^{τ}- instead of N^{π}-protected His isomer will be the predominant derivative derived under the standard side chain protection conditions due to the disadvantageous steric effect of the latter. His-N^{π}-protection is normally achieved through the alkylation of preformed N^{τ}-protected His derivative.[21] Some His derivatives protected on N^{π}-moiety, e.g., Fmoc-His(π-Bum)-OH and Fmoc-His(π-Bom)-OH (Fig. 11.15) are available for peptide synthesis in order to suppress the notoriously high racemization affecting N^{τ}-protected His counterpart at the step of carboxylate activation and coupling process.

Removal of His(π-Bum)-OH side chain protecting group can be realized by TFA whereas His could be regenerated from His(π-Bom)-OH by HF, TFMSA or hydrogenolysis,[118] rendering them applicable for Fmoc and Boc chemistry, respectively. Utilization of Fmoc-His(π-Bum)-OH as building block in an effort to assemble those peptides that are susceptible to racemization upon His incorporation has been unequivocally verified to be much more effective to suppress the racemization side reaction than Fmoc-His(τ-Trt)-OH and Fmoc-His(τ-Bum)-OH.[20] The predominant drawbacks of His-N^{τ}-Bum protection lie in the fact that the quantitative removal of Bum is sluggish, relative to Trt protecting group, and the formation of formaldehyde as a degradative byproduct upon N^{τ}-Bum cleavage. The released formaldehyde could induce undesired modifications on susceptible residues, e.g., Cys and Trp, or initiate other crosslinking processes (see also Section 7.4.3.1). These limitations could, nevertheless, be attenuated by the rational utilization of Cys[119] or hydroxylamine derivative such as methoxyamine hydrochloride[20] as scavengers to quench the released formaldehyde in order to shield the sensitive residues from formaldehyde-induced irreversible modification. In another investigation, Z-His(π-Pac)-OH has exhibited substantial advantages over the Z-His(τ-Pac)-OH counterpart in terms of reducing His racemization.[120] This result also explicitly verifies the efficiency of the strategy to minimize His racemization by selectively shielding the π-position on His imidazolyl side chain.

In view of the intrinsic difficulties with regard to the selective masking of the N^{π} moiety the more practical tactic is to protect the N^{τ} by a bulky or electron-withdrawing group in order to minimize the autocatalysis effect of the unshielded His-N^{π} during His activation and the consequent racemization. N^{τ}-Trt, N^{τ}-Mtt, N^{τ}-Mmt, and N^{τ}-Boc protected Histidine can be applied to Fmoc-mode peptide synthesis[121] while N^{τ}-Tos,[122] N^{τ}-Doc,[123] and N^{τ}-Dnp[124] protected Histidine derivatives can be used in Boc chemistry.

11.3.3 Coupling Reagent

Since racemization of an amino acid is triggered once its backbone carboxylate group is activated, the manners of amino acid activation will understandably regulate the inclination of the racemization. Elaboration of racemization suppression could hence be achieved by meticulous coupling reagent screening.

The impact of the amino acid backbone carboxylate activation on racemization shall be analyzed with a dialectical approach. On one hand, an intensive activation of the carboxylate on amino acid/peptide could facilitate the acylation of the reciprocal amino group, and the risk of racemization of the activated amino acid/peptide will be lowered accordingly by a prompt consumption of the activated species upon conversion to the corresponding amide derivative; on the other hand, the excessive-activation on the carboxylate from the subjected amino acid/peptide might also intensify racemization since the enhanced leaving propensity of the activation group could facilitate the oxazolone formation or enolation/direct H^{α} abstraction which eventually results in racemization as a consequence. Meticulous fine-tuning of activation strategy is hence practical to balance these two seemingly paradoxical facets.

Since the pathway of racemization and susceptibility to this side reaction are both amino acid and peptide sequence specific, it should be prudent to draw general conclusions from certain specific investigations regarding the correlation between the choice of coupling reagents and their racemization suppression effects.

11.3.3.1 Amino Acid Azides

In spite of its arguably obsolete status as an activated amino acid species, employment of amino acid azide as a building block for peptide stepwise assembly or segment condensation is regarded as a low-racemization strategy.[125] Since the activation effect of azide on the addressed carboxylate group is relatively mild compared with other activation manners, the susceptibility of the affected amino acid to oxazolone formation is attenuated which diminishes the racemization extent.[126] Nevertheless, under certain conditions particularly in the presence of excess of base[51] the azide activation strategy could also lead to racemization. The amount[127] as well as type of auxiliary base[128] in azide activation procedure can play a significant role to the extent of racemization that the concerned amino acid has to suffer.

11.3.3.2 Amino Acid Halides

Amino acid fluoride exhibits superior attributes to the corresponding acid chloride counterpart in that they are relatively more resistant to oxazolone formation, and therefore less racemization could occur to the amino acid fluoride-mediated peptide assembly. Meanwhile, amino acid fluorides are provided with sufficiently high reactivity in difficult couplings, e.g., the coupling reaction between secondary alcohol and cyclohexyl amino acid,[129] as well as other challenging couplings.[130,131] Moreover, another unique property of acid fluoride is that it is capable of undergoing efficient aminolysis with the substoichiometric amount of base or even in the absence of base.[132] This intrinsic property of amino acid fluoride can substantially restrict the occurrence of base-catalyzed oxazolone formation from active amino acid derivative and the subsequent racemization.

Unlike its Fmoc counterpart, Boc- and Z-amino acid fluoride is readily prepared in the form of crystalline with significantly less inclination to undergo NCA formation.[133] Highly racemization-susceptible amino acids such as phenylglycine could be incorporated into the target peptide chain in the manner of Boc- or Z-Phg-F with no significant racemization (<1%).[133] The activated species of *N,N*-Bis(Boc) or (*N*-Boc, *N*-Z)-amino acids that preclude oxazolone formation and are therefore to some extent unaffected by oxazolone-mediated racemization are nonetheless inappropriate for acylating reactions due to their significant steric hindrances. Most of the canonical coupling tactics, e.g., active ester, mixed anhydride, and carbodiimide activation are frequently ineffective to quantitatively activate and assemble *N,N*-Bis(alkoxycarbonyl) amino acid into the target peptide chain.[134] However, *N,N*-Bis(Boc) amino acid fluoride could elude this limitation and be utilized as an effective acylating agent with low susceptibility to racemization via oxazolone intermediate pathway.[107]

In spite of the extraordinarily high inclination to undergo oxazolone formation and subsequent racemization, amino acid chloride could be generated *in situ* to circumvent this inherent limitation. It has been exhibited that the BTC/collidine mediated amino acid couplings were successfully employed in the SPPS of Omphalotin A[135] and Cyclosporin O[136] derivatives that are rich in *N*-alkyl amino acids, taking advantage of the *in situ* formation of amino acid chloride. No detectable to very limited amount of racemization was generated even at the most difficult coupling steps between consecutive *N*-alkyl amino acids such as acylation of *N*-Me-Val-resin by Fmoc-*N*-Me-Val-OH. These investigations have verified the feasibility of the *in situ* formation of amino acid chloride mediated by BTC, and its applicability for *N*-alkyl amino acid couplings manifested by coupling efficiency enhancement and racemization suppression. It is indicated that NMP is reactive toward BTC to form the chloroiminium reminiscent of an active intermediate species generated in the Vilsmeier–Haack reaction, and this side reaction will induce amino acid racemization as a consequence.[137] BTC-mediated *in situ* formation of amino acid chloride should, therefore, be conducted in inert solvents like THF.

11.3.3.3 Halophosphonium Salts

Halophosphonium coupling reagents, e.g., BroP and PyBroP seem to intensify the racemization in the process of ordinarily amino acid acylation compared with coupling reactions regulated by phosphonium salts like PyBOP. Conversely, when *N*-alkyl amino acids are to be acylated, or in case both amino and carboxyl components of the coupling reactions are *N*-alkyl amino acids, halophosphonium coupling reagents are able to exhibit an evidently superior property with respect to racemization suppression relative to PyBOP. For instance, it is observed that the coupling of Z-*N*-Me-Val-OH to H-*N*-Me-Val-OMe by PyBOP or HBPyU results in an unacceptable extent of racemization (18 and 13%, respectively) whereas BroP, PyBroP, PyClop or HPyClU-mediated coupling reactions generate almost no racemization.[138]

11.3.3.4 Uronium Salts

Among various uronium salts HATU (Fig. 11.16) is regarded as one of the most excellent coupling reagents. It exhibits outstanding performances both in facilitating amino acid coupling reactions into completion and suppressing racemization in this process.[139] Its utility in industrial peptide manufacturing is, nevertheless, substantially limited due to its prohibitively high price and its hazardous explosive property.[140]

As a sort of trade-off between a HOAt- and HOBt-type uronium coupling reagent, HCTU (Fig. 11.16) has found its application in peptide synthesis with good performances with respect to coupling efficiency[141] and racemization suppression. HCTU outperforms HBTU in terms of the capability to restrain racemization in the process of amino acid couplings.[142]

Uronium salts derived from HODhbt structure, e.g., HDTU[142] and TDBTU[143] (Fig. 11.16) are effective coupling reagents to reduce the extent of amino acid racemization, and this attribute could be more evidently reflected in the process of amino acid couplings that are particularly susceptible to racemization. The advantageous characterizations of HODhbt-derived uronium coupling reagents in terms of amino acid racemization suppression are basically in line with the performance of HODhbt as a coupling additive to alleviate racemization relative to other hydroxylamine derivatives such as HOBt.[144] The HODhbt-type of uronium coupling reagents have not been widely utilized in peptide synthesis in spite of sporadic report,[145] predominantly due to the inherent ring opening side reaction that leads to the formation of azide side product and peptide chain termination as a consequence (see also Section 5.7).

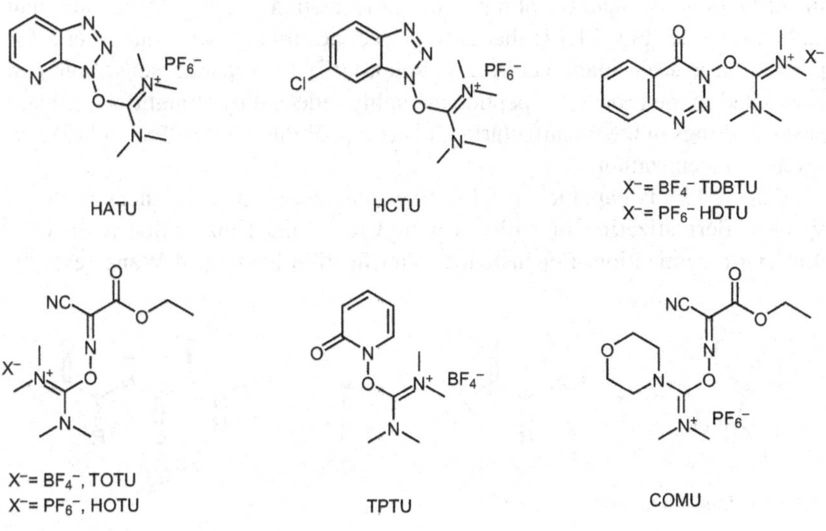

FIGURE 11.16 Selected uronium salt coupling reagents.

Oxyma as an amino acid coupling additive has gained distinct attention and application in peptide synthesis. It exhibits outstanding performances, and becomes an excellent replacement for HOBt and other coupling additives.[146] Correspondingly Oxyma has derivatized a plethora of coupling reagents that have been utilized in peptide synthesis. TOTU and HOTU (Fig. 11.16) are representatives of this group, and have been employed to direct amino acid coupling under certain conditions. TOTU exhibits improved property in terms of restraining racemization occurring during peptide segment condensation compared with BOP or TBTU.[147] COMU (Fig. 11.16), another Oxyma-derived uronium salt, has been verified to be less explosive, and possesses superior racemization suppression capability relative to HOBt-based coupling reagents. The racemization suppressive performance of COMU is comparable to those of HOAt-derived coupling reagents.[148]

Oxo-pyridine type of uronium coupling reagents, e.g., TPTU (Fig. 11.16) has also found its application in peptide synthesis. TPTU is recommended for the peptide segment condensation that is susceptible to racemization as well as *N*-methyl amino acid couplings. It exhibits superior properties to TBTU in this connection.[143]

11.3.3.5 UNCA

Urethane-protected amino acid *N*-carboxyanhydride (UNCA) is regarded as an activated urethane protected amino acid species.[149] The commercially available UNCA derivatives are generally stable crystalline compounds that can be stored for reasonably long periods of time.[150] The advantages of utilization of UNCA as an active species of amino acid could be reflected in the following facets: (1) high reactivity of UNCA that could be rationally exploited for the difficult amino acid coupling;[151] (2) no formation of byproducts other than carbon dioxide (Fig. 11.17) that is in perfect accordance with the criteria for peptide large-scale manufacturing in solution;[152] (3) separate activation step is avoided in the process of peptide assembly reflected by straightforward and easy handlings in the manufacturing; (4) compatibility with SPPS; and (5) low levels of racemization.

Fmoc-NCA is capable of exhibiting outstanding performances with respect to derivatization of hydroxymethyl resin and Rink amide resin with minimal racemization. For instance, esterification loading of Wang resin by

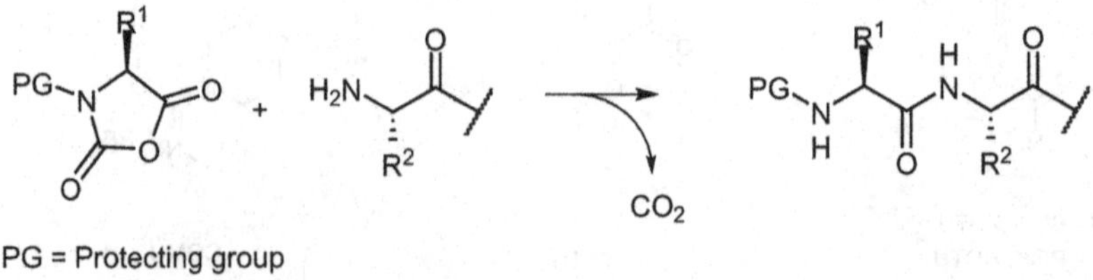

FIGURE 11.17 **UNCA mediated amino acid coupling.**

Fmoc-His(Trt)-OH, that is notoriously known for high racemization inclination, with various coupling reagents (e.g., DCC/DMAP, CIP, BOI, DIC/DMAP and PyBroP) resulted in poor yields and high extent of racemization whereas Fmoc-His(Trt)-NCA in combination with NMM as base produced 70–100% yields and almost undetectable racemization (<0.3%) in this process.[153]

11.3.3.6 Miscellaneous Coupling Reagents

DEPBT, a phosphate-type coupling reagent that consists of HODhbt moiety, has found applications in peptide synthesis particularly in the event of racemization-prone coupling reactions. The proposed mechanism of amino acid activation and coupling by DEPBT is demonstrated in Fig. 11.18.[154]

Thanks to the inherent property of HODhbt ester DEPBT is preferably utilized in stepwise synthesis or segment condensation where loss of chiral integrity upon carboxylate activation poses to be an issue with ordinary coupling reagents. DEPBT could exhibit superior property with respect to restrict racemization under certain circumstances compared with PyBroP, HATU, HBTU and BOP.[154] In an investigation focusing on the coupling of chiral PNA DEPBT outperforms TDBTU, HBTU and even HATU in terms of suppressing the racemization of the subjected chiral PNA building block during its activation and coupling.[155] The utility of DEPBT, however, is somewhat limited by the distinctive side reaction that is attributed to HODhbt ester moiety (see also Section 5.7).

Coupling reagents derived from CDMT (2-chloro-4,6-dimethoxy[1,3,5]triazine), e.g., DMTMM have also been utilized in peptide synthesis (Fig. 11.19) thanks to its significant cost advantages[156] and ability to suppress amino acid racemization.[157] DMTMM exhibits excellent performances with regard to sterically hindered peptide synthesis, and under certain circumstances outperforms

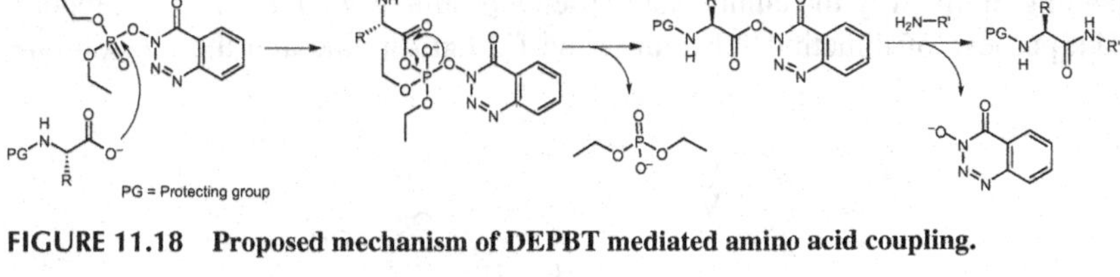

FIGURE 11.18 Proposed mechanism of DEPBT mediated amino acid coupling.

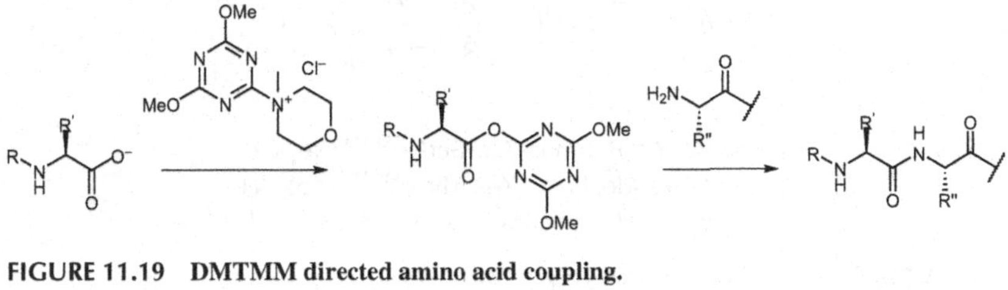

FIGURE 11.19 DMTMM directed amino acid coupling.

HBTU/HOBt, BOP-Cl and CDMT in terms of racemization and triazinylation side reaction, respectively.[158] Another DMTMM advantage is reflected by its capability to regulate amino acid coupling reactions in alcoholic or even aqueous solvents with good yields without suffering significantly from alcoholysis and hydrolysis compared with DCC-mediated analogous reactions.[159]

11.3.4 Coupling Tactics

11.3.4.1 Pseudoproline

It is known that N^{α}-urethane-protected proline is substantially less susceptible to racemization upon carboxylate activation compared with other chiral amino acids due to the fact that N^{α}-urethane-protected proline is not prone to form the corresponding oxazolone intermediate. As a consequence, proline racemization via oxazolone formation pathway is largely restrained. This distinctive property of proline has been exploited to come up with the concept of pseudoproline derived from amino acid precursors bearing β-hydroxyl substituent, e.g., Ser and Thr, or β-sulfhydryl amino acid, e.g., Cys. The utility of pseudoproline building blocks has been intensively developed.

The benefits of the pseudoproline strategy in peptide synthesis are predominantly evinced by the following two aspects: Firstly, the incorporation of pseudoproline unit into the target peptide chain might be able to evidently modify the secondary structure of the affected peptide chain. For instance, β-sheet conformation that is attributed to intra- or intermolecular aggregation might be disrupted by the insertion of the pseudoproline moiety into the peptide chain.[160] The rigidity of the peptide chain might be substantially alleviated, and the efficiencies of the following amino acid couplings can be consequently enhanced. In view of the inherent difficulty of pseudoproline acylation on solid support, this moiety is normally preformed in solution synthesis in the manner of the corresponding dipeptide Fmoc-Xaa-Ser/Thr/Cys($\Psi^{Me,Me}$ Pro)-OH (Fig. 11.20), bypassing in this way the cumbersome pseudoproline acylation on solid support. The purpose of dimethyl substitution on *C*2 oxazolidine and the thiazolidine

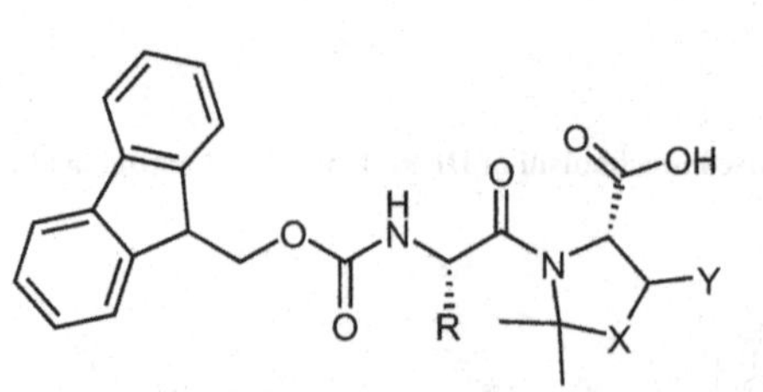

X = O, Y = H, Fmoc-Xaa-Ser($\psi^{Me,Me}$ Pro)-OH
X= O, Y= Me, Fmoc-Xaa-Thr($\psi^{Me,Me}$ Pro)-OH
X= S, Y = H, Fmoc-Xaa-Cys($\psi^{Me,Me}$ Pro)-OH

FIGURE 11.20 Fmoc-Xaa-Ser/Thr/Cyc($\psi^{Me,Me}$ Pro)-OH.

ring is to increase the acid lability of the corresponding pseudoproline derivative so that the regeneration of Ser/Thr/Cys from their pseudoproline precursors could be readily completed by TFA treatment.[160] Pseudoproline dipeptides have been successfully utilized in numerous cases of difficult peptide synthesis, and their performances will not be elaborated herein.

Secondly, the introduction of pseudoproline strategy evidently enhances the applicability of peptide segment condensation methodology for the construction of large peptides from the corresponding reciprocal peptide fragments. It is known that peptide segment condensation tactics are normally restricted to the *N*-component peptide segment that is located by a Gly or Pro at its *C*-terminus. Otherwise high susceptibility of racemization on the *C*-terminal residue in the condensation process might invalidate the synthetic strategy. The absence of an imperative Gly or Pro residue at the appropriate location in the target peptide sequence will inevitably restrict the utility of peptide segment condensation strategy to construct large peptide from its fragment precursors. This inherent limitation, is nevertheless, at least partially overcome by the introduction of pseudoproline methodology. Ser, Thr, and Cys can be regarded as the synthon of the requested proline residue, and the choice of the proper site for the segment condensation is thus broadened. Not only Gly/Pro, but also Ser/Thr/Cys could serve as the junction residue of the segment condensation. For the latter group they could be incorporated into the peptide segment in the form of the corresponding pseudoproline. This synthetic strategy could not only increase the solubility of the subjected segment and render it more applicable for the following condensation, but also, more importantly, tremendously reduce the risk of the *C*-terminal amino acid to suffer from racemization at the condensation step.[160]

In summary, pseudoproline methodology drastically enriches the repertoire of synthetic strategies for a large peptide that is less amenable for standard peptide preparation. The introduction of a solubilizing pseudoproline moiety into the target peptide sequence could significantly increase the efficacies of the following amino acid couplings. The employment of pseudoproline as a peptide building block can also reinforce the adaptability of peptide segment condensation tactics for the assembly of a large peptide by circumventing the nuisance of racemization that will probably interfere with an ordinary segment condensation that is void of Gly or Pro at the *C*-terminal position on the *N*-component peptide segment.

11.3.4.2 Natural Chemical Ligation

NCL offers a complementary synthetic strategy to the peptide segment condensation methodologies with respect to large peptide construction. One of the most noteworthy features of NCL relative to conventional convergent condensation lies in the fact that the *C*-terminal residue on the *N*-component fragment could be exempted from racemization in the ligation process. The restriction that the *C*-terminal residue is preferably occupied by Gly or Pro

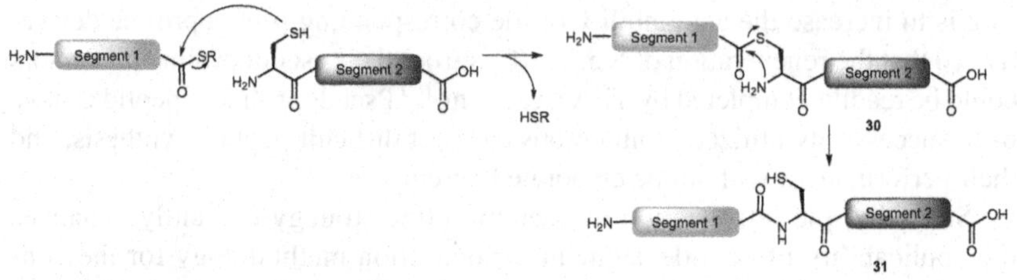

FIGURE 11.21 Mechanism of natural chemical ligation.

in the conventional chemical segment condensation is therefore bypassed. In fact, systematic feasibility tests of model peptide construction by means of NCL with 20 proteinogenic amino acids on the *C*-terminal position of the *N*-component fragment have been conducted, and the liability of the NCL strategy has been verified by this systematic study.[161] The ligation kinetics differs nevertheless substantially due to the steric hindrance of the side chains on the *C*-terminal thioester residues.

The mechanism of peptide segment condensation via NCL is elucidated in Fig. 11.21. The *N*-component fragment participates in the ligation process in the manner of peptide thioester, and the *C*-component fragment with side chain-unprotected Cys residue located at its *N*-terminus. The natural chemical ligation process starts with a reversible transthioesterification between the thioester moiety, and the sulfhydryl side chain on the Cys residue at neutral pH[162] in a chemoselective manner, giving rise to a thioester intermediate **30**, which undergoes rapid intramolecular *S*→*N* acyl rearrangement via a favorable five-member ring transition state to form the native peptide bond between the *N*- and *C*-component fragment and generate the target peptide product **31**.[163]

The advantages of NCL over conventional ligation strategy are reflected in many aspects. Firstly, NCL reactions usually proceed smoothly without major side reactions. Secondly, NCL is normally conducted between the side chain unprotected fragments in buffered aqueous solution in a highly chemical selective manner, thus bypassing the solubility limitation that normally precludes the ligation under high concentration conditions. Moreover, NCL can be conducted in the presence of denaturing agents, e.g., urea or sodium dodecylsulfonate (SDS) without noticeable interferences, which might substantially increase the solubility of the subjected peptide fragments[164] and accelerate the kinetics of the concerned condensation reaction.

NCL is per se a racemization-free process. However, loss of chiral integrity could occur at the step of peptide thioester preparation from its corresponding carboxylate precursor.[165] A plethora of peptide thioester preparative strategies has been hitherto developed to stress this intrinsic limitation. Many of them take advantage of intramolecular *O*→*S* acyl shift[166] that precludes the necessity of peptide *C*-terminal activation for thioester formation, thus reinforcing the applicability of the NCL strategy for the construction of complicated peptides. Besides, peptide thioester could also be readily derived by means of SPPS facilitated by resin loaded with Fmoc-chemistry-compatible alkanesulfonamide "safety-catch" linker.[167]

11.3.5 Solvent

In light of the mechanism of amino acid racemization in peptide synthesis indicated in Figs 11.1–11.3 it could be inferred that racemization in these processes will be affected by the attributes of solvent, predominantly from the aspect of solvent polarity since all these racemization processes proceed via a charged intermediate. Indeed the choice of organic solvent for peptide synthesis exerts significant impacts on the outcomes of amino acid racemization. It has been clearly exhibited by a variety of investigations that amino acid racemization in the activation and coupling process could be reduced by the addition of DCM to the routinely utilized DMF and NMP as cosolvent for the concerned reactions. This finding is applicable to the amino acid couplings mediated by HATU, DCC,[168] DIC,[169] UNCA,[170] BOP, PyBOP and TBTU.[171] It is to note that the choice of solvent for peptide synthesis influences both the kinetics of amino acid activation/coupling and the level of racemization. Sometimes the effects thereof might be counteractive. Generally speaking synthesis of a short peptide in solution could be suitable for DCM or $CHCl_3$.[172] For some racemization-susceptible synthesis, e.g., peptide segment condensation[168] or Fmoc-Cys(X)-OH activation[9] DCM/DMF mixture as solvent mixture is beneficial to suppress the configuration inversion of the *C*-terminus amino acid from the *N*-component peptide fragment. One of the disadvantages related with the utilization of DCM as organic solvent for peptide synthesis is reflected by the relatively low solubility of some of the coupling reagents. The addition of fluorinated alcohols, e.g., TFE and HFIP to $CHCl_3$, has been applied in solution phase peptide fragment condensation[173] as well as SPPS.[168] Its advantageous effects with regard to racemization suppression have been evidently manifested in certain cases.[174]

Although it is hard to generalize the correlation between amino acid racemization and the components of organic solvents, it is advisable to screen the appropriate solvent or solvent mixtures in combination with other parameters in case the extent of the racemization of certain amino acids in the process of peptide synthesis reaches unacceptable levels.

11.3.6 Base

Tertiary organic bases are normally needed to facilitate the couplings of amino acids, in case the amino groups to be acylated are in the form of salt, or coupling reagents such as phosphonium or uronium salts are opted to regulate the subjected amino acid activation. These auxiliary tertiary bases actually are endowed with dual effects in the process of amino acid activation and coupling. On one side the base is capable of catalyzing and accelerating the target reaction; on the other side, the involvement of bases in the process of amino acid activation will induce and intensify the amino acid racemization regulated by the mechanism of the oxazolone intermediate formation or direct H^{α} abstraction. As exhibited in Figs 11.1 and 11.3, the base directly contributes to the occurrence of racemization in the process of amino acid activation. Hence, the choice of tertiary bases will significantly influence the extent of amino acid racemization in peptide synthesis.

Basically speaking, tertiary bases with relatively weaker basicity and larger steric hindrance will be helpful to diminish the extent of amino acid racemization upon activation. DIEA is the most widely utilized organic base for peptide assembly. In case of severe amino acid racemization tertiary bases other than DIEA should be employed in an effort to decrease the extent of amino acid racemization. NMM could exhibit moderately to evidence improved performance in terms of racemization suppression under certain circumstances compared with DIEA[9,175] whereas sometimes the effect might even be reversed.[9] Thanks to its relatively weaker basicity ($pK_a = 7.4$) and distinct steric effect 2,4,6-collidine is frequently utilized in place of DIEA or NMM for the racemization-prone amino acid coupling. It has been verified in abundant cases that 2,4,6-collidine is able to substantially reduce the extent of the racemization of susceptible amino acids such as Cys,[9] compared with DIEA and NMM. Likewise, 2,4,6-collidine is also evidently effective to alleviate racemization occurring to the *C*-terminal amino acid in the process of convergent peptide segment condensation.[175] 2,4,6-collidine outperforms other applicable tertiary bases, e.g., DIEA, NMM, NMI, and dimethylpyridine (lutidine) in terms of suppressing amino acid racemization during carboxylate activation.[175] On the other hand, the amount of base applied in the amino acid coupling could also impact on the outcomes of the racemization. Less base input would help reduce the extent of amino acid racemization.[9,175,176]

Rational screening of the organic tertiary base in combination with an appropriate choice of coupling reagents and solvents is applicable to minimize amino acid/peptide racemization in the process of carboxylate activation.

11.3.7 Amino Acid Activation Mode

Amino acid preactivation is a strategy that is frequently adopted in peptide synthesis, particularly in large-scale peptide manufacturing, in order to minimize peptide *N*-terminal endcapping by excess coupling reagents introduced by a high local concentration upon the coupling reagent addition. Preactivation of the to-be-incorporated amino acid by stoichiometric, or even slightly substoichiometric coupling reagents, could consume the coupling reagents in advance and avoid the undesired derivatization on the to-be-acylated amino groups. Although established as a standard process for peptide manufacturing the preactivation of amino acid will elongate the existence period of the activated species of N^{α}-protected amino acid, this will consequently increase the risk of the concerned amino acid to suffer from racemization via either oxazolone formation or a direct H^{α} abstraction mechanism.[168,177] It is proven that the avoidance of amino acid preactivation could reduce the extent of racemization[9] since the formed amino acid active species will be consumed *in situ* by the unacylated N^{α} group, that minimizes their existence period and reduces the risk of racemization as a consequence. Although it is not recommended to skip the preactivation step in large-scale peptide manufacturing, for special cases in which an amino acid is substantially susceptible to racemization the *in situ* activation mode could be tested and implemented in the manufacturing process in case no undesired N^{α}-endcapping is induced.

11.3.8 Temperature

Although peptide synthesis at elevated temperatures has been tested in a variety of investigations and obtained success,[178] the correlated higher risk of the concerned amino acid to suffer from racemization cannot be ignored. Indeed some racemization processes seem to be less affected by temperature under certain circumstances,[73,168] nonetheless, it has been well characterized by a plethora of works that the elevated temperature results in an enhanced level of amino acid racemization,[179] and in case of severe racemization, amino acid activation at reduced temperature could be adopted in order to minimize this side reaction.[180] Preactivation could also be conducted at lower temperatures in the circumstances when racemization of the concerned amino acid in the process of ambient temperature preactivation is severe.

11.3.9 Cu(II) Salt Additive

It has been reported that Cu(II) salt, e.g., $CuCl_2$, could be utilized as an additive in the carbodiimide coupling reagent-mediated amino acid activation/coupling reaction in order to minimize the extent of the racemization side reaction occurring in this process.[181,182] This synthetic strategy is even capable of effectively alleviating racemization of the *N*-methyl amino acid that is substantially susceptible to racemization.[43] Moreover, DIC/$Cu(OBt)_2$ has been successfully employed as a coupling reagent/additive in the preparation of a peptide with the reversed direction ($N \rightarrow C$)[183] which is normally significantly prone to suffer from racemization at the step of peptide *C*-terminal carboxylate activation step. The inherent disadvantage of $CuCl_2$-involved carbodiimide-regulated amino acid coupling is reflected by the unnegligible yield drop in some cases.[43] Preformed Cu(II) salt, e.g., $Cu(OBt)_2$ and $Cu(OAt)_2$ could substantially suppress amino acid racemization even with prolonged preactivation treatment.[184] These Cu(II) salt additives could be utilized together with carbodiimide, phosphonium or uronium coupling reagents.

REFERENCES

1. (a) Lee DL, Powers J-PS, Pflegerl K, Vasil ML, Hancock REW, Hodges RS. *J Pept Res.* 2004;6:69–84. (b) McInnes C, Kondejewski LH, Hodges RS, Sykes BD. *J Biol Chem.* 2000;275:14287–14294. (c) Chalmers DK, Marshall GR. *J Am Chem Soc.* 1995;117: 5927–5937.
2. (a) Khosla MC, Hall MM, Smeby RR, Bumpus FM. *J Med Chem.* 1973;16:829–832. (b) Kondejewski LH, Jelokhani-Niaraki M, Farmer SW, et al. *J Biol Chem.* 1999;274: 13181–13192.
3. Goodman M, McGahren WJ. *Tetrahedron.* 1967;23:2031–2050.
4. Griehl C, Kolbe A, Merkel S. *J Chem Soc Perkin Trans.* 1996;2:2525–2529.
5. (a) Benoiton NL, Chen FMF. *Can J Chem.* 1981;59:384–389. (b) Jones JH, Witty MJ. *J Chem Soc.* 1979;1:3203–3206:Perkin Trans.
6. Bodanszky M. *Pept Res.* 1992;5:134–139.
7. Friedman M, Agric J. *Food Chem.* 1999;47:3457–3479.
8. Bodanszky M, Bodanszky A. *Chem Commun (London).* 1967;591–593.
9. Han Y, Albericio F, Barany G. *J Org Chem.* 1997;62:4307–4312.
10. Harding SJ, Jones JH, Sabirov AN, Samukov VV. *J Pept Sci.* 1999;5:368–373.
11. Geiger T, Clarke S. *J Biol Chem.* 1987;262:785–794.
12. Bada JL. *Methods Enzymol.* 1984;106:98–115.
13. Radkiewicz JL, Zipse H, Clarke S, Houk KN. *J Am Chem Soc.* 1996;118:9148–9155.
14. Manning JM, Moore S. *J Biol Chem.* 1968;243:5591–5597.
15. Kaiser K, Benner R. *Limnol Oceanogr Methods.* 2005;3:318–325.
16. Liardon R, Ledermann S, Ott U. *J Chromatogr.* 1981;203:385–395.
17. Merrifield RB. *J Am Chem Soc.* 1963;85:2149–2154.
18. Sieber P, Riniker B. *Tetrahedron Lett.* 1987;28:6031–6034.
19. Veber DF. In: Walter R, Meinenhofer J, eds. *Peptides: Chemistry, Structure and Biology.* Ann Arbor, MI: Ann Arbor Science; 1975:307.
20. Mergler M, Dick F, Sax B, Schwindling J, Vorherr TH. *J Pept Sci.* 2001;7:502–510.
21. Brown T, Jones JH, Richards JD. *J Chem Soc Perkin Trans.* 1982;1:1553–1561.
22. Kemp DS. Gross E, Meinenhofer J, eds. *The Peptides: Analysis, Synthesis, Biology*, Vol. 1. New York: Academic Press; 1979:315.
23. Kovacs J, Mayers GL, Johnson RH, Cover RE, Ghatak UR. *J Org Chem.* 1970;35: 1810–1815.
24. Musiol H-J, Siedler F, Quarzogo D, Moroder L. *Biopolymers.* 1994;34:1553–1562.
25. Behrendt R, Schenk M, Musiol H-J, Moroder L. *J Pept Sci.* 1999;5:519–529.
26. Sieber P. *Tetrahedron Lett.* 1987;28:6147–6150.
27. Atherton E, Benoiton NL, Brown E, Sheppard RC, Williams BJ. *J Chem Soc Chem Commun.* 1981;336–337.
28. (a) Atherton E, Hardy PM, Harris DE, Matthews BH. In: Giralt E, Andreu D, eds. *Peptides 1990.* Leiden: ESCOM; 1991:243. (b) Rietman BH, Peters RFR, Tesser GI. *Recl Trav Chim Pays-Bas.* 1995;114:1–5.
29. Spiro RG. *Adv Protein Chem.* 1973;27:349–467.
30. Kornfeld R, Konfeld S. *Ann Rev Biochem.* 1985;54:631–664.
31. Van den Steen P, Rudd PM, Dwek RA, Opdenakker G. *Crit Rev Biochem Mol Biol.* 1998;33:151–208.
32. Dondoni A, Marra A. *Chem Rev.* 2000;100:4395–4422.
33. Weiss JB, Lote CJ. *Nature New Biol.* 1971;234:25–26.

34. (a) Benjamin GD. *Chem Rev*. 2002;102:579–602. (b) Jung K-H, Möller M, Schmidt RR. *Chem Rev*. 2000;100:4423–4442. (c) Herzner H, Reipen T, Schultz M, Kunz H. *Chem Rev*. 2000;100:4495–4583.
35. (a) Kottenhahn M, Kessler H. *Liebigs Ann Chem*. 1991;727–744. (b) Andrews DM, Seale PW. *Int J Pept Protein Res*. 1993;42:165–170.
36. Bejugam M, Flitsch SL. *Org Lett*. 2004;6:4001–4004.
37. Di Fenza A, Tancredi M, Galoppini C, Rovero P. *Tetrahedron Lett*. 1998;39:8529–8532.
38. Bada JL, Shou M-Y. In: Hare PE, Hoering TC, King K, eds. *Biochemistry of Amino Acids*. New York: Wiley; 1980:235.
39. Grant GS, Reddy GV. *J Org Chem*. 1989;54:4529–4535.
40. Zhang Y, Muthana SM, Farnsworth D, et al. *J Am Chem Soc*. 2012;134:6316–6325.
41. McDermott JR, Benoiton NL. *Can J Chem*. 1973;51:2562–2570.
42. McDermott JR, Benoiton NL. *Can J Chem*. 1973;51:2555–2561.
43. Nishiyama Y, Tanaka M, Saito S, Ishizuka S, Mori T, Kurita K. *Chem Pharm Bull*. 1999;47:576–578.
44. Davies JS, Mohammed K. *J Chem Soc Perkin Trans*. 1981;1:2982–2990.
45. Sureshbabu VV, Narendra N. In: Hughes AB, ed. *Amino Acids, Peptides and Proteins in Organic Chemistry. Vol. 4. Protection Reactions, Medicinal Chemistry and Combinatorial Synthesis*. Weinheim: Wiley; 2011:7.
46. Sewald N. *Angew Chem Int Ed*. 2002;41:4661–4663.
47. Polese A, Formaggio F, Crisma M, et al. *Chem Eur J*. 1996;2:1104–1111.
48. Leitgeb B, Szekeres A, Manczinger L, Vágvölgyi C, Kredics L. *Chem Biodivers*. 2007;4:1027–1051.
49. Sammes PG, Weller DJ. *Synthesis*. 1995;1205–1222.
50. Wipf P, Heimgartner H. *Helv Chim Acta*. 1986;69:1153–1162.
51. Weygand F, Prox A, König W. *Chem Ber*. 1966;99:1451–1460.
52. Williams RM, Hendrix JA. *Chem Rev*. 1992;92:889–917.
53. Kohli RM, Walsh CT, Burkart MD. *Nature*. 2002;418:658–661.
54. Smith GG, Sivakua T. *J Org Chem*. 1983;48:627–634.
55. Elsawy MA, Hewage C, Walker B. *J Pept Sci*. 2012;18:302–311.
56. Shen B, Makley DM, Johnston JN. *Nature*. 2010;465:1027–1033.
57. Berry DJ, DiGiovanna CV, Metrick SS, Murugan R. *Arkivoc*. 2001;i:201–226.
58. Coin I. *J Pept Sci*. 2010;16:223–230.
59. Nielsen J, Lyngsø LO. *Tetrahedron Lett*. 1996;37:8439–8442.
60. Frérot E, Coste J, Pantaloni A, Dufour M-N, Jouin P. *Tetrahedron*. 1991;47:259–270.
61. Patgiri A, Witten MR, Arora PS. *Org Biomol Chem*. 2010;8:1773–1776.
62. Mergler M, Nyfeler R, Tanner R, Gosteli J, Grogg P. *Tetrahedron Lett*. 1988;29:4009–4012.
63. Van Nispen JW, Polderdijk JP, Greven HM. *Recl Trav Chim Pays-Bas*. 1985;104:99–100.
64. Cabrele C, Langer M, Beck-Sickinger AG. *J Org Chem*. 1999;64:4353–4361.
65. Granitza D, Beyermann M, Wenschuh H, et al . *J Chem Soc Chem Commun*. 1995;2223–2224.
66. Albericio F, Bailén MA, Chinchilla R, Dodsworth DJ, Nájera C. *Tetrahedron*. 2001;57: 9607–9613.
67. Rodríguez H, Suarez M, Albericio F. *J Pept Sci*. 2010;16:136–140.
68. Matsushita T, Hinou H, Fumoto M, et al. *J Org Chem*. 2006;71:3051–3063.
69. Seo J, Michaelian N, Owens SC, et al. *Org Lett*. 2009;11:5210–5213.
70. Brandt M, Gammeltoft S, Jensen J. *Int J Pept Res Ther*. 2006;12:349–357.
71. Bacsa B, Horváti K, Bõsze S, Andreae F, Kappe CO. *J Org Chem*. 2008;73:7532–7542.
72. Palasek SA, Cox ZJ, Collins JM. *J Pept Sci*. 2007;13:143–148.

73. Souza MP, Tavares MFM, Miranda MTM. *Tetrahedron*. 2004;60:4671–4681.
74. Gracia C, Isidro-Llobet A, Cruz LJ, et al. *J Org Chem*. 2006;71:7196–7204.
75. Lloyd-Williams P, Albericio F, Giralt E. *Tetrahedron*. 1993;49:11065–11133.
76. Kaiser T, Nicholson GJ, Kohlbau HJ, Woelter W. *Tetrahedron Lett*. 1996;37:1187–1190.
77. Tesser GI, Balvert-Geers IC. *Int J Pept Protein Res*. 1975;7:295–305.
78. Samukov VV, Sabirov AN, Pozdnyakov PI. *Tetrahedron Lett*. 1994;35:7821–7824.
79. Albericio F. *Pept Sci. 55*. 2000;123–139.
80. Wunderlin R, Minakakis P, Tun-Kyi A, Sharma SD, Schwyzer R. *Helv Chim Acta*. 1985;68:1–11.
81. Boon PJ, Tesser GI. *Int J Pept Protein Res*. 1985;25:510–516.
82. Harjunpää I, Kuusela P, Smoluch MT, Silberring J, Lankinen H, Wade D. *FEBS Lett*. 1999;449:187–190.
83. Harding SJ, Heslop I, Jones JH, Wood ME. In: Maia HLS, ed. *Proceedings of the Twenty-Third European Peptide Symposium*. Leiden: ESCOM; 1995:189.
84. Carreno C, Méndez ME, Kim Y-D, et al. *J Peptide Res*. 2000;56:63–69.
85. Zervas L, Borovas D, Gazis E. *J Am Chem Soc*. 1963;85:3660–3666.
86. Poduška K. *Collect Czech Chem Commun*. 1968;33:3779–3789.
87. Zervas L, Hamalidis C. *J Am Chem Soc*. 1965;87:99–104.
88. Kessler W, Iselin B. *Helv Chim Acta*. 1966;49:1330–1344.
89. Mukaiyama T, Ueki M, Maruyama H, Matsueda R. *J Am Chem Soc*. 1968;90:4490–4491.
90. Moroder L, Gemeiner M, Goehring W, Jaeger E, Thamm P, Wünsch E. *Biopolymers*. 1981;20:17–37.
91. Bednarek MA, Bodanszky M. *Int J Pept Protein Res*. 1995;45:64–69.
92. Juillerat M, Bargetzi JP. *Helv Chim Acta*. 1976;59:855–866.
93. Bodanszky M, Meienhofer J, du Vigneaud V. *J Am Chem Soc*. 1960;82:3195–3198.
94. Sureshbabu VV, Narendra N. In: Hughes AB, ed. *Amino Acids, Peptides and Proteins in Organic Chemistry. Vol. 4. Protection Reactions, Medicinal Chemistry, Combinatorial Synthesis*. Weinheim: Wiley-VCH; 2011:10–11.
95. Carpino LA, Ionescu D, El-Faham A, et al. *Tetrahedron Lett*. 1998;39:241–244.
96. Fukuyama T, Jow C-K, Cheung M. *Tetrahedron Lett*. 1995;36:6373–6374.
97. Seitz O. *Angew Chem Int Ed*. 1998;37:3109–3111.
98. Weinreb SM, Demko DM, Lessen TA, Demers JP. *Tetrahedron Lett*. 1995;19:2099–2102.
99. Leggio A, Belsito EL, de Marco R, Liguori A, Perri F, Viscomi MC. *J Org Chem*. 2010;75:1386–1392.
100. Miller SC, Scanlan TS. *J Am Chem Soc*. 1998;120:2690–2691.
101. Barlos K, Mamos P, Papaioannou D, Patrianakou S, Sanida C, Schäfer W. *Liebigs Ann Chem*. 1987;1025–1030.
102. de la Torre B, Marcos MA, Eritja R, Albericio F. *Lett Pept Sci*. 2001;8:331–338.
103. Alsina J, Giralt E, Albericio F. *Tetrahedron Lett*. 1996;37:4195–4198.
104. Barlos K, Papaioannou D, Patrianakou S, Tsegenidis T. *Liebigs Ann Chem*. 1986;1950–1955.
105. Sim TB, Rapoport H. *J Org Chem*. 1999;64:2532–2536.
106. Gunnarsson K, Grehn L, Ragnarsson U. *Angew Chem Int Ed*. 1988;100:411–412.
107. Šavrda J, Chertanova L, Wakselman M. *Tetrahdron*. 1994;50:5309–5322.
108. Carpino LA, Mansour E-SME, El-Faham A. *J Org Chem*. 1993;58:4162–4164.
109. Weigelt S, Huber T, Hofmann F, et al. *Chem Eur J*. 2012;18:478–487.
110. Nepomniaschiy N, Grimminger V, Cohen A, DiGiovanni S, Lashuel HA, Brik A. *Org Lett*. 2008;10:5243–5246.
111. Tedebark U, Meldal M, Panza L, Bock K. *Tetrahedron Lett*. 1998;39:1815–1818.

112. Lundquist JT, Pelletier JC. *Org Lett*. 2001;3:781–783.
113. DalPozzo A, Ni M, Muzi L, et al. *J Org Chem*. 2002;67:6372–6375.
114. Tornøe CW, Davis P, Porreca F, Meldal M. *J Pept Sci*. 2000;6:594–602.
115. Kovacs J, Cortegiano H, Cover RE, Mayers GL. *J Am Chem Soc*. 1971;93:1541–1543.
116. Fujiwara Y, Akaji K, Kiso Y. *Chem Pharm Bull*. 1994;42:724–726.
117. Moroder L, Musiol H-J, Schaschke N, Chen L, Hargittai B, Barany G. In: Goodman M, Toniolo C, Moroder L, Felix A, eds. *Methods in Organic Chemistry. Vol. E22a. Synthesis of Peptides and Peptidomimetics*. Stuttgart and New York: Georg Thieme Verlag; 2002:390–391.
118. Isidro-Llobet A, Álvarez M, Albericio F. *Chem Rev*. 2009;109:2455–2504.
119. Kumagaye KY, Inui T, Nakajima K, Kimura T, Sakakibara S. *Pept Res*. 1991;4:84–87.
120. Jones JH, Ramage WI. *J Chem Soc Chem Commun*. 1978:472–473.
121. Barlos K, Chatzi O, Gatos D, Stravropoulos G, Tsegenidis T. *Tetrahedron Lett*. 1991;32:475–478.
122. Kusunoki M, Nakagawa S, Seo K, Hamana T, Fukuda T. *Int J Pept Protein Res*. 1990;36:381–386.
123. Kalström A, Undén A. *Chem Commun*. 1996:959–960.
124. Gesquière J-C, Najib J, Letailleur T, Maes. P, Tartar A. *Tetrahedron Lett*. 1993;34:1921–1924.
125. Klausner Y, Bodanszky M. *Synthesis*. 1974:549–559.
126. Determann H, Wieland T. *Justus Liebigs Ann Chem*. 1963;670:136–140.
127. Kemp DS, Wang SW, Busby III G, Hugel G. *J Am Chem Soc*. 1970;92:1043–1055.
128. Sieber P, Brugger M, Rittel W. In: Nesvadba H, ed. *Peptides 1971*. Amsterdam: North-Holland; 1973:49.
129. Mayer SC, Joullié MM. *Syn Commun*. 1994;24:2367–2377.
130. Milton SCF, de Lisle Milton RC, Kates SA, Glabe C. *Lett Pept Sci*. 1996;6:151–156.
131. Wenschuh H, Beyermann M, Krause E, et al. *J Org Chem*. 1994;59:3275–3280.
132. Wenschuh H, Beyermann M, El-Faham A, Ghassemi S, Carpino LA, Bienert M. *J Chem Soc Chem Commun*. 1995:669–670.
133. Carpino LA, Mansour E-SME, Sadat-Aalaee D. *J Org Chem*. 1991;56:2611–2614.
134. Gunnarsson K, Ragnarsson U. *Acta Chem Scand*. 1990;44:944–951.
135. Thern B, Rudolph J, Jung G. *Angew Chem Int Ed*. 2002;41:2307–2309.
136. Thern B, Rudolph J, Jung G. *Tetrahedron Lett*. 2002;43:5013–5016.
137. Falb E, Yechezkel T, Salitra Y, Gilon C. *J Pept Res*. 1999;53:507–517.
138. Coste J, Frérot E, Jouin P. *Tetrahedron Lett*. 1991;32:1967–1970.
139. Carpino LA. *J Am Chem Soc*. 1993;115:4397–4398.
140. Wehrstedt KD, Wandrey PA, Heitkamp D. *J Hazard Mater*. 2005;126:1–7.
141. Sabatino G, Mulinacci B, Alcaro MC, Chelli M, Rovero P, Papini AM. *Lett Pept Sci*. 2002;9:119–123.
142. Di Fenza A, Rovero P. *Lett Pept Sci*. 2002;9:125–129.
143. Knorr R, Trzeciak A, Bannwarth W, Gillessen D. *Tetrahedron Lett*. 1989;30:1927–1930.
144. König W, Geiger R. *Chem Ber*. 1970;103:2024–2033.
145. Hiebl J, Alberts DP, Banyard AF, et al. *J Pept Res*. 1999;54:54–65.
146. Subirós-Funosas R, Prohens R, Barbas R, El-Faham A, Albericio F. *Chem Eur J*. 2009;15:9394–9403.
147. König W, Breipohl G, Pokorny P, Birkner M. In: Giralt E, Andreu D, eds. *Peptides 1990*. Leiden: ESCOM; 1991:143.
148. El-Faham A, Subirós-Funosas R, Prohens R, Albericio F. *Chem Eur J*. 2009;15:9404–9416.
149. Fuller WD, Goodman M, Naider FR, Zhu Y-F. *Pept Sci*. 1996;40:183–205.
150. Konopi ska D, Siemion IZ. *Angew Chem Int Ed Engl*. 1967;6:248.

151. Auvin-Guette C, Frérot E, Coste J, Rebuffat S, Jouin P, Bodo B. *Tetrahedron Lett.* 1993;34:2481–2482.
152. Zhu Y-F, Fuller WD. *Tetrahedron Lett.* 1995;36:807–810.
153. Zhu Y-F, Blair RK, Fuller WD. *Tetrahedron Lett.* 1994;35:4673–4676.
154. Li H, Jiang X, Ye Y, Fan C, Romoff T, Goodman M. *Org Lett.* 1999;1:91–94.
155. Tedeschi T, Corradini R, Marchelli R, Pushl A, Nielsen PE. *Tetrahedron Asymmetr.* 2002;13:1629–1636.
156. Falchi A, Giacomelli G, Porcheddu A, Taddei M. *Synlett.* 2000;275–277.
157. Kunishima M, Kitao A, Kawachi C, et al. *Chem Pharm Bull.* 2002;50:549–550.
158. Shieh W-C, Chen Z, Xue S, et al. *Tetrahedron Lett.* 2008;49:5359–5362.
159. Kunishima M, Kawachi C, Hioki K, Terao K, Tani S. *Tetrahedron.* 2001;57:1551–1558.
160. Wöhr T, Wahl F, Nefzi A, et al. *J Am Chem Soc.* 1996;118:9218–9227.
161. Hackeng TM, Griffin JH, Dawson PE. *Proc Natl Acad Sci USA.* 1996;96:10068–10073.
162. Haase C, Seitz O. *Angew Chem Int Ed.* 2008;47:1553–1556.
163. Hackenberger CPR, Schwarzer D. *Angew Chem Int Ed.* 2008;47:10030–10074.
164. Valiyaveetil FI, MacKinnon R, Muir TW. *J Am Chem Soc.* 2002;124:9113–9120.
165. Nagalingam AC, Radford SE, Warriner SL. *Synlett.* 2007:2517–2520.
166. (a) Botti P, Villain M, Manganiello S, Gaertner H. *Org Lett.* 2004;6:4861–4864. (b) Warren JD, Miller JS, Keding SJ, Danishefsky SJ. *J Am Chem Soc.* 2004;126:6576–6578. (c) Tofteng AP, Jensen KJ, Hoeg-Jensen T. *Tetrahedron Lett.* 2007;48:2105–2107.
167. (a) Shin Y, Winas KA, Backes BJ, Kent SBH, Ellman JA, Bertozzi CR. *J Am Chem Soc.* 1999;121:11684–11689. (b) Ingenito R, Bianchi E, Fattori D, Pessi A. *J Am Chem Soc.* 1999;121:11369–11374.
168. Carpino LA, El-Faham A, Albericio F. *Tetrahedron Lett.* 1994;35:2279–2282.
169. Haver AC, Smith DD. *Tetrahedron Lett.* 1993;34:2239–2242.
170. Romoff TT, Goodman M. *J Pept Res.* 1997;49:281–292.
171. Benoiton NL, Lee YC, Steinaur R, Chen FMF. *Int J Pept Protein Res.* 1992;40:559–566.
172. (a) Kamenecka TM, Danishefsky SJ. *Angew Chem Int Ed.* 1998;37:2993–2995. (b) Jou G, González I, Albericio F, Lloyd-Williams P, Giralt E. *J Org Chem.* 1997;62:354–366. (c) Sin N, Meng L, Auth H, Crews CM. *Bioorg Med Chem.* 1998;6:1209–1217.
173. Nishiuchi Y, Inui T, Nishio H, et al. *Proc Natl Acad Sci USA.* 1998;95:13549–13554.
174. Riniker B, Flörsheimer A, Fretz H, Sieber P, Kamber B. *Tetrahedron.* 1993;49:9307–9320.
175. Carpino LA, El-Faham A. *J Org Chem.* 1994;59:695–698.
176. Forest M, Fournier A. *Int J Pept Protein Res.* 1990;35:89–94.
177. Schnölzer M, Alewood P, Jones A, Alewood D, Kent SBH. *Int J Pept Protein Res.* 1992;40:180–193.
178. Tam JP, Lu Y-A. *J Am Chem Soc.* 1995;117:12058–12063.
179. Coste J, Frérot E, Jouin P. *J Org Chem.* 1994;59:2437–2446.
180. Jarrett JT, Lansbury Jr PT. *Tetrahedron Lett.* 1990;31:4561–4564.
181. Miyazawa T, Otomatsu T, Fukui Y, Yamada T, Kuwata S. *Int J Pept Protein Res.* 1992;39:237–244.
182. Ryadnov MG, Klimenko LV, Mitin YV. *J Pept Res.* 1999;53:322–328.
183. Thieriet N, Guibé F, Albericio F. *Org Lett.* 2000;2:1815–1817.
184. Nest WVD, Yuval S, Albericio F. *J Pept Sci.* 2001;7:115–120.

Chapter 12

Side Reactions in Peptide Phosphorylation

Posttranslational modifications of proteins by means of phosphorylation, sulfation or glycosylation endow the affected proteins with distinctive physiological and biochemical functions and/or reinforcement of certain significant protein conformation. Principally these protein modifications could be realized by chemical derivatization on target functional groups.

Protein phosphorylations, for instance, are frequently applied in nature to regulate enzyme activities[1,2] and to create recognition areas for the desirable aggregation of multiprotein complexes.[3,4] These events are utilized in a variety of signal transduction pathways and controls of cell cycles. Aberrations in protein phosphorylation and dephosphorylation processes are associated with diseases such as diabetes[5] and cancer.[6] Study of these mechanisms constitutes a major field in modern biochemistry and medicine. Furthermore, phosphorylated proteins also play important roles with their distinctive calcium phosphate moieties, and consequently serve as nucleators or regulators of biomineralization.[7]

Most of the protein phosphorylation processes occur on hydroxyl-bearing amino acid residues serine and threonine and to a lesser extent, on the phenolic side chain of tyrosine. Generally speaking, methodology of the chemical synthesis of phosphopeptides falls into two broad categories: (1) utilization of preformed phosphoamino acids as building blocks to assemble the target phosphopeptides regulated by Fmoc or Boc chemistry; (2) postassembly modifications of the substrate hydroxyl substituents on the unprotected Ser/Thr/Tyr residues from the addressed peptide precursors in order to convert them globally to the corresponding phosphate counterparts.[8] Nevertheless, a variety of side reactions could be aroused in the processes of peptide phosphorylation, some of them are characteristic to phosphopeptide synthesis.

12.1 FORMATION OF *H*-PHOSPHONATE SIDE PRODUCT

Global phosphorylation is regarded as an important strategy for the chemical synthesis of phosphopeptides. Peptide precursors with side chain-unprotected Ser/Thr/Tyr residues are subjected to global phosphorylation treatment, either in solution[9] or on solid support,[10] and are uniformly converted to their

Side Reactions in Peptide Synthesis. http://dx.doi.org/10.1016/B978-0-12-801009-9.00012-4

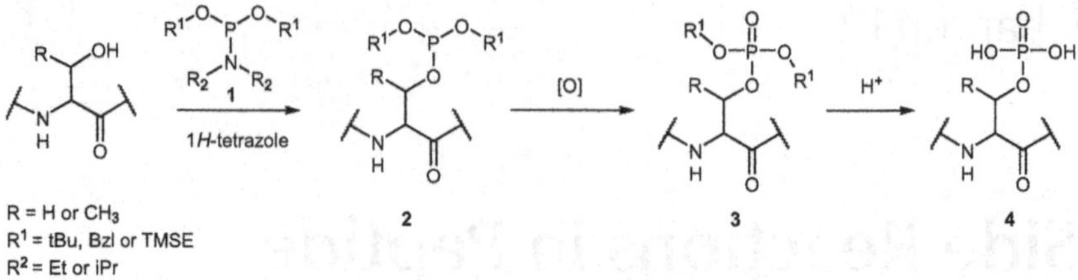

FIGURE 12.1 **Scheme of phosphopeptide synthesis via phosphitylation and oxidation treatment.**

phosphorylated counterparts. Basically, processes of peptide global phosphorylation could be divided into two consecutive synthetic steps: phosphitylation and oxidation. The concerned hydroxyl-containing phosphorylation substrates are firstly converted to their dialkyl phosphite triester derivative **2** at the phosphitylation step by phosphoramidite **1** treatment facilitated by acidic catalyst, e.g., 1*H*-tetrazole (Fig. 12.1). The most commonly applicable phosphoramidite are: $(tBuO)_2PNR_2$, $(BzlO)_2PNR_2$ and $(TMSEO)_2PNR_2$ (R = Et, iPr).[11,12] The generated phosphite triester **2** is subsequently oxidized by tBuOOH, mCPBA or aqueous iodine to the corresponding protected phosphate **3**, which undergoes acidolytic deprotection to give rise to the final phosphopeptide **4**.

The global peptide phosphorylation process is nevertheless inherently accompanied with the formation of *H*-phosphonate side product with a −16 amu molecular weight difference relative to the target phosphopeptide product.[13,14] The structure of the concerned impurity has been unambiguously elucidated as *H*-phosphonate derivative.[15] It has been assumed that the formation of *H*-phosphonate is due to the insufficient oxidation of the phosphite triester intermediate after phosphitylation step since it was observed that the deliberate omission of the oxidation after phosphitylation would significantly intensify the *H*-phosphonate formation.[13] It was discovered later on, however, that this side reaction is actually attributed to the elongated exposure of the phosphite triester intermediate to the excess acidic 1*H*-tetrazole at the phosphitylation step, resulting in the rearrangement to the *H*-phosphonate diester derivative.[16,17]

The extent of *H*-phosphonate formation would be evidently enhanced if the oxidation step following phosphitylation is delayed. This observation also justifies the assertion that *H*-phosphonate formation precedes the addition of the oxidant.[17] It has been revealed that the formation of *H*-phosphonate side product is subject to the following three conditions:[17] (1) an extended delay of the oxidation after phosphitylation; (2) exposure of the phosphite triester intermediate to water; and (3) exposure of the phosphite triester intermediate to acid. Rapid evolution of isobutylene has been observed upon heating tri-*tert*-butyl phosphite derivative which eventually leads to the quantitative formation of the corresponding di-*tert*-butyl phosphonate counterpart.[18] Based on this

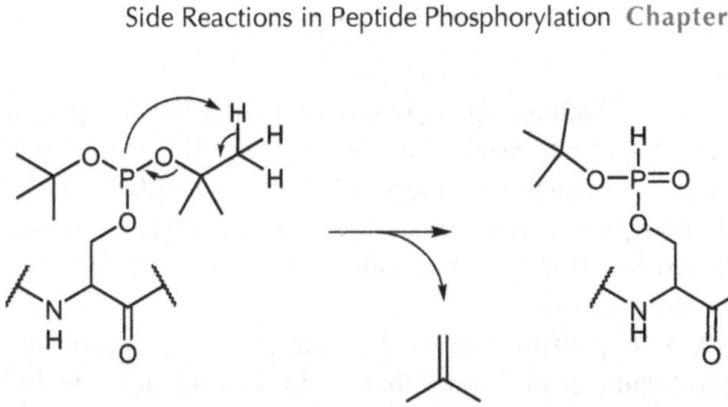

FIGURE 12.2 Proposed mechanism of *H*-phosphonate formation from di-*tert*-butyl phosphite.

observation it was inferred that *H*-phosphonate formation is inherently attributed to the lability of *tert*-butyl phosphite derivative to thermolytic or acidolytic degradation provoked by the presence of the tetrazole derivative. The proposed mechanism of *H*-phosphonate formation is indicated in Fig. 12.2. Indeed, it has been confirmed by a number of studies that benzyl and halobenzyl phosphite is much less susceptible to *H*-phosphonate formation than their *tert*-butyl or TMSE phosphite counterparts.[16,17]

An alternative mechanism of *H*-phosphonate formation involves the participation of two water molecules. This is in compliance with the observation that water could facilitate *H*-phosphonate formation, and the rate of *H*-phosphonate formation from tripropyl phosphite is second order in reactant water, and first order in reactant phosphite. Moreover, experiments conducted in ^{18}O-enriched water proved that a phosphorous–oxygen bond on tripropyl phosphite was broken in the process of phosphite hydrolysis.[19] A complex transition state comprising one molecule of phosphite and two molecules of water is firstly formed that subsequently undergoes phosphite hydrolysis. As a consequence, the derived hydrolytic product in this process rapidly tautomerizes to the more stable *H*-phosphonate form (Fig. 12.3).

The *H*-phosphonate formation at the step of peptide global phosphorylation could affect serine, threonine and tyrosine peptides.[20] It has been justified by a variety of studies that *H*-phosphonate formation is directly attributed to the

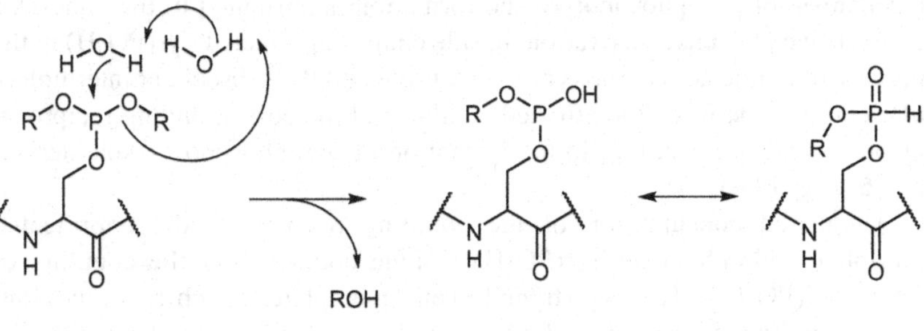

FIGURE 12.3 Proposed mechanism of water-involved *H*-phosphonate formation.

acidic conditions to which phosphite is subjected after the phosphitylation step in the presence of acidic tetrazole compounds. Actually, peptide phosphonate compounds could be intentionally prepared simply through the TFA treatment of the peptide phosphite precursors derived from phosphitylation of the side chain unprotected Ser/Thr/Tyr containing peptides by appropriate phosphoramidite and tetrazole catalyst.[21]

Conversion of *H*-phosphonate to the corresponding phosphate could be realized by the treatment of *H*-phosphonate by I_2 oxidation.[8] In light of this reaction it was recommended to treat the phosphite intermediate prepared at dibenzylphosphoramidite-mediated phosphitylation step by I_2 oxidation as to circumvent the *H*-phosphonate formation side reaction.[16] Alternatively, phosphopeptide could also be synthesized from its *H*-phosphonate precursor through I_2 oxidation.[22] The detrimental effects of I_2 on oxidizable residues, e.g., Cys, Met, and Trp should be nonetheless taken into account when this kind of synthetic strategy is concerned. Besides, the prompt oxidation treatment of phosphite intermediate directly after phosphitylation step is crucial for suppressing the formation of the *H*-phosphonate side product. The holding time between phosphitylation and oxidation steps should be kept minimal. This concern is particularly applicable to large-scale production.

12.2 FORMATION OF PYROPHOSPHATE SIDE PRODUCT

In light of the inherent advantages of utilization of side chain-unprotected phosphotyrosine $Tyr(PO_3H_2)$ over monoprotected $Tyr(PO_3R,H)$ (R = alkyl) and diprotected $Tyr(PO_3R_2)$ counterparts, the former has been successfully employed as a building block for the preparation of phosphotyrosine peptides.[13] These merits are more distinctively reflected for the synthesis of small- to medium-length simple phosphotyrosine peptides.

The incorporation of side chain-unprotected phosphotyrosine residues, in spite of the mentioned advantages, could not be regarded as problem-free due to various intrinsic side reactions. One of the most evident challenges in this connection is the pyrophosphotyrosine formation.[23] The side product generated in this process might have a molecular weight of [2M-18] (M refers to the molecular weight of the target phosphotyrosine-containing peptide). The mechanism of pyrophosphotyrosine formation is attributed to the undesired competitive phosphate activation on side chain-unprotected $Tyr(PO_3H)$ in the process of amino acid carboxylate activation, and the subsequent intermolecular coupling between the affected residue and the corresponding reciprocal phosphotyrosine, resulting in the formation of pyrophosphotyrosine derivative **5** (Fig. 12.4).

Undesired consumption of the coupling reagent by the unprotected phosphate side chain on $Tyr(PO_3H)$ residue could render the coupling of Fmoc-$Tyr(PO_3H)$-OH sluggish, and even incomplete, which requests elongated coupling times and/or double couplings. It was figured out that the

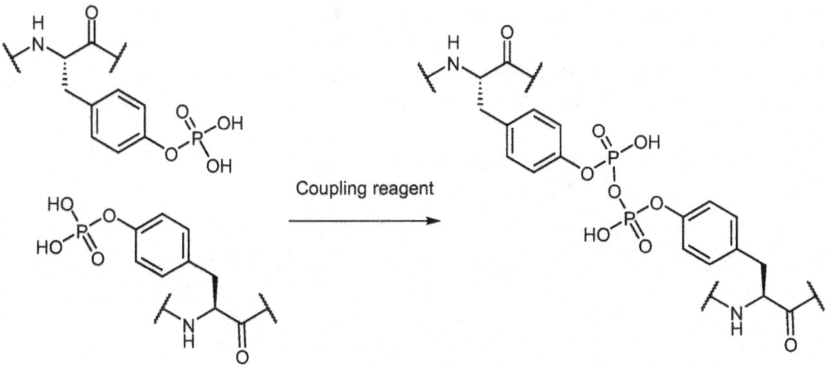

FIGURE 12.4 Scheme of pyrophosphotyrosine formation.

susceptibility of unprotected phosphate side chain to the undesired activation is governed by the type of the coupling reagents. Hence meticulous screening of the activation conditions could diminish the extent of pyrophosphate formation. However, the most effective strategy to restrain this side reaction is the substitution of the unprotected phosphotyrosine by its phosphate monobenzyl ester counterpart. The shielding effect of benzyl substituent on the subjected phosphate moiety can maximally inhibit the activation of the phosphate group by the coupling reagent, and consequently the formation of the pyrophosphate side product.

REFERENCES

1. Burnett G, Kennedy EP. *J Biol Chem.* 1954;211:969–980.
2. Cozzone AJ. *Ann Rev Microbiol.* 1988;42:97–125.
3. Yaffe MB. *Nat Rev Mol Cell Biol.* 2002;3:177–186.
4. Ingham RJ, Colwill K, Howard C, et al. *Mol Cell Biol.* 2005;25:7092–7106.
5. Kenner KA, Anyanwu E, Olefsky JM, Kusari J. *J Biol Chem.* 1996;271:19810–19816.
6. Dobrusin EM, Fry DW. *Ann Rep Med Chem.* 1992;27:169–178.
7. Marsh ME, Sass RL. *Biochemistry.* 1984;23:1448–1456.
8. McMurray JS, Coleman DRJIV, Wang W, Campbell ML. *Biopolymers.* 2001;60:3–31.
9. Perich JW, Johns RB. *Aust J Chem.* 1990;43:1623–1632.
10. Perich JW, Johns RB. *Tetrahedron Lett.* 1988;29:2369–2372.
11. de Bont HBA, Liskamp RMJ, O'Brian CA, Erkellens C. *Int J Pept Protein Res.* 1989;33:115–123.
12. Andrews DM, Kitchin J, Seale PW. *Int J Pept Protein Res.* 1991;38:469–475.
13. Ottinger EA, Shekels LL, Bernlohr DA, Barany G. *Biochemistry.* 1993;32:4354–4361.
14. Poteur L, Trifilieff E. *Lett Pept Sci.* 1995;2:271–276.
15. Austin B, Hawkes J. In: Epton R, ed. *Innovation and Perspectives in Solid Phase Synthesis.* Birmingham: Mayflower Worldwide; 1994:437–440.
16. Perich JW. *Lett Pept Sci.* 1998;5:49–55.
17. Xu Q, Ottinger EA, Solé NA, Barany G. *Lett Pept Sci.* 1996;3:333–342.

18. Mark V, van Wazer JR. *J Org Chem*. 1964;29:1006–1008.
19. Aksnes G, Aksnes D. *Acta Chem Scand*. 1964;18:1623–1628.
20. Hoffmann R, Wachs WO, Berger RG, et al. *Int J Pept Protein Res*. 1995;45:26–34.
21. Kuyl-Yeheskiely E, Tromp CM, van der Marel GA, van Boom JH. *Tetrahedron Lett*. 1987;28:4461–4464.
22. Kupihár Z, Kele Z, Tóth GK. *Org Lett*. 2001;3:1033–1035.
23. García-Echeverría C. *Lett Pept Sci*. 1995;2:93–98.

Chapter 13

Cys Disulfide-Related Side Reactions in Peptide Synthesis

Disulfide moiety is ubiquitously located in many peptides and plays important roles in the establishment and reinforcement of the peptide/protein overall structure, as well as the regulation of the activity of the peptide/protein. Due to the inherent instability of the disulfide bond toward a variety of conditions it could undergo divergent side reactions, e.g., thiol-disulfide exchange that reversibly modifies the pattern of the disulfide bonds and hence the activity of the affected peptide/protein molecule. On top of this process disulfide bridge moiety could also suffer from other degradation such as hydrolysis, β-elimination or relatively infrequent α-degradation, leading to the formation of degradative side products like trisulfide or lanthionine derivatives.

13.1 DISULFIDE SCRAMBLING VIA THIOL-DISULFIDE EXCHANGE

Thiol-disulfide exchange constitutes a core pathway for disulfide formation in living systems, providing a cornerstone of catalyzed protein disulfide-bond formation in all organisms from prokaryotes to eukaryotes.[1] A group of enzymes denoted as disulfide oxidoreductases catalyze the process of thiol-disulfide exchange reactions *in vivo* among which protein disulfide isomerase (PDI) was one of the firstly identified entities.[2] Depending on the redox environment and substrate protein properties, PDI could regulate the processes such as disulfide bond formation, disulfide bond reduction and disulfide bond isomerization via the thiol-disulfide exchange pathway.[3]

Actually, many chemical synthetic peptide disulfide derivatives are prepared by means of decent exploitation of the corresponding thiol-disulfide exchange process, e.g., disulfide formation with dithiasuccinoylglycine,[4] 2,2′-dithiodipyridine,[5] 5,5′-dithiobis(2-nitrobenzoic acid), and di[3-nitro(2-pyridyl)]disulfide.[6]

In spite of the versatility of the thiol-disulfide exchange synthetic strategy in the context of disulfide bond formation this process could induce undesired disulfide bond scrambling side reactions in peptide synthesis, or during the storage of the disulfide bond-containing peptide/protein materials. The root cause of disulfide scrambling is principally attributed to the uncontrolled reactions

Side Reactions in Peptide Synthesis. http://dx.doi.org/10.1016/B978-0-12-801009-9.00013-6

FIGURE 13.1 Thiol-disulfide exchange-induced disulfide bond scrambling.

between the free thiolate derivatives and disulfide moiety, giving rise to the formation of a new disulfide and a thiolate (Fig. 13.1).

Disulfide scrambling through thiol-disulfide exchange is in principle an S_N2 reaction.[7] Normally only thiolate could regulate such a process. In an acidic milieu where thiolate is converted to the corresponding thiol derivative via protonation the potential disulfide scrambling side reaction would be largely suppressed. The inclination of the occurrence of disulfide scrambling is understandably partially decided by the reactivity of the concerned thiolate which inherently depends on the dual properties of its thiol precursor with respect to its basicity and nucleophilicity.[8,9] Leaving out the steric hindrance effect the thiol derivative with a higher p*K*a value is expected to display a stronger nucleophilic property. However, as the actual reactant in the thiol-disulfide exchange is believed to be thiolate anion, the thiol derivative with a lower p*K*a value will be more prone to undergo deprotonation and get involved in the referred S_N2 reaction. Taking these two contradictory factors into account the most reactive thiol compounds toward thiol-disulfide exchange are those with p*K*a values close to the pH of the reaction solution.[8,9] It has been observed that base could catalyze this reaction when the p*K*a value of the participated thiol reactant is higher than the pH of the solution since the extent of thiolate formation is decided by the solution pH.[10]

Preparations of disulfide bond-rich peptides by means of chemical synthesis frequently induce various extents of disulfide scrambling side reactions.[11] For example, the unmatched thiolate moiety from one peptide could initiate the thiol-disulfide exchange with a disulfide bond from another molecule or the same peptide, giving rise to a mismatched S—S bond that covalently oligomerizes individual peptide chains (Fig. 13.2).

A redox reagent such as GSH/GSSG-directed cysteine oxidation could promote the generation of the thermodynamically most stable native cystine connectivity via reshuffling of the disulfide bridge networks. However, the same mechanism of thiol-disulfide exchange might also result in the degradation

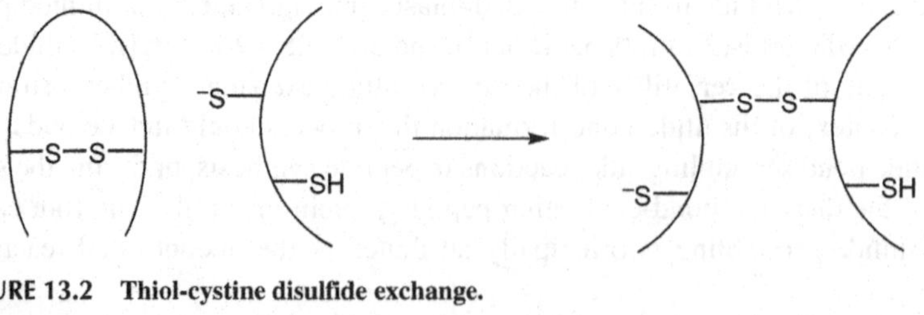

FIGURE 13.2 Thiol-cystine disulfide exchange.

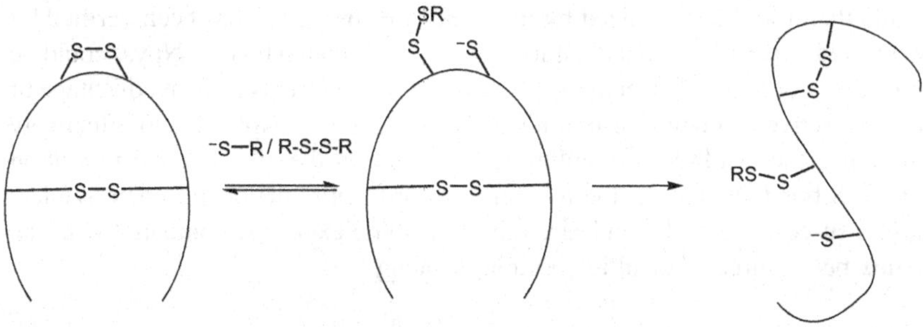

FIGURE 13.3 **Redox buffer-induced disulfide bond scrambling.**

of the established disulfide bond and the formation of a new mismatched one (Fig. 13.3). Besides redox buffer-mediated S—S bond formation air oxidation of the free cysteines in the presence of a preformed disulfide bond could also lead to disulfide scrambling via the thiol-disulfide exchange pathway.[12]

Since disulfide scrambling side reactions frequently affect disulfide bond-rich peptides, the chemical synthesis of these compounds should be exerted with the utmost caution. Following the first disulfide bond formation between two cysteines, additional sequential inter- or intra-chain disulfide bond formation would have to be conducted in the presence of the preformed disulfide bridge. The potential thiol-disulfide exchange could hence not be overlooked under such circumstances.

Regioselective stepwise establishment of multiple disulfide networks is regarded as an appropriate strategy to circumvent the potential thiol-disulfide exchange side reactions. For example, site-specific activation of the Cys residue in the form of Cys(Npys) and its subsequent reaction with the free complementary Cys would form the target disulfide in a regioselective manner. The driving force of disulfide bond formation between free Cys and Cys(Npys) is the low p*Ka* and high leaving tendency of aromatic 3-nitropyrdine-2-thiol that is eventually tautomerized to the inert 3-nitropyridine-2(1*H*)-thione[13] upon displacement from its Cys(Npys) active precursor (Fig. 13.4).

In light of the distinctive character that the unsymmetrical disulfide bridging between Cys and Cys(Npys) could take place in slightly acidic conditions, the undesired thiol-disulfide exchange reactions prevailing in the basic environments

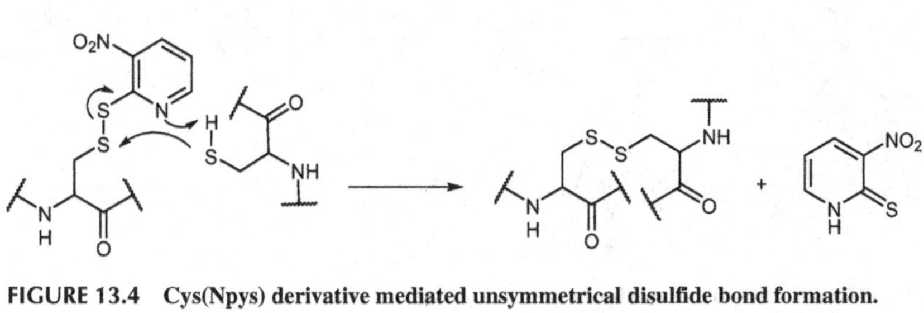

FIGURE 13.4 **Cys(Npys) derivative mediated unsymmetrical disulfide bond formation.**

could therefore be minimized by this synthetic strategy. It has been verified by dedicated experiments that disulfide formation regulated by Cys(Npys) could be effectively conducted even in the pH range of 3.5–6.[14] This evident advantage of regioselective formation of multidisulfide through Cys(Npys) could effectively minimize the disulfide scrambling taking place in the air oxidative formation of S—S bond. Moreover, the tolerance toward acidic pH environment renders it less susceptible to the undesired thiol-disulfide exchange compared with the redox-buffer directed disulfide bridging strategy.

13.2 DISULFIDE DEGRADATION AND CONSEQUENT TRISULFIDE AND LANTHIONINE FORMATION

13.2.1 Disulfide Degradation Pattern

Under basic conditions labile peptide disulfide moiety could suffer from degradation via pathways like hydrolysis, β-elimination or relatively infrequent α-degradation, leading to the formation of degraded fragments such as cysteine thiolate, dehydroalanine, persulfide derivatives, HS^-, thioketone, and so forth.[15–17] The formation of these degradative intermediates could trigger further tandem side reactions as a consequence, e.g., trisulfide or lanthionine formation characterized by a sulfur intrusion and extrusion, respectively,[18] or the regeneration of cysteine as well as the formation of cysteine sulfenic acid, sulfinic acid and dehydroalanine adducts.

Generally speaking there are three patterns of cystine disulfide degradation: (1) cystine homolytic hydrolysis, (2) cystine β-elimination, and (3) cystine α-elimination.

It is believed that the direct nucleophilic attack of hydroxide on disulfide moiety could take place unless the referred reaction is prohibited by the negative charge in proximity to the concerned S—S bond, or if there is an highly acidic proton on C^{α} or C^{β} that will favor β-elimination or α-elimination pathway of disulfide degradation, respectively.[19] The scheme of cystine homolytic hydrolysis is described in Fig. 13.5. The disulfide compound undertakes

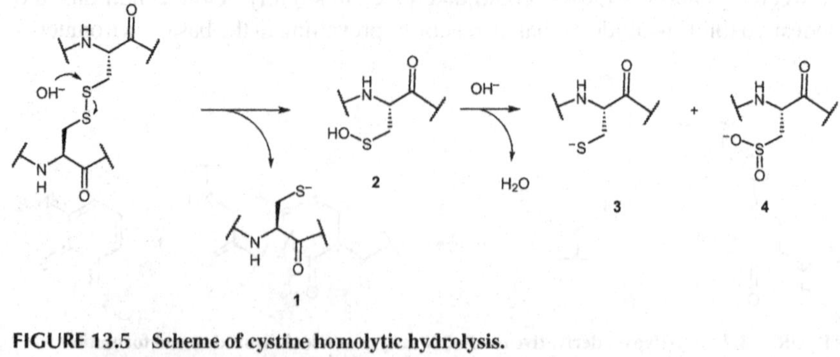

FIGURE 13.5 Scheme of cystine homolytic hydrolysis.

the nucleophilic attack of hydroxide and is degraded to cysteine thiolate **1** and cysteine sulfenic acid derivative **2**. The latter will be disproportionate to the corresponding cysteine thiolate **3** and cysteine sulfinic acid **4**.[20] For instance, an aromatic disulfide 2,2′-dinitro-5,5′-dithiodibenzoic acid, denoted as Ellman's reagent bearing no reactive hydrogen atom will opt to undergo a rapid hydrolytic decomposition to generate a thiolate and a sulfinic acid under basic conditions.[21]

Normally the cystine moiety in the parental peptide is acylated on its *N*-terminus and aminolysed on the *C*-terminus. The existence of two electron-withdrawing groups attached (-CONH- and -NHCO-) substantially enhances the acidity of the H^{α} on the concerned Cys[22] and consequently favors the disulfide degradation via β-elimination pathway.[23] The mechanism of cystine β-elimination is illustrated in Fig. 13.6. The acidic H^{α} on cystine was firstly pulled by base and the affected peptide is consequently decomposed to dehydroalanine **5** and cysteine persulfide **6**. The fates of instable persulfide are somehow in dispute. It was alleged that persulfide derivatives are readily decomposed to the corresponding thiolate **7** upon releasing elemental sulfur.[24] Contrarily it was suggested that persulfide **6** could liberate hydrogen sulfide anion HS^- in the presence of hydroxide, and be converted to the corresponding sulfenic acid **8** that is subsequently disproportionated to the commensurate thiolate **9** and sulfinic acid derivative **10**.[16]

Cystine-containing peptide could also, to a much lesser extent, be susceptible to an α-elimination degradation.[16] In this process, the H^{β} on the concerned disulfide is subtracted by base, and the affected disulfide bond is subsequently degraded into the corresponding thiolate **11** and thioaldehyde **12**. The latter is further converted to the corresponding aldehyde **13** by OH^- (Fig. 13.7).

13.2.2 Trisulfide Formation

In spite of the prevalence of disulfide cystine moiety in natural peptides and proteins, other patterns of polysulfide cystine analogues do exist in natural products. Trisulfide structures, for instance, could be found in esperamicins[25] and calicheamicins[26] as a class of antitumor antibiotics. Nevertheless, the intentional synthesis of trisulfide derivatives does not fall into the contents of this section. On the other hand, the formation of trisulfide impurities have been detected in some protein derivatives such as human IgG2 antibody,[27] superoxide dismutase,[28] a truncated form of interleukin-6 mutein,[29] and rhGH expressed from *Escherichia coli*.[30,31]

The formation of trisulfide derivatives might be confused with the oxidation of cysteine to sulfinic acid or double Met(O) formation by means of MS analysis. However, with high resolution mass spectrometry[32] it is practically realizable to differentiate the subtle distinction of molecular weight between a single S atom (31.9721 Da) and two O atoms (31.9898 Da).[27] Trisulfide impurities are normally eluted later than their disulfide counterparts from RP-HPLC due to the increased hydrophobicity attributed by the incorporated sulfur atom.[33]

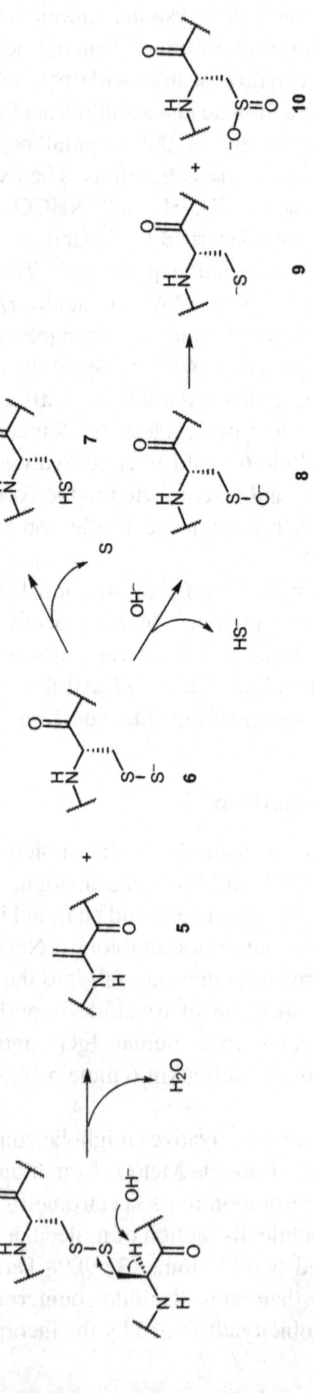

FIGURE 13.6 Scheme of cystine peptide degradation via β-elimination pathway.

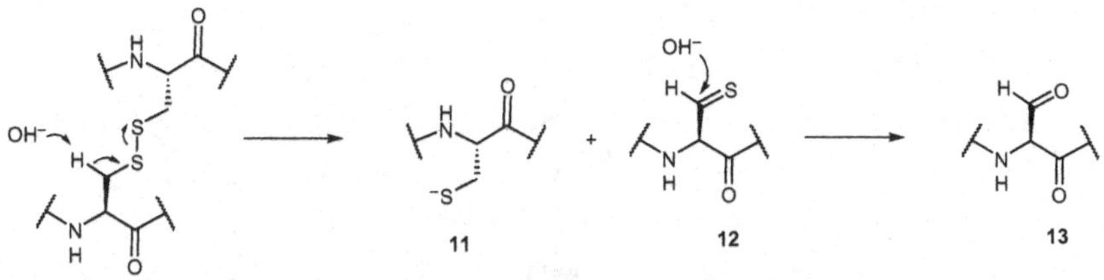

11 12 13

FIGURE 13.7 **Scheme of cystine peptide degradation via α-elimination pathway.**

The trisulfide bond is not stable in the basic environment, and it is not compatible with reagents utilized to cleave the N^{α}-Fmoc protecting group.[34] Treatment of the trisulfide derivative by reducing agents like DTT could regenerate the free thiol derivatives.[28] Cysteine- or glutathione-mediated trisulfide reduction would also liberate the free cysteine residues from its trisulfide precursor in an evident pH-dependent manner (low pH environment would retard the reduction kinetics).[27,30]

The mechanism of trisulfide formation is controversial. As is elucidated in Section 13.2.1, there are diverse patterns of disulfide degradation. Consensus has been reached that the affected disulfide-containing peptides suffer firstly from fragmentation by means of β-elimination, and subsequently give rise to a dehydroalanine and a cysteine persulfide (see also Fig. 13.6).[35] It is proposed[30,36] that the released active anionic cysteine persulfide would subsequently function with the free cysteine residue either from the same molecule or another peptide to generate the intramolecular or intermolecular trisulfide bridge, respectively (Fig. 13.8). It has also been hypothesized that H_2S released upon disulfide degradation[30,37] or during the fermentation process for the protein formation regulates the trisulfide formation since the treatment of proteins, e.g., rhGH with H_2S containing solution leads to an enhanced level of trisulfide formation.[38]

Trisulfide formation is not an uncommon side reaction in the chemical synthesis of disulfide peptides. It was reported that the utilization of solid-phase bound Ellman's reagent derivatives, e.g., solid-phase $(Npys)_2$ **14** for the regioselective formation of cystine moiety could lead to trisulfide or even tetrasulfide impurities, prevailing preferentially in acidic conditions.[39] It was proposed that the mechanism of this side reaction is probably due to the formation of the solid phase trisulfide intermediate **15** derived via C—S bond scission that is deviated from the routine thiolate-disulfide exchange reaction. The subsequent function of free thiol with compound **15** could result in different sulfide derivatives, e.g., disulfide, trisulfide and tetrasulfide (Fig. 13.9).

Other oxidants, either in the form of polymer-supported derivatives such as oligomethionine sulfoxide resin, commercially available CLEAR-OX™ resin[6] (Fig. 13.10) or free entity like Dts-Gly, have also been reported to foster the undesired trisulfide formation during the process of cysteine oxidation to the corresponding disulfide target products.[40]

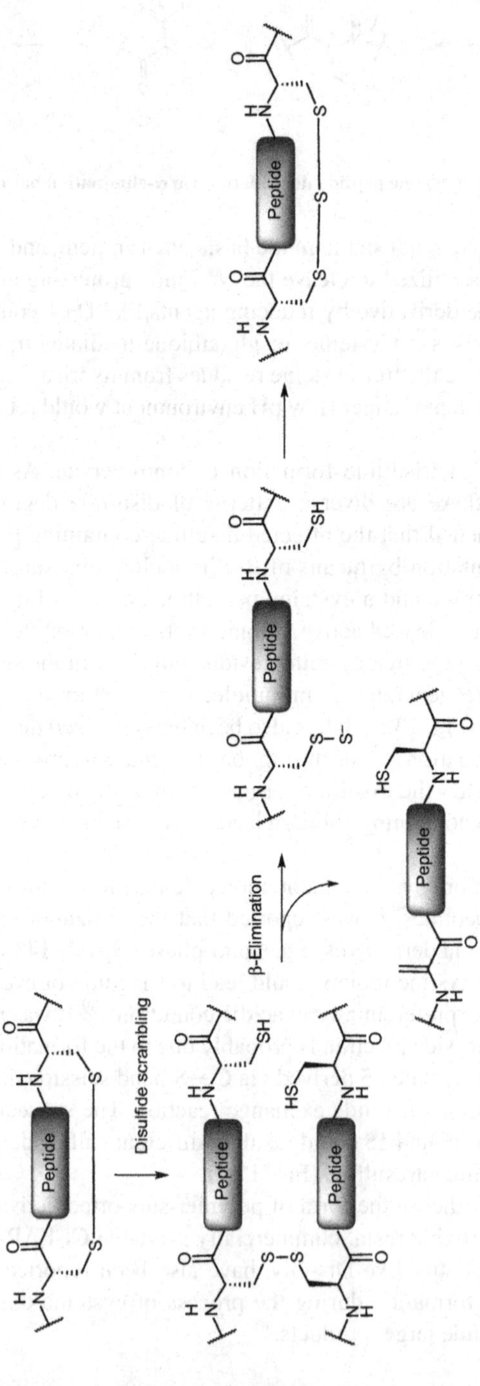

FIGURE 13.8 Proposed mechanism of trisulfide formation via disulfide β-elimination and the function between the degraded persulfide and thiol.

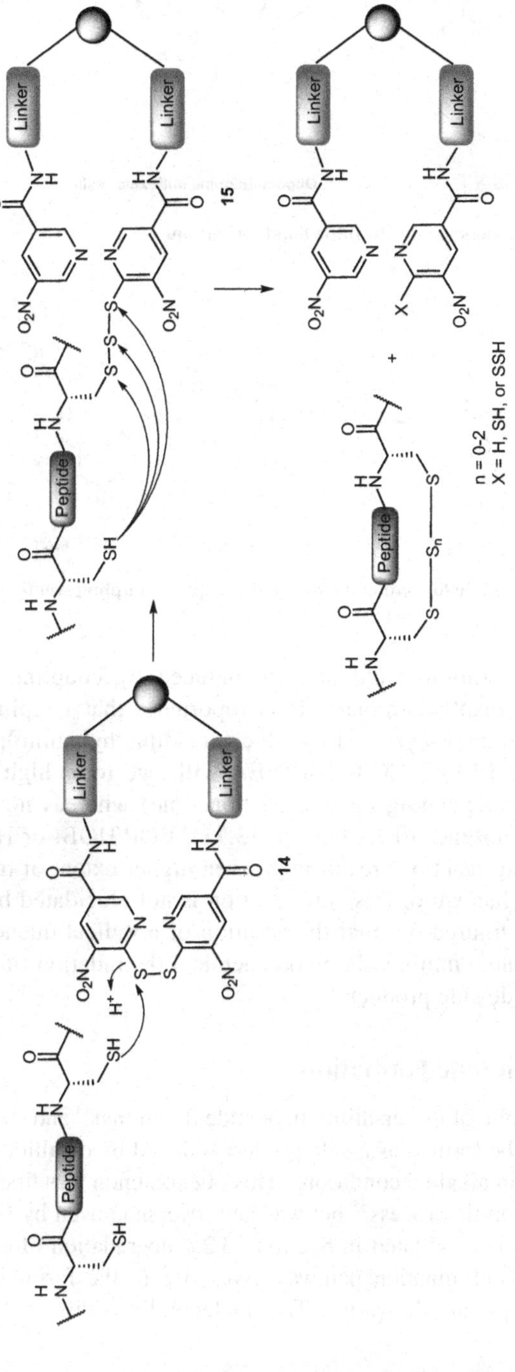

FIGURE 13.9 **Proposed mechanism of solid phase (Npys)$_2$-induced tri- and tetra-sulfide formation.**

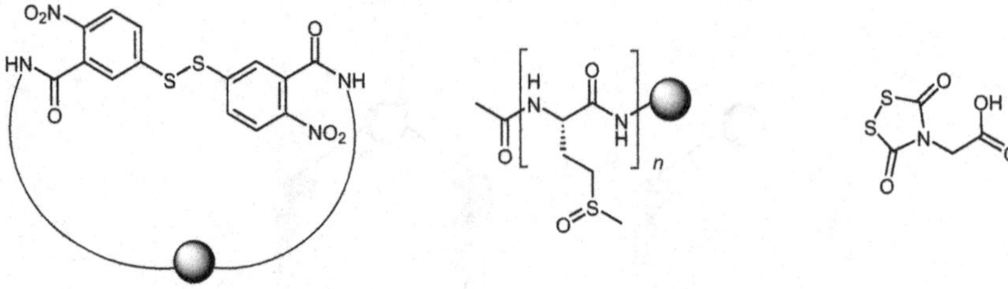

FIGURE 13.10 **Oxidants for the disulfide bond formation.**

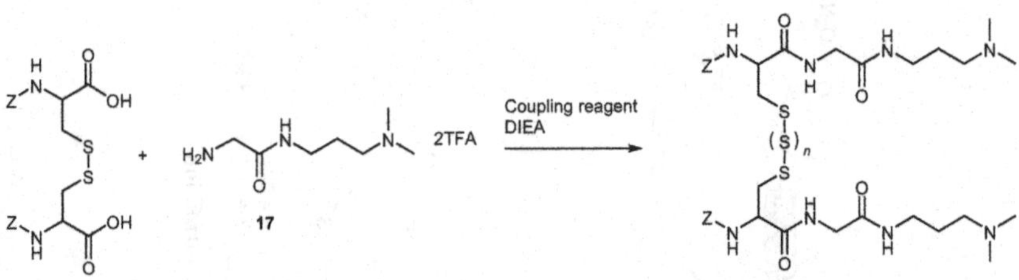

FIGURE 13.11 **Trisulfide formation during cystine peptide coupling reaction.**

Trisulfide generation could also be induced by coupling reagents in the presence of disulfide moiety. It is reported[41] that coupling of *N,N′*-bis(benzyloxycarbonyl)-cystine **16** to glycine(3-dimethylamino)propylamide ditrifluoroacetate **17** by DCC/HOBt/DIEA will lead to as high as 10% trisulfide impurity (depending on the reaction time) whereas no trisulfide is generated in the absence of DCC (Fig. 13.11). BOP/HOBt or HBTU/HOBt mediated coupling reactions result in an even higher extent of trisulfide formation. The mechanism of this side reaction is not elucidated by the investigation but it is figured out that the addition of a radical quenching agent, e.g., hydroquinone[42] inhibits the photochemical degradation of cystine that generates trisulfide side products.

13.2.3 Lanthionine Formation

Lanthionine, in spite of its versatility in peptide functional[43] and structural studies,[44] might also be formed as a side product induced by disulfide desulfurization degradation in alkaline conditions. This phenomenon was firstly explained by disulfide hydrolytic process[45] but was later overshadowed by β-elimination mechanism.[35] As is elucidated in Section 13.2.1 degradation of cystine disulfide moiety via β-elimination pathway gives rise to the formation of dehydroalanine and cysteine derivative. The nucleophilic addition of the thiolate

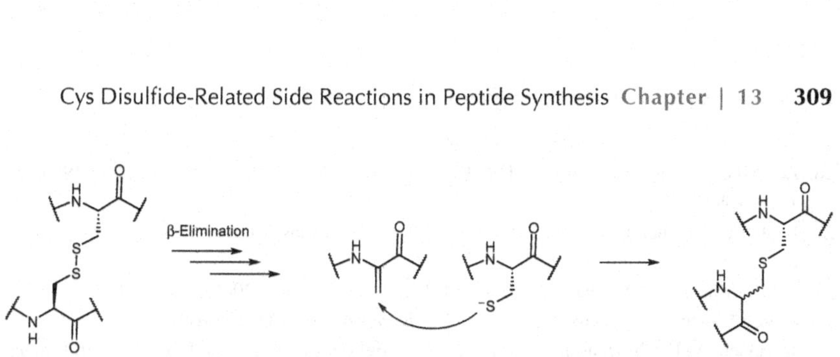

FIGURE 13.12 **Mechanism of lanthionine formation through cystine disulfide degradation.**

substituent on cysteine to the dehydroalanine olefin side chain through Michael-addition mechanism[18] will generate lanthionine moiety (Fig. 13.12). Besides thiolate on cysteine side chain, dehydroalanine could also accommodate nucleophilic attacks from other amino acids, e.g., ornithine, lysine, histidine, and tryptophan to generate the ornithinoalanine, lysinoalanine, and corresponding histidine- and tryptophan-adduct.[15]

REFERENCES

1. Sevier CS, Kaiser CA. *Nat Rev Mol Cell Biol.* 2002;3:836–847.
2. Golberger RF, Epstein CJ, Anfinsen CB. *J Biol Chem.* 1963;238:628–635.
3. Hatahet F, Ruddock LW. *Antioxid Redox Signal.* 2009;11:2807–2850.
4. Chen L, Barany G. *Lett Pept Sci.* 1996;3:283–292.
5. Maruyama K, Nagasawa H, Suzuki A. *Peptides.* 1999;20:881–884.
6. Annis I, Chen L, Barany G. *J Am Chem Soc.* 1998;120:7226–7238.
7. (a) Wilson JM, Bayer RJ, Hupe DJ. *J Am Chem Soc.* 1977;99:7922–7926. (b) Houk J, Whitesides GM. *J Am Chem Soc.* 1987;109:6825–6836.
8. Whitesides GM, Lilburn JE, Szajevski RP. *J Org Chem.* 1977;42:332–338.
9. DeCollo TV, Lees WJ. *J Org Chem.* 2001;66:4244–4249.
10. (a) Rothwarf DM, Scheraga HA. *Proc Natl Acad Sci USA.* 1992;89:7944–7948. (b) Keire DA, Strauss E, Guo W, Noszal B, Rabenstein DL. *J Org Chem.* 1992;57:123–127.
11. Browning JL, Mattaliano RJ, Chow EP, et al. *Anal Biochem.* 1986;155:123–128.
12. Ponsati B, Giralt E, Andreu D. *Tetrahedron.* 1990;46:8255–8266.
13. Lunkenheimer W, Zahn H. *Justus Liebigs Ann Chem.* 1970;740:1–17.
14. Bernatowicz MS, Matsueda R, Matsueda GR. *Int J Pept Protein Res.* 1986;28:107–112.
15. Nashef AS, Osuga DT, Lee HS, Ahmed AI, Whitaker JR, Feeney RE. *J Agric Food Chem.* 1977;25:245–251.
16. Florence TM. *Biochem J.* 1980;189:507–520.
17. Trivedi MV, Laurence JS, Siahaan TJ. *Curr Protein Pept Sci.* 2009;10:614–625.
18. Galande AK, Trent JO, Spatola AF. *Biopolymers.* 2003;71:534–551.
19. Danehy JP, Parameswaran KN. *J Org Chem.* 1968;33:568–572.
20. Nagahara N. *Amino Acids.* 2011;41:59–72.
21. Danehy JP, Elia VJ, Lavelle CJ. *J Org Chem.* 1971;36:1003–1005.
22. Cecil R, McPhee JR. *Adv Protein Chem.* 1959;14:255–389.
23. Schneider JF, Westley J. *J Biol Chem.* 1969;244:5735–5744.
24. (a) Bergmann M, Stather F. *Z Physiol Chem.* 1926;152:189–201. (b) Nicolet BH. *J Am Chem Soc.* 1931;53:3066–3072. (c) Tarbell DS, Harnish DP. *Chem Rev.* 1951;49:1–90.
25. Golik J, Clardy J, Dubay G, et al. *J Am Chem Soc.* 1987;109:3461–3462.

26. Lee MD, Manning JK, Williams DR, Kuck NA, Testa RT, Borders DB. *J Antibiot.* 1989;42:1070–1087.
27. Pristatsky P, Cohen SL, Krantz D, Acevedo J, Ionescu R, Vlasak J. *Anal Chem.* 2009;81:6148–6155.
28. Okado-Matsumoto A, Guan Z, Fridovich I. *Free Radic Bio Med.* 2006;41:1837–1846.
29. Breton J, Avanzi N, Valsasina B, et al. *J Chromatogr A.* 1995;709:135–146.
30. Jespersen AM, Christensen T, Klausen NK, Nielsen PF, Sorensen HH. *Eur J Biochem.* 1994;219:365–373.
31. Canova-Davis E, Baldonado IP, Chloupek RC, et al. *Anal Chem.* 1996;68:4044–4051.
32. Russell DH, Edmondson RD. *J Mass Spectrom.* 1997;32:263–276.
33. Erlanson DA, Wells JA. *Tetrahedron Lett.* 1998;39:6799–6802.
34. Chen L, Zoulikova I, Slaninova J, Barany G. *J Med Chem.* 1997;40:864–876.
35. Federici G, Duprè S, Matarese RM, Solinas SP, Cavallini D. *Int J Pept Protein Res.* 1977;10:185–189.
36. Windisch V, Deluccia F, Duhau L, et al. *J Pharm Sci.* 1997;86:359–364.
37. Thakur SS, Balaram P. *J Am Soc Mass Spectrom.* 2009;20:783–791.
38. Hemmendorrf B, Castan A, Persson A. Method for the production of recombinant peptides with a low amount of trisulfides. US Patent 7,232,894. June 19, 2007.
39. Annis I, Barany G. In: Fields GB, Tam JP, Barany G, eds. *Peptides for the New Millennium.* Dordrecht: Kluwer Academic; 2000:96.
40. Ronga L, Verdié P, Sanchez P, et al. *Amino Acids.* 2013;44:733–742.
41. Parmentier B, Moutiez M, Tartar A, Sergheraert C. *Tetrahedron Lett.* 1994;35:3531–3534.
42. Asquith RS, Hirst L. *Biochim Biophis Acta.* 1969;184:345–357.
43. (a) Schnell N, Entian K-D, Schneider U, et al. *Nature.* 1988;333:276–278. (b) Nutt RF, Veber DF, Saperstein RJ. *J Am Chem Soc.* 1980;102:6539–6545.
44. (a) Rosenfield Jr RE, Parthasarathy R. *J Am Chem Soc.* 1974;96:1925–1930. (b) Knerr PJ, van der Donk WA. *J Am Chem Soc.* 2013;135:7094–7097.
45. Donovan JW, White TM. *Biochemistry.* 1971;10:32–38.

Chapter 14

Solvent-Induced Side Reactions in Peptide Synthesis

Effective solvation of peptide resin is one of the most important prerequisites for efficient peptide assembly on the solid support during SPPS. The criticality of the roles of solvents in peptide synthesis is not the focus of this chapter. Conversely, various side reactions in peptide synthesis inherently triggered by organic solvents will be elaborated hereinafter.

14.1 DCM-INDUCED SIDE REACTION

Dichloromethane is an ordinary organic solvent frequently utilized in peptide synthesis. Under normal circumstances DCM maintains highly stable and will not cause significant side reactions in peptide synthesis. However, exceptions exist where DCM functions with certain reagents and interferes with the target reactions.

In spite of the numerously reported utilization of piperidine/DCM solution for Fmoc removal in peptide and other ordinary organic synthesis,[1] this strategy is nonetheless not recommendable especially for large-scale peptide production. It has been indicated in the literature that DCM is capable of reacting with primary, secondary and tertiary aliphatic amines.[2] Practically DCM could function with piperidine to generate amine salt that gradually crystallizes in the solution.[3] The mechanism of the reaction between DCM and piperidine is proposed in Fig. 14.1. It is believed that the DCM firstly undergoes an S_N2 substitution by piperidine, forming 1-(chloromethyl)piperidinium chloride **1**, which is rapidly converted to 1,1'-methylenebis(piperidinium) dichloride **2** by another molecule of piperidine, presumably via 1-methylenepiperidinium chloride **3** intermediate.[4] The precipitated solids observed in piperidine/DCM could be either 1,1'-methylenebis(piperidinium) chloride **2** or piperidinium chloride **4**.

The reaction between DCM and piperidine could consume the piperidine intended for the target reactions, resulting probably in incomplete Fmoc removal.[3] Moreover, the formation of the crystals during the piperidine/DCM-directed reaction might necessitate extra handling steps in peptide production, and is particularly unadaptable to automatic instruments. The reaction between DCM and piperidine is exothermal, and this effect is particularly adverse to large-scale peptide productions.

Side Reactions in Peptide Synthesis. http://dx.doi.org/10.1016/B978-0-12-801009-9.00014-8

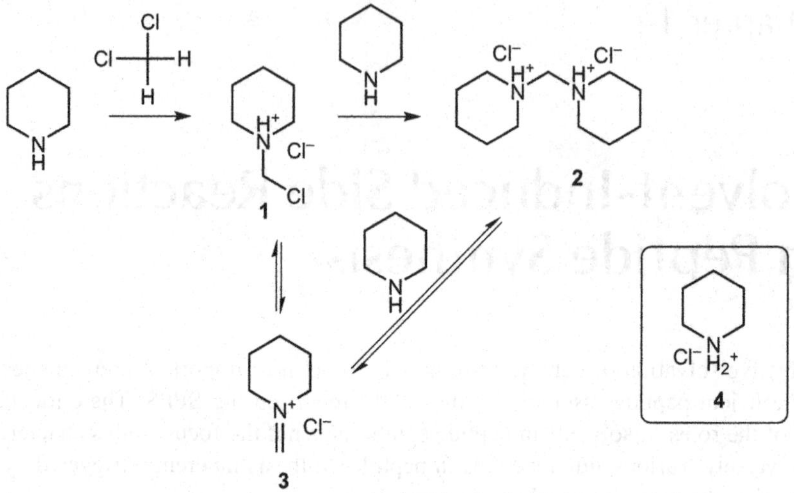

FIGURE 14.1 **Reactions between DCM and piperidine.**

Besides secondary amine like piperidine, DCM is capable of functioning with tertiary amine[5] such as pyridine and TEA. It is known that pyridine is frequently mixed with DCM in reactions such as alcohol protection and acylation.[6] However, pyridine/DCM solution will be slowly evolved at an ambient temperature into crystal identified as 1,1'-methylenebis(pyridinium) dichloride. DMAP undergoes similar conversion as pyridine but exhibits enhanced kinetics.[5]

In light of the aforementioned observations, it is recommended to avoid the utilization of DCM as the solvent for amines like piperidine, pyridine and DMAP. Caution is supposed to be exercised particularly when the concerned bases are charged in equimolar quantities to the corresponding substrates. The consumption of the amine reactants by DCM under such circumstances will evidently interfere with the target reaction. Generally speaking, DCM is not recommended as a solvent for a S_N1 and a S_N2 reaction[7] and the employment of piperidine/DCM solution should be avoided as much as possible in industrial production.

14.2 DMF-INDUCED SIDE REACTION

DMF is one of the most frequently utilized organic solvents in peptide synthesis. It could, nonetheless, induce a variety of side reactions in peptide production due to its innate properties.

14.2.1 DMF-Induced *N*-Formylpiperidine Formation

In Fmoc-mode peptide synthesis, acidic coupling additives such as HOBt are frequently charged into piperidine/DMF solution to facilitate Fmoc deblocking while suppressing the occurrence of side reactions like aspartimide formation[8]

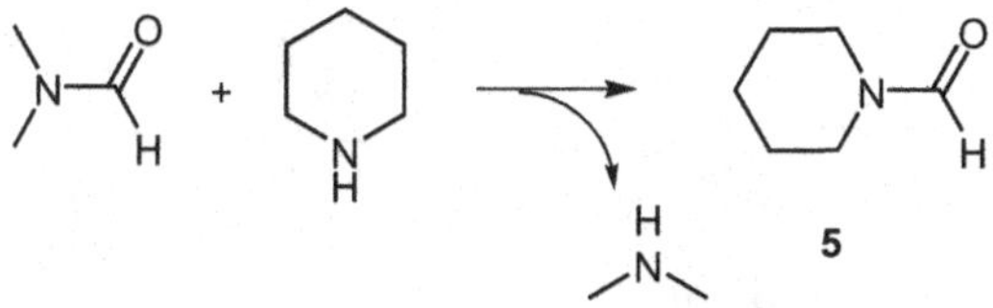

FIGURE 14.2 Formation of *N*-formylpiperidine from degradation of DMF by piperidine.

that are predominately induced by base treatment in this process. HOBt/piperidine/DMF stock solution is sometimes prepared in advance. However, the compatibility of DMF with piperidine in the presence of HOBt is often overlooked, indicated by the darkening of the HOBt/piperidine/DMF stock solution upon storage (turning yellow). In light of the NMR analysis (Yang, Y., unpublished results) it could be concluded that DMF/piperidine is converted to *N*-formylpiperidine **5** and dimethylamine (Fig. 14.2) at ambient temperature in the presence of HOBt, and this process is evidently accelerated at higher temperature, e.g., 50°C. Formation of *N*-formylpiperidine at room temperature without HOBt is also possible, but the decomposition kinetics are evidently slowed down relative to that when HOBt is present.

N-Formylpiperidine can be used as a polar aprotic solvent with enhanced hydrocarbon solubility that is greater than DMF.[9] *N*-Formylpiperidine serves also as an efficient formylation agent[10] that reacts with aryl-, alkyl- and vinyl-lithium, or Grignard compounds, and results in the formation of the corresponding aldehydes upon acidic workup. The exact consequences of *N*-formylpiperidine formation on peptide synthesis are currently unknown. Nonetheless, the deterioration of piperidine/DMF/HOBt solution would alter the stoichiometry of the affected reaction.

14.2.2 DMF-Induced Formylation Side Reactions

DMF degradation is a major source of peptide formylation side reactions in peptide synthesis. This phenomenon is frequently directly attributed to the decomposition of DMF under various conditions. Formic acid, which is one the degradative products of DMF, or DMF itself could function as formyl donor, and trigger formylation during peptide synthesis. Moreover, peptide formylation induced by DMF might follow the mechanism analogous to the Vilsmeier–Haack reaction. These side reactions have been elucidated in Section 7.3.3 and will not be reiterated herein.

14.2.3 DMF-Induced Acid Chloride Formation Side Reactions

It is known that acid chlorides could be obtained from the reaction of phosgene, thionyl chloride, oxalyl chloride, phosphorus trichloride, phosphorus oxychloride or phosphorous pentachloride with the corresponding carboxylic acids.[11] These reactions could generally be catalyzed by DMF.[12] The mechanism of DMF-catalyzed acid chloride formation by oxalyl chloride is illustrated

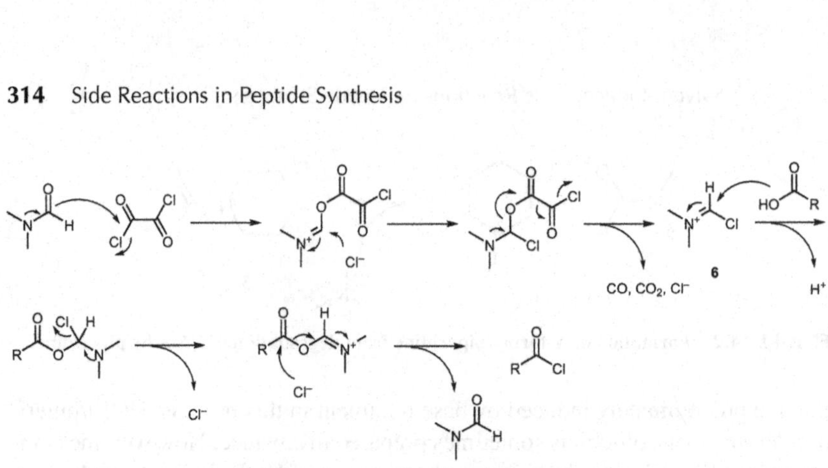

FIGURE 14.3 Mechanism of DMF-catalyzed acid chloride formation from carboxylic acid and oxalyl chloride.

in Fig. 14.3. The process follows a pathway akin to Vilsmeier–Haack reaction via a chloroiminium intermediate **6**,[13] that reacts with the carboxylic acid to give rise to the corresponding acid chloride and regenerates DMF. The reaction mechanism involving thionyl chloride and phosphorous pentachloride are analogous to that of oxalyl chloride.

In spite of the evident catalytic effect DMF-mediated acid chloride formation sometimes interferes with the target reactions. NCA preparation, for example, might suffer from the residual DMF in the reaction system and the undesired acid chloride formation as a consequence. NCA preparation is frequently conducted by the slow bubbling of slight excesses of phosgene through a hot suspension of the amino acid in an inert solvent (Fuchs–Farthing method).[14] *N*-Chloroformyl amino acid intermediate **7** formed in the process of NCA preparation undergoes a cyclization step and is converted to the target NCA derivative (Fig. 14.4).

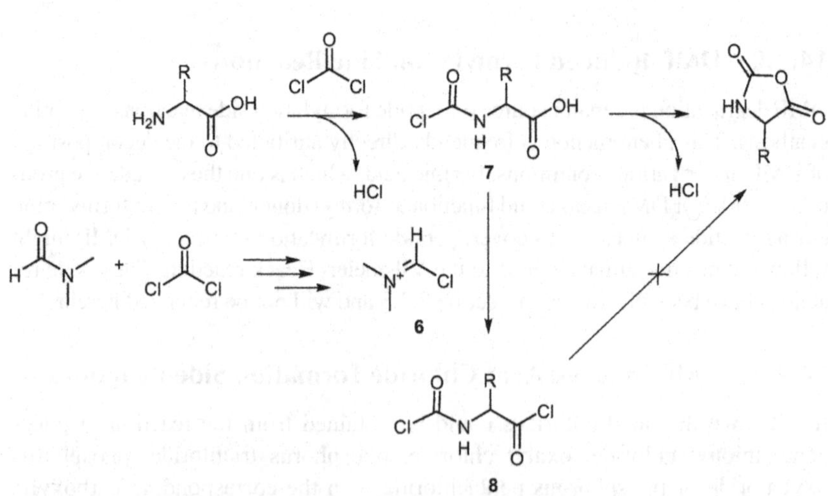

FIGURE 14.4 Phosgene-mediated NCA formation and DMF-induced side reaction in this process.

However, if the carboxyl group on **7** is transformed to the corresponding acid chloride by the active chloroiminium derivative **6** derived from the process analogous to Vilsmeier–Haack reaction, *N*-chloroformyl amino acid chloride derivative **8** will be generated as a consequence. Very low content of DMF in the reaction system is capable of facilitating this side reaction to a significant extent. If *N*-chloroformyl amino acid chloride derivative **8** is formed, it will not be converted to the target NCA.[15] It is, therefore, necessary to avoid the contamination of the solvents used for NCA manufacturing by DMF in order to prevent the undesired formation of *N*-chloroformyl amino acid chloride that terminates the NCA formation.

14.3 METHANOL/ETHANOL-INDUCED SIDE REACTIONS

14.3.1 Methanol-Induced Esterification Side Reactions

Methanol is utilized in peptide production mostly as resin rinsing solvent or eluent solution for RP-HPLC and sometimes as solvents for instrument cleaning. The existence of methanol could trigger undesired peptide esterification side reactions in the presence of an acid catalyst. This is often the case when the target peptide undergoes RP-HPLC purification with methanol as eluent and TFA as buffer. The collected fractions are subsequently subjected to the concentration workup, and significant extent of methyl esterification on the Asp/Glu carboxylate side chain and/or peptide *C*-terminal carboxylate backbone could be triggered in this process. Moreover, if the peptide resin after SPPS assembly is rinsed by methanol, and if the rinsing methanol is not sufficiently removed in the subsequent drying process, the survived methanol might be able to induce the formation of methyl ester impurities in the following TFA-mediated peptide cleavage and side chain global deprotection reaction. The content of methanol in the peptide intermediate should be, therefore, controlled under such circumstances especially for those derivatives that are susceptible to esterification side reactions. This kind of methanol-induced esterification side reaction has been illustrated in Section 8.1.2.

14.3.2 Methanol-Induced *N*-Alkylation Side Reactions in Catalytic Hydrogenation

Another common side reaction induced by alcohol takes place in the process of catalytic hydrogenation in which Z protecting groups are removed in the presence of Pd catalysts. Hydrogenolysis is generally carried out in alcohols (e.g., MeOH, iPrOH, tBuOH), AcOH, DMF, NMP or other applicable solvents. When the catalytic hydrogenation is carried out in alcohol, particular methanol, it is notable that oxygen should be removed from the system meticulously, since methanol, ethanol and benzyl alcohol could be converted to the corresponding aldehydes in the presence of palladium catalyst and oxygen.[16] The generated aldehyde could function with the amino functionalities released from the catalytical hydrogenation to give Schiff-bases, which are reduced *in situ* to the corresponding *N*-alkyl amines by catalytic hydrogenation.[17] Di-methylated amine derivatives could be generated in this process as a consequence.[18] This side reaction could also affect Lys-N^{ε}.[19]

14.4 ACETONITRILE-INDUCED SIDE REACTION

Acetonitrile is one of the most frequently utilized eluents in reverse phase chromatographic purification of peptides, partially thanks to its low viscosity, high chemical stability and strong eluting power. Moreover, it has also found widespread applications as a polar aprotic solvent in organic synthesis.

Acetonitrile is produced mainly as a byproduct of acrylonitrile manufacture via Sohio process by means of propylene ammoxidation.[20] In the acrylonitrile production with the aforementioned process hydrogen cyanide is released as a byproduct.[21] Pure acetonitrile is recovered by distillation from the waste before the treatment. If residual hydrogen cyanide survives the intermediary purification steps it will contaminate the acetonitrile. Moreover, in spite of its significant chemical stability acetonitrile does suffer from decomposition when heated or reacted with acid or oxidizing agents. The pyrolysis or chemical degradation of acetonitrile will also lead to the formation of hydrogen cyanide.[22,23] It is know that cyanide could function with carbonyl derivatives, e.g., ketones or aldehydes, by means of nucleophilic addition to generate the corresponding cyanohydrin derivatives (Fig. 14.5).

It has been detected in certain experiments that an impurity with a +27 amu molecular weight increase could be generated by stressing the target peptide in acetonitrile (Yang, Y., unpublished results). The content of the concerned impurity will be augmented upon the addition of NaOH. This undesired side reaction is thoroughly suppressed in the event of replacement of acetonitrile by methanol. Even though no direct proof was obtained to correlate the occurrence of the referred +27 amu impurity with the possible existence of hydrogen cyanide in acetonitrile, it is assumed that the product might suffer from the nucleophilic attack from the residual hydrogen cyanide accumulated in acetonitrile, especially taking into consideration the existence of the susceptible ketone carbonyl moiety in the target peptide molecule that serves as the receptor for cyanide addition. For the majority of

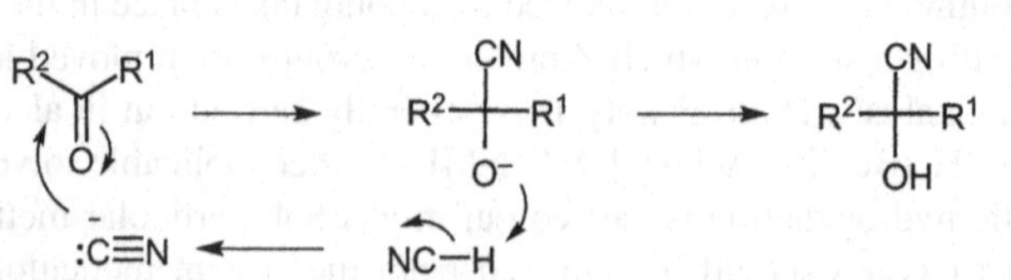

FIGURE 14.5 Mechanism of cyanohydrin formation via nucleophilic addition of cyanide to carbonyl derivative.

peptides the aforementioned potential side reaction triggered by acetonitrile might be trivial, however, the severity of this undesired phenomenon could be enhanced under certain conditions where factors such as acetonitrile quality, existence of susceptible functional group, pH, temperature and reaction time act synergistically.

Another acetonitrile-induced side reaction is originated from acetonitrile hydrolysis which is catalyzed by both acid[24] and base.[25] This reaction proceeds in two consecutive steps in which acetonitrile is firstly hydrolyzed to acetamide and the latter is further converted to acetic acid.[26] Formation of acetamide under basic condition is directly influenced by the water content in the system (Yang, Y., unpublished results) and its occurrence might affect the product quality if acetamide is not sufficiently removed by lyophilization or other workup handlings. On the other side, acetic acid is potentially capable of triggering acetylation side reaction in appropriate conditions. The content of acetic acid is generally listed in the acetonitrile material specification in an effort to release the acetonitrile as a raw material for the corresponding production.

14.5 ACETONE-INDUCED SIDE REACTION

Thanks to its distinctive attributes of high volatility as well as high miscibility with water and many other organic solvents, acetone is frequently utilized as the solvent of choice for instrument cleaning in the laboratory and in chemical/pharmaceutical manufacturing. However, if acetone is not sufficiently removed from the system, the residual acetone molecules might be capable of initiating some undesired modifications on the target peptides.

One of the side reactions induced by acetone has been introduced in Section 7.4.4. Acetone could function with the peptide *N*-terminal His residue on its imidazolyl side chain group to form *N*-(prop-1-en-2-yl) moiety with a +40 amu increase of the molecular weight relative to the target peptide derivative.

The modification of acetone on peptide molecules is not limited to the *N*-terminal His residues. Amino acid residues other than His that occupy the *N*-terminus of a certain peptide are also susceptible to acetone-mediated derivatization. It has already been reported that Oxytocin would be evidently inactivated by acetone due to the formation of an acetone-Oxytocin adduct.[27] It has been confirmed by various analyses that Oxytocin is firstly converted by acetone to an unstable Schiff-base intermediate that is subsequently rearranged to its stabilized 2,2-dimethyl-4-imidazolidinone ring structure.[28]

Generalized from the observation of Oxytocin modification by acetone, it is discovered that peptides with an *N*-terminal amino group and a primary amide bond between the first and second amino acid residues are prone to be affected by this kind of acetone modification. The process presumably experiences a Schiff-base intermediate which is subsequently rearranged to its corresponding

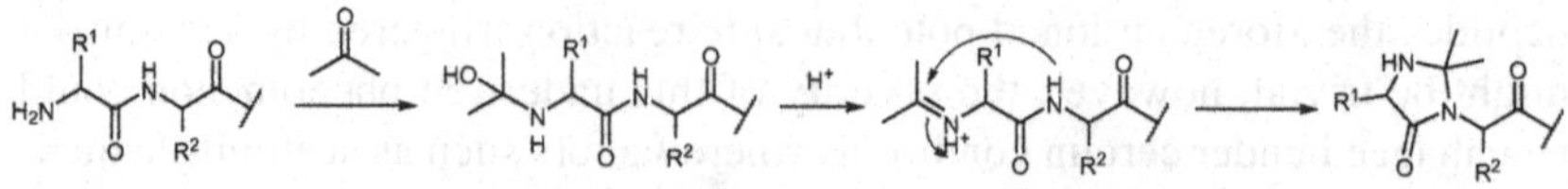

FIGURE 14.6 Imidazolidinone formation through acetone-mediated peptide *N*-terminus modification.

imidazolidinone counterpart.[29,30] This process is illustrated in Fig. 14.6. The nucleophilic addition of peptide N^{α} on acetone carbonyl moiety leads to the formation of a carbinolamine derivative **9** that undergoes dehydration subsequently and is converted to the corresponding Schiff-base **10**. Due to the intrinsic instability the generated Schiff-base **10** intermediate is tautomerized to the stable imidazolidinone derivative **11** upon an intramolecular nucleophilic attack from the backbone amide nitrogen on the iminium moiety.

Not only peptide but also relevant compounds such as amino acid amide, 2-aminobezoic acid and ampicillin could suffer from this kind of acetone modification in a similar manner.[30] Other carbonyl derivatives like formaldehyde, acetaldehyde, benzaldehyde and cyclohexanone[30,31] could direct analogous reactions as well. Acetone-induced imidazolidinone formation is reflected by the characteristic +40 amu molecular weight increase, and this side reaction could be readily characterized by MS and NMR analyses.

Thanks to the high chemoselectivity between an aminooxy functional group and a carbonyl group,[32] aminooxy peptides have found wide applications in the domains of oxime ligation with the complementary peptide aldehydes.[33] A variety of side reactions have nevertheless been identified in the synthesis and purification of aminooxy peptides.[34,35] Ironically, one of the major undesired side reactions affecting the aminooxy group is its conjugation with irrelevant aldehyde and ketones that leads to the formation of stable oxime, in a similar manner to the target oxime ligation process.[36] Due to its high reactivity toward aldehydes and ketones aminooxy peptides can react with traces of acetone or formaldehyde, even with those from the softeners of the plastic tubes, not only at the reaction steps but also in the process of HPLC purification and even upon storage.[37] The formation of oxime impurity between aminooxy peptide and acetone is illustrated in Fig. 14.7.

In view of the high reactivity of aminooxy peptides toward acetone it is necessary to avoid the contact of aminooxy peptides with even traces of acetone/aldehyde. Solutions like working under inert atmosphere devoid of acetone,

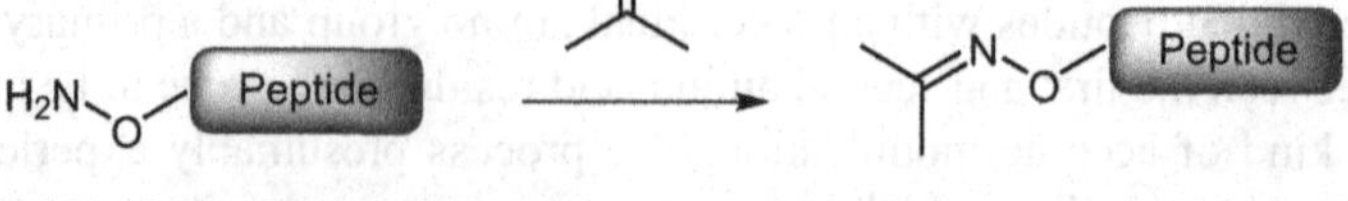

FIGURE 14.7 Oxime formation between aminooxy peptide and acetone.

utilization of freshly distilled diethyl ether,[38] shortening work-up process, utilization of methanol in place of acetonitrile for HPLC purification, storage of concerned aminooxy peptides at low temperature in proper containers, are favorable with regards to suppress the occurrence of this side reaction. The addition of carbonyl scavengers such as (aminooxy)acetic acid to the reaction systems in which undesired oxime formations could potentially take place is also a practical solution to restrict this side reaction.[37]

It is also to be noted that the mixing of concentrated H_2O_2 with acetone should be avoided since this operation could lead to the formation of explosive acetone peroxide derivatives.[39]

14.6 MTBE-INDUCED SIDE REACTION

Ethers are frequently utilized in peptide manufacturing to precipitate peptide crude after the peptide side chain global deprotection step. Thanks to its enhanced resistance to peroxide formation[40] as well as advantages with regards to safety concerns, MTBE is preferred to other ether derivatives such as diethyl ether, especially in peptide industrial production.

In spite of the aforementioned benefits MTBE might be able to induce some side reactions in peptide manufacturing at the peptide precipitation step. It is already known that MTBE could serve as the alkylating agent for the *tert*-butylation of various aromatic derivatives such as *p*-cresol,[41] phenol,[42] aniline,[43] and hydroquinone[44] in the presence of strong acid catalysts, presumably via the isobutene intermediate as the source of *tert*-butylation.[45,46] In peptide manufacturing strong acids like TFMSA are sometimes utilized for the removal of peptide side chain protecting groups. It has been discovered[47] that MTBE-mediated workup under such occasion might lead to the undesired *tert*-butylation side reaction particularly when the addressed product has abundant alkylation-susceptible residues such as Trp, Tyr, Cys, Met or PNA nucleobases. These *tert*-butylated impurities are suppressed upon the substitution of MTBE with diethyl ether as the antisolvent for the concerned peptide precipitation. This observation implies to some extent the incompatibility of MTBE with strong acids, especially in the presence of alkylation-susceptible substrates. The extent of the concerned *tert*-butylation side reaction induced by MTBE acidolysis is seemingly correlated to the acidity of the applied acid, and when TFA is employed without other strong auxiliary acids (e.g., TFMSA or TMSOTf) the severity of *tert*-butylation will be evidently alleviated.

14.7 TFE-INDUCED SIDE REACTION

Due to its distinctively superior peptide solubilization attribute TFE has found widespread application as a solvent or as a component of a solvent mixture for peptide synthesis,[48] especially recommended in an effort to disrupt β-sheet aggregation.[49] Moreover, TFE is intensively employed in peptide structure investigations due to its helix-inducing effects.[50] As a reagent, TFE is utilized in peptide synthesis to cleavage the side-chain protected peptide from CTC resin.[51]

TFE could be manufactured by various processes such as catalytic hydrogenation of trifluroacetamide,[52] or trifluoroacetate,[53] or by reduction of trifluoroacetic acid with metal hydrides.[54] Due to the high varieties of TFE manufacturing processes diverse impurity profiles could be relevant, and some active impurities could interfere with the target reactions. It is reported that a significant extent of trifluoroacetylation side reactions on Lys-N^{ε} is triggered by stressing the concerned peptide in 2-phenylethylamine TFE solution for a couple of hours.[55] This phenomenon was attributed to the contamination of TFE by TFA or trifluoroacetate which in turn leads to the formation of reactive trifluoroethyl trifluoroacetate, an analog to ethyl trifluoroacetate that is employed to mediate trifluoroacetylation modification on the amino group.[56] In another investigation, it was discovered that peptide stressed in pure TFE is partially converted to a derivative with a +26 amu molecular weight increase (Yang, Y., unpublished results). The chromatographic property of this compound is identical to that of the original peptide treated by acetaldehyde. The occurrence of this side reaction is presumably induced by the residual acetaldehyde impurity in the concerned TFE solvent, which leads to the formation of a Schiff-base side product.

REFERENCES

1. (a) Albericio F, Barany G. *Int J Pept Protein Res*. 1987;30:206–216. (b) Androutsou M-E, Saifeddine MS, Hollenberg MD, Matsoukas J, Agelis G. *Amino Acids*. 2010;38:985–990. (c) Shaginian A, Rosen MC, Binkowski BF, Belshaw PJ. *Chem Eur J*. 2004;10:4334–4340.
2. (a) Hansen SH, Nordholm L. *J Chromatogr A*. 1981;204:97–101. (b) Wright DA, Wulff CA. *J Org Chem*. 1970;35:4252. (c) Almarzoqi B, George AV, Issacs NS. *Tetrahedron*. 1986;42:601–607.
3. Albericio F, Kneib-Cordonier N, Biancalana S, et al. *J Org Chem*. 1990;55:3730–3743.
4. Mills JE, Maryanoff CA, McComsey DF, Stanzione RC, Scott L. *J Org Chem*. 1987;52:1857–1859.
5. Rudine AB, Walter MG, Wamser CC. *J Org Chem*. 2010;75:4292–4295.
6. Ruecker C. *Chem Rev*. 1995;95:1009–1064.
7. Reichardt C. *Solvents and Solvent Effects in Organic Chemistry*. VCH Publisher; 1988:424.
8. Mergler M, Dick F, Sax B, Stähelin C, Vorherr T. *J Pept Sci*. 2003;9:518–526.
9. Scriven EFV, Murugan R. Pyridine and pyridine derivatives. 4th ed. *Kirk-Othmer Encyclopedia of Chemical Technology*. John Wiley & Sons, Inc.; 2005:30.
10. Olah GA, Arvanaghi M. *Angew Chem Int Ed Engl*. 1981;20:878–879.
11. (a) Pizey JS, ed. *Synthetic Reagents*, Vol. 1. New York: Wiley; 1974:321–357. (b) Pearson AJ, Roush WR, eds. *Handbook of Reagents for Organic Synthesis: Activating Agents and Protecting Groups*. New York: Wiley; 1999:370–373. (c) Knapp S, Gibson FS. *Organic Syntheses*, Vol. IX. New York: Wiley; 1998:516–521. (d) Antell MF. In: Patai S, ed. *The Chemistry of Acyl Halides*. London: Interscience; 1972:40–44.
12. Bruckner R. *Advanced Organic Chemistry: Reaction Mechanism*. San Diego: Harcourt/Academic; 2002:239.
13. Vilsmeier A, Haack A. *Ber Dtsch Chem Ges A/B*. 1927;60:119–122.

14. Farthing AC. *J Chem Soc*. 1950:3213–3217.
15. Blacklock TJ, Shuman RF, Butcher JW, et al. In: Pascoe WE, ed. *Catalysis of Organic Reactions*. New York: Marcel Dekker Inc; 1992:47–50.
16. Benoiton NL. *Int J Pept Protein Res*. 1993;41:611.
17. Filira F, Biondi L, Gobbo M, Rocchi R. *Tetrahedron Lett*. 1991;32:7463–7464.
18. Bowman RE, Stroud HH. *J Chem Soc*. 1950:1342–1345.
19. Mazaleyrat J-P, Xie J, Wakselman M. *Tetrahedron Lett*. 1992;33:4301–4302.
20. Pollak P, Romeder G, Hagedorn F, Gelbke HP. *Nitrile. Ullmann's Encyclopedia of Industrial Chemistry*; Weinheim: Wiley-VCH Verlag GmbH & Co. KGaA; 2000:252.
21. Weissermel K, Arpe H-J. *Industrial Organic Chemistry*. Weinheim: Verlag Chemie; 1978: 268–269.
22. Reynolds JEF, Prasad AB. *Martindale: The Extra Pharmacopoeia*. London: Pharmaceutical Press; 1982.
23. Lifshitz A, Moran A, Bidani S. *Int J Chem Kinet*. 1987;19:61–79.
24. Barbosa LAMM, van Santen RA. *J Catal*. 2000;191:200–217.
25. Ma J, Zhang X, Zhao N, Xiao F, Wei W, Sun Y. *J Mol Struct Theochem*. 2009;911:40–45.
26. Schaefer FC. In: Rappoport Z, ed. *The Chemistry of Cyano Group*. New York: Interscience; 1970:[Chapter 6].
27. Yamashiro D, Aanning HL, du Vigneaud V. *Proc Natl Acad Sci USA*. 1965;54:166–171.
28. Hruby VJ, du Vigneaud V. *J Am Chem Soc*. 1969;91:3624–3626.
29. (a) Hruby VJ, Yamashiro D, du Vigneaud V. *J Am Chem Soc*. 1968;90:7106–7110. (b) Cardinaux F, Brenner M. *Helv Chim Acta*. 1973;56:339–347.
30. Panetta CA, Pesh-Imam M. *J Org Chem*. 1972;37:302–304.
31. Nicholls R, de Jersey J, Worrall S, Wilce P. *Int J Biochem*. 1992;24:1899–1906.
32. Ulrich S, Boturyn D, Marra A, Renaudet O, Dumy P. *Chem Eur J*. 2014;20:34–41.
33. Cremer GA, Bureaud N, Lelièvre D, Piller V, Piller F, Delmas A. *Chem Eur J*. 2004;10: 6353–6360.
34. Buré C, Levièvre D, Delmas A. *Rapid Commun Mass Spectrom*. 2000;14:2158–2164.
35. Decostaire IP, Levièvre D, Zhang HH, Delmas AF. *Tetrahedron Lett*. 2006;47:7057–7060.
36. Canne LE, Ferré-D'Amaré AR, Burley SK, Kent SBH. *J Am Chem Soc*. 1995;117:2998–3007.
37. Mezö G, Szabó I, Kertész I, et al. *J Pept Sci*. 2011;17:39–46.
38. Carulla N, Woodward C, Barany G. *Bioconjugate Chem*. 2001;12:726–741.
39. Oxley JC, Brady J, Wilson SA, Smith JL. *J Chem Health Saf*. 2012;19:27–33.
40. Winterberg M, Schulte-Körne E, Peters U, Nierlich F, *Methyl* Tert-Butyl *Ether. Ullmann's Encyclopedia of Industrial Chemistry*, Vol. 23. Weinheim: Wiley-VCH Verlag GmbH & Co. KGaA; 2010:119–130.
41. Yadav GD, Pujari AA, Joshi AV. *Green Chem*. 1999;1:269–274.
42. Yadav GD, Doshi NS. *Appl Catal A*. 2002;236:129–147.
43. Yadav GD, Doshi NS. *J Mol Catal A Chem*. 2003;194:195–209.
44. Yadav GD, Doshi NS. *Catal Today*. 2000;60:263–273.
45. Cunill F, Tejero J, Izquierdo JF. *Appl Catal*. 1987;34:341–351.
46. Tejero J, Cunill F, Manzano S. *Appl Catal*. 1988;38:327–340.
47. de la Torre BG, Andreu D. *J Pept Sci*. 2008;14:360–363.
48. (a) Kuroda H, Chen Y-N, Kimura T, Sakakibara S. *Int J Pept Protein Res*. 1992;40:294–299. (b) Giralt E, Eritja R, Pedroso E. *Tetrahedron*. 1986;42:691–698.
49. Yamashiro D, Blake J, Li CH. *Tetrahedron Lett*. 1976;17:1469–1472.
50. (a) Goodman M, Listowsky I, Masuda YFB. *Biopolymers*. 1963;1:33–42. (b) Walgers R, Lee TC, Cammers-Goodwin A. *J Am Chem Soc*. 1998;120:5073–5079.

51. Bollhagen R, Schmiedberger M, Barlos K, Grell E. *J Chem Soc Chem Commun*. 1994:2559–2560.
52. Gillman H, Jones RG. *J Am Chem Soc*. 1948;70:1281–1282.
53. Anello LG, Cunningham WJ. US Patent 3,390,191 A. Issued Mar 24, 1964.
54. Siegemund G, Schwertfeger W, Feiring A, et al. *Fluorine Compounds, Organic. Ullmann's Encyclopedia of Industrial Chemistry*, Vol. 15. Weinheim: Wiley-VCH Verlag GmbH & Co. KGaA; 2000:444–494.
55. Rizo J, Albericio F, Giralt E, Pedroso E. *Tetrahedron Lett*. 1992;33:397–400.
56. Xu D, Prasad K, Repic O, Blacklock TJ. *Tetrahedron Lett*. 1995;36:7357–7360.

Appendix I

Molecular Weight Deviation of Peptide Impurity

Average Δmass	Modification	Proposed side reaction scheme	References
−98	β-Elimination of phosphopeptide	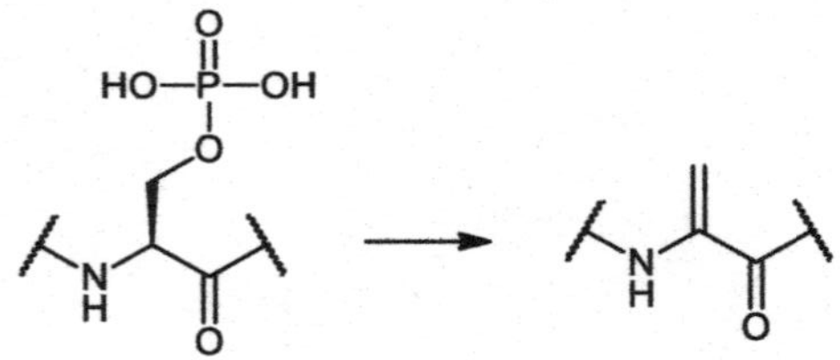	[1]
−80	Peptide dephosphorylation	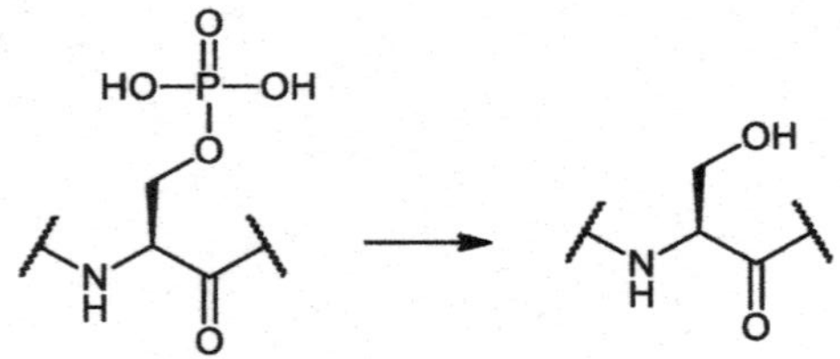	[2]
−80	Cys-induced intramolecular nucleophilic substitution on bromoacetyl moiety and monosulfide formation	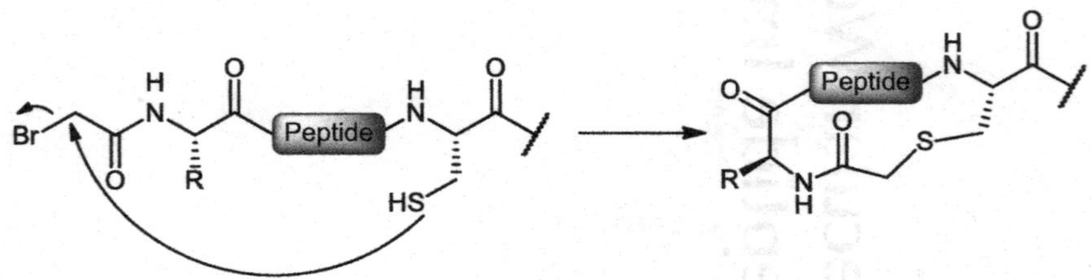	[3]

−48	Degradation of Met to allylglycine	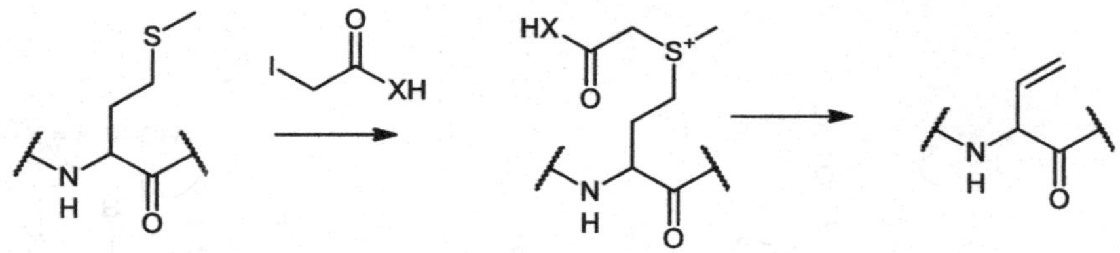 X = O or NH	[4]
−42	Conversion of Arg to Orn via deguanidination	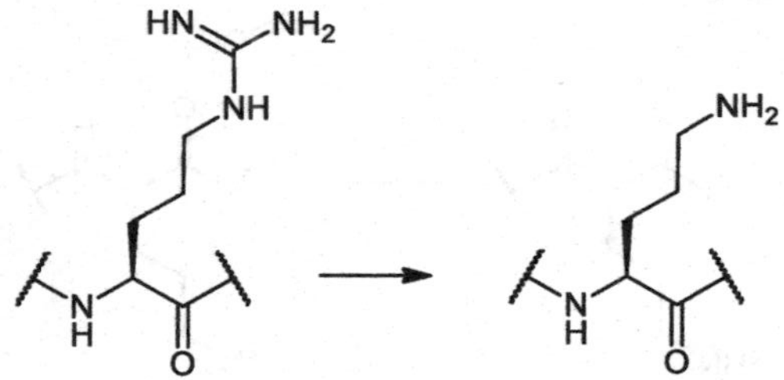	[5]
−34	Cysteine β-elimination	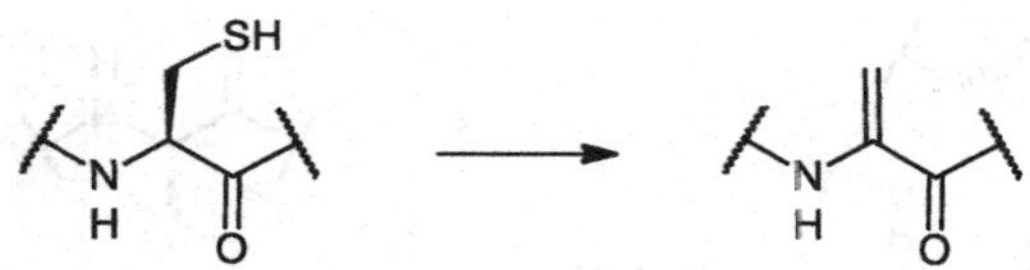	[6]

(Continued)

Average Δmass	Modification	Proposed side reaction scheme	References
−32	Disulfide desulfurization	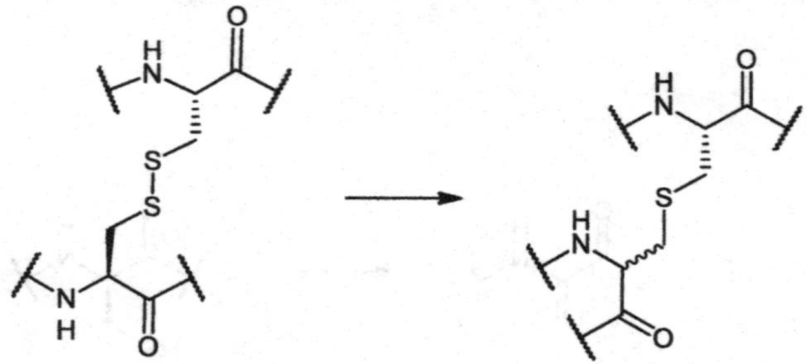	[7]
−26	Reduction of Nva(N_3) to Orn	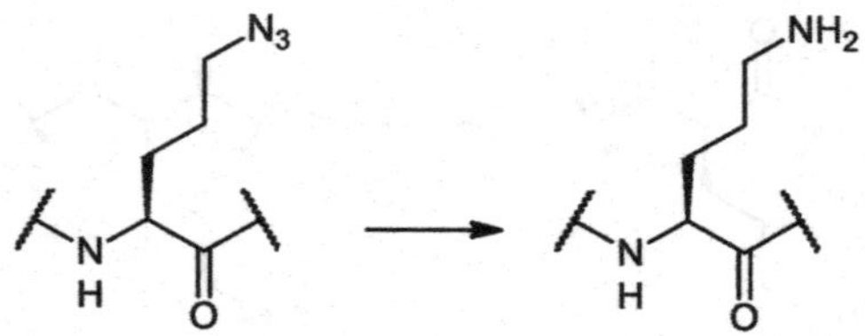	[8]
−20	Oxazole formation from Ser/Thr	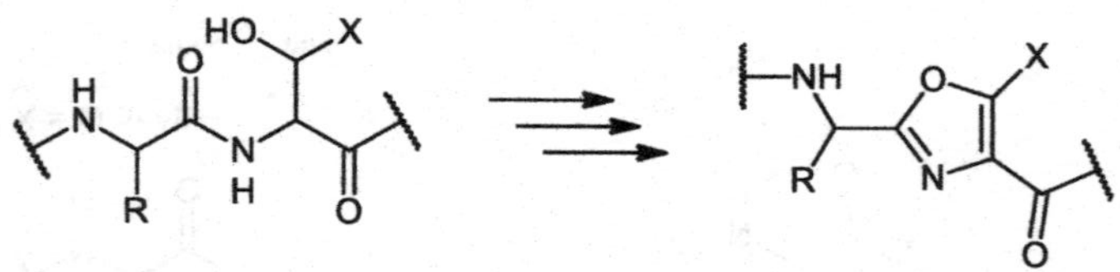 X = H or CH_3	[9]

−18	Pyroglutamate formation from Glu	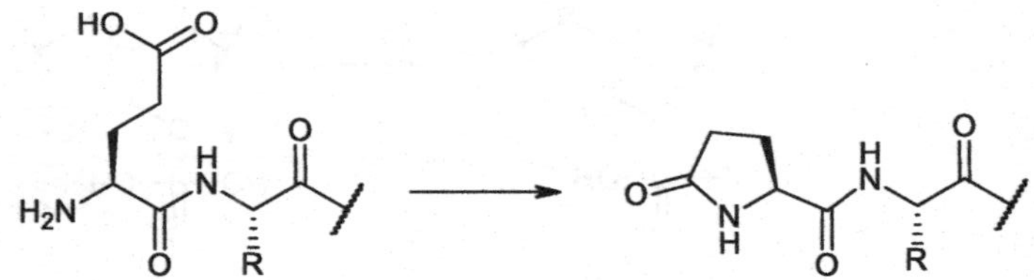	[10]
−18	Aspartimide/Glutarimide formation from Asp/Glu	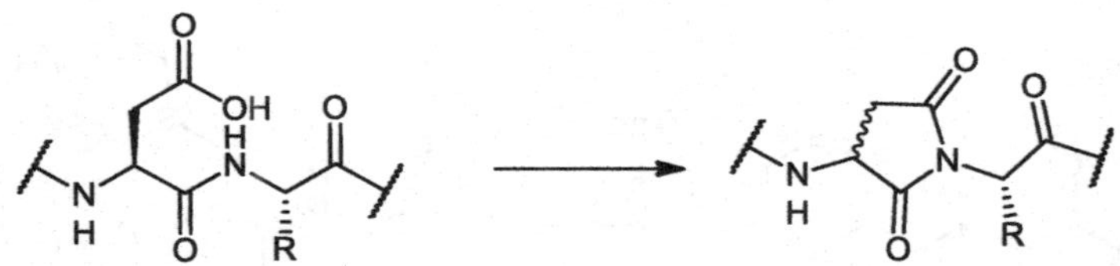	[11,12]
−18	β-Elimination of Ser	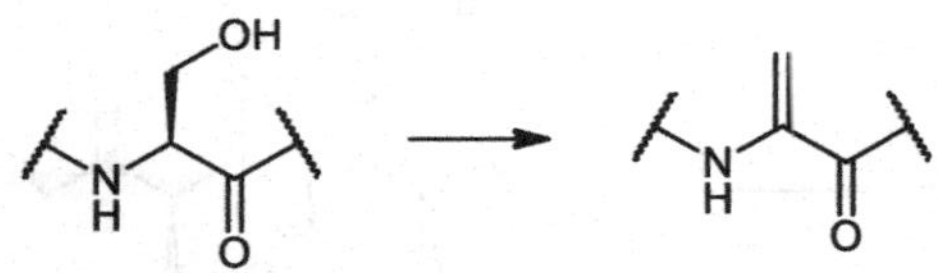	
−18	Dehydration of Asn/Gln	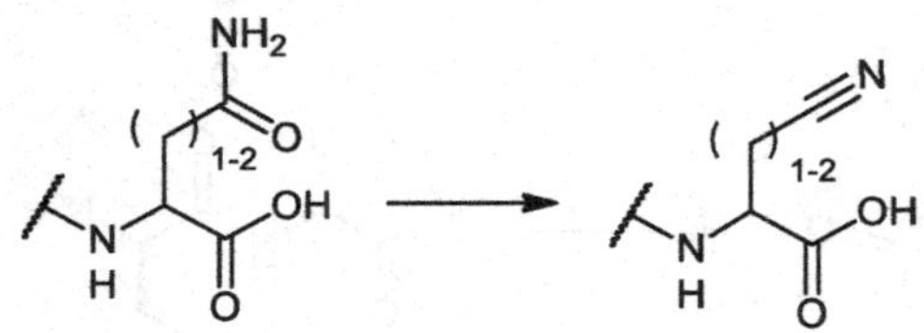	[13]

(Continued)

Average Δmass	Modification	Proposed side reaction scheme	References
−17	Pyroglutamate formation from Gln	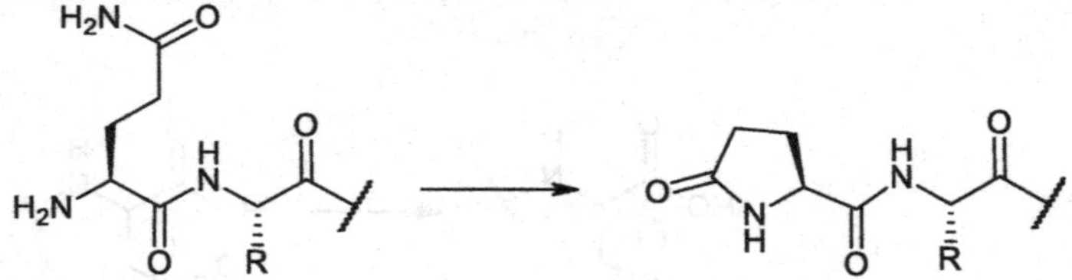	[14]
−17	Aspartimide/Glutarimide formation from Asn/Gln		[15]
−17	Cyclization of *N*-terminal Cys(Cam)	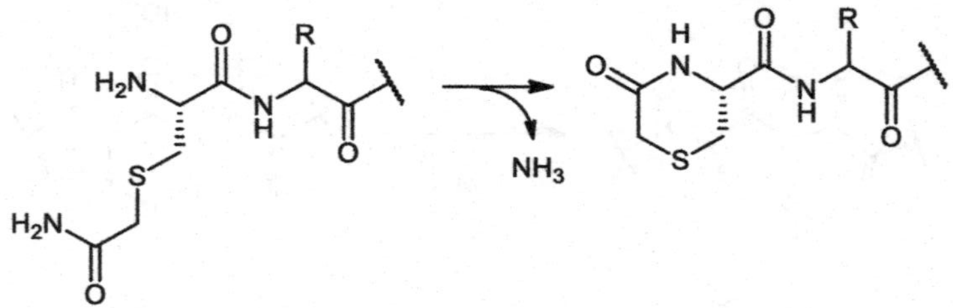	[16]
−16	*H*-phosphonate formation	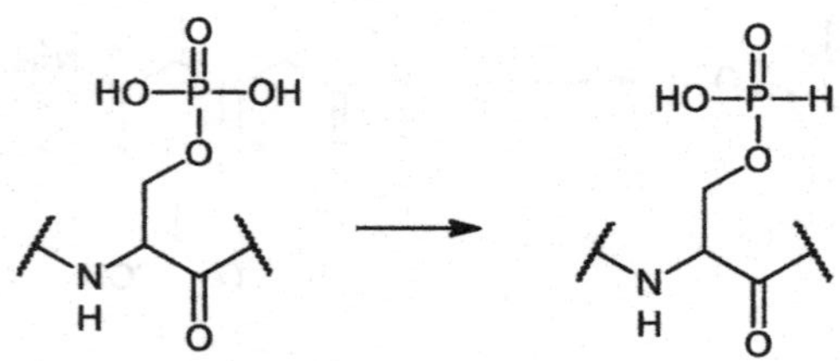	[17]

−14	Thioanisole-induced Tyr (Me) demethylation	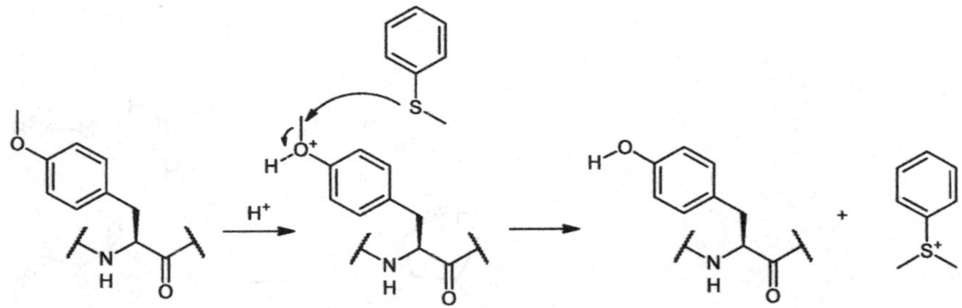	[18]
−2	Cysteine oxidation to Cystine	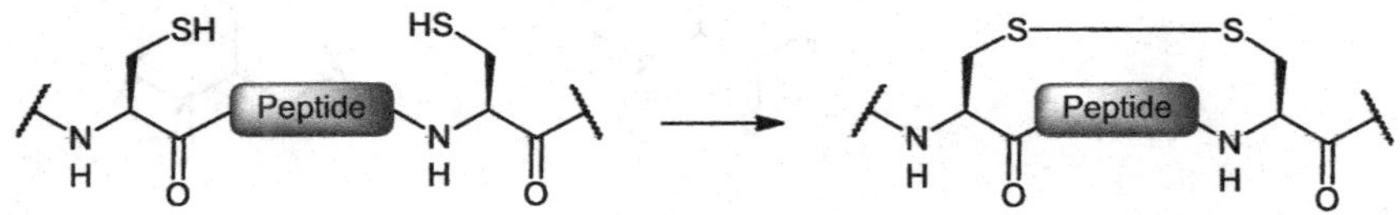	[19]
+1	Asn/Gln hydrolysis	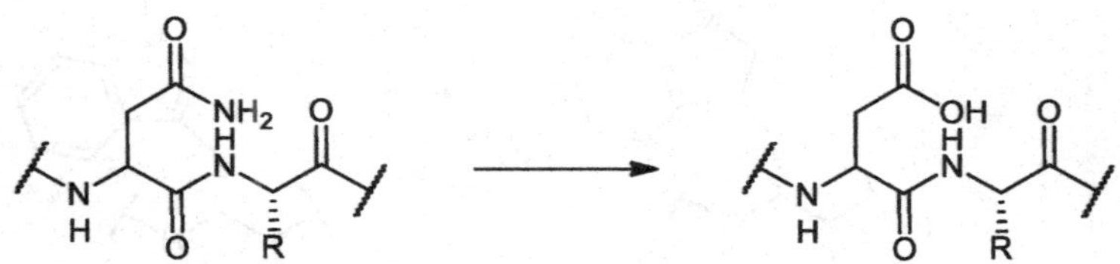	[20]
+1	Peptide amide hydrolysis		

(Continued)

Average Δmass	Modification	Proposed side reaction scheme	References
+2	Cystine reduction		
+2	Trp reduction	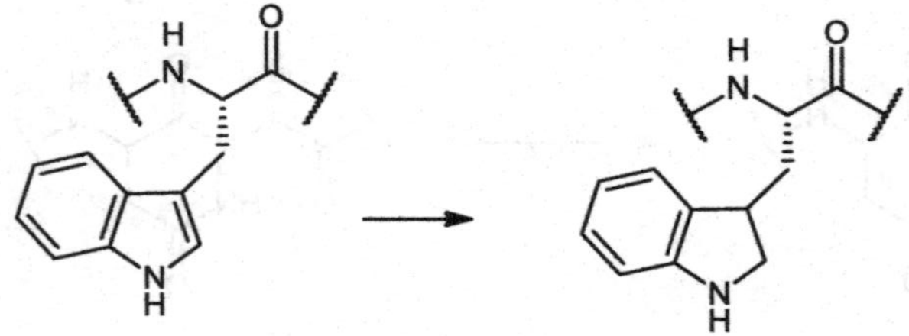	[21]
+4	Oxidation of Trp to Kynurenine	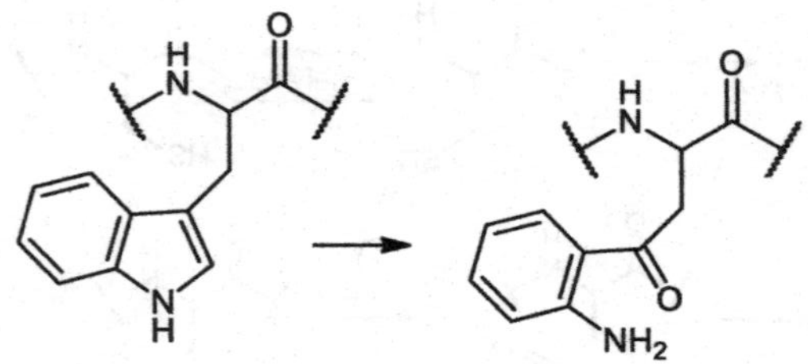	[22]
+12	Imine formation on amino-containing peptide	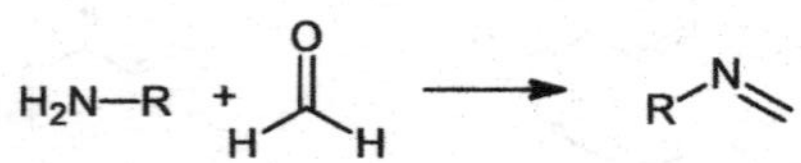	[23]

+12	Imidazolin-4-one formation on peptide *N*-terminus	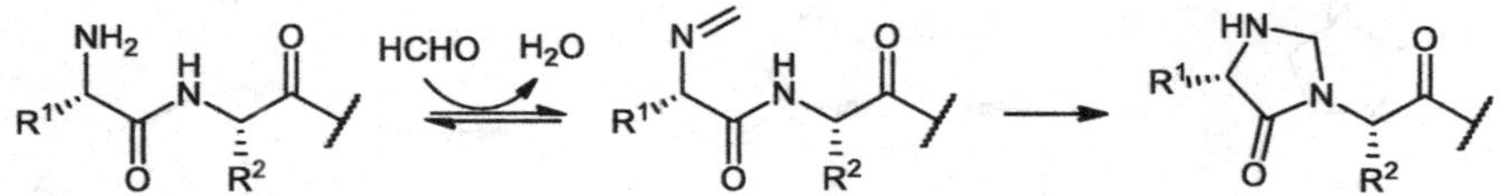	[24]
+12	Formaldehyde-induced crosslinking of peptide *N*-terminal Cys, Trp, Lys(Nma)	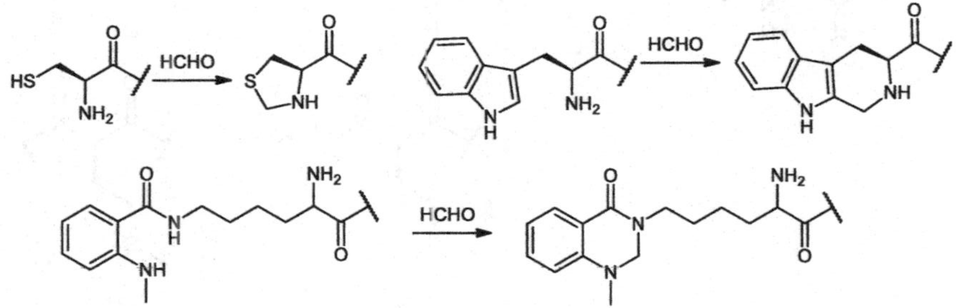	[25]
+14	Methylation of amino group		[26]
+14	Methylesterfication on carboxyl group		
+15	Unsymmetric urea formation from acyl azide	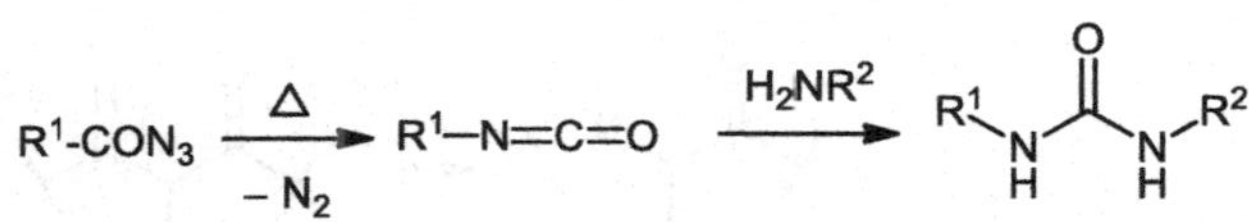	[27]

(Continued)

Average Δmass	Modification	Proposed side reaction scheme	References
+16	Oxidation of Cys to Cysteine sulfenic acid	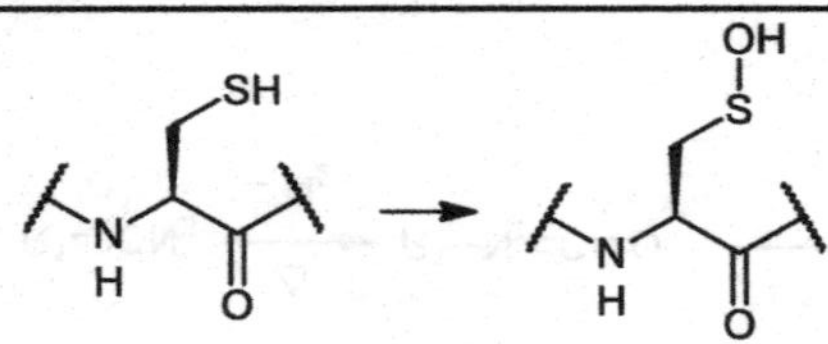	[28]
+16	Oxidation of Trp to Oia (Oxindolylalanine)	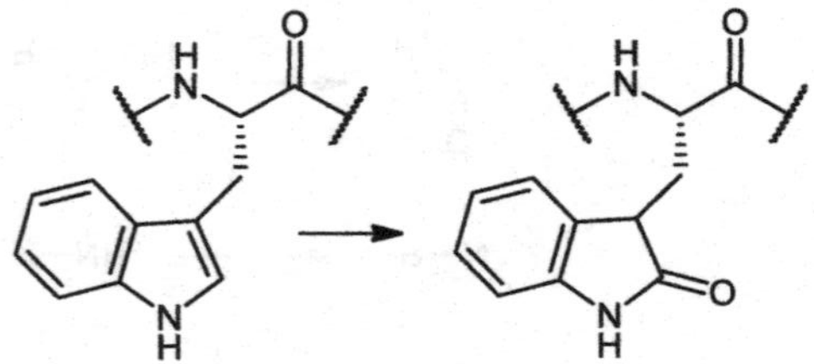	[29]
+16	Oxidation of Met to Met sulfoxide	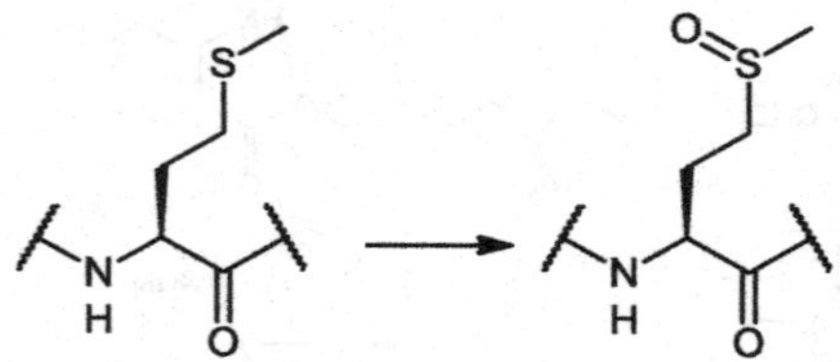	[30]
+16	Oxidation of His to 2-oxo-His	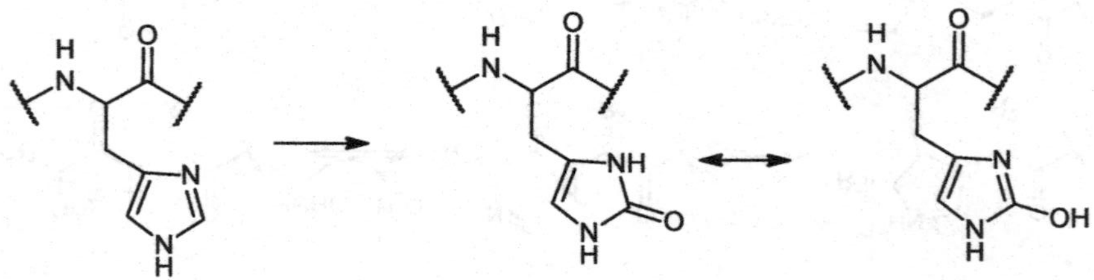	[31]

+25	Met cyanilation	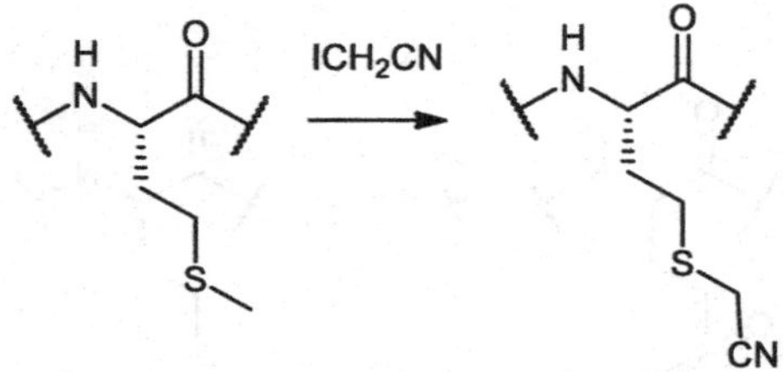	[32]
+25	Cys cyanilation	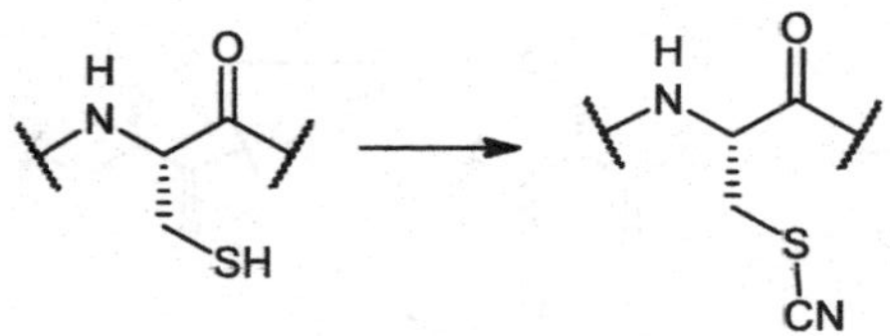	[33]
+26	Schiff-base formation from amino group and acetaldehyde		
+26	Oxime formation between aminoxy peptide and acetal-dehyde		[34]
+26	Hydantoin formation	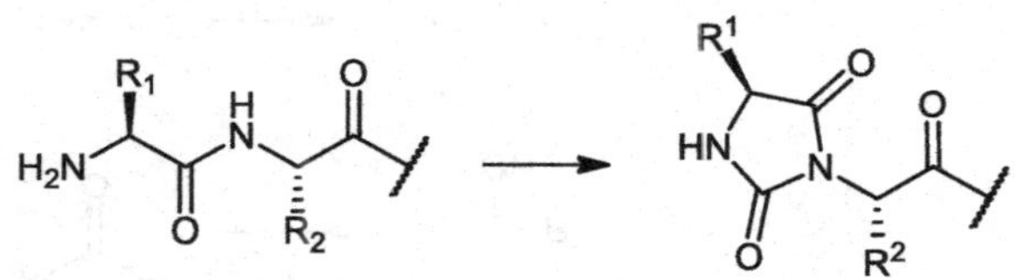	[35]

(Continued)

Average Δmass	Modification	Proposed side reaction scheme	References
+27	Cyanohydrin formation		
+28	Peptide formylation at N^{α}, Lys-N^{ε}, Trp-N^{in} or His-N^{im}	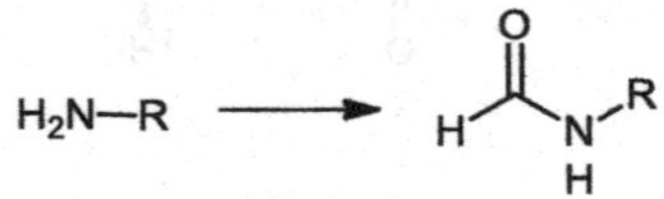	[36–40]
+28	Carboxylate ethylation		[41]
+28	*N*-Dimethylation		[26]
+32	Trisulfide formation		[42,43]
+32	Oxidation of Met to Met sulfone	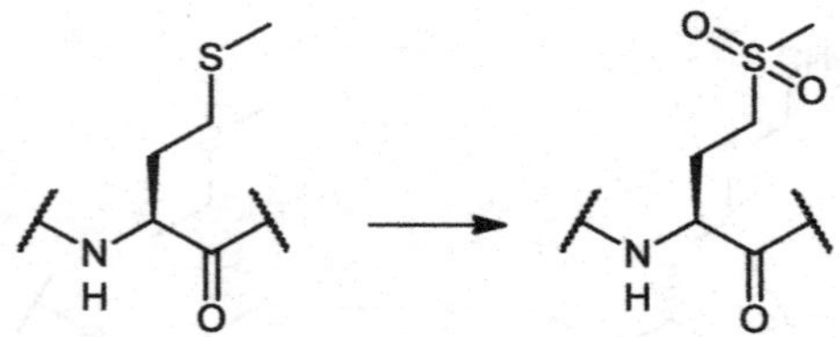	[44]

+32	Oxidation of Cys to Cysteine sulfinic acid	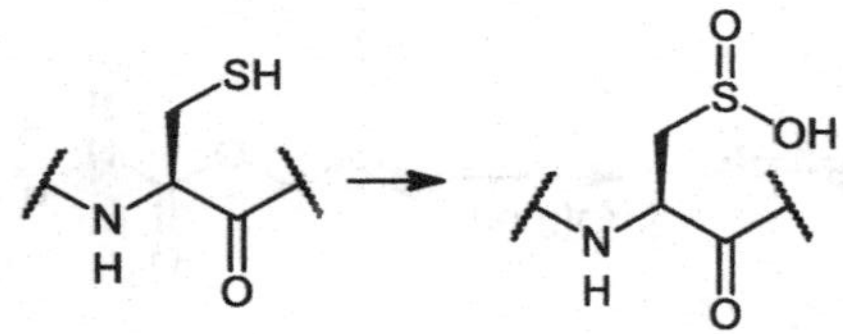	[45]
+32	Oxidation of Trp to *N*-formylkynurenine	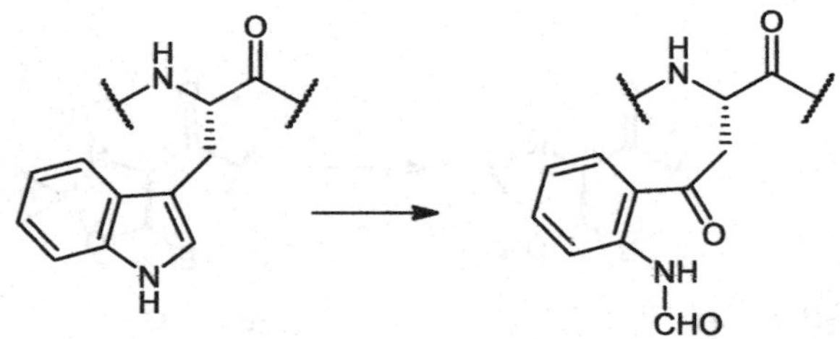	[46]
+34	Chlorination of Tyr	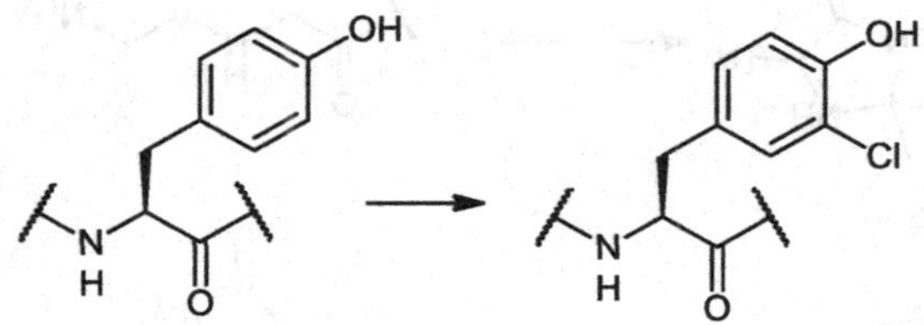	[47]
+40	Oxime formation between aminoxy peptide and acetone		[34,41]

(Continued)

Average Δmass	Modification	Proposed side reaction scheme	References
+40	Enamination of His imidazolyl side chain by acetone	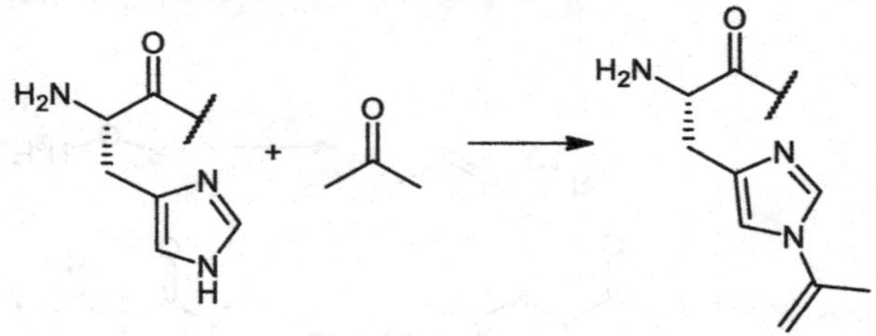	
+40	Acetone induced peptide *N*-terminal imidazolidinone formation	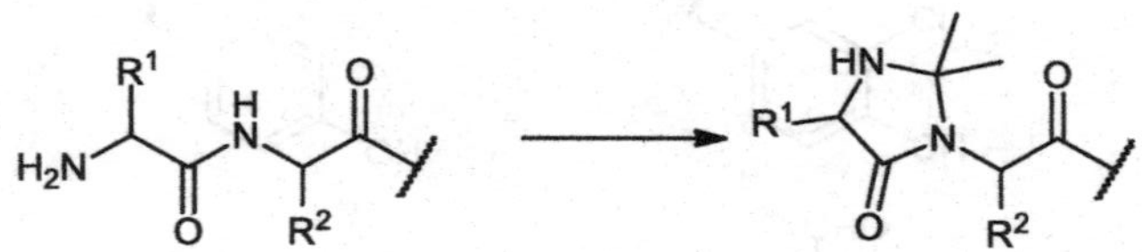	[48]
+40	Insufficient de-protection of Ser/Thr($\Psi^{Me,Me}$ Pro)	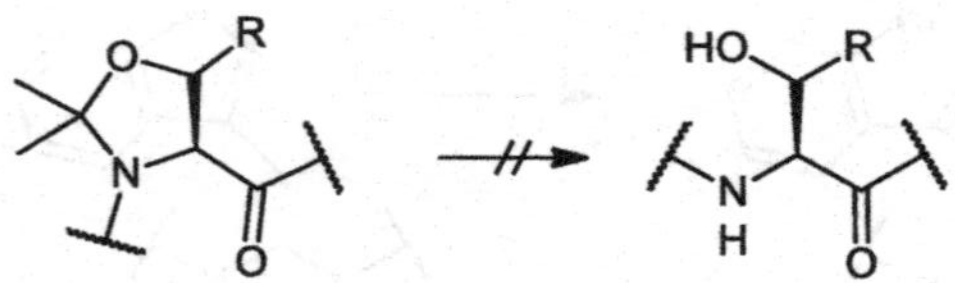 R = H or CH_3	
+40	Allylation	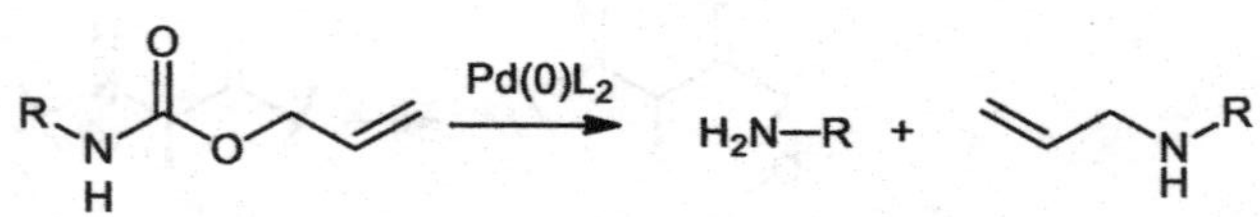 L = Ligand	[49]

+41	Amidine formation		[50–52]
+42	Acetylation on N^{α}, Lys-N^{ε}, *O*-Ser/Thr	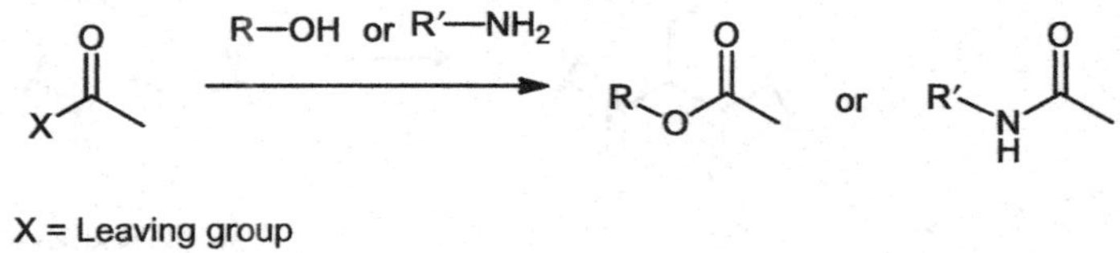	[53]
+44	Trp carbamate	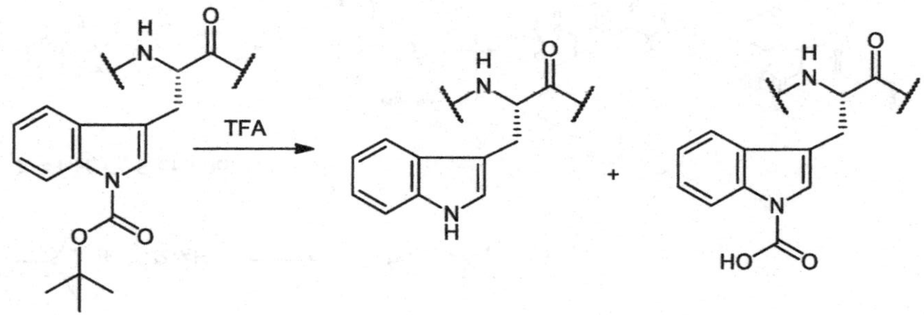	[54]
+48	Oxidation of Cys to Cys sulfonic acid	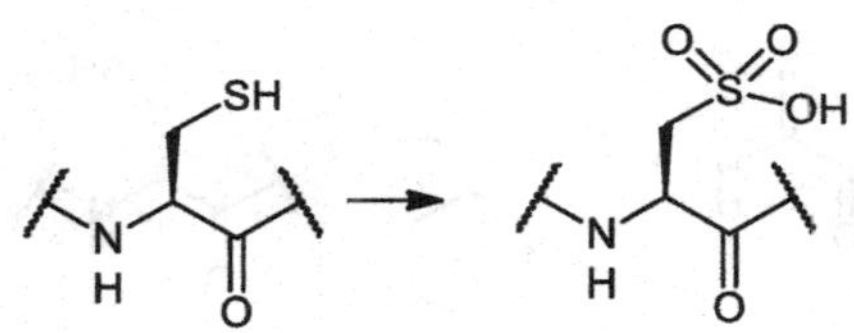	[55]

(Continued)

Average Δmass	Modification	Proposed side reaction scheme	References
+51	Cys beta-elimination and piperidide adduct formation	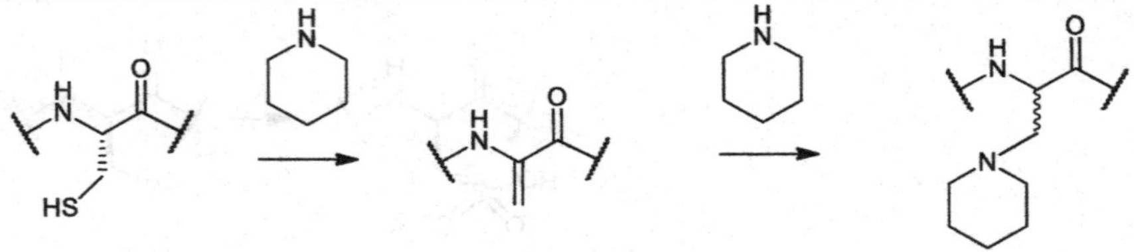	
+56	*tert*-butylation on nucleophilic amino acid or insufficient removal of tBu protecting group	 X = NH, O, S or indolyl	
+67	Asp-piperidide	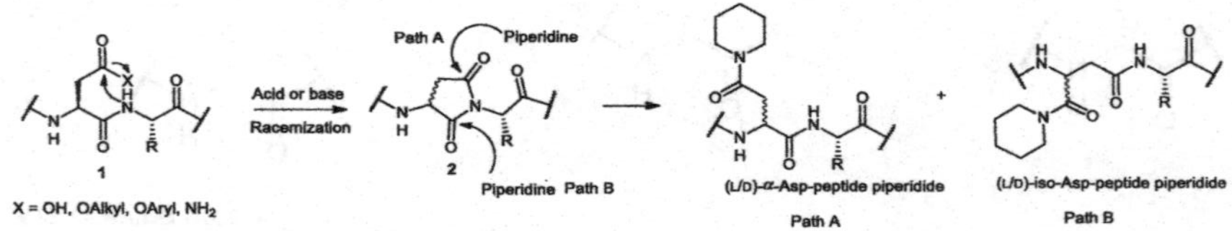	
+71	Acetamidomethylated Ser/Thr	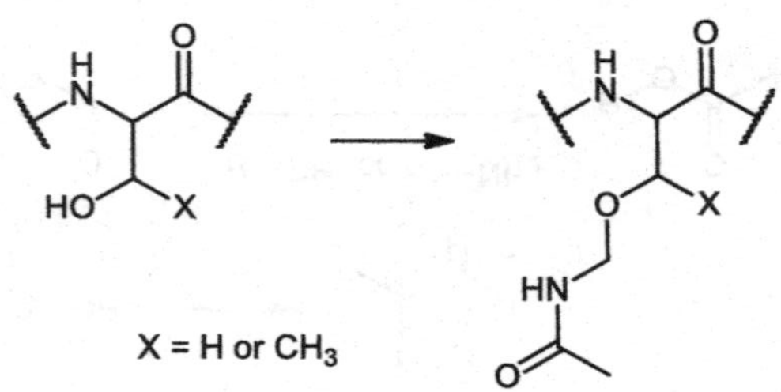	

+71	*endo*-β-alanine	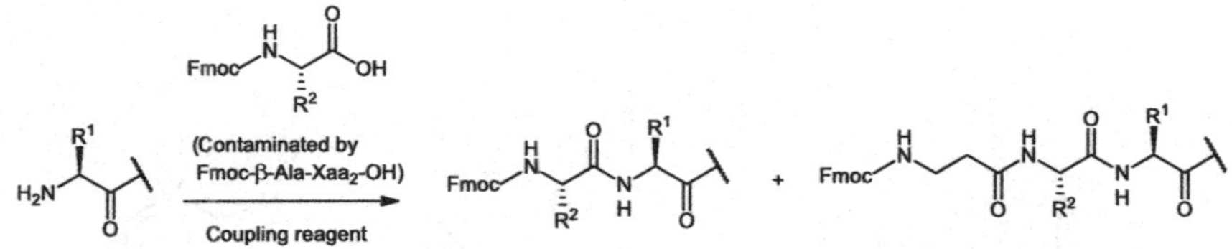	[56]
+72	Michael addition of mercaptan to acrylic acid	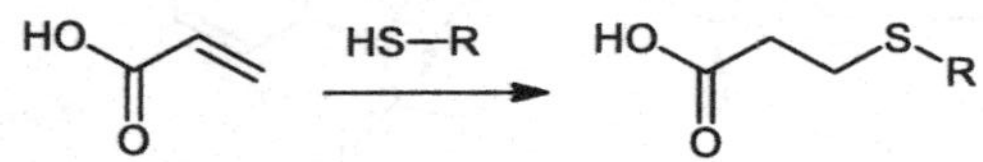	[57]
+74	Esterification of Asp/Glu by glycerol	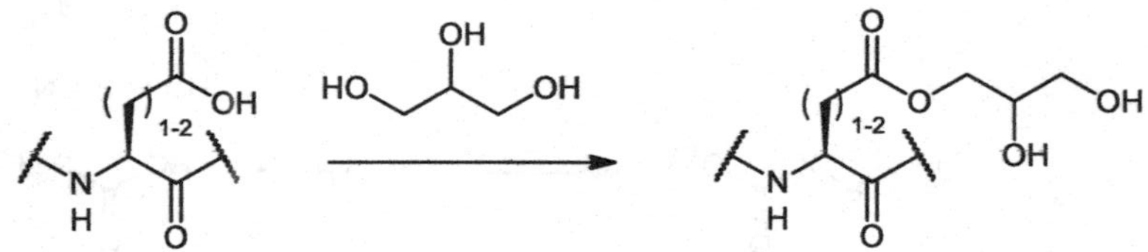	[58]
+76	EDT-mediated dithioketal adduct on ketone	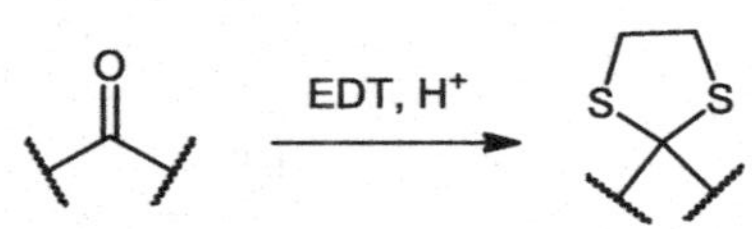	[59]
+80	Sulfonation	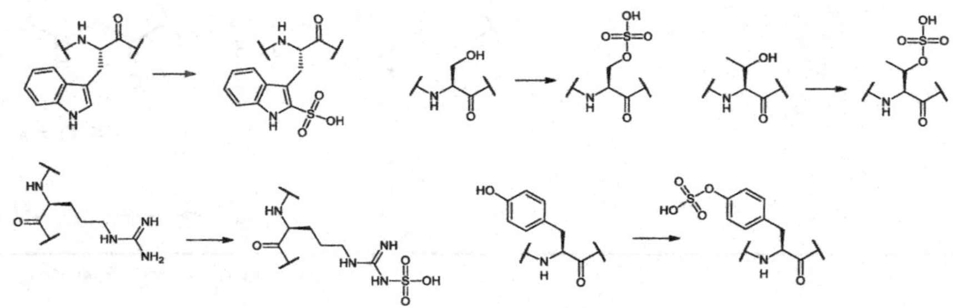	[60–62]

(Continued)

Average Δmass	Modification	Proposed side reaction scheme	References
+90	Benzylation	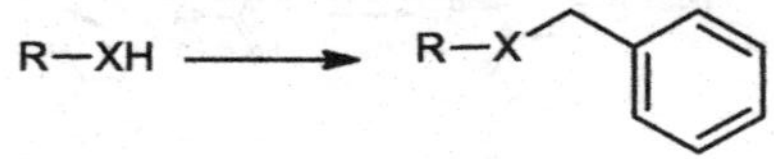 X = O or S	[63]
+92	Cys-EDT adduct	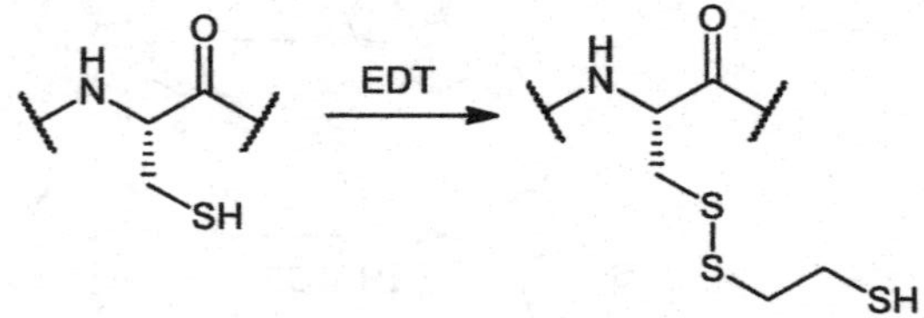	
+96	Trifluoroacetylation of amino or hydroxyl group	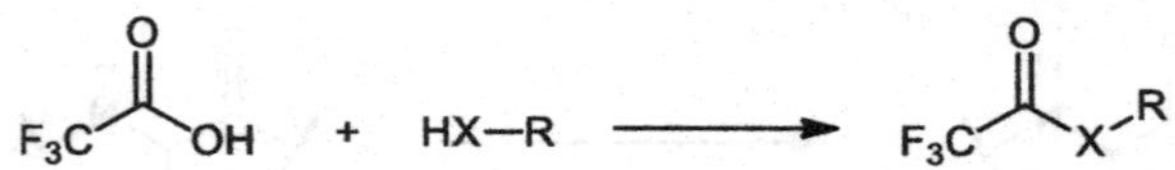 N = O or NH	[64]
+98	Guanidinium formation on amino group	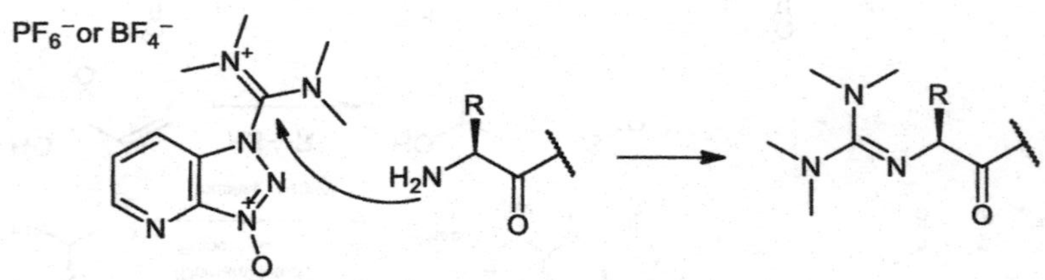	[65]

+106	4-Hydroxylbenzylation	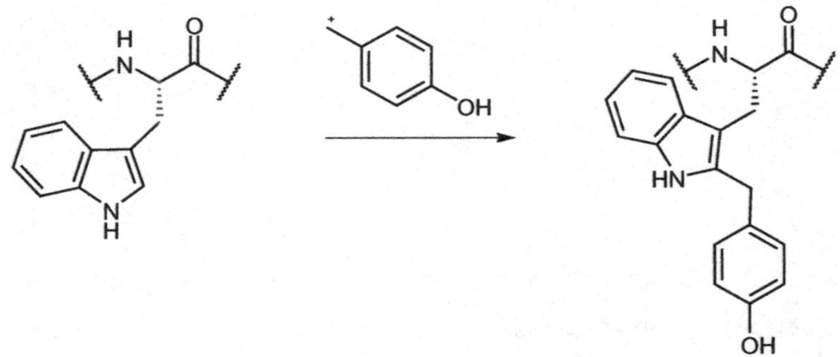	[66]
+117	Methionine alkylation dy DODT	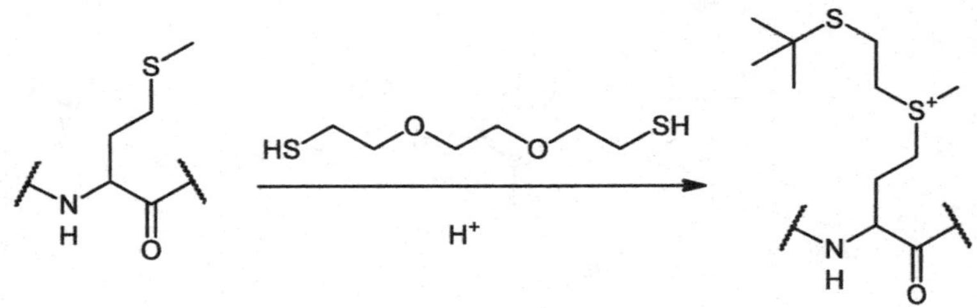	[67]
+126	DIC-mediated guanidinium/hydantoin formation on N^{α}	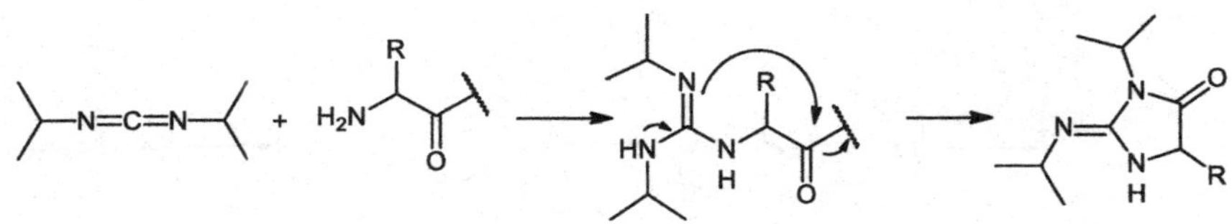	[68]

(Continued)

Average Δmass	Modification	Proposed side reaction scheme	References
+145	HOOBt/carbodiimide-induced N^{α}-2-azidobenzoyl endcapping	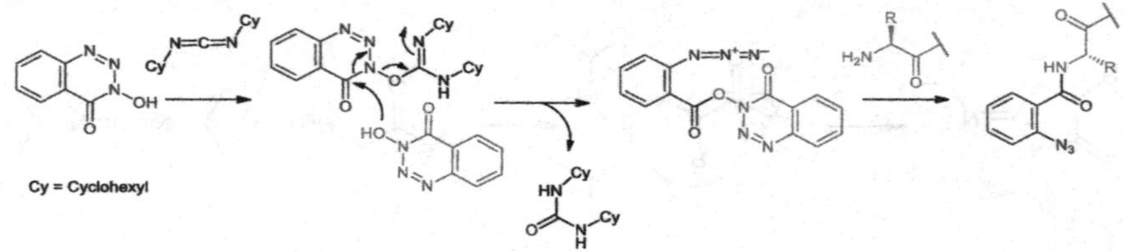	[69]
+148	Cys-EDT-tBu adduct	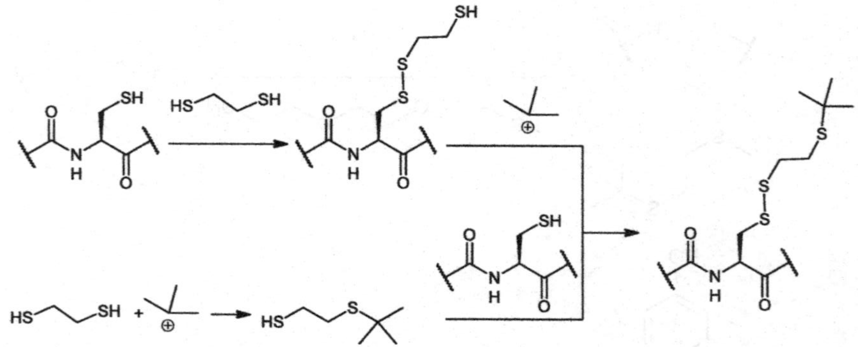	

+163	Degraded rink amide MBHA linker-induced Trp alkylation	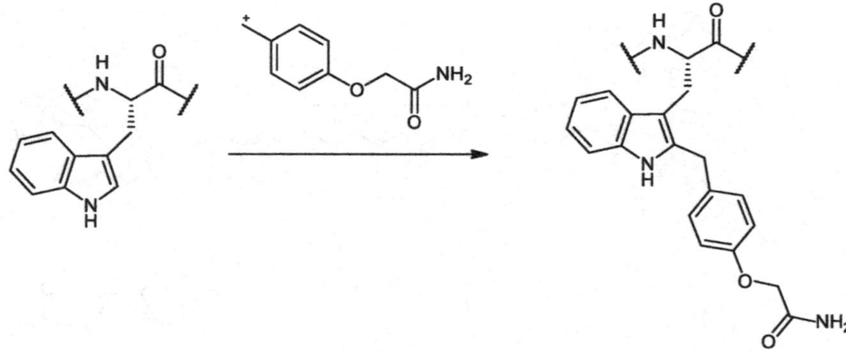	[70]
+172	Trp EDT-TFA adduct	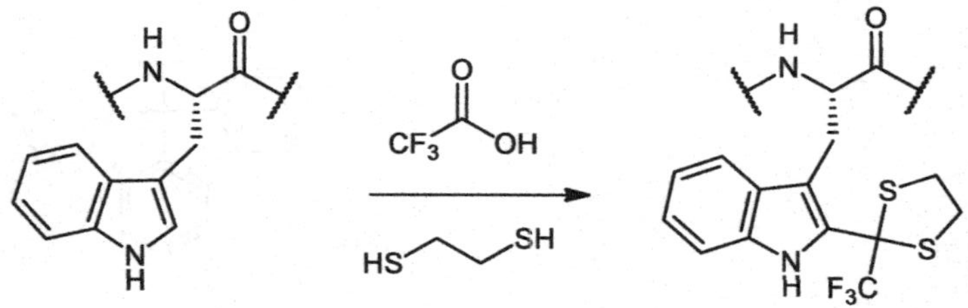	[60]
+178	N^{α}-Fm endcapping	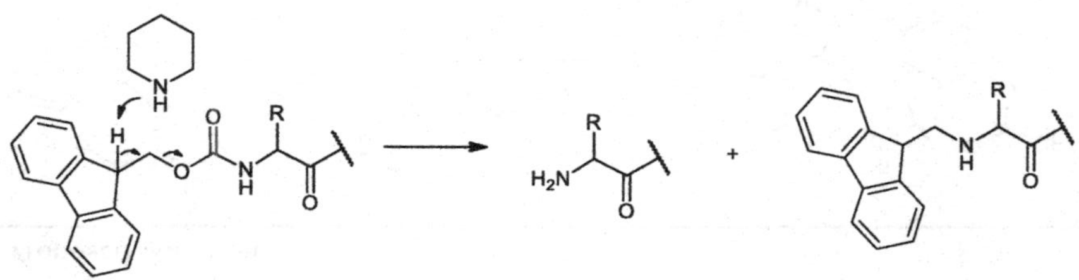	[71]

(Continued)

Average Δmass	Modification	Proposed side reaction scheme	References
+202	4-Trifluoroacetyoxy-benzylation of Trp	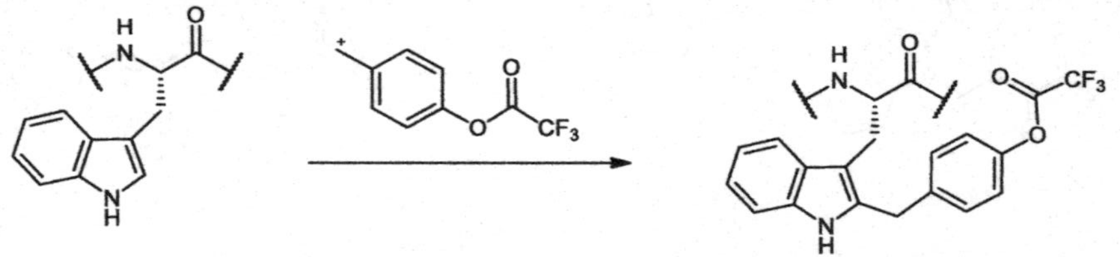	[70]
+242	Trtylation	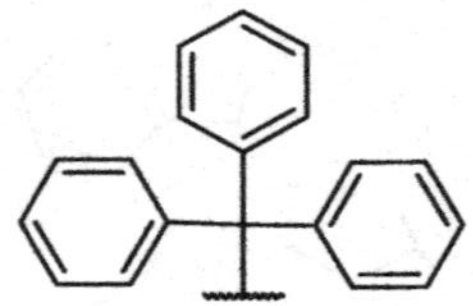	
+252	Pbf derivatization	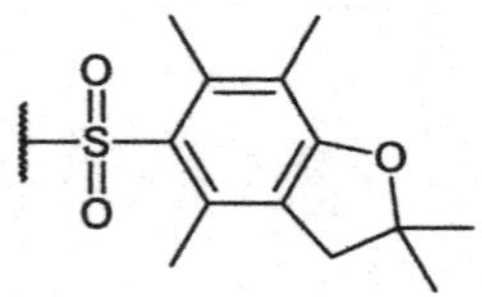	
+265	Degraded Pal linker-induced Trp alkylation	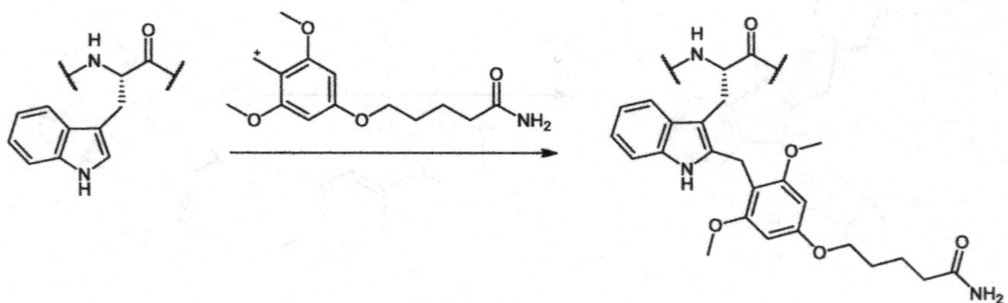	[70]

REFERENCE

1. Attard TJ, O'Brien-Simpson NM. *Int J Pept Ther*. 2009;15:69–79.
2. Luedtke NW, Schepartz A. *Chem Commun*. 2005:5426–5428.
3. Robey FA. In: Pennington MW, Dunn BM, eds. *Methods in Molecular Biology. Vol. 35, Peptide Synthesis Protocols*. Totowa, New Jersey: Humana Press; 1994:81-82.
4. Jones MD, Merewether LA, Clogston CL, Lu HS. *Anal Biochem*. 1993;216:135–146.
5. Rink H, Sieber P, Raschdorf F. *Tetrahedron Lett*. 1984;25:621–624.
6. Lukszo J, Patterson D, Albericio F, Kates SA. *Lett Pept Sci*. 1996;3:157–166.
7. Federici G, Duprè S, Matrarese RM, Solina SP, Cavallini D. *Int J Pept Protein Res*. 1977;10: 185–189.
8. Schneggenburger PE, Worbs B, Diederichsen U. *J Pept Sci*. 2010;16:10–14.
9. Paulus T, Riemer C, Beck-Sickinger AG, Henle T, Klostermeyer H. *Eur Food Res Technol*. 2006;222:242–249.
10. Schilling S, Wasternack C, Demuth HU. *Biol Chem*. 2008;389:983–991.
11. Ryakhovsky VV, Khachiyan GA, Kosovova NF, Isamiddinova EF, Ivanov AS. *Beilstein J Org Chem*. 2008;4(39).
12. Zhu J, Marchant RE. *J Pept Sci*. 2008;14:690–696.
13. Stroup AN, Cole LB, Dhingra MM, Gierasch LM. *Int J Pept Protein Res*. 1990;36:531–537.
14. Schilling S, Wasternack C, Demuth HU. *Biol Chem*. 2008;389:983–991.
15. Sandmeier E, Hunziker P, Kunz B, Sack R, Christen P. *Biochem Biophys Res Commun*. 1999;261:578–583.
16. Geohegan KF, Hoth LR, Tan DH, Borzilleri KA, Withka JM, Boyd JG. *J Proteome Res*. 2002;1:181–187.
17. Ottinger EA, Shekels LL, Bernlohr DA, Barany G. *Biochemistry*. 1993;32:4354–4361.
18. Kiso Y, Nakamura S, Ito K, et al. *J Chem Soc Chem Commun*. 1979:971–972.
19. Allison WS. *Acc Chem Res*. 1976;9:293–299.
20. Capasso S, Mazzarella L, Sica F, Zagari A, Salvadori S. *J Chem Soc Perkin Trans*. 1993;2: 679–682.
21. Pearson DA, Blanchette M, Baker ML, Guindon C. *Tetrahedron Lett*. 1989;30:2739–2742.
22. Takikawa O, Yoshida R, Kido R, Hayaishi O. *J Biol Chem*. 1986;261:3648–3653.
23. da Silva RA, Estevam IHS, Bieber LW. *Tetrahedron Lett*. 2007;48:7680–7682.
24. Fowles LF, Beck E, Worrall S, Shanley BC, de Jersey J. *Biochem Pharmacol*. 1996;51: 1259–1267.
25. Taichi M, Kimura T, Nishiuchi Y. *Int J Pept Res Ther*. 2009;15:247–253.
26. Eschweiler W. *Chem Ber*. 1905;38:880–882.
27. Schnabel E. *Justus Liebigs Ann Chem*. 1962;659:168–184.
28. Seo YH, Carroll KS. *Proc Natl Acad Sci USA*. 2009;106:16163–16168.
29. Simat TJ, Steinhart H. *J Agric Food Chem*. 1998;46:490–498.
30. Witkop B. *Adv Protein Chem*. 1962;16:221–321.
31. Li S, Schöneich C, Borchardt RT. *Biotechnol Bioeng*. 1995;48:490–500.
32. Flavell RR, Huse M, Goger M, Trester-Zedlitz M, Kuiyan J, Muir TW. *Org Lett*. 2002;4: 165–168.
33. Tang H-Y, Speicher DW. *Anal Biochem*. 2004;334:48–61.
34. Mezö G, Szabó I, Kertész I, et al. *J Pept Sci*. 2011;17:39–46.
35. Zhang H-C, McComsey DF, White KB, et al. *Bioorg Med Chem Lett*. 2001;11:2105–2109.
36. Effenberger F, Mück AO, Bessey E. *Chem Ber*. 1980;113:2086–2099.
37. Lelièvre D, Turpin O, El Kazzouli S, Delmas A. *Tetrahedron*. 2002;58:5525–5533.

38. Vilsmeier A, Haack A. *Ber Dtsch Chem Ges A/B*. 1927;60:119–122.
39. Ohno M, Tsukamoto S, Makisumi S, Izumiya N. *Bull Chem Soc Jpn*. 1972;45:2852–2855.
40. Viville R, Scarso A, Durieux JP, Loffet A. *J Chromatogr*. 1983;262:411–414.
41. Simpson DM, Beynon RJ. *J Proteome Res*. 2010;9:444–450.
42. Jespersen AM, Christensen T, Klausen NK, Nielsen PF, Sorensen HH. *Eur J Biochem*. 1994;219:365–373.
43. Windisch V, Deluccia F, Duhau L, et al. *J Pharm Sci*. 1997;86:359–364.
44. Fujii N, Sasaki T, Funakoshi S, Irie H, Yajima H. *Chem Pharm Bull*. 1978;26:650–653.
45. Reddie KG, Carroll KS. *Curr Opin Chem Biol*. 2008;12:746–754.
46. Berlett BS, Stadtman ER. *J Biol Chem*. 1997;272:20313–20316.
47. Kantouch A, Abdel-Fattah SH. *Chem Zvesti*. 1971;25:222–230.
48. Hruby VJ, du Vigneaud V. *J Am Chem Soc*. 1969;91:3624–3626.
49. Minami I, Ohashi Y, Shimizu I, Tsuji J. *Tetrahedron Lett*. 1985;26:2449–2452.
50. van der Veken P, Dirksen EHC, Ruijter E, et al. *ChemBioChem*. 2005;6:2271–2280.
51. Hauel NH, Nar H, Priepke H, Ries U, Stassen J-M, Wienen W. *J Med Chem*. 2002;45: 1757–1766.
52. Joshi BK, Ramsey B, Johnson B, et al. *J Pharm Sci*. 2010;99:3030–3040.
53. Macdonald JM, Haas AL, London RE. *J Biol Chem*. 2000;275:31908–31913.
54. Franzén H, Grehn L, Ragnarsson U. *J Chem Soc Chem Commun*. 1984:1699–1700.
55. Aversa MC, Barattucci A, Bonaccorsi P, Giannetto P. *Curr Org Chem*. 2007;11:1034–1052.
56. Isidro-Llobet A, Just-Baringo X, Ewenson A, Álvarez M, Albericio F. *Pept Sci*. 2007;88: 733–737.
57. Lamthanh H, Roumestand C, Deprun C, Ménez A. *Int J Pept Protein Res*. 1993;41:85–95.
58. Xing G, Zhang J, Chen Y, Zhao Y. *J Proteome Res*. 2008;7:4603–4608.
59. Breslav M, Becker J, Naider F. *Tetrahedron Lett*. 1997;38:2219–2222.
60. Sieber P. *Tetrahedron Lett*. 1987;28:1637–1640.
61. Jaeger E, Remmer HA, Jung G, et al. *Biol Chem Hoppe-Seyler*. 1993;374:349–362.
62. Beck-Sickinger AG, Schnorrenberg G, Metzger J, Jung G. *Int J Pept Protein Res*. 1991;38: 25–31.
63. Tiefenbrunn TK, Dawson PE. *Protein Sci*. 2009;18:970–979.
64. Quibell M, Turnell W, Johnson T. *J Chem Soc Perkin Trans*. 1993;1:2843–2849.
65. Albericio F, Bofill JM, El-Faham A, Kates SA. *J Org Chem*. 1998;63:9678–9683.
66. Cironi P, Tulla-Puche J, Barany G, Albericio F, Álvarez M. *Org Lett*. 2004;6:1405–1408.
67. Harris PWR, Kowalczyk R, Yang S-H, Williams GM, Brimble MA. *J Pept Sci*. 2014;20: 186–190.
68. DeTar DF, Silverstein R, Rogers Jr FF. *J Am Chem Soc*. 1966;88:1024–1030.
69. König W, Geiger R. *Chem Ber*. 1970;103:2034–2040.
70. Guy CA, Fields G. *Methods Enzymol*. 1997;289:67–83.
71. Carpino LA. *Acc Chem Res*. 1987;20:401–407.

Appendix II

List of Abbreviations

AAA	amino acid analysis
Abu	aminobutanoic acid
Ac	acetyl
Acm	acetamidomethyl
ACN	acetonitrile
AcOH	acetic acid
ACTH	adrenocorticotropic hormone
Ada	adamantyl
Adoc	adamantyloxycarbonyl
Ag(TFMSO)	silver(I) trifluoromethanesulfonate
ε-Ahx	6-aminohexanoic acid
Aib	2-aminoisobutyric acid
Ala	alanine
All	allyl
Alloc	allyloxycarbonyl
AM	aminomethyl
4-AMP	4-(aminomethyl)pyridine
amu	atomic mass unit
Aoa	aminooxyacetic acid
Aph	amino phenylalanine
Arg	arginine
Asi	aspartimide
Asn	asparagine
Asp	aspartic acid
AZT	azidothymidine
Boc	*tert*-butoxycarbonyl
BOI	2-[(1*H*-benzotriazol-1-yl)oxy]1,3-dimethylimidazolidinium hexafluorophosphate
Bom	benzyloxymethyl
Bpa	*p*-benzoylphenylalanine
Bpoc	2-(4-biphenylyl)isopropyloxycarbonyl
BroP	bromotris(dimethylamino)phosphonium hexafluorophosphate
BTC	bis(trichloromethyl)carbonate
Bum	*tert*-butoxymethyl

Bzl	benzyl
Cam	carbamoylmethyl
CBD	cellulose-binding domain
CDI	1,1′-carbonyldiimidazole
CDMT	2-chloro-4,6-dimethoxy-1,3,5-triazine
CHA	cyclohexylamine
Chx	cyclohexyl
CIP	2-chloro-4,5-dihydro-1,3-dimethyl-1*H*-imidazolium hexafluorophosphate
CLEAR®	cross-linked ethoxylate acrylate resin
COMU	1-[(1-(cyano-2-ethoxy-2-oxoethylideneaminnooxy)-dimethylamino-morpholinomethylene)] methanaminium hexafluorophosphate
CPME	cyclopentyl methyl ether
CTC	2-chlorotrityl chloride
Cys	cysteine
Da	dalton
Dapa	2,3-diaminopropionic acid
DBF	dibenzofulvene
DBU	1,8-diazabicyclo[5.4.0]undec-7-ene
DCC	*N,N′*-dicyclohexylcarbodiimide
DCM	dichloromethane
Dcpm	dicyclopropylmethyl
DCU	*N,N′*-dicyclohexylurea
DDQ	2,3-dichloro-5,6-dicyano-1,4-benzoquinone
Ddz	α,α-dimethyl-3,5-dimethoxybenzyloxycarbonyl
DEPBT	3-(diethoxyphosphoryloxy)-1,2,3-benzotriazin-4(3*H*)-one
DIC	*N,N′*-diisopropylcarbodiimide
DIEA	diisopropylethylamine
DiOia	dioxindolylalanine
DKP	diketopiperazine
DMAc	dimethylacetamide
DMAP	4-dimethylaminopyridine
Dmb	2,4-dimethoxybenzyl
DMB	1,3-dimethoxybenzene
DMF	*N,N*-dimethylformamide
DMS	dimethyl sulfide
DMSO	dimethyl sulfoxide
DMTMM	4-(4,6-dimethoxy-1,3,5-trazine-2-yl)-4-methylmorpholinium chloride
DNP	2,4-dinitrophenol
Doc	2,4-dimethylpent-3-yloxycarbonyl
Dod	4,4′-dimethoxydityl
DODT	3,6-dioxa-1,8-octanedithiol
DoE	design of experiments
DOPA	3,4-dihydroxyphenylalanine
DPPA	diphenyl phosphorazidate
DSC	*N,N′*-disuccinimidyl carbonate
Dts	*N*-dithiasuccinoyl
DTT	dithiothreitol

Dyn A	dynorphin A
E. coli	*Escherichia coli*
EBT	1,1′-ethylidenebis[tryptophan]
EDC	1-ethyl-3-(3-dimethylaminopropyl)carbodiimide
EDOT	3,4-ethylenedioxy-2-thienyl
EDT	1,2-ethanedithiol
eNOS	endothelial nitric-oxide synthase
equiv.	equivalent
FAB MS	fast atom bombardment mass spectrometry
FITC	fluorescein isothiocyanate
Fm	9-fluorenylmethyl
Fmoc	9-fluorenylmethyloxycarbonyl
For	formyl
Glc	glycosyl
Gln	glutamine
Glu	glutamic acid
Gly	glycine
GSH	glutathione
GSSG	glutathione disulfide
HATU	*N*-[(dimethylamino)-1*H*-1,2,3-triazolo-[4,5-*b*]pyridin-1-ylmethylene]-*N*-methylmethanaminium hexafluorophosphate *N*-oxide
HBPyU	*O*-(benzotriazol-1-yl)-*N*,*N*,*N*′,*N*′-bis(tetramethylene)uronium hexafluorophosphate
HBTU	*O*-(benzotriazol-1-yl)-*N*,*N*,*N*′,*N*′-tetramethyluronium hexafluorophosphate
HCTU	1-[bis(dimethylamino)methylene]-5-chlorobenzotriazolium 3-oxide hexafluorophosphate
HDTU	*O*-(4-oxo-3,4-dihydro-1,2,3-benzotriazin-3-yl)-*N*,*N*,*N*′*N*′-tetramethyluronium hexafluorophosphate
HFIP	hexafluoroisopropanol
His	histidine
Hmb	2-hydroxyl-4-methoxybenzyl
HMPA	hydroxymethylphenoxyacetic acid
HMPT	hexamethylphosphoramide
HOAt	1-hydroxy-7-aza-benzotriazole
HOBt	1-hydroxy-1*H*-benzotriazole
HODhbt	3-hydroxy-1,2,3-benzotriazin-4(3*H*)-one
HOOBt	3-hydroxy-1,2,3-benzotriazin-4(3*H*)-one
Hor	hydroorotyl
HOSu	*N*-hydroxysuccinimide
HOTU	*O*-[(ethoxycarbonyl)cyanomethylenamino]-*N*,*N*,*N*′,*N*′-tetramethyluronium hexafluorophosphate
Hyl	hydroxylysine
Hyp	4-hydroxyproline
IC	ionic chromatography
IgG	immunoglobulin G
Ile	isoleucine
Im	imidazole

iPrOH	isopropanol
Kyn	kynurenine
LC	liquid chromatography
Leu	leucine
LPPS	liquid-phase peptide synthesis
Lys	lysine
MALDI-TOF-MS	Matrix-Assisted Laser Desorption/Ionization Time of Flight Mass Spectrometry
MBHA	methylbenzhydrylamine
MBT	mercaptobenzothiazole
mCPBA	m-chloroperoxybenzoic acid
Me	methyl
MeSub	2-methoxy-5-dibenzosuberyl
Met	methionine
MIM	1-methyl-3-indolylmethyl
MIS	1,2-dimethylindole-3-sulfonyl
Mmt	monomethoxytrityl
Mob	4-methoxybenzyl
Moz	*p*-methoxybenzyloxycarbonyl
MS	mass spectrometry
Msc	2-(methylsulfonyl)ethoxycarbonyl
MTBE	methyl *tert*-butyl ether
Mtr	4-methoxy-2,3,6-trimethylbenzenesulfonyl
Mts	mesityl-2-sulfonyl
Mtt	4-methyltrityl
NBD-Cl	4-chloro-7-nitrobenzo-2-oxa-1,3-diazole
Nbz	*N*-acyl-benzimidazolinone
NCA	*N*-carboxyanhydride
NCL	natural chemical ligation
NFK	*N*-formylkynurenine
Nle	Norleucine
Nma	2-(methylamino)-benzoyl
NMI	*N*-methylimidazole
NMM	*N*-methylmorpholine
NMP	*N*-methyl-2-pyrrolidone
NMR	nuclear magnetic resonance
Nps	2-nitrophenylsulfanyl
Npys	3-nitro-2-pyridylsulfanyl
Nsc	2-(4-nitrophenylsulfonyl)ethoxycarbonyl
Nva	norvaline
OBO	4-methyl-2,6,7-trioxabicyclo[2.2.2]oct-1-yl
ODie	2,3,4-trimethylpent-3-yl
ODmab	4-{*N*-[1-(4,4-dimethyl-2,6-dioxocyclohexylidene)-3-methylbutyl]-amino}benzyl ester
ODT	1,8-octanedithiol
Oia	oxindolylalanine
OMpe	3-methylpen-3-yl ester
***o*Nbs**	2-nitrobenzenesulfonyl

ONp	4-nitrophenyloxy
OPhFl	9-phenyl-fluoren-9-yl ester
OPp	2-phenylisopropyl ester
OPyBzh	4-pyridyl-diphenylmethyl ester
Orn	ornithine
OTcm	tricyclohexylmethyl ester
OTim	triisopropylmethyl ester
Oxyma	ethyl 2-cyano-2-(hydroxyimino)acetate
PAC	4-(hydroxymethyl)phenoxyacetic acid handle
Pac	phenacyl
PAL	peptide amide linker
Palm	palmitoyl
Pam	phenylacetamidomethyl
Pbf	2,2,4,6,7-pentamethyldihydrobenzofuran-5-sulfonyl
PDI	protein disulfide isomerase
PEG/PS	polyethylene glycol-polystyrene graft support
PEGA	polyethylene glycol polyamide copolymer
Pfp	pentafluorophenyl
PG	protecting group
Phe	phenylalanine
Phg	phenylglycine
Pip	pipecolic acid
Pmc	2,2,5,7,8-pentamethylchroman-6-sulfonyl
PNA	peptide nucleic acid
***p*Nbs**	4-nitrobenzenesulfonyl
pNZ	*p*-nitrobenzyloxycarbonyl
Pro	proline
PS	polystyrene
PyAOP	(7-azabenzotriazol-1-yloxy)tripyrrolidinophosphonium hexafluorophosphate
PyBOP	(benzotriazol-1-yloxy)tripyrrolidinophosphonium hexafluorophosphate
PyBroP	bromotripyrrolidinophosphonium hexafluorophosphate
PyCloP	chlorotripyrrolidinophosphonium hexafluorophosphate
2-PySH	2-mercaptopyridine
QC	quality control
rhGH	recombinant human growth hormone
ROS	reactive oxygen species
RP-HPLC	reverse phase high performance liquid chromatography
SAR	structure-activity relationship
SASRIN	super acid sensitive resin
SDS	sodium dodecyl sulfate
Ser	serine
SES	2-(trimethylsilyl)ethylsulfonyl
SPPS	solid-phase peptide synthesis
StBu	*tert*-butylthio
Sub	5-dibenzosuberyl
Suben	ω-5-dibenzosuberenyl
TAEA	tris(2-aminoethyl)amine

TBAF	tetrabutylammonium fluoride
TBTU	*N,N,N′,N′*-tetramethyl-*O*-(benzotriazol-1-yl)uronium tetrafluoroborate
tBu	*tert*-butyl
tBuOH	*tert*-butanol
TCEP	tris(2-carboxyethyl)phosphine
TDBTU	*O*-(3,4-dihydro-4-oxo-1,2,3-benzotriazin-3-yl)-*N,N,N′,N′*-tetramethyluronium tetrafluoroborate
TEA	triethylamine
TES	triethylsilane
TFA	trifluoroacetic acid
tfa	trifluoroacetate
TFE	trifluoroethanol
TFMSA	trifluoromethanesulfonic acid
TfOH	trifluoromethanesulfonic acid
THF	tetrahydrofuran
Thr	threonine
Thz	thiazolidine-4-carboxylic acid
TIS	triisopropylsilane
$Tl(tfa)_3$	thallium (III) trifluoroacetate
TMAH	tetramethylammonium hydroxide
Tmob	trimethoxybenzyl
TMSBr	trimethylsilyl bromide
TMSCl	trimethylsilyl chloride
TMSE	2-(trimethylsilyl)ethyl
TMSOTf	trimethylsilyl trifluoromethanesulfonate
Tos	tosyl
TOTU	*O*-[(ethoxycarbonyl)cyanomethylenamino]-*N,N,N′,N′*-tetramethyluronium tetrafluoroborate
TPTU	*O*-(1,2-dihydro-2-oxo-1-pyridyl-*N,N,N′,N′*-tetramethyluronium tetrafluoroborate
Trp	tryptophan
Trt	trityl
Tyr	tyrosine
UNCA	urethane-protected amino acid *N*-carboxyanhydride
UV	ultraviolet
Val	valine
XAL	xanthenylamide acid labile linker
Xan	9*H*-xanthen-9-yl
Z	benzyloxycarbonyl

Subject Index

A

B

G

H

I

K

L

M

N

O

Q

R

U

V

W

X

Y

Z

Printed in the United States
By Bookmasters